Engineering

an introduction to
a creative profession

THIRD EDITION

GEORGE C. BEAKLEY, P.E.   **H. W. LEACH, P.E.**

Professor of Engineering, Arizona State University   *Consulting Engineer*

With Contributions by
J. KARL HEDRICK

Associate Professor of Mechanical Engineering, Massachusetts Institute of Technology

Macmillan Publishing Co., Inc.
NEW YORK

Collier Macmillan Publishers
LONDON

Macmillan Publishing Co., Inc.
866 Third Avenue, New York, New York 10022

Collier Macmillan Canada, Ltd.

Library of Congress Cataloging in Publication Data

Beakley, George C
    Engineering.

    Edition for 1951, by H. W. Leach and G. C. Beakley,
published under title: Elementary problems in engineering; edition for 1960, by H. W. Leach and G. C. Beakley,
published under title: Engineering, the profession and
elementary problem analysis.
    Includes index.
    1. Engineering—Vocational guidance.   2. Engineering
mathematics.   3. Engineering design.   I. Leach, H. W.,
joint author.   II. Leach, H. W.   Elementary problems in
engineering.   III. Title.
TA147.B4   1976   620'.0023   76-13591
ISBN 0-02-307190-7

Printing: 1 2 3 4 5 6 7 8   Year: 7 8 9 0 1 2 3

Preface

Today's world is a world of change, challenge, and opportunity. Never before have so many compelling technological problems occupied positions of prominence in man's system of values. The engineer's role in this environment is more important than ever before in history, and the student who chooses engineering as a career should realize that, perhaps more than any other, his profession will shape the destiny of civilizations yet unborn. Challenge and opportunity? Yes . . . but not without a consciousness for and a realization of the moral and ethical responsibilities that should accompany the emergence of new processes, designs, and systems. Recognizing the importance of an early commitment toward these ends, this work has been prepared as an informational and motivational instrument for the use of those who are interested in preparing themselves today for the solution of tomorrow's problems.

The first edition of this textbook was experimental in the sense that it was multipurpose in scope, rather than singular in purpose, as has been the case traditionally with many other introductory engineering textbooks. Its nine printings over a brief span of time attest to its popularity for use in (*a*) informational courses that introduce the student to the profession of engineering, (*b*) engineering problems and analysis courses that give the student some practice in engineering problem solving, and (*c*) introduction to engineering design courses that give the student a sense of personal involvement in undertaking real engineering tasks. Some teachers have preferred to organize courses that draw material from each of these areas.

The second edition expanded further the options available to the teacher and student, and increased emphasis was placed upon the expanding role of the technician and technologist in modern industry. This edition was accorded even greater adopter response than the first edition.

In this third edition the authors have attempted to retain those qualities of the text most often praised and to update and improve it throughout. They are most appreciative of the suggestions received from the more than 300 schools who have used the first two editions, and they are very desirous of learning the opinions of those who read this new edition—both students and faculty—concerning its utility and serviceability in meeting the needs for which it has been written. Improvements that are suggested will be incorporated in later editions.

In order to give the student an understanding of the types of problems that are likely to be encountered in the practice of engineering, a large number of varied situations are described in problems throughout the book. Some of the problems are straightforward, and appropriate data are given to permit a unique solution. In other problems, data are given in general terms, sometimes with insufficient data so that the student must add information, and sometimes with an overabundance of data from which the student must select what he needs. In all cases, the problems are designed to introduce the student to the realm of engineering study, to offer work with engineering concepts, and to provide situations in which decisions must be made where a number of choices exist.

The problem-solving method introduced in the authors' previous textbooks has been expanded to include more of the creative phases of engineering. It is not enough just to manipulate numbers in engineering work; the engineer must be able to see applications of scientific principles, to develop designs that are based upon abstract principles, and to assume the lead in formulating innovative solutions to unfamiliar problems. Every effort is made here to motivate the student to think imaginatively and constructively, and also to present material that will provide the best introduction to a career in engineering or technology.

For 28 years one of the text's most acclaimed chapters has been "The Slide Rule." With the rapid development of microelectronics, this material has now been replaced by Chapter 10, "Using Electronic Hand Calculators." As before, hundreds of problems have been included in this chapter that may be used for practice. Others are included for homework or class assignment. It is the authors' belief that students should be thoroughly familiar with both English and metric unit systems. Therefore Chapter 12 has been expanded to include a thorough treatment of the metric SI system of units, and many of the problems in Chapter 13 are described in metric terms. Chapter 14 uses only the SI metric unit system.

A completely new chapter, "The Modeling of Engineering Systems," gives the student an insight into the analysis of engineering systems and emphasizes the commonality that exists between the engineering disciplines.

The format of this text is such that the student can record points of emphasis and certain class notes and questions directly as they occur without the use of extra notepaper. Marking pens may be used for highlighting sentences or phrases, and the use of colored ink for underlining is recommended. For the student to get the most out of the text he should "live in it." If he does, the book will have a "lived-in" appearance.

In preparing this book the authors have been aided not only by reviews and criticisms of previous editions but also by comments and suggestions from numerous engineering professors and practicing engineers and technologists in industry throughout the United States. The professional colleagues of the authors have been most influential in giving this book a unique blend of the academic and industrial viewpoints.

In particular the authors wish to thank Professors Harold Nelson and a number of faculty colleagues for their review of portions of the manuscript. Professor William Anderson prepared much of the material in Chapter 10, "Using Electronic Hand

Calculators," Professor J. Karl Hedrick prepared the majority of the material for Chapter 14, "The Modeling of Engineering Systems;" and George C. Beakley III prepared the material in Chapter 16 relating to the planning of engineering projects. The authors gratefully acknowledge these contributions as well as those of the many organizations who have supplied illustrative materials for inclusion in the text.

The cover design, an electronic-computer-produced artform, was used originally for the first and second editions. It is the work of Donald Robbins and Leigh Hendricks of the Sandia Corporation, Albuquerque, New Mexico, who have made it available for this specific use. The typing and proofreading of the manuscript were masterfully accomplished by Esther F. Taylor.

To the many others who have given the authors the benefit of their experience by making recommendations as to format and content of this new third edition, we wish to express our sincere gratitude.

G. C. B.
H. W. L.

Acknowledgments

Contents

Part One

Engineering
—the profession
in review

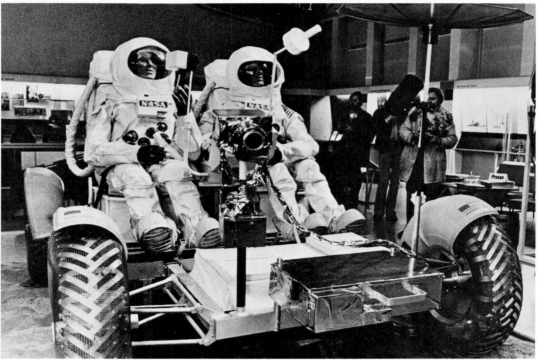

The mode of propulsion—via individual electrical wheel hub motors—used in the design of the 50 million dollar American lunar vehicle in 1971, was first used in 1900 by Ferdinand Porsch to propel his racing car.

1

Historical perspectives of a developing profession

When did engineering begin? Who were the first engineers? What were the objectives of work by the early engineers? Answers to these questions and others concerning the beginning of engineering appear in the fragments of historical information available to us. In fact the beginnings of civilization and the beginnings of engineering are coincident. As early man emerged from caves to make homes in communities, he adapted rocks and sticks as tools to aid him. Simple as these items may seem to us today, their useful employment suggests that the creative ideas which emerged in the minds of early man were developed into useful products to serve the recognized needs of the day. Some served as tools in the struggle for existence of an individual or group, and others were used for protection against wild animals or warlike neighbors. Early engineering was therefore principally either civil or military.

Down through the ages, the engineer has been in the forefront as a maker of history. His material accomplishments have had as much impact on world history as any political, economic, or social development. Sometimes his accomplishments have stemmed from the pressures of need from evolving civilizations. At other times his abilities to produce and meet needs have led the way for civilizations to advance. In general, engineers do the things required to serve the needs of the people and their culture.

Basically, the role of the engineer has not changed through the centuries. His job is to take knowledge and make practical use of it. He converts scientific theory into useful application, and in so doing, he provides for man's material needs and well-being. From era to era, only the objectives that he has pursued, the techniques of solution that he has used, and the tools of analysis at his disposal have changed.

It is helpful to review the past to gain insight to the driving forces of science and to learn of the men who developed and applied these principles. A review also will reveal certain facts concerning the discovery and use of fundamental scientific principles. Primarily, science builds its store of knowledge on facts which, once determined, are available from then on for further discovery. This principle is in contrast to the arts, since, for example, the ability of one person to produce a beautiful painting does not make available to others his skills in producing paintings.

Outstanding characteristics of engineers through the centuries have been a willingness to work and an intellectual curiosity about the behavior of things. Their queries about "Why?," "How?," "With what?," and "At what cost?" have all served to stimulate an effort to find desirable answers to many types of technological problems.

Another characteristic associated with engineers is the ability to "see ahead." The engineer must have a fertile imagination, must be creative, and must be ready to accept new ideas. Whether an engineer lived at the time of construction of the pyramids or has only recently graduated in nuclear engineering, these characteristics have been an important part of his intellectual makeup.

The following sections present a brief picture of the development of engineering since the dawn of history and outline the place that the engineer has held in various civilizations.

The beginnings of engineering: 6000 B.C.—3000 B.C.

The beginning of engineering probably occurred in Asia Minor or Africa some 8000 years ago. About this time, man began to cultivate plants, domesticate animals, and build permanent houses in community groups. With the change from a nomadic life came requirements for increased food production. Among the first major engineering projects were irrigation systems to promote crop growing. Increased food production permitted time for men to engage in other activities. Some became rulers, some priests, and many became artisans, whom we may call the first engineers.

Early achievements in this era included methods of producing fire at will, melting certain rocklike materials to produce copper and bronze tools, invention of the wheel and axle, development of a system of symbols for written communication, origination of a system of mathematics, and construction of irrigation works.

Early records are so fragmentary that only approximate dates can be given for any specific discovery, but evidence of the impact of early engineering achievements is readily discernible. For example, in setting up stable community life in which land was owned, men had to provide both for irrigation and for accurate location and maintenance of boundaries. This necessity stimulated the development of surveying and of mathematics. The moving of earth to make canals and dams required computations, and to complete the work the efforts of many men had to be organized and directed. As a result, a system of supervisors, foremen, and workers was established that formed the beginnings of a class society.

In this society, craftsmen became a distinct group producing useful items such as pottery, tools, and ornaments that were desired by others. As a result, trade and commerce were stimulated and roads were improved. Some 5000 years ago man first used the wheel and axle to make two-wheeled carts drawn by animals.

In order to record the growing accumulation of knowledge about mathematics and engineering, the early engineer needed a system of writing and some type of writing material. In the Mesopotamian region, soft clay was used on which cuneiform characters were incised. When baked, the clay tile material was used for permanent documents, some of which are legible even today (see Illustration 1-1). In the Nile

Illustration 1-1
Mesopotamia, often called the cradle of civilization, also may be said to have begun engineering. Excavations have revealed extensive architecture, irrigation systems, roads, and land planning. In this picture is shown a party of surveyors using tools for measurement which, for the period, were remarkably accurate.

Map of caravan routes, mountains, cities, and water.

Clay tablet of a city plan.

City planning and building.

Irrigation systems were extensive.

BABYLON AND SURROUNDING AREA

Mesopotamia, often called the "Cradle of civilization," could also be said to have nurtured engineering in its infancy. Clay tablets, such as the ones shown on this page, have been unearthed which show city plans, irrigation, water supply systems, and what appear to be road maps. Although no engineering tools have been discovered among the remains of ancient Mesopotamia, the evidence unearthed of their remarkable architectural construction indicates that they used measuring tools, which, even though primitive, aided in producing engineering of a high degree for this period. Their cities, with their water supply, irrigation systems, and road networks, were among the wonders of the ancient world.

Many outstanding contributions of mathematics were made by the Mesopotamians. It has been proven that they had knowledge of the sexagesimal system, in which they divided the circle into 360 degrees, the hour into 60 minutes and the minute into 60 seconds.

Valley, a paperlike material called *papyrus* was made from the inner fibers of a reed. In other parts of Asia Minor, treated skins of animals were used to form parchment. Occasionally, slabs of stone or wood were used as writing materials. The type of writing that developed was strongly influenced by the writing material available. For example, the incised characters in soft clay differed significantly from the brush strokes used in writing on papyrus.

In engineering work, a source of energy is necessary. This requirement led to the enslavement and use of numbers of humans as primary sources of energy. The construction of all early engineering works, whether Oriental, Mediterranean, or American Indian, was accomplished principally by human labor. It was not until near the end of the period of history known as the Middle Ages that mechanical sources of power were developed.

Engineering in early civilizations: 3000 B.C.–600 B.C.

After about 3000 B.C., enough records were made on clay tablets, on papyrus and parchment, on pottery, and as inscriptions on monuments and temples to provide us with information about ancient civilization. These records show that urban civilizations existed in Egypt, Mesopotamia, and the Indus Valley, and that a class society of craftsmen, merchants, soldiers, and government officials was a definite part of that civilization.

In Mesopotamia, clay tablets have been uncovered which show that Babylonian engineers were familiar with basic arithmetic and algebra. From these writings we know that they routinely computed areas and volumes of land excavations. Their number system, based on 60 instead of 10, has been handed down through the centuries to us in our measures of time and angle. Their buildings were constructed principally of baked brick. Primitive arches were used in some of their early hydraulic works. Bridges were built with stone piers carrying wooden stringers for the roadway. Some roads were surfaced with a naturally occurring asphalt, a construction method that was not used again until the nineteenth century.

It was in Egypt that some of the world's most remarkable engineering was performed (see Illustration 1-3). Beginning about 3000 B.C. and lasting for about 1000 years, the Pyramid Age flourished in Egypt. The first pyramids were mounds covered with stone, but the techniques progressed rapidly until the Great Pyramid was begun about 2900 B.C. Stones for the structures were cut by workmen laboriously chipping channels in the native rock, using a ball made of a harder rock as a tool. By this method, blocks weighing 15 tons or more were cut for use in building. The Egyptian engineers apparently used only the lever, the inclined plane, the wedge, and the wheel in their construction efforts (see Illustration 1-4).

Although early construction tools were primitive, the actual structures, even by today's standards, are outstanding examples of engineering skill in measurement and layout. For example, the base of the Great Pyramid is square within about 1 inch in a distance of 756 feet, and its angles are in error by only a few minutes despite the fact that the structure was built on a sloping rocky ledge.

The Egyptian engineers and architects held a high place in the Pharaoh's court. Imhotep, a designer of one of the large pyramids, was so revered for his wisdom and ability that he was included as one of the Egyptian gods after his death. Not only were the Egyptian engineers skilled builders, they were also skilled in land measurement. Annual overflows of the Nile River obliterated many property lines and a resurvey of the valley was frequently necessary. Using geometry and primitive measuring equipment, they restored markers for land boundaries after the floods receded.

The Egyptians also were skilled in irrigation work. Using a system of dikes and canals, they reclaimed a considerable area of desert. An ancient engineering contract to build a system of dikes about 50 miles long has recently been discovered.

Although the skill and ingenuity of the Egyptian engineers were outstanding, the culture lasted only a relatively short time. Reasons which may account for the failure to maintain leadership are many, but most important was the lack of pressure to continue development. Once the engineers formed the ruling class, little influence could be brought to bear to cause them to continue their creative efforts. Since living conditions were favorable after an agricultural system was established, little additional engineering was required. The lack of urgency to do better finally stifled most of the creativity of the engineers and the civilization fell into decay.

Illustration 1-3

In Egypt the science of measurement and construction developed rapidly. The pyramids are engineering marvels both in design and construction. Papyrus scrolls show that the Egyptians had knowledge of the triangle and were able to compute areas and volumes.

Resetting boundaries after the Nile floods.

Early geometric application.

In ancient Egypt warfare and strife delayed the development of engineering; however, with the unification of Upper and Lower Egypt, the science of measurement and construction made rapid progress. Buildings, city planning, and irrigation systems show evidence of this development. Good judgment and reasonable engineering design resulted in sound and durable structures. The Pyramids are engineering marvels both in design and construction.

That the Egyptians advanced mathematics is attested to by papyrus scrolls, dating back to 1500 B.C., which show that the Egyptians had knowledge of the triangle and were able to compute areas and volumes. They also had a device to obtain the azimuth from the stars.

The annual floods of the Nile afforded ample practice in measurement surveying. This may well have been the first example of the importance of resurveys. The rope used as a measure was first soaked in water, dried, and then coated heavily with wax to ensure constant length. Probably some crude surveying instruments were devised, but none have been found.

Illustration 1-4
The pyramids of Egypt exemplify man's desire to create and build enduring monuments.

Science of the Greeks and Romans: 600 B.C.—A.D. 400

The history of engineering in Greece had its origins in Egypt and the East. With the decline of the Egyptian civilization, the center of learning shifted to the island of Crete and then about 1400 B.C. to the ancient city of Mycenae in Greece.

To the engineers of Mycenae were passed not only the scientific discoveries of the Egyptians but also a knowledge of structural building materials and a language that formed the basis of the early Greek language. These engineers subsequently developed the corbeled arch and made wide use of irrigation systems.

The Greeks of Athens and Sparta borrowed many of their developments from the Mycenaean engineers. In fact, the engineers of this period were better known for the intensive development of borrowed ideas than for creativity and invention. Their water system, for example, modeled after Egyptian irrigation systems, showed outstanding skill in the use of labor and materials, and these Greeks established technical procedures that have endured for centuries (see Illustration 1-5).

Greece was famous for its outstanding philosophers. Significant contributions

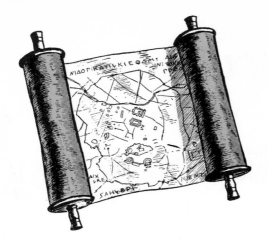

Illustration 1-5
The Greeks constructed many buildings of unusual beauty which show a high degree of engineering skill and architectural design. Their cities had municipal water supplies that required dams and aqueducts to bring water from the mountains. This picture shows a builder laying out a building foundation, using a divided circle, a plumb bob, and a knotted rope.

Hydraulics provide public water

Aqueducts, tunnels and highways

The outstanding progress made by the Ancient Grecians in architecture and mathematics and their contribution to the advancement of engineering demand our admiration.

Aristotle contended that the world was a spheroid. He stated that observations of the various stars showed the circumference of the earth to be about 400,000 *stadia* (4600 miles).

Erathosthenes of Cyrene observed that the sun's rays, when perpendicular to a well at Alexandria, cast a shadow equal to one fiftieth of a circle at Syene (Aswan) 500 miles away. Thus he established that the circumference of the earth was 50 times 500 miles or 25,000 miles.

The Greeks constructed many buildings and structures of large size, which show engineering skill and excellent architectural design. One tunnel, which was built to bring water to Athens, measured 8 feet by 8 feet and was 4200 feet in length. The construction of such a tunnel necessitated extremely accurate alignment both on the surface and underground.

> When looms weave by themselves man's slavery will end.
> —Aristotle

were made by men such as Plato, Aristotle, and Archimedes. In the realm of abstract thought, they perhaps have never been equaled, but at that time extensive use of their ideas was retarded because of the belief that verification and experimentation, which required manual labor, were fit only for slaves. Of all the contributions of the Greeks to the realm of science, perhaps the greatest was the discovery that nature has general laws of behavior which can be described with words.

The best engineers of antiquity were the Romans. Within a century after the death of Alexander, Rome had conquered many of the eastern Mediterranean countries, including Greece. Within two more centuries Rome had dominion over most of the known civilized areas of Europe, Africa, and the Middle East. Roman engineers liberally borrowed scientific and engineering knowledge from the conquered countries for use in warfare and in their public works. Although in many instances they lacked originality of thought, Roman engineers were superior in the application of techniques (see Illustration 1-6).

From experience Rome had learned the necessity for establishing and maintaining a system of communications to hold together the great empire. Thus Roman roads became models of engineering skills. By first preparing a deep subbase and then a compact base, the Romans advanced the technique of road construction so far that some Roman roads are still in use today. At the peak of Roman sovereignty, the network of roads comprised over 180,000 miles stretching from the Euphrates Valley to Great Britain.

In addition, Roman engineers were famous for the construction of aqueducts and bridges. Using stone blocks in the constructing of arches, they exhibited unusual skill. An outstanding example of this construction is the famous Pont du Gard near Nîmes, France, which is 150 feet high and over 900 feet long. It carries both an aqueduct and a roadway.

By the time of the Christian era, iron refining had developed to the extent that iron was being used for small tools and weapons. However, the smelting process was so inefficient that over half of the metallic iron was lost in the slag. Except in the realm of medicine, no interest was being shown in any phase of chemistry.

Despite their outstanding employment of construction and management techniques, the Roman engineers seemed to lack the creative spark and imagination necessary to provide the improved scientific processes required to keep pace with the expanding demands of a far-flung empire. The Romans excelled in law and civil administration but were never able to bring distant colonies fully into the empire. Finally, discontent and disorganization within the empire led to the fall of Rome to a far less cultured invader.

Engineering in the middle ages: fifth to sixteenth centuries

After the fall of Rome, scientific knowledge was dispersed among small groups, principally under the control of religious orders. In the East, an awakening of

Illustration 1-6

The rise of the Roman Empire was attributed to the application of engineering principles to military tactics. This picture shows a construction party as they build a section of the famous Roman highways. Notice the heavy foundations which exist to this day.

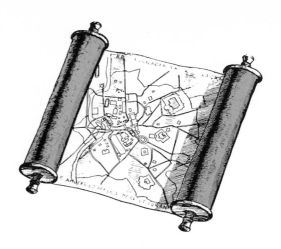

Scientific approach to navigational problems.

Piers and arches, a product of geometry

ROME
AND PART OF THE ROMAN EMPIRE AT ITS HEIGHT

The Romans excelled in the building of aqueducts. Many of these carried water for great distances with perfect grade and alignment. The key design in this type of construction was the arch, which was also used in bridges, tunnels, buildings, and other construction.

Evidence of the Romans' knowledge and understanding of basic geometric principles is further shown by their river and harbor construction and the scientific approach to navigational problems.

Sanitary systems, paved roads, magnificent public buildings, water supply systems, and other public works still in evidence today stand as monuments to the Roman development of engineering as a key to the raising of the standard of living.

The rise of the Roman Empire was attributable to the application of engineering principles applied to military tactics. The invincibility of the Roman legions was the result not only of the valor of the fighting men but also, and perhaps more strongly, to the genius of the Roman military engineers.

> To be ignorant of the lives of the most celebrated men of antiquity is to continue in a state of childhood all our days.
> —Plutarch, ca. A.D. 106

technology began among the Arabs but little organized effort was made to carry out any scientific work. Rather, it was a period in which isolated individuals made new discoveries or rediscovered earlier known scientific facts.

It was during this time that the name *engineer* first was used. Historical writings of about A.D. 200 tell of an *ingenium,* an invention, which was a sort of battering ram used in attacks on walled defenses. Some thousand years later, we find that an *ingeniator* was the man who operated such a device of war—the beginning of our modern title, *engineer.*

Several technical advances were made late in this period. One important discovery involved the use of charcoal and a suitable air blast for the efficient smelting of iron. Another advance was made when the Arabs began to trade with China and a process of making paper was secured from the Chinese. Within a few years the Arabs had established a paper mill and were making paper in large quantities. With the advent of paper, communication of ideas began to be reestablished. Also in Arabia, the sciences of chemistry and optics began to develop. Sugar refining, soap making, and perfume distilling became a part of the culture. The development of a method of making gun powder, probably first learned from China about the fourteenth century, also had rapid and far-reaching results.

After centuries of inaction, the exploration of faraway places began again, aided greatly by the development of a better compass. With the discovery of other cultures and the uniting of ideas, there gradually emerged a reawakening of scientific thought.

With the growth of Christianity, an aversion arose to the widespread use of slaves as primary sources of power. This led to the development of waterwheels and windmills and to a wider use of animals, particularly horses, as power sources.

About 1454, Gutenberg, using movable type, produced the first books printed on paper. This meant that the knowledge of the ages, which previously had been recorded laboriously by hand, now could be disseminated widely and in great quantities. Knowledge, which formerly was available only to a few, now was spread to scholars everywhere. Thus the invention of paper and the development of printing served as fitting climaxes to the Middle Ages.

Seldom has the world been blessed with a genius such as that of Leonardo da Vinci (1452–1519). Although still acclaimed today as one of the greatest of all artists, his efforts as an engineer, inventor, and architect are even more impressive. Long after his death his designs of a steam engine, machine guns, a camera, conical shells, a submarine, and a helicopter have been proven to be workable.

Galileo (1564–1642) was also a man of great versatility. He was an excellent writer, artist, and musician, and he is also considered one of the foremost scientists of that period. One of his greatest contributions was his formulation of what he considered to be the scientific method of gaining knowledge.

The revival of science:
seventeenth and eighteenth centuries

Following the invention of printing, the self-centered medieval world changed rapidly. At first, the efforts to present discoveries of Nature's laws met with opposition and in some cases even hostility. Slowly, however, freedom of thought was permitted and a new concept of *testing to evaluate a hypothesis* replaced the early method of establishing a principle solely by argument.

Four men in this period made discoveries and formulated laws which have proved to be of great value to engineering. They were Boyle, who formulated a law relating pressures and volumes of gases; Huygens, who investigated the effects of gravitational pull; Hooke, who experimented with the elastic properties of materials; and Newton, who is famous for his three laws of motion. All the early experimenters were hampered by a lack of a concise vocabulary to express their ideas. Because of this many of the principles were expressed in a maze of wordy statements.

During this period, significant advancements were made in communication and transportation. Canals and locks were built for inland water travel and docks and harbors were improved for ocean commerce. Advances in ship design and improved methods of navigation permitted a wide spreading of knowledge that formerly had been isolated in certain places.

The search for power sources to replace human labor continued. Water power and wind power were prime sources, but animals began to be used more and more. About this time, the first attempts to produce a steam engine were made by Papin and Newcomen. Although these early engines were very inefficient, they did mark the beginning of power from heat engines.

An important industry was made possible in this period by the development of spinning and weaving machinery by such men as Jurgen, Hargreaves, Crampton, and Arkwright. This period also marked a general awakening of science after the Dark Ages. Individual discoveries, although usually isolated, found their way into useful products within a short period of time because of the development of printing and the improvements in communication.

The basic discoveries in this era were made by men who were able to reject old, erroneous concepts and search for principles that were more nearly in accord with Nature's behavior. Engineers in any age must be equally discerning if their civilization is to advance.

Beginnings of modern science:
nineteenth century

Early in the nineteenth century, two developments provided an impetus for further technological discoveries. The two developments were the introduction of a method, developed by Henry Cort, of refining iron and the invention of an efficient steam engine by James Watt (Illustration 1-7). These developments provided a source of iron for machinery and power plants to operate the machinery.

As transportation systems began to develop, both by water and by land, a network

Illustration 1-7
This engraving of James Watt, a Scotch engineer, first appeared in 1860. It portrays an experimental design which was to become the forerunner of the modern condensing steam engine. Watt had been given a Newcomen engine to repair, in 1764, and, noting its extreme wastefulness of fuel, set about the task of building a better machine.

of railroads and highways was built to tie together the major cities in Europe and in the United States.

In this period, the awakening of science and engineering truly had begun. Now, although people were slow to accept new ideas, knowledge was not rejected as it had been in earlier centuries. Colleges began to teach more and more courses in science and engineering, and it was here that the fuse was lighted for an explosion of discoveries in the twentieth century.

One of the most important reasons for the significant development of technology in this period was the increasingly close cooperation between science and engineering. It began to become more and more evident that discoveries by research scientists could be used to develop new articles for commerce. Industry soon began to realize that money spent for research and development eventually returned many times its value.

Twentieth-century technology

As the twentieth century came into being, a number of inventions emerged that were destined to have far-reaching effects on our civilization. The automobile began to be more widely used as better roads were made available. The inventions of Edison

Illustration 1-8

Nations the world over acknowledge their indebtedness to Thomas Alva Edison, inventor, and Charles Proteus Steinmetz, electrical engineer, for their significant inventions relating to electricity.

(Illustration 1-8) and DeForest of electrical equipment and electron tubes started the widespread use of power systems and communication networks. Following the demonstrations by the Wright brothers that man could build a machine that would fly, aircraft of many types developed rapidly (Illustration 1-9).

These inventions, typical of many basic discoveries that were made early in the century, exemplify the spirit of progress of this period. So fast has been the pace of discovery, with one coming on the heels of another, that it is difficult to evaluate properly their relative importance, although we certainly can realize their impact on our way of life. However, in a number of instances the practicality of an engineering

> I know this world is ruled by Infinite Intelligence. It required Infinite Intelligence to create it and it requires Infinite Intelligence to keep it on its course. Everything that surrounds us—everything that exists—proves that there are Infinite Laws behind it. There can be no denying this fact. It is mathematical in its precision.
> —Thomas Alva Edison
>
> No man really becomes a fool until he stops asking questions.
> —Charles P. Steinmetz

Illustration 1-9

Orville and Wilbur Wright, U.S. inventors and aviation pioneers, achieved in 1903 the world's first successful powered, sustained, and controlled flights of an airplane.

invention has been demonstrated many years in advance of its implementation (Illustration 1-10).

Until late in the nineteenth century, engineering as an applied science was divided

Illustration 1-10

The first public demonstration of television as an adjunct to the telephone took place on April 7, 1927, when Herbert Hoover, then Secretary of Commerce, and other officials in Washington, D.C., spoke "face-to-face" with officials of the Bell Laboratories in New York City. In the ensuing 50 years the concept of a video telephone system has progressed through several evolutionary designs, represented by the models in the foreground.

into two principal groups, civil and military. Mining and metallurgy was the first group to be recognized as a separate branch, and the American Institute of Mining and Metallurgical Engineers was founded in 1871. In 1880 the American Society of Mechanical Engineers was founded, and in 1884 the American Institute of Electrical Engineers (now the Institute of Electrical and Electronics Engineers) was founded. In 1908 the American Institute of Chemical Engineers was founded and since then a number of other branch societies have been founded with objectives peculiar to specialized fields of engineering endeavor.

An outstanding characteristic of this century is the increased use of power. Albert Einstein was one of the world's most acclaimed physicists. His statement of the equivalence of mass and energy, $E = mc^2$, made possible the emergence of nuclear power (Illustration 1-11). In 1940 it was estimated that the total energy generated in the United States would be the equivalent in "muscle-power energy" of 153 slaves working for every American man, woman, and child in the country. Today a similar calculation would show that about 500 "slaves" are available to serve each person.[1] This is a considerable advance from the days of the Egyptians and Greeks.

Following World War II, the political, economic, and scientific disorganization in the world caused the emigration of many outstanding educators, scientists, and engineers to the United States. Here they have been able to expand their knowledge and skills and to aid generally in advancing our own understanding of the basic natural laws on which the improved techniques of the future will be based.

[1] *The Humble Way,* Third Quarter 1970, p. 10.

Illustration 1-11
Albert Einstein, father of the nuclear age, emphasized the importance of simplicity in all matters. He believed that harmony will result if people act in accordance with principles founded on consciously clear thinking and experience.

Engineering today

Broadly speaking, modern engineering had its beginnings about the time of the close of the Civil War. Within the last century, the pace of discovery has been so rapid that it can be classed as a period within itself. In these modern times, engineering endeavor has changed markedly from procedures used in the time of Imhotep, Galileo, or Ampere. Formerly, engineering discovery and development were accomplished principally by individuals. With the increased store of knowledge available and the widening of the field of engineering to include so many diverse branches, it is usual to find groups or teams of engineers and scientists working on a single project. Where formerly an individual could absorb and understand practically all of the scientific knowledge available, now the amount of information available is so vast that an individual can retain and employ at best only a part of it.

Since 1900 the ratio of engineers and scientists in the United States in comparison to the total population has been steadily increasing. Predictions based upon past increases seem to indicate the following:

Year	Ratio of U.S. engineers and scientists to population
1900	1 to 1800
1950	1 to 190
1960	1 to 130
1980	1 to 65
2000	1 to 35

If this is the case, there will be an even greater increase in technological advance in the next 20 years than there has been in the past 20 years.

Within the past two decades, four technological developments have produced profound changes in our way of life. These developments are nuclear power, the electronic digital computer, interplanetary space navigation, and microelectronics. These concepts are still in their early stages of development, but historians of the future may well refer to our time by such terms as the *nuclear age,* the *computer age,* or the *age of space travel* (Illustration 1-12). The engineer has been a principal developer of these concepts because of the need for their capabilities. The ocean offers great possibilities for technological exploration and perhaps even greater rewards for civilization than has space exploration (Illustration 1-13).

In this age, as in any age, the engineer must be creative and must be able to visualize what may lie ahead. He must possess a fertile imagination and a knowledge of what others have done before him. As Sir Isaac Newton is reputed to have said, "If I have seen farther than other men, it is because I have stood on the shoulders of giants." The giants of science and engineering still exist. All any person must do to increase his field of vision is to climb up on their shoulders.

> My interest is in the future because I am going to spend the rest of my life there.
> —Charles F. Kettering

Illustration 1-12
Man's safe exploration of the moon will long stand as one of his greatest engineering achievements.

Problems on history of engineering

1-1. Prepare a chart as a series of columns, showing happenings and their approximate dates in a vertical time scale for various civilizations, beginning about 3000 B.C. and extending to about A.D. 1200.

Chinese	Middle East	Egyptian	Greek	Roman	Western Europe

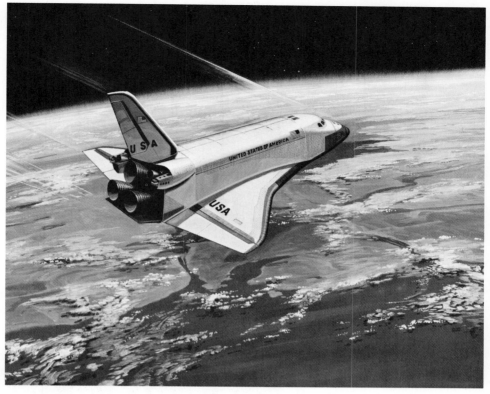

Illustration 1-13
The space shuttle orbiter will carry satellites into space for placement in orbit. It not only has the ability to maneuver in space like a spacecraft, but also to enter the earth's atmosphere and land in a manner similar to a jet transport.

1-2. Prepare a brief essay on the possible circumstances surrounding the discovery that an iron needle when rubbed on a lodestone and then supported on a bit of wood floating on water will point to the north.

1-3. Determine from historical references the approximate number of years that major civilizations existed as important factors in history.

1-4. What were the principal reasons for the lack of advancement of discovery in Greek science?

1-5. Explain why the development of a successful horsecollar was a major technological advancement.

1-6. Draw to some scale a typical cross section of the "Great Wall" of China, and estimate the volume of rock and dirt required per mile of wall.

1-7. Trace the development of a single letter of our alphabet from its earliest known symbol to the present.

1-8. Describe the details of preparing papyrus from reed-like plants which grew in Egypt.

1-9. Describe the patterns of behavior and accomplishments of ancient engineers that seem to have made successful civilizations, and to have prolonged their existence.

1-10. Prepare lists of prominent persons who contributed outstanding discoveries and developments to civilization during the period from A.D. 1200 to A.D. 1900 in the fields of (*a*) mathematics, (*b*) astronomy, (*c*) electricity, (*d*) mechanics, and (*e*) light.

Person	Date	Major contribution

1-11. List the ten most significant engineering achievements of the twentieth century.

1-12. Beginning with 3000 B.C., list the 25 most significant engineering achievements.

1-13. Based upon your knowledge of world history, describe the probable changes that might have occurred had the airplane not been invented until 1970.

1-14. Describe the precision with which the pyramids of Egypt were constructed. How does this precision compare with that of modern office buildings of more than 50 stories in height?

1-15. Trace the development of the power-producing capability of man from 3000 B.C. to the present.

Illustration 1-14
Our ability to achieve a better future will in large measure be determined by our understanding of the past.

2

Serving the needs of a constantly changing society

The earth and its inhabitants form a complex system of constantly changing inter-relationships. From the time of creation until a few thousand years ago the laws of nature programmed the actions of all living things in relation to each other and to their environments. From the beginning, however, change was ever present. For example, as glaciers retreated, new forests grew to reclaim the land, ocean levels and coastlines changed, the winding courses of rivers altered, lakes appeared, and fish and animals migrated, as appropriate, to inhabit the new environments. Changes in climate and/or topography always brought about consequential changes in the distribution and ecological relationships of all living things—plants and animals alike. Almost invariably these changes brought about competitive relationships between the existing and migrating species. In this way certain competing forms were forced to adapt to new roles, while others became extinct. As a result of this continual change in the ecological balance of the earth over eons of time, there currently exist some 1,300,000 different kinds of plants and animals that make their homes in rather specific locations. Only a few, such as the cockroach, housefly, body louse, and house mouse, have been successful in invading a diversity of environments—because they chose to follow man in his travels. Presumably even these would be confined to specific regions if man did not exist on the earth.

Primitive man was concerned with every facet of his environment and had to be acutely aware of many of the existing ecological interrelationships. For example, he made it his business to know those places most commonly frequented by animals that he considered to be good to eat or whose skins or pelts were valued for clothing. He distinguished between the trees, plants, and herbs and he knew which would provide him sustenance. Although by today's standards of education men of earlier civilizations might have been classified as "unlearned," they were certainly not ignorant. The Eskimo, for example, knew long ago that his sled dogs were susceptible to the diseases

> The air, the water and the ground are free gifts to man and no one has the power to portion them out in parcels. Man must drink and breathe and walk and therefore each man has a right to his share of each.
> —James Fenimore Cooper, *The Prairie*, 1827

of the wild arctic foxes, and the Masai of East Africa have been aware for centuries that malaria is caused by mosquito bites.[1]

For thousands of years after man first inhabited the earth, populations were relatively small and, because man was mobile, the cumulative effect of his existence upon the ecology of the earth was negligible. If, perchance, he did violence to a locality (for example, caused an entire forest to be burned), it was a relatively simple matter for him to move to another area and to allow time and nature to heal the wound. Due to the expanded world population this alternative is no longer available to him.

From century to century man has continued to add to his store of knowledge and understanding of nature. In so doing he has advanced progressively from a crude nomadic civilization, where he used what he could find useful to him in nature, to one that was sustained by domestication and agriculture, where he induced nature to produce more of the things that he wanted, and currently to one in which he is endeavoring to use technologies of his own design to control the forces of nature.

Man's existence is a part of, not independent of, nature—specifically it is most concerned with that part of nature that is closest to the surface of the earth, known as the *biosphere*. This is the wafer-thin skin of air, water, and soil comprising only a thousandth of the planet's diameter and measuring less than 8 miles thick.[2] It might be said to be analogous to the skin of an apple. Within this relatively narrow space, however, it encompasses the entire fabric of life as we know it—from virus to field mouse, man, and whale. Most of the life forms live within a domain extending from the top $\frac{1}{2}$ mile of the ocean to 2 miles above the earth's surface, although some creatures do inhabit the extreme boundaries of the biosphere. A number of processes of nature provide the biosphere with a delicate balance of characteristics that are necessary to sustain life. The life cycles of all living things, both fauna and flora, are interdependent and inextricably interwoven to form a delicately balanced *ecological system* that is as yet not completely understood by man. We do know, however, that not only is every organism affected by the environment of the "world" in which it lives, but it also has some effect on this environment.[3] The energy necessary to operate this system comes almost entirely from the sun, and is utilized primarily through the processes of photosynthesis and heat. Any changes that man exerts on any part of the system will affect its precarious balance and cause internal adjustment of either its individual organisms, its environment, or both. The extent and magnitude of the modifications that man has exerted on this ecological system have increased immeasurably within the past few years, particularly as a consequence of his rapid population growth. Certain of these modifications are of particular concern to the engineer.

Even during the period of the emergence of agriculture and domestication of

[1] Peter Farb, *Ecology*, Time, New York (1963), p. 164.
[2] Robert C. Cook (ed.), "The Thin Slice of Life," *Population Bulletin*, Vol. XXIV, No. 5, p. 101.
[3] Marston Bates, "The Human EcoSystem," *Resources and Man*, W. F. Freeman, San Francisco (1969), p. 25.

Illustration 2-1
There are few areas to be found today that have been spared the ravages of man.

animals, man began to alter the ecological balance of his environment. Eventually some species of both plants and animals became extinct, while the growth of others was stimulated artificially. All too often man was not aware of the extent of the consequences of his actions and more particularly of the irreversibility of the alterations and imbalances that he had caused in nature's ecosystem. His concerns have more often been directed toward subduing the earth than replenishing it. Over the period of a few thousand years nature's law of "survival of the fittest" was gradually. replaced by man's law of "survival of the most desirable." From man's short-term point of view, this change represented a significant advantage to him. Where once he was forced to gather fruit and nuts and to hunt animals to provide food and clothing for himself and his family—a life filled with uncertainty, at best—now he could simplify his food-gathering processes by increasing the yield of crops such as wheat,

And God blessed them, and God said unto them, be fruitful, and multiply, and *replenish* the earth, and *subdue* it: and have dominion over the fish of the sea, and over the fowl of the air, and over every living thing that moveth upon the earth.
—Genesis 1:28

There is another design that is far better. It is the design that nature has provided. . . . It is pointless to superimpose an abstract, man-made design on a region as though the canvas were blank. It isn't. Somebody has been there already. Thousands of years of rain and wind and tides have laid down a design. Here is our form and order. It is inherent in the land itself—in the pattern of the soil, the slopes, the woods—above all, in the patterns of streams and rivers.
—William H. Whyte, *The Last Landscape*

We shape our buildings; and forever afterwards our buildings shape us.
—Sir Winston Churchill; Speech in House of Commons, October 1943

Illustration 2-2
Man's continued misuse of the world's resources can only lead to a depletion of the essentials for life.

corn, rice, and potatoes. In addition certain animals such as cows, sheep, goats, and horses were protected from their natural enemies and, in some instances, their predators were completely annihilated. Such eradication seemed to serve man's immediate interests, but it also eliminated nature's way of maintaining an ecological balance. Man, in turn, was also affected by these changes. At one time only the strongest of his species survived, and the availability or absence of natural food kept his population in balance with the surroundings. With domestication these factors have become less of a problem and "survival of the fittest" no longer governs his increase in numbers. In general, today both strong and weak live and procreate. Because of this condition the world's population growth has begun to mount steadily and *alarmingly* because of the manifold problems that accompany large populations and for which solutions are still to be found.

In many respects the young engineer of today lives in a world that is vastly different from the one known to his grandfather or great grandfather. Without question he enjoys a standard of living that is unsurpassed in the history of mankind; and yet, in spite of the significant agricultural and technological advances that have been made in this generation, over half the world's population still lives in perpetual hunger. Famine, disease, and war continue to run rampant throughout portions of the earth. In addition, the residue from his own technology continues to mount steadily. Nevertheless his population growth continues unabated, further compounding these problems.

The population explosion —a race to global famine

Until comparatively recent times the growth of the human race was governed by the laws of nature in a manner similar to that controlling the growth of all other living things. As man's culture changed from nomadic to agrarian to technological, however, he began to alter nature's population controls significantly. Through control of disease and pestilence his average life span has been extended over three times. His ability to supply his family consistently with food and clothing has also been improved immeasurably. Combinations of these two factors have caused his population to increase in geometric progression: 2–4–8–16–32–64–128, and so on.[4] Initially it took hundreds of thousands of years for a significant change to occur in the world population. However, as the numbers became larger, and particularly in more recent times as man's life span began to lengthen as a result of his gaining some control over starvation, disease, and violence, the results of geometric growth began to have a profound effect. Where it has taken an estimated 2 million years for the world population to reach slightly over 3 billion persons, it will take only 35 years to add the next 3 billion *if present growth rates remain unchanged.* The significance of this problem is illustrated by Figure 2-1.

As in the case of man's altering his ecological environment, he has not always been wise enough to anticipate all the various effects of his changes. In the case of his own propagation he has managed to introduce "death control," but birth rates have continued to climb, particularly in the underdeveloped countries. It is estimated that the average annual increase in world population in 1650 was only 0.3 per cent.[5] In 1900 this annual growth rate had increased to 0.9 per cent; in 1930, 1.0 per cent; in 1960, 2.0 per cent; and in 1970, over 2.1 per cent. Unless these rates decline significantly, which does not now seem likely, a worldwide crisis is fast approaching.

The earth's land area, only 10 per cent of which appears to be arable, is fixed and

[4] T. R. Malthus, 1798, *An Essay on the Principle of Population As It Affects the Future Improvements of Mankind,* facsimile reprint in 1926 for J. Johnson. Macmillan, London.
[5] Joseph M. Jones, *Does Overpopulation Mean Poverty?,* Center for International Economic Growth, Washington, D.C. (1962), p. 13.

Figure 2-1

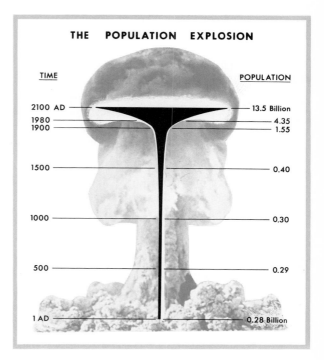

THE POPULATION EXPLOSION

TIME		POPULATION
2100 AD		13.5 Billion
1980		4.35
1900		1.55
1500		0.40
1000		0.30
500		0.29
1 AD		0.28 Billion

Illustration 2-3
Traditionally one person signifies loneliness; two persons—companionship; three persons—a crowd. In more recent times a new concept has been added: multitudes signify pollution, frustration, and depression.

unexpandable, and a shortage of food and water is already an accepted fact of life in many countries. Although conditions in some slum areas of the United States are very bad, they bear little resemblance to many areas of the world where people grovel in filth and live little better than animals. A recent visit to India is described as follows:

> I have understood the population explosion intellectually for a long time. I came to understand it emotionally one stinking hot night in Delhi a couple of years ago. My wife and daughter and I were returning to our hotel in an ancient taxi. The seats were hopping with fleas. The only functional gear was third. As we crawled through the city, we entered a crowded slum area. The temperature was well over 100, and the air was a haze of dust and smoke. The streets seemed alive with people. People eating, people washing, people sleeping. People visiting, arguing, and screaming. People thrusting their hands through the taxi window, begging. People defecating and urinating. People clinging to buses. People herding animals. People, people, people, people. As we moved slowly through the mob, hand horn squawking, the dust, noise, heat, and cooking fires gave the scene a hellist aspect.[6]

Over one third of the human beings now living on the earth are starving and another one third are ill fed. The underdeveloped countries of the world are incapable of producing enough food to feed their populations. This deficiency is 16 million tons of food each year and will grow to a staggering 88 million tons of food per year by 1985. For these people to be fed an adequate diet, the current world food production would have to double by 1982, *which appears to be an impossible task*. The quantity of food available is not the only problem; it must also be of the proper quality. For example, in Central Africa every other baby born dies before the age of five *even though food is generally plentiful*. They die from a disease known as *kwashiorkor*, which is caused by a lack of sufficient protein in the diet. The magnitude of the problem continues to grow as the world population increases. More population

[6] Paul R. Ehrlich, *The Population Bomb*, Ballantine, New York (1968), p. 15.

Illustration 2-4
*One half of the people
in the world had food
to eat today . . .*

means more famine. It also means more crowding, more disease, less sanitation, more waste and garbage, more pollution of air, water, and land, and ultimately . . . the untimely death of millions, or possibly, the end of human life on earth.

In the United States, just as in the underdeveloped countries of the world, the population has tended to migrate to cities and some of the cities have grown to monstrous size, often called megalopolises. These treks have been brought about by the widespread use of mechanized agriculture and the impoverishment of the soil. With these migrations special problems have arisen. No longer is a city dweller self-sufficient in his ability to provide sustenance, shelter, and security for his family. Rather, his food, water, fuel, and power must all be brought to him by others and his wastes of every kind must be taken away. Most frequently his work is located many miles from his home and his reliance upon a transportation system becomes critical.

Illustration 2-5
*. . . the other one half face
starvation.*

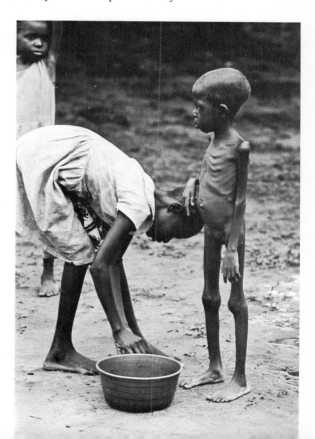

Illustration 2-6
Wheat is one of the world's most important food crops because each grain contains a large amount of protein. Even though engineering designs and advances in agricultural science have made possible a substantially increased production, the disparity between food supply and world population continues to increase.

He finds himself vulnerable to every kind of public emergency, and the psychological pressures of city life often lead to mental illness or escape into the use of alcohol and drugs. The incidence of crime increases, and his clustering invites a more rapid spread of disease and pestilence than ever. In general his cities are enormous consumers of electrical and chemical energy and producers of staggering amounts of wastes and pollution. Today 70 per cent of the people in our country live on 1 per cent of the land, and the exodus from the countryside continues.

What are the implications for the engineer of these national and international sociological crises? As a member of a profession that is committed to the *solving* of man's physical and economic problems, the engineer—of all professionals—is perhaps the most concerned. However, the complexity of his task is threefold. First, he must learn all he can about the extent and causes of the technological problems that exist; second, he must direct his energies and abilities to solving them; and third, he

> For this is the word that the Lord has spoken.
> The earth dries up and withers,
> The whole world withers and grows sick;
>   the earth's high places sicken,
> and the earth itself is desecrated by the feet of those who live in it,
> because they have broken the laws, disobeyed the statutes
> and violated the eternal covenant.
> —Isaiah xxiv:3–5 (New English Translation)

must continually assess the manifold effects of his designs to ensure that they tend to restore, rather than depreciate, the equilibrium to the ecological system of nature in those cases where it has become unbalanced.

Just as the engineer must understand how the physical laws of nature govern the universe, so also must he recognize that nature's laws governing the procession and diversity of life on this planet are equally valid and unyielding. The remainder of this chapter will consider the severity and complexity of several critical problems that confront today's society and, more particularly, the engineer's role in designing for their solution.

Our polluted planet

In the last few years the average American has become aware that our "spaceship earth" is undergoing many severe and detrimental ecological changes which may take hundreds of years to restore. Unfortunately he is not always able to distinguish effects that are of a temporary nature from those with long-term consequences. Frequently his most damaging actions to the environment are either of an incremental or visually indistinguishable nature, and for this reason he participates willingly in them. In some measure man's reactions are dulled by the slowness of deterioration. This is somewhat analogous to the actions of a frog that will die rather than jump out, when placed in a bucket of water that is being *slowly* heated. In contrast, if the frog is pitched into a bucket of boiling water, he will immediately jump out and thereby avoid severe injury. The engineer in particular must learn to understand such "cause-and-effect" relationships so that his designs will enhance the orderly and natural development of life.

The air environment

The atmosphere, which makes up the largest fraction of the biosphere, is a dynamic system that absorbs continuously a wide range of solids, liquids, and gases from both natural and man-made sources. These substances often travel through the air, disperse, and react with one another and with other substances, both chemically and physically. Eventually most of these constituents find their way into a depository, such as the ocean, or to a receptor, such as a man. Some, however, such as helium, escape from the biosphere. Others, such as carbon dioxide, may enter the atmosphere faster than they enter a receptor and thus gradually accumulate in the air.[7]

Clean, dry air contains 78.09 per cent nitrogen by volume, and 20.94 per cent oxygen. The remaining 0.97 per cent is composed of a gaseous mixture of carbon dioxide, helium, argon, krypton, nitrous oxide, and xenon, as well as very small amounts of some other organic and inorganic gases whose amounts vary with time and place in the atmosphere. Through both natural and man-made processes that

---

[7] *Cleaning Our Environment: The Chemical Basis for Action,* American Chemical Society, Washington, D.C. (1969), p. 23.

Illustration 2-7
Complacency leads to industrial pollution.

exist upon the earth, varying amounts of contaminants continuously enter the atmosphere. That portion of these substances which interacts with the environment to cause toxicity, disease, aesthetic distress, physiological effects, or environmental decay has been labeled by man as a pollutant. In general, the actions of people are the primary cause of pollution and, as population increases, the attendant pollution problems also increase proportionally. This is not a newly recognized relationship, however. The first significant change in man's effect on nature came with his deliberate making of a fire. No other creature on earth starts fires. Prehistoric man built a fire in his cave home for cooking, heating, and to provide light for his family. Although the smoke was sometimes annoying, no real problem existed with regard to pollution of the air environment. However, when his friends or neighbors visited him and also built fires in the same cave, even prehistoric man recognized that he then had an *air pollution problem*. Some nineteenth-century cities with their hundreds of thousands of smoldering soft-coal grates coughed amid a thicker and deadlier smog than any modern city can concoct.[8] Today the natural terrain that surrounds large

[8]Tom Alexander, "Some Burning Questions About Combustion," *Fortune,* February 1970, p. 130.

> A recent scientific analysis of New York City's atmosphere concluded that a New Yorker on the street took into his lungs the equivalent in toxic materials of 38 cigarettes a day.
> —Robert Rienow and Leona Train Rienow, *Moment in the Sun*

Table 2-1   Sources of air pollutants in the United States [millions of tons/year]*

Source	Totals	Per cent of totals	Carbon monoxide	Sulfur oxides	Hydro-carbons	Nitrogen oxides	Particulate matter	Other
Motor vehicles	86	60	66	1	12	6	1	†
Industry	25	17	2	9	4	2	6	2
Power plants	20	14	1	12	†	3	3	
Space heating	8	6	2	3	1	1	1	†
Refuse disposal	4	3	1	†	1	†	1	†
Totals	143	100	72	26	18	12	12	4

* Adapted from *Your Right to Clean Air,* The Conservation Foundation, Washington, D.C. (1970), p. 15.
† Less than 1 per cent.

cities is recognized as having a significant bearing on the air pollution problem. However, this is not an altogether new concept either. Historians tell us that the present Los Angeles area, which in recent years has become a national symbol of comparison for excessive smog levels,[9] was known as the "Valley of Smokes" when the Spaniards first arrived.[10] In recent years air pollution has become a problem of world concern.

In the United States the most common air pollutants are carbon monoxide, sulfur oxides, hydrocarbons, nitrogen oxides, and particles (Table 2-1). Their primary sources are motor vehicles, industry, electrical power plants, space heating, and refuse disposal, with approximately 60 per cent of the bulk being contributed by motor vehicles and 17 per cent by industry. It seems possible that America's streets will contain twice as many automobiles as the current 100 million by the year 2000. This is certainly a foreboding prospect *unless improved engineering designs are able to alleviate the situation.* Certainly restoring the quality of the atmosphere ranks as one of the most difficult and challenging tasks of our generation. As with the determined effort of the 1960's to explore the moon, an engineering venture of this magnitude requires a coordinated national effort of great magnitude. With every man, woman, and child in the United States producing an average of 1400 pounds per year of air pollutants, the problem is certainly one of serious proportions.

It has been found that the significantly increasing volume of particulate matter entering the atmosphere scatters the incoming sunlight. This reduces the amount of heat that reaches the earth and tends to reduce its temperature. The decreasing mean global temperature of recent years has been attributed to the rising concentrations of

[9] The term *smog* was coined originally to describe a combination of smoke and fog, such as was common in London when coal was widely used for generating power and heating homes. More recently it has come to mean the accumulation of photochemical reaction products which result largely from the action of the radiant energy of the sun on the emissions of internal combustion engines (automobile exhaust).
[10] Henry C. Wohlers, *Air Pollution—The Problem, the Source, and the Effects,* Drexel Institute of Technology, Philadelphia (1969), p. 1.

> Environment pollution . . . now affects the whole earth. Smog produced in urban and industrial areas is hovering over the countryside and beginning to spread over the oceans . . . cities will not benefit much longer from the cleansing effects of the winds for the simple reason that the wind itself is contaminated.
> —*The New York Times,* January 6, 1969

airborne particles in the atmosphere.[11] A counteracting phenomenon, commonly referred to as the "greenhouse effect," is caused by the increasing amounts of carbon dioxide found in the atmosphere. Although carbon dioxide occurs naturally as a constituent of the atmosphere and is not normally classified as an air pollutant, man does generate an abnormally large amount of it in those combustion processes that utilize coal, oil, and natural gas. The presence of water vapor in the atmosphere, and to a lesser extent carbon dioxide and ozone, acts in a manner similar to the glass in a greenhouse. Light from the sun arrives as short-wavelength radiation (visible and ultraviolet) and is allowed to pass through it to heat the earth, but the relatively long wavelength infrared radiation (heat radiation) that is emitted by the earth is absorbed—thereby providing an unnatural and additional heating effect to the earth. It has been estimated that if the carbon dioxide content in the atmosphere continues to increase at the present rate, the mean global temperature could rise enough to cause undesirable side effects.

Air pollution can cause death, impair health, reduce visibility, bring about vast economic losses, and contribute to the general deterioration of both our cities and countryside. It is therefore a matter of grave importance that engineers of all disciplines consciously incorporate in their designs sufficient constraints and safeguards to ensure that they do not contribute to the pollution of the atmosphere. In addition, they must apply their ingenuity and problem-solving abilities to eliminating air pollution where it exists and restoring the natural environment. However, it is equally important that they use their expertise in these matters to assist the general public, and particularly the political leadership of the country, to appraise realistically remedial measures that should be implemented. All too often the public is inclined to listen to vocal self-proclaimed "experts" from one faction or another who historically

[11] Reginald E. Newell, "The Global Circulation of Atmospheric Pollutants," *Scientific American,* January 1971, p. 40.

Illustration 2-8
One solution to the air pollution problem. Surely there must be a more desirable alternative!

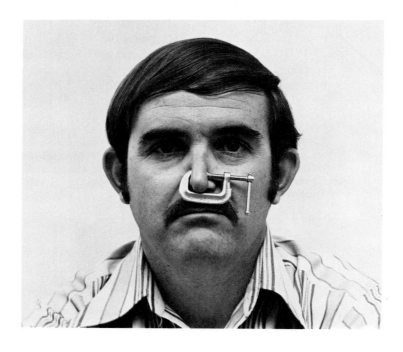

prey on the drama of the moment to bear their tidings of gloom and doom for all civilization.

The quest for water quality

Water is the most abundant compound to be found upon the face of the earth and, next to air, it is the most essential resource for man's survival. The per capita use of water in the United States is 1000 gallons per day and this demand continues to grow. Early man was most concerned with the quality (purity) of his drinking water, and even he was aware that certain waters were contaminated and could cause illness or death. In addition, modern man has found that he must be concerned also with the quantity of the water available for his use. An abundant supply of relatively pure water is no longer available in most areas. Today water pollution, *the presence of toxic or noxious substances or heat in natural water sources,* is considered to be one of the most pressing social and economic issues of our time. Just how bad water pollution can become was dramatically illustrated a few years ago when the oily, chocolate-brown Cuyahoga River in Cleveland, Ohio, caught fire and blazed for several days, nearly destroying two railroad bridges that spanned the river. This once-beautiful river was so filled with municipal and industrial wastes that even the leeches and sludge worms normally found only in polluted rivers could not survive, and it was, in effect, nothing more than a flammable sewer.[12] Unfortunately, many other water bodies in the United States are no less polluted, and now approximately one half the people in the nation "drink their own treated sewage." Unfortunately a number of city water treatment plants merely remove the particulate matter and disinfect the available water with chlorine to kill bacteria, since they were not originally designed to remove pesticides, herbicides, and other organic and inorganic chemicals that may be present.[13] This problem has become acute in a number of areas as hundreds of new contaminants have been discovered in streams and lakes: bacteria and viruses, detergents, municipal sewage, acid from mine drainage, pesticides and weed killers, radioactive substances, phosphorus from fertilizers, trace amounts of metals and drugs, and other organic and inorganic chemicals. As the population continues to increase, the burden assumed by the engineer to design more comprehensive water treatment plants also mounts. Indeed, the well-being of entire communities may depend upon the engineer's design abilities, because it is now a recognized condition of population increase that "everyone cannot live upstream."

The processes of nature have long made use of the miraculous ability of rivers and lakes to "purify themselves." After pollutants find their way into a water body they are subject to dilution, or settling, action of the sun, and to being consumed by

[12] Gene Bylinsky, "The Limited War on Water Pollution," *The Environment,* Harper & Row, New York (1970), p. 20.
[13] Ibid., p. 25.

Modern man . . . has asbestos in his lungs, DDT in his fat, and strontium 90 in his bones.
—*Today's Health,* April 1970

beneficial bacteria. The difficulty arises when man disturbs the equilibrium of the ecosystem by dumping large amounts of his organic wastes into a particular water body, thereby intensifying the demand for purification. In time the body of water cannot meet the demand, organic debris accumulates, anaerobic areas develop, fish die, and putrification is the result. This process also occurs in nature but it may take many hundreds of years to complete the natural processes of deterioration. Man can alter nature's time scale appreciably.

Lake Erie, the smallest of the Great Lakes, provides a classic example of what can happen when man ignores his responsibility to protect a natural resource.[14] *Eutrophication,* a term used to denote the process of nutrient enrichment by which lakes fill and die, has overtaken the lake prematurely, and because of this it is estimated to have aged hundreds of years since 1925. These consequences, however, should come as no surprise. The lake is fed by a number of heavily polluted rivers, including the Cuyahoga and Buffalo rivers. In addition, over 9.6 billion gallons of industrial waste and 1.5 billion gallons of sewage are dumped into the lake *each day* from adjoining states.[15] Among the most damaging pollutants in this lake are those that stimulate the growth of algae—tiny, green plants—which then multiply until they become large

[14] Victor W. Wigotsky, "Engineering and the Urban Crisis," *Design News,* December 7, 1970, p. 31.
[15] "Eat, Drink, and Be Sick," *Medical World News,* September 26, 1969, p. 32.

Illustration 2-9
As a direct result of good engineering design, the average drinking water quality in the United States is the best of any country in the world.

Illustration 2-10
*Pollution does not
necessarily need to
accompany poverty,
but it most frequently
does.*

green mats that literally clog and stifle the lake. When the algae die, their decay depletes the water of its life-giving oxygen, choking forms of desirable aquatic life, including fish.[16]

In spite of the conditions described above, Lake Erie is not a dead body of water. Its waters continue to be used as the primary source of drinking water for a number of cities, including Cleveland, Toledo, Loraine, Erie, and Buffalo. Also, extensive use is made of its water for commercial fishing, agricultural uses, industrial cooling water, recreation, and wildlife.[17]

The presence of radioactive wastes and excess heat are relatively new types of pollution to water bodies, but they are of no less importance for the engineer to take into account in his designs. All radioactive materials are biologically injurious. Therefore, radioactive substances that are normally emitted by nuclear power plants are suspected of finding their way into the ecological food chain where they could cause serious problems. For this reason all radioactive wastes should be isolated from the biological environment during the "life" of the isotopes (as much as 1000 years in some cases). The "heat" problem arises because nuclear electrical power generating plants require great quantities of water for cooling. Although the heated discharge water from a nuclear power plant is approximately the same temperature as that from a fossil-fuel power plant, the quantity has been increased by approximately 40 per cent. The warmer water absorbs less oxygen from the atmosphere, and this accelerates the normal rate of decomposition of organic matter. This unnatural heat also unbalances the life cycles of fish who, being cold blooded, cannot regulate their body temperatures correspondingly. With the prospect of the number of nuclear power plants doubling by 1980, this problem will loom larger than ever.

Engineering designs of the future must take into account all these factors to ensure for all the nation's inhabitants a water supply that is both sufficient in quantity and

[16] Senator Gaylord Nelson, *Congressional Record,* February 26, 1970, p. S2444.
[17] Cy A. Adler, *Ecological Fantasies,* Delta, New York (1973), p. 113.

unpolluted in quality. This is not only the engineer's challenge, but his responsibility as well.

Solid waste disposal

We are living in a most unusual time—a time where possibly the most valuable tangible asset that a person could own would be a "bottomless hole." Never before in history have so many people had so much garbage, refuse, trash, and other wastes to dispose of, and at the same time never before has there been such a shortage of "dumping space." The proliferation of refuse, however, is only partly attributable to the population explosion. A substantial portion of the blame must be assumed by an affluent society that is careless with its increasing purchasing power and which has demonstrated a decided distaste for secondhand articles.[18] Table 2-2 lists the primary solid waste constituents of the United States.

The most popular method of solid waste disposal has long been "removal from the immediate premises." For centuries man has been aware of the health hazards

[18] Alex Hershaft, "Solid Waste Treatment," *Science and Technology,* June 1969, p. 35.

Table 2-2 Solid waste of the United States

Category	Composition		Percentage distribution	Estimated production, 10^6 tons/year
Refuse				
Garbage	Animal and vegetable kitchen wastes		15	↑
Rubbish	Dry household, commercial, and industrial waste	Paper	28	
		Yard waste	14	230
		Glass and *metal*	10	
		Other	10	
Municipal waste	Construction waste and street sweepings		23	↓
Industrial wastes	Industrial processing scrap and by-products			120
Scrap metal	Automobiles, machinery, and major appliances			20
Sewage residue	Grit, sludge, and other residue from sewage treatment plants			30
Mining displacement	Overburden and gangue			1200
Agriculture waste	Manure, animal carcasses, crop and logging debris			2400

*Drawn from compilation by Alex Hershaft, "Solid Waste Management," Grumman Aerospace Corporation, Bethpage, N.Y. (1970).

that accompany the accumulations of garbage. Historians have recorded that a sign at the city limits of ancient Rome warned all persons to transport their refuse outside the city or risk being fined. Also, it has been recognized that in the Middle Ages the custom of dumping garbage in the streets was largely responsible for the proliferation of disease-carrying rats, flies, and insects that made their homes in piles of refuse.[19]

On the average, each person in the United States has established the following record with regard to the generation of solid waste, refuse, and garbage:

1920	2.75	pounds per person per day
1970	5.5	pounds per person per day
1980	8.0	pounds per person per day (estimated)

From New York to Los Angeles, cities throughout the nation are rapidly depleting their disposal space, and there is considerable concern that too little attention has been given to what is fast becoming one of man's most distressing problems—solid waste disposal. Why is trash becoming such a problem? The answer seems to lie partially in man's changing value system. Just a generation or two ago thrift and economy were considered to be important tenets of American life, and few items with any inherent value were discarded. Today, we live in the era of "the throwaway." More and more containers of all types are being made of nondecomposable plastic or glass, or nonrustable aluminum, and everything from furniture to clothing is being made from disposable paper products which are sold by advertising that challenges purchasers to "discard when disenchanted." In 1970 the American public threw away approximately 50 billion cans, 30 billion bottles, 35 million tons of paper, 5 million tons of plastic, and 100 million automobile tires. The problem of disposing of 7

[19] Ibid., p. 34.

Illustration 2-12
What contributions do you make to the country's mounting crisis in solid waste disposal?

million junked automobiles each year is also becoming a problem that can no longer be ignored—particularly when this volume is expected to climb to over 10 million per year by 1980. The truth is that most "consumers" in America have instead become "users and discarders," but this fact has not been taken into account.

Unfortunately most current waste disposal practices make no attempt to recover any of the potential values that are in solid wastes. The methods of disposing of refuse in most common use today are dumping and burning in the open, sanitary land fills,

Illustration 2-13
The character of a nation is revealed by examining its garbage.

Illustration 2-14
Pollution is a personal matter.

burial in abandoned mines or dumping at sea, and grinding in disposal systems followed by flushing into sewers. Some edible waste, such as garbage, is fed to hogs.

Since almost all the products of the engineer's design ingenuity will eventually be discarded due to wear or obsolescence, it is imperative that consideration for disposal be given to each design *at the time that it is first produced.* In addition, the well-being of society as we know it appears to depend in some measure upon the creative design abilities of the engineer to devise new processes of recycling wastes, and either the changing of the physical form of wastes or the manner of their disposal, or both. Such designs must be accomplished within the constraints of economic considerations and without augmenting man's other pollution problems: air, water, and sound. Basically, the solution of waste accumulation is a matter of attitude, ingenuity, and economics—all areas in which the engineer can make significant contributions.

The rising crescendo of unwanted sound

A silent world is not only undesirable but impossible to achieve. Man's very nature is psychologically sensitive to the many sounds that come to his ears. For example, he is pleased to hear the gurgle and murmur of a brook or the soothing whispering wind as it filters through overhead pine trees, but his blood is likely to chill if he recognizes the whirring buzz of a rattlesnake or hears the sudden screech of an automobile tire as it slides on pavement. He may thrill to the sharp bugle of a far-off hunting horn, but his thoughts often tend to lapse into dreams of inaccessible places as a distant train whistle penetrates the night.[20] Yes, sounds have an important bearing on man's sense

[20] *Noise: Sound Without Value,* Committee on Environmental Quality of the Federal Council for Science and Technology, Washington, D.C. (1968), p. 1.

Illustration 2-15
Man is affected psychologically by the sounds that he hears.

of well-being. While the average person's ears continue to alert him to impending dangers, their sensitivity is far less acute than in generations past, particularly for those who lived in less densely populated areas. It is said that, even today, aborigines living in the stillness of isolated African villages can easily hear each other talking in low conversational tones at distances as great as 100 yards, and that their hearing acuity diminishes little with age.[21] Even as man's technology has brought hundreds of thousands of desirable and satisfying innovations, it has also provided the means for the retrogression of his sense of hearing—for deafness caused by a deterioration of the microscopic hair cells that transmit sound from the ear to the brain. It has been found that prolonged exposure to intense sound levels will produce permanent hearing loss, and it matters not that such levels may be considered pleasing. (Some people purport to enjoy "rock" music concerts at sound levels exceeding 110 decibels,[22] Figure 2-2). Today noise-induced hearing loss looms as one of America's major health hazards.

Noise is generally considered to be any annoying or unwanted sound. Noise (like sound) has two discernible effects on man. One causes a deterioration of his sensitivity of hearing, and the other affects his "psychological state of mind." The adverse effects of noise have long been recognized as a form of environmental pollution. Julius Caesar was so annoyed by noise that he banned chariot driving at night, and, prior to the Civil War, studies in England reported substantial hearing losses among

[21] The Editors of Fortune, *The Environment,* Harper & Row, New York (1970), p. 136.
[22] A *decibel* (abbreviated db) is a unit of measure of sound intensity, or pressure change on the ear.

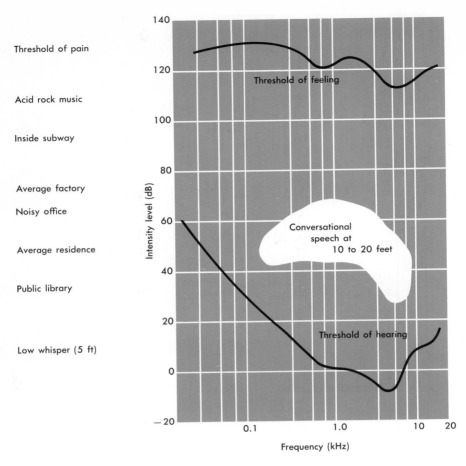

Figure 2-2 Average sound levels.
[Adapted from graphical presentations in Wesley E. Woodson and Donald W. Conover, *Human Engineering Guide for Equipment Designers,* second edition, (Berkley: University of California Press, 1964), p. 4–10.]

blacksmiths, boilermakers, and railroad men.[23] It has only been in recent years, however, that noise has been recognized as a health hazard.[24]

It is estimated that the average background noise level throughout the United States has been doubling each ten years. At this rate of increase, living conditions will be intolerable within a few years. Such a crescendo of sound results from the steady increase of population and the concomitant growth of the use of power on every hand—from the disposal in the kitchen, to the motorcycle in the street, to power tools in the factory. Buses, jet airliners, television sets, stereos, dishwashers, tractors, mixers, waste disposers, air conditioners, automobiles, jackhammers, power lawn mowers, vacuum cleaners, and typewriters are but a few examples of noise producers that are deemed desirable to today's high standard of living but which may very well also

[23] Aram Glorig, *Noise As a Public Health Hazard,* W. Dixon Ward and James E. Fricke (eds.), The American Speech and Hearing Association, Washington, D.C. (1969), p. 105.
[24] Andrew D. Hosey and Charles H. Powell (eds.), *Industrial Noise,* U.S. Department of Health, Education, and Welfare, Washington, D.C. (1967).

prevent man from fully enjoying the fruits of his labors, unless the sound levels at which they operate are altered significantly.

Except in the case of minimizing aircraft noise, the United States lags far behind many countries in noise prevention and control. Virtually all man-made noise can be suppressed, and the same engineer who formulates the idea for a new type of kitchen aid or designs an improved family vehicle must also be capable of solving the acoustical problems that are associated with his designs. In this regard he is responsible to generations yet unborn for the consequences of his actions.

Man's insatiable thirst for energy

In man's earliest habitation of the earth he competed for energy with other members of the earth's ecological environment. Initially his energy requirements were primarily satisfied by food—probably in the range of 2000 kilocalories (100 thermal watts) per person per day. However, as he has been able to make and control the use of fire, domesticate the plant and animal kingdoms, and initiate technologies of his own choosing, his per capita consumption of energy has increased appreciably. Today in the United States, man's thirst for energy (or *power,* which is the time rate use of energy) exceeds 10,000 thermal watts per capita per day,[25] which is about 100 times the

[25] M. King Hubbert, *Resources and Man,* W. H. Freeman, San Francisco (1969), p. 237.

Illustration 2-16
If the ear were to shatter or bleed profusely when subjected to abuse from intense or prolonged noise, we might be more careful of its treatment.

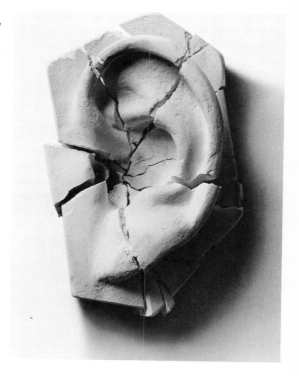

average of underdeveloped nations.[26] This demand has followed an exponential pattern of growth similar to the growth of the world population *except that the annual rate of increase for nonnutrient energy utilization is growing at a rate* (approximately 4 per cent per year) *considerably in excess of the world's growth in population* (approximately 2 per cent per year). This is brought about by man's appetite for more gadgets, faster cars and airplanes, heavier machinery, and so on.

The principal sources of the world's energy prior to about A.D. 1200 were solar energy, wood, wind, and water. At about this time in England it was discovered that certain "black rocks" found along the seashore would burn. From this there followed in succession the mining of coal[27] and the exploration of oil and natural gas reservoirs. More recently nuclear energy has emerged as one of the most promising sources of power yet discovered. The safe management and disposal of radioactive wastes, however, continue to present problems for the engineer.

Figure 2-3 provides a record of the history of energy consumption in the United States since 1800, and represents a prediction of how the continually rising demand might expand in the future.[28] Of course the future is unknown, and such a prediction of our energy sources for the year 2000 and beyond is mere conjecture. It depends to a large extent upon the background and experience of the predictor. External factors will also intervene. For example, today America is facing a daily use deficiency of energy that borders on the critical. This condition has been brought to public attention because changing world political conditions have curtailed oil imports, and

[26] Fred Singer, "Human Energy Production as a Process in the Biosphere," *Scientific American,* September 1970, p. 183.

[27] There are evidences that coal was used in China, Syria, Greece, and Wales as early as 1000 to 2000 B.C.

[28] L. P. Gaucher, "Energy Sources in the United States," *Journal of Solar Energy Society,* Vol. 9, No. 3 (1965), p. 122.

Illustration 2-17
The use of solar energy to heat and cool homes, schools, and industrial buildings requires a revision of traditional building designs.

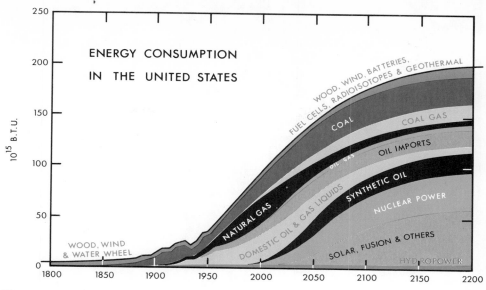

WOOD, WIND, BATTERIES, FUEL CELLS, RADIOISOTOPES & GEOTHERMAL

COAL

COAL GAS

OIL IMPORTS

OIL GAS

SYNTHETIC OIL

NUCLEAR POWER

NATURAL GAS

DOMESTIC OIL & GAS LIQUIDS

WOOD, WIND & WATER WHEEL

SOLAR, FUSION & OTHERS

HYDROPOWER

10^{15} B.T.U.

Figure 2-3

alternative sources had not been developed to supply the deficiency. However, the problem has been developing over a long period of time. The total energy consumption in the United States has more than doubled since 1950, and the average annual rate of energy consumption has grown from 3.5 to 4.5 per cent in the same period. With 6 per cent of the world's population, the United States today consumes one third of the world's energy.[29]

As with other societies of the past that have faced similar difficulties,[30] the United States must initiate energy-saving practices, shift energy consumption to sources other than oil, expand the production of existing principal energy sources (oil, gas, coal, and uranium), and develop several new energy sources.[31] At the current rate of usage, 180 $(10)^{15}$ Btu's of energy will be required by the turn of the century. Table 2-3 lists the

[29] *Exploring Energy Choices,* Ford Foundation (1974), p. 1.
[30] "Man's Age-Old Struggle for Energy," *EXXON USA,* Vol. XII, No. 4 (1973).
[31] H. A. Bethe, "The Necessity of Fusion Power," *Scientific American,* January 1976, p. 21.

Table 2-3   United States energy consumption

Present energy sources	Btu's of energy used annually × 10^{15}		Possible future energy sources	Estimated Btu's of energy × 10^{15} that can be produced annually by 2000 A.D.
Petroleum	34.7		Nuclear fusion	Not probable prior to 2000 A.D.
Natural gas	23.6		Solar radiation	20.
Coal	13.5		Wind power	2.
Hydropower	2.9		Sea thermal gradients	0.5
Uranium	0.85		Tidal energy	0.1
	75.55	1973 total	Organic wastes	1.
			Geothermal energy	3.
			Fuel from wastes	4.

Illustration 2-18

Some parts of the country, particularly in the West, can harness geothermal energy sources that exist beneath the surface of the earth in the form of hot rock, hot water, or steam.

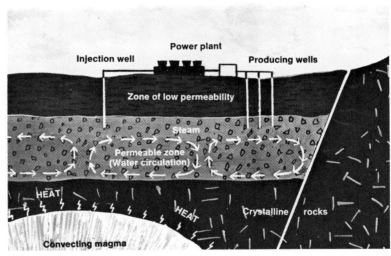

use that was made in 1973 of present energy sources and suggests levels of utilization that might be made of some alternative sources by the year 2000 A.D.

In the United States, the
Use of Power is Distributed
Approximately as follows:[32]

		Estimated Distribution in 2000 A.D.	
Household	17 per cent	Household	9 per cent
Commercial	13 per cent	Commercial	8 per cent
Industrial	28 per cent	Industrial	27 per cent
Transportation	25 per cent	Transportation	22 per cent
Energy Processing	15 per cent	Energy Processing	30 per cent
Other	2 per cent	Other	4 per cent

[32] "Exploring Energy Needs," The Ford Foundation, 1974, p. 45.

48   *Part One: Engineering—the profession in review*

Illustration 2-19
In certain parts of the country the energy of the winds can be harnessed and the resultant electrical power either stored or transmitted to more desirable use locations.

Considering the fact that currently the five most common air pollutants (carbon monoxide, sulfur oxides, hydrocarbons, nitrogen oxides, and solid particles) are primarily by-products of the combustion of fossil fuels, it behooves the engineer to design and utilize energy sources that are as free from such pollution-causing wastes as possible. It would appear that in the long run the earth can tolerate a significant increase in man's continuous release of energy (perhaps as much as 1000 times the current U.S. daily consumption—or more) without deleterious effect. Such increases would, of course, be necessary to accommodate a constantly increasing population. However, extrapolations and statements of this type concerning the future are meaningless unless the short-range problems—the problems of today—are solved. Our society has invested the engineer with a responsibility for leadership in this regard, and he must not fail.

Go—Go—Go

This year American motorists have travelled over 1 trillion miles on the nation's highways—an equivalent distance of over 2 million round trips to the moon. More than one half of this travel has been in urban areas, Figure 2-4, where for the most part the physical layouts—the planning, the street design, and basic service systems—were created over 100 years ago.[33] As the population of the nation continues to

[33] Alvin M. Weinberg and R. Philip Hammond, "Limits to the Use of Energy," *American Scientist,* August 1970, p. 413.

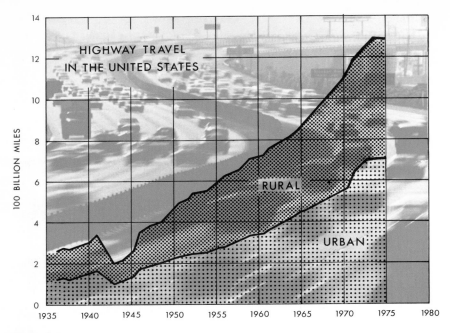

Figure 2-4

shift to the urban areas, many of the *frustrating* problems of today will become *unbearable* in the future. Since 1896, when Henry Ford built his first car, the mores of the nation have changed gradually from an attitude of "pioneer independence" to a state of "apprehensive dependence"—to the point where one's possession of a means of private transportation is now considered to be a *necessity*.

Over three fourths of the families in the United States own at least one automobile, and over one fourth can boast of owning two or more. However, due to inadequate planning this affluence has brought its share of problems for all concerned. Beginning about 3500 B.C. and until recent times roads and highways were used primarily as trade routes for the transport of commerce among villages, towns, and cities. The Old Silk Trade Route that connected ancient Rome and Europe with the Orient, a distance of over 6000 miles, was used extensively for the transport of silk, jade, and other valuable commodities. The first really expert road builders, however, were the Romans, who built networks of roads throughout their empire to enable their soldiers to move more quickly from place to place. In this country early settlers first used the rivers, lakes, and oceans for transportation, and the first communities were located at easily accessible points. A few crude roads were constructed, but until 1900 the railroad was generally considered to be the most satisfactory means of travel, particularly where long distances were involved. With the advent of the automobile, individual desires could be accommodated more readily, and many road systems were improvised to connect the railroad stations with frontier

> In the next 40 years, we must completely renew our cities. The alternative is disaster. Gaping needs must be met in health, in education, in job opportunities, in housing. And not a single one of these needs can be fully met until we rebuild our mass transportation systems.
> —Lyndon Baines Johnson, 1968

Illustration 2-20
All too frequently we become concerned about the dwindling supply of oil products only *when we are personally affected.*

settlements. At first these roads existed mainly so farmers could market their produce, but subsequent extensions were the direct result of public demands for an improved highway system. People in the cities wanted to visit the countryside (Illustration 2-21) and people in the outlying areas were eager to "get a look at the big city." Within a few years we became a *mobile* people, but the road and highway system in use today was designed primarily to accommodate the transfer of goods rather than large volumes of people. Because of this, many of these "traffic arteries" are not in the best locations, nor of the most appropriate designs to satisfy *today*'s demands. Thus, attempts to *drive* to work, *drive* downtown to shop, or take a leisurely *drive* through the countryside on a Sunday afternoon are more likely than not apt to be "experiences in frustration" (Illustration 2-22). Vehicle parking is also becoming a critical problem (Illustration 2-23).

Illustration 2-21
*Automobile travel and
parking problems of
yesterday.*

Most cities have made only half-hearted attempts to care for the transportation needs of their most populous areas. Although those owning automobiles do experience annoying inconveniences, those without automobiles suffer the most—especially the poor, the handicapped, the secondary worker, the elderly, and the young. Too often the public transit services that do exist are characterized by excessive walking distances to and from stations, poor connections and transfers, infrequent service, unreliability, slow speed and delays, crowding, noise, lack of comfort, and a lack of information for the rider's use. Moreover, passengers are often exposed to dangers to their personal safety while awaiting service. Certainly not to be minimized are the more than 4 million injuries and the 52,000 fatalities that result annually from motor vehicle accidents. (For perspective, since 1963 highway fatalities have exceeded more than ten times the total loss of American lives in the Vietnam War.)

Traditionally people have moved into a locality, built homes, businesses, and schools and then demanded that adequate transportation facilities be brought to

Illustration 2-22
Automobile travel today.

Illustration 2-23
Parking problems today.

them. We may now live in an era where this independence is no longer feasible, but rather people eventually may be required to settle around previously designed transportation systems. Engineers can provide good solutions for all these problems *if they are allowed to do so by the public.* However, there will be a cost for each improvement—whether it be a better vehicle design, computerized control of traffic flow, redesigned urban bus systems, rapid transit systems (Illustration 2-24), highway guideway systems for vehicles, or some other entirely new concept. In some instances city, state, or federal taxes must be levied; in others, the costs must be borne by each person who owns private transportation. Certainly the quality of urban life depends upon a unified commitment to this end.

The challenge of crime

Crime, one form of social pollution, is increasing rapidly in the United States in particular and throughout the world in general. The rate of increase in this country can be attributed variously to the population explosion, the increasing trend to urbanization, the changing composition of the population (particularly with respect to such factors as age, sex, and race), the increasing affluence of the populace, the diminishing influence of the home, and the deterioration of previously accepted value systems, mores, and standards of morality. A recent survey of the National Opinion Research Center of the University of Chicago indicates that the actual amount of crime in the United States is known to be several times that reported. Figure 2-5 provides a comparison of recent increases in the seven forms of crime that are considered to be most serious in this country. A brief examination of these data indicates that the rate of increase of crime is now several times greater than the rate of

Illustration 2-24
Rapid transit systems have proved their usefulness in a number of major metropolitan areas.

Figure 2-5 Reported crimes in the United States.

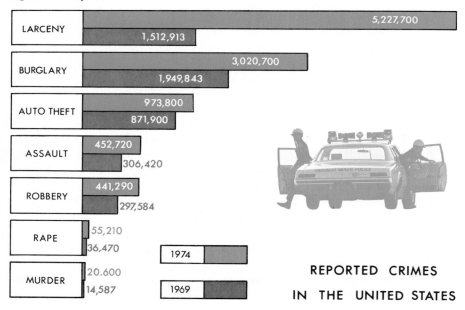

	1974	1969
LARCENY	5,227,700	1,512,913
BURGLARY	3,020,700	1,949,843
AUTO THEFT	973,800	871,900
ASSAULT	452,720	306,420
ROBBERY	441,290	297,584
RAPE	55,210	36,470
MURDER	20,600	14,587

REPORTED CRIMES
IN THE UNITED STATES

increase of the population. In fact, crime is becoming such a serious social issue as to challenge the very fabric of our American way of life.

Not all people react in the same way to the threat of crime. Some are inclined to relocate their residences or places of business; some become fearful, withdrawn, and antisocial; some are resentful and revengeful; and a large percentage become suspicious of particular ethnic groups whom they believe to be responsible. A number, of course, seize the opportunity to "join in," and they adopt crime as an "easy way" to get ahead in life. The majority, however, merely display moods of frustration and bewilderment. In all cases the consequential results are detrimental to everyone concerned, because a free society cannot long endure such strains on public and private confidences, nor tolerate the continual presence of fear within the populace.

Traditionally the detection, conviction, punishment, and even the prevention of crime have been functions of local, state, or federal agencies. Only in rare instances has private enterprise been called upon to assist in any significant way, and certainly there has been no concentrated effort to bring to bear on these situations the almost revolutionary advances that have been made in recent years in engineering, science, and technology. Rather, a few of the more spectacular developments have been modified or adapted for police operations or surveillance (Illustration 2-25).

Illustration 2-25
The digital data transmission equipment is one contribution the engineer has made to aid in the suppression of crime.

What is needed, and needed now, is a delineation of the vast array of problems that relate to the prevention, detection, and punishment of crime, with particular attention being directed toward achieving *general* rather than *specific* solutions. In this way technological efforts can be concentrated in those areas where they are most likely to be productive. The engineer can make a significant contribution in this endeavor, and, in fact, we may have reached the time when such attention will be required if the wave of lawlessness now sweeping the country is to be stemmed.

Other opportunities and challenges

The discussions in this chapter are succinct and not intended to be complete in signaling the manifold and varied problems and challenges that confront the engineer *today*. Rather, an attempt has been made to point out specific areas of opportunity for the engineer and to discuss how these may relate to the well-being of society today and to that of future generations. Of necessity, many very important challenges have not been discussed, such as the mounting congestion caused by the products of communication media and the threatening inundation of existing information-processing systems, ocean exploration with all of its varied technical problems and yet almost unlimited potential as a source of material, the expertise that the engineer can contribute to the entire field of health care and biological and medical advance (Illustration 2-26), and the attendant social problems that are closely related to urbanization and population growth—such as mass migration, metropolitan planning, improved housing, and unemployment caused by outmoded work assignment.

It is axiomatic that technological advance always causes sociocultural change. In this sense the engineers and technologists who create new and useful designs are also "social revolutionaries."[34] After all, it was they who brought about the obsolescence of slave labor, the emergence of transportation machines that allowed redistribution of the population, the radio and television sets that provide "instant communication," and every convenience of liberation for the housewife—from mixers, waste disposers, dishwashers, ironers, and dryers to frozen foods. Frequently, society is not prepared to accept such abrupt changes—even though it is generally agreed that they are for the overall betterment of mankind. Because of this the engineer has a dual responsibility to society. He not only must continue to bring about improvements for the benefit of society, but he must exert every possible means to acquaint society with its

[34] Melvin Kranzberg, "Engineering: A Force for Social Change," *Our Technological Environment: Challenge and Opportunity,* American Society for Engineering Education, Washington, D.C. (1971).

> The machine can free man or enslave him, it can make of this world something resembling a paradise or a purgatory. Men have it within their power to achieve a security hereto dreamed of only by the philosophers, or they may go the way of the dinosaurs, actually disappearing from the earth because they fail to develop the social and political intelligence to adjust to the world which their mechanical intelligence has created.
> —William G. Carleton

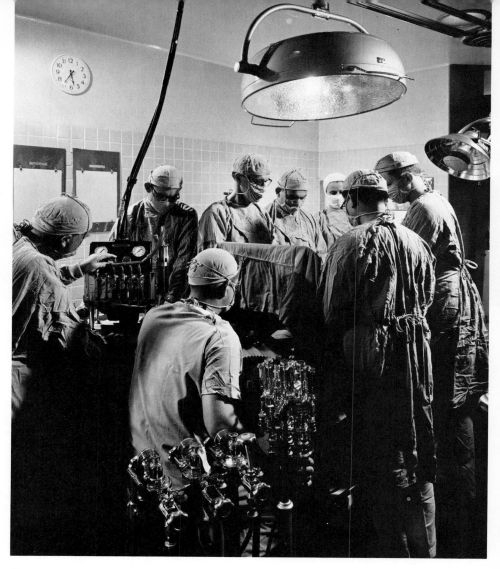

Illustration 2-26
Millions of people are alive today because of the availability of technical support systems that have been made possible by biomedical engineering design.

responsibility for continual change. Without such an active voice in community and governmental affairs, irrational forces and misinformation can prevail.

An environment of change

Man has always lived in an environment of change. As he has been able to add to his store of technical knowledge, he has also been able to change his economic structure and his sociological patterns. For centuries the changes that took place during a

Illustration 2-27
The quality of work of many professionals (such as physicians and attorneys) is evaluated primarily by individuals. The engineer's designs, on the other hand, are evaluated by the critical jury of public opinion.

lifetime were hardly discernible. Beginning about 1600, the changes became more noticeable; and today technological change is literally exploding at an exponential rate. Although a description of this accelerated expansion by empirical means will suffice for general purposes, it is interesting to contemplate one's future if a growth curve relationship such as $k = a(i)^t$ is followed (Figure 2-6).

In Figure 2-6, engineering and scientific knowledge is assumed to be doubling every 15 to 20 years. Experience with other growth curves of this nature indicates that at some point a threshold will be reached and the rate will begin to decline. However, in considering the expansion of technology, no one can say with certainty when this slowing down is most likely to occur. National and international factors certainly must be taken into account.

We have many ways of measuring the increase in engineering knowledge and in the number of engineers. For example, the world supply of engineering and scientific manpower has been following a consistent growth pattern since the mid-seventeenth century. With this supply doubling approximately every 15 years, it is easy to see that approximately 90 per cent of the engineers and scientists who have ever lived *are alive today.* The total number of scientific journals founded, the number of doctorates granted in engineering, the growth of a tomato vine, the growth of scientific discoveries, the growth of the U.S. population, and many other relationships which vary

> The man who graduates today and stops learning tomorrow is uneducated the day after.
> —Newton D. Baker

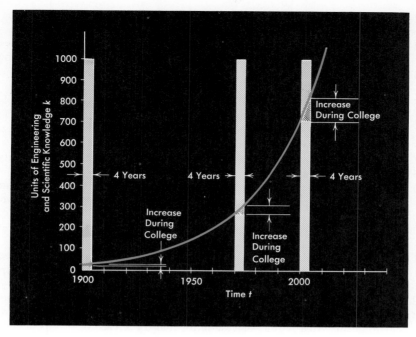

Figure 2-6 Growth of engineering and science.

with time also tend to follow an exponential growth pattern.[35] Of course, they do not all grow at the same rate.

Similar factors are working to provoke changes in educational goals and patterns. In 1900, for example, the engineering student studied for four years to earn his baccalaureate degree, and he saw relatively little change take place in his technological environment during this period. Today, however, due to the accelerated growth pattern of engineering and scientific knowledge, many significant changes will have taken place between his freshman and senior years in college. In fact, complete new industries will be bidding for the services of the young graduate that were not even in existence at the time he began his freshman year of college. This is particularly true of the engineering student who continues his studies for a master's or a doctorate. It is also interesting to contemplate that, at the present rate of growth, engineering and scientific knowledge will have doubled within 20 years after graduation. This places a special importance on continuing lifetime studies for all levels of engineering graduates.

Problems

2-1. Describe one instance in which the ecological balance of nature has been altered unintentionally by man.

2-2. Plot the rate of population growth for your state since 1900. What is your prediction of its population for the year 2000?

[35] Derek J. De Solla Price, *Little Science, Big Science,* Columbia University Press, New York (1963), p. 11.

2-3. What can the engineer do that would make possible the improvement of the general "standards of living" in your home town?

2-4. Investigate world conditions and estimate the numbers of people who need some supplement to their diet. How can the engineer help to bring about such help?

2-5. From a technological and economic point of view, what are the fundamental causes of noise in buildings?

2-6. Borrow a sound-level meter and investigate the average sound level in decibels of (*a*) a busy freeway, (*b*) a television "soap opera," (*c*) a college classroom lecture, (*d*) a library reading room, (*e*) a home vacuum cleaner, (*f*) a riverbank at night, (*g*) a "rock" combo, (*h*) a jack hammer, (*i*) a chain saw, and (*j*) a kitchen mixer.

2-7. Which of the air pollutants appear to be most damaging to man's longevity? Why?

2-8. Explain the "greenhouse effect."

2-9. Investigate how the "smog intensity level" has changed over the past ten years for the nearest city of over 100,000 population. With current trends, what level would you expect for 1985?

2-10. Describe some effects that might result from a continually increasing percentage of carbon dioxide in the atmosphere.

2-11. Investigate the methods used in purifying the water supply from which you receive your drinking water. Describe improvements that you believe might be made to improve the quality of the water.

2-12. What are the apparent sources of pollution for the water supply serving your home?

2-13. Seek out three current newspaper accounts where man has caused pollution of the environment. What is your suggestion for remedy of each of these situations?

2-14. Investigate the problems that might be caused by increasing the average water temperature of the nearest river 20°F.

2-15. What means is currently being used to dispose of solid waste in the city where you live? Would you recommend some other procedure or process?

2-16. Estimate the amount of energy consumed by the members of your class in one year.

2-17. Considering the expanding demand for energy throughout the world, list ten challenges that require better engineering solutions.

2-18. What are the five most pressing problems that exist in your state with regard to transportation? Suggest at least one engineering solution for each.

2-19. List five new engineering designs that are needed to help suppress crime.

2-20. In the United States, what are the most pressing communications problems that need solving?

2-21. List five general problems not discussed in this chapter that need engineering solutions.

2-22.[36] The Marginal Chemical Corp. is a small company by Wall Street's standards, but it is one of the biggest employers and taxpayers in the little town in which it has its one and only plant. The company has an erratic earnings record, but production has been trending up at an average of 6 per cent a year—and along with it, so has the pollution from the plant's effluents into the large stream that flows by the plant. This stream feeds a large lake that has become unfit for bathing or fishing.

The number of complaints from town residents about this situation has been rising, and you, as a resident of the community and the plant's senior engineer, also have become increasingly concerned. Although the lake is a gathering place for the youth of the town, the City Fathers have applied only token pressure on the plant to clean up. Your immediate superior, Mr. Jones, the plant manager, has other worries because the plant has been caught in a cost/price squeeze and is barely breaking even.

After a careful study, you propose to Jones that, to have an effective pollution-abatement system, the company must make a capital investment of $1 million. This system will cost another $100,000 per year in operating expenses (e.g., for treatment chemicals, utilities, labor, laboratory support). Jones' reaction is:

[36] Problems 2-22 through 2-25 are reprinted by special permission from *Chemical Engineering*, November 2, 1970, pp. 88–93, copyright ©, by McGraw-Hill, Inc., New York, N.Y. 10036.

"It's out of the question. As you know, we don't have an extra million around gathering dust—we'd have to borrow it at 10 per cent interest per year and, with the direct operating expenses, that means it would actually cost us $200,000 a year to go through with your idea. The way things have been going, we'll be lucky if this plant *clears* $200,000 this year, and we certainly can't raise prices. Even if we had the million handy, I'd prefer to use it to expand production of our new pigment; that way, it would give us a better jump on our competitors and on overseas competition. You can create a lot of new production—and new jobs—for a million dollars. This town needs new jobs more than it needs crystal-clear lakes, unless you want people to fish for a living. Besides, even if we weren't putting anything in the lake, it still wouldn't be crystal clear—there would still be all sorts of garbage in it."

During further discussion, the only concessions you can get from Jones is that you can spend $10,000 so that one highly visible (but otherwise insignificant) pollutant won't be discharged into the stream, and that if you can come up with an overall pollution-control scheme that will pay for itself via product recovery, he will give it serious consideration. You feel that the latter concession does not offer much hope, because not enough products with a ready market appear to be recoverable.

If you were this engineer, what do you think you *should* do? Consider the alternatives below.

a. Report the firm to your state and other governmental authorities as being a polluter, and complain about the laxness of city officials (even though the possible outcome might be your dismissal, or the company deciding to close up shop).

b. Go above Jones' head (i.e., to the president of the company). If he fails to overrule Jones, quit your job, and then take step *a*.

c. Go along with Jones on an interim basis, and try to improve the plant's competitive position via a rigorous cost-reduction program so that a little more money can be spent on pollution control in a year or two. In the meantime, do more studies of product-recovery systems, and keep him aware of your continued concern with pollution control.

d. Relax, and let Jones tell you when to take the next antipollution step. After all, he has managerial responsibility for the plant. You have not only explained the problem to him, but have suggested a solution, so you have done your part.

e. Other action (explain).

2-23. *d.* You are the division manager of Sellwell Co.—a firm that has developed an inexpensive household specialty that you hope will find a huge market among housewives. You want to package this produce in 1-gallon and ½-gallon sizes. A number of container materials would appear to be practical—glass, aluminum, treated paper, steel, and various types of plastics. A young engineer whom you hired recently and assigned to the manufacturing department has done a container-disposal study that shows that the disposal cost for 1-gallon containers can vary by a factor of three—depending upon the weight of the container, whether it can be recycled, whether it is easy to incinerate, whether it has good landfill characteristics, and so on.

Your company's marketing expert believes that the container material with the highest consumer appeal is the one that happens to present the biggest disposal problem and cost to communities. He estimates that the sales potential would be at least 10 per cent less if the easiest-to-dispose-of, salvageable container were used, because this container would be somewhat less distinctive and attractive.

Assuming that the actual costs of the containers were about the same, to what extent would you let the disposal problem influence your choice? Would you:

(1) Choose the container strictly on its marketing appeal, on the premise that disposal is the community's problem, not yours (and also that some communities may not be ready to use the recycling approach yet, regardless of which container material you select).

(2) Choose the easiest-to-dispose-of container, and either accept the sales penalty, or try to overcome it by stressing the "good citizenship" angle (even though the marketing department is skeptical about whether this will work).

(3) Take the middle road, by accepting a 5 per cent sales penalty to produce a container that is midway on the disposability scale.

(4) Other action (explain).

b. Do you think the young engineer who made the container-disposal study (but who is not a marketing expert) has any moral obligation to make recommendations as to which container to use? Explain your position.

2-24. Stan Smith, a young engineer with two years of experience, has been hired to assist a senior engineer in the evaluation of air and water pollution problems at a large plant—one that is considering a major expansion that would involve a new product. Local civic groups and labor unions favor this expansion, but conservation groups are opposed to it.

Smith's specific assignment is to evaluate control techniques for the effluents in accordance with state and federal standards. He concludes that the expanded plant will be able to meet these standards. However, he is not completely happy, because the aerial discharge will include an unusual by-product whose effects are not well known, and whose control is not considered by state and federal officials in the setting of standards.

In doing further research, he comes across a study that tends to connect respiratory diseases with this type of emission in one of the few instances where such an emission took place over an extended time period. An area downwind of the responsible plant experienced a 15 per cent increase in respiratory diseases. The study also tends to confirm that the pollutant is difficult to control by any known means.

When Smith reports these new findings to his engineering supervisor, he is told that by now the expansion project is well along, the equipment has been purchased, and it would be very expensive and embarrassing for the company suddenly to halt or change its plans.

Furthermore, the supervisor points out that the respiratory-disease study involved a different geography of the country and, hence, different climatic conditions, and also that apparently only transitory diseases were increased, rather than really serious ones. This increase might have been caused by some unique combination of contaminants, rather than only the one in question, and might not have occurred at all if the other contaminants had been controlled as closely as they will be in the new facility.

If Smith still believes that there is a reasonable possibility (but not necessarily certainty) that the aerial discharge would lead to an increase in some types of ailments in the downwind area, should he:

a. Go above his superior, to an officer of the company (at the risk of his previously good relationship with his superior).

b. Take it upon himself to talk to the appropriate control officials and to pass their opinions along to his superior (which entails the same risk).

c. Talk to the conservation groups and (in confidence) give them the type of ammunition they are looking for to halt the expansion.

d. Accept his superior's reasoning (keeping a copy of pertinent correspondence so as to fix responsibility if trouble develops).

e. Other action (explain).

2-25. Jerry Williams is a chemical engineer working for a large diversified company on the East Coast. For the past two years, he has been a member—the only technically trained member—of a citizens' pollution-control group working in his city.

As a chemical engineer, Williams has been able to advise the group about what can reasonably be done about abating various kinds of pollution, and he has even helped some smaller companies to design and buy control equipment. (His own plant has air and water pollution under good control.) As a result of Williams' activity, he has built himself considerable prestige on the pollution-control committee.

Recently, some other committee members started a drive to pressure the city administration into banning the sale of phosphate-containing detergents. They have been impressed by reports in their newspapers and magazines on the harmfulness of phosphates.

Williams believes that banning phosphates would be misdirected effort. He tries to explain that although phosphates have been attacked in regard to the eutrophication of the Great Lakes, his city's sewage flows from the sewage-treatment plant directly into the ocean. And he feels that nobody has shown any detrimental effect of phosphate on the ocean. Also, he is aware that there are conflicting theories on the effect of phosphates, even on the Great Lakes (e.g., some theories put the blame on nitrogen or carbon rather than phosphates, and suggest that some phosphate substitutes may do more harm than good).

In addition, he points out that the major quantity of phosphate in the city's sewage comes from human wastes rather than detergent.

Somehow, all this reasoning makes little impression on the backers of the "ban phosphates" measure. During an increasingly emotional meeting, some of the committee men even accuse Williams of using stalling tactics in order to protect his employer who, they point out, has a subsidiary that makes detergent chemicals.

Williams is in a dilemma. He sincerely believes that his viewpoint makes sense, and that it has nothing to do with his employer's involvement with detergents (which is relatively small, anyway, and does not involve Williams' plant). Which step should he now take?

a. Go along with the "ban phosphates" clique on the grounds that the ban won't do any harm, even if it doesn't do much good. Besides, by giving the group at least passive support, Williams can preserve his influence for future items that really matter more.

b. Fight the phosphate foes to the end, on the grounds that their attitude is unscientific and unfair, and that lending it his support would be unethical. (Possible outcomes: his ouster from the committee or its breakup as an effective body.)

c. Resign from the committee, giving his side of the story to the local press.

d. Other action (explain).

2-26. For present energy sources predict the Btu's of energy that predictably might be produced by 2000 A.D.

3

The technological team

Just over a decade ago our civilization was privileged to participate in and help fulfill one of man's oldest desires—to walk on the moon. Indeed it was as engineer–astronaut Neil Armstrong described so aptly, "One small step for man, one giant leap for mankind." For centuries, exploring the surface of the moon had been one of man's cherished dreams. Moreover, the mathematical principles and laws of nature governing whether such a voyage could be possible had been known in scientific circles for many years. Yet it all seemed to be an impossible dream. The lacking ingredient was a broad interdisciplinary technology *of undreamed complexity* that would be capable of fulfilling the requirements of a lunar exploration. And, of course, without a national commitment of purpose to complete such a voyage successfully, the dream would still be unfulfilled.

Today we live in a technological world where such people as housewives, farmers, home builders, accountants, teachers, and physicians carry out their daily tasks in a manner totally different than their counterparts did only 50 years ago. In the case of each worker, one or more *technological teams* have been instrumental in producing new and improved devices, designs, and systems that have eased the task and improved the overall quality of life. These team members are known in our society today as scientists, engineers, technologists, technicians, and craftsmen. In earlier cultures, the attributes of the scientist–engineer–craftsman were most frequently embodied in one person who had the basic knowledge, interest, and skill in the full technological spectrum of activities. Leonardo da Vinci was such a person. Today, partially as a result of the tremendous increase in population that has occurred and somewhat as a result of the overall increased complexity of technology, occupational specialization in a separate career field is more desirable for the majority of people. Thus, it is now more likely that a high school graduate would plan an educational–work-experience program to become a scientist, or an engineer, or a technologist, and

Craftsman	Work involves repetitive and manipulative skills requiring physical dexterity. Rarely supervises the work of others.
Technician	Performs routine equipment checks and maintenance. Carries out plans and and designs of engineers. Sets up scientific experiments. Seldom supervises others.
Technologist	Applies engineering principles for industrial production, construction, and operation. Works with engineering design components. Occasionally supervises others.
Engineer	Identifies and solves problems. An innovator in applying principles of science to produce economically feasible designs. Frequently supervises others.
Scientist	Searches for new knowledge concerning the nature of man and the universe. Infrequently involved in supervisory work.

Figure 3-1 Occupational spectrum.

so on, rather than some combination. A purpose of this chapter is to further clarify the types of work that each member of the technological team will likely perform, and to list some of the personality traits and aptitudes that are indicative of success for each career field. Figure 3-1 depicts graphically this occupational spectrum.

We must recognize that as individuals we do not always fit neatly into one specific area of the technological occupational spectrum. Rather, our interests and aptitudes may bridge across one or two of the work areas—or quite naturally across the entire spectrum. If so, good! However, educational programs leading to the various career choices are usually quite different in their composition. In most schools they strengthen the interests and preferences depicted in Figure 3-2.

In a general sense Figure 3-2 indicates how such factors as degree of aptitude for manipulating mathematical expressions and the degree of satisfaction that is derived from manual artistry will influence the probability of lasting interest in these various career fields.

During the last 100 years, by whatever names they may have been called, members of the technological team have been largely responsible for the development in this country of the highest standard of living that the world has ever known. This cooperative effort has made possible the solution of practical problems on a scale never accomplished before in all recorded history. Each member of the team has been

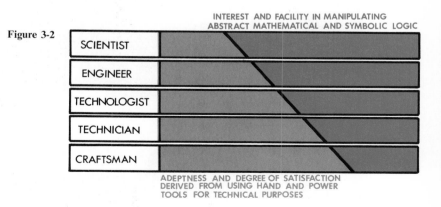

Figure 3-2

INTEREST AND FACILITY IN MANIPULATING
ABSTRACT MATHEMATICAL AND SYMBOLIC LOGIC

SCIENTIST

ENGINEER

TECHNOLOGIST

TECHNICIAN

CRAFTSMAN

ADEPTNESS AND DEGREE OF SATISFACTION
DERIVED FROM USING HAND AND POWER
TOOLS FOR TECHNICAL PURPOSES

important in this effort. However, the craftsman was the first to appear on the American scene.

The craftsman

In colonial times the master craftsman held a position of considerable esteem in the community. Usually he operated his own shop, determined the designs, set standards, and directed the apprentice workmen and other artisans who worked for him. The period of apprenticeship was usually about seven years, and individual skills were mastered by constant repetition. Paul Revere, an outstanding master craftsman of the colonial era, was skilled in the casting and working of metal, particularly silver and copper.

Since the products produced were usually sold, the average craftsman was sensitive to the style, marketability, and serviceability required by his customers. He was, therefore, accustomed to working within definite limits. Examples of craftsmen of this period were silversmiths, weavers, cabinetmakers, potters, blacksmiths, and candlestick makers. They were the backbone of early American industry.

As American industry became more industrialized, the need for master craftsmen skills diminished and the role of the craftsman changed. Machines were developed to do many of the routine jobs, and individual workers found themselves to be more captives of the machines than the reverse. Also, as factories grew a worker no longer was responsible for making a product from start to finish. Now he specialized in a

Illustration 3-1
Welding craftsmen are important contributors to the successful completion of many engineering designs.

single task—one of a series of operations—and he performed this task over and over to achieve the desired production. When this happened, the pride of accomplishment that is so necessary to motivate the craftsman disappeared. Because of this trend many individuals who aspired to be true craftsmen moved into other areas of work.

Today craftsmen continue to work in a variety of technologically related areas. These do provide the same intense feeling of pride and satisfaction for the individual as did the domestic-type craft work of several decades ago. Among these are carpenters, welders, tool and die makers, pattern makers, and precision machinists. A high school education is needed for entry into these specialty areas. In most cases this is followed by an intensified period of skill training that may vary from 6 to 18 months.

The technician

The engineering technician is interested in working with equipment and in assembling the component parts of designs that are designed by others. In this respect he is an experimentalist—an Edison-type thinker rather than an Einstein- or Steinmetz-type thinker. He prefers to assemble, repair, or make improvements in technical equipment by learning its characteristics rather than by studying the scientific or engineering basis for its original design.

A technician possesses many of the skills of a craftsman, and he is personally able to effect required physical changes in engineering hardware. Because he works directly with such equipment he is often able to suggest possible answers to difficult problems that have not been thought of previously.

The technician will probably work in a laboratory, out in the field on a construction job, or perhaps troubleshoot on a production line, rather than, for example, work at a desk. Of course, some technicians do their work in drafting rooms or offices under the direction of engineers. They also are frequently found in research laboratories, where they render effective service in repairing equipment, setting up experiments, and accumulating scientific data. They are very important in this role because it is often the case that scientists themselves are not too adept or interested in carrying out tasks of this type.

In construction work, technicians also play key roles. They are needed to accomplish tasks in surveying, to make estimates of material and labor costs, to be responsible occasionally for the coordination of skilled labor and the work of subcontractors, and for the delivery of materials to the job sites.

A substantial number of engineering technicians are employed in the manufacturing and electronics industry and by public utilities. They may carry out standard calculations, serve as technical salesmen, make estimates of costs, or assist in preparing service manuals, such as for electronic equipment and plant operation and maintenance. They install and maintain, make checks on, and frequently modify electrical and mechanical equipment. As a group they are important problem-solving-oriented persons whose interests are directed more to the practical than to the theoretical. Technicians learn fundamental scientific theory and master the mathematical topics most useful in analyzing and solving problems. The majority of their studies, however, are more practically oriented.

The technician's education typically requires two full years of collegiate-level study. Generally this work is taken in a technical institute or community college and leads to an associate degree in technology. In many instances this school work is transferable to a senior academic institution and may be applied as credit toward a bachelor of engineering technology program. In most cases, however, these courses would not be applicable toward a degree in engineering or science.

Illustration 3-2
The technician is an experimentalist. Here a mechanical technician examines a ceramic sensor that monitors engine exhaust gases and serves as the input element in a feedback control for a closed-loop fuel-metering system.

The technologist

The engineering technologist is the most recent of the technological team to appear, having emerged in the last two decades. Engineering technology is that part of the technological spectrum which requires the application of scientific and engineering knowledge and methods combined with technical skills in support of engineering activities. Technologists work in the occupational domain between the craftsman-technician and the engineer. Their areas of interest typically are less theoretical and mathematically based than those of the engineer. Some of these are construction, operation, maintenance, and production. Also, as a group they are more inclined to be known as design organizer–producers, rather than design innovators.

Ordinarily the engineering technologist will work more with the development of various design components of systems that have been designed and developed by engineers. In research he is valuable in liaison work. Also, he is well suited to assume a variety of technical supervisory and management roles in industry.

Engineering technologists are graduates of baccalaureate degree programs in technology. The programs exist in two forms. One is a separate and distinctive

Illustration 3-3
Technologists must have a fundamental understanding of the theoretical principles involved as well as a facility for working with the associated engineering equipment.

four-year curriculum that emphasizes the solution of practical engineering problems. The other has similar objectives but is a two-year extension of the two-year associate degree program for engineering technicians.

The engineer

The engineer is a person who enjoys changing the status quo to gain an improvement. In this role he performs two functions. First, he recognizes, identifies, and defines problems and deficiencies, or conditions that need improvement. Second, he couples his reservoir of knowledge and skill in innovative thought to produce one or more acceptable solutions.

Thus, as a personal characteristic the engineer should possess a persistently inquiring and creative mind—yet one that is receptive to producing realistic innovative action. He must have a fundamental understanding of the laws of nature and of mathematics, and be able to recognize and interpret their application in real situations. Although these characteristics are most often associated with those who provide our nation technological leadership in commerce and industrial production, they are also highly desired qualities to attain for those who want to follow careers in management, medicine, and law.

The baccalaureate degree in engineering is required for entry into industry, and the masters degree (a total of five years of college) is preferred for many types of work.

> Thousands of engineers can design bridges, calculate strains and stresses, and draw up specifications for machines, but the great engineer is the man who can tell *whether* the bridge or the machine should be built at all, *where* it should be built, and *when*.
> —Eugene G. Grace

Illustration 3-4
Engineers create custom, special-purpose *designs that have never existed previously, like the lunar vehicle pictured here.*

Illustration 3-5
The primary objective of the scientist is to discover new knowledge. Here a research physicist uses a low-energy electron diffraction technique to study the arrangement of atoms entering into chemical reactions on the surface of crystals.

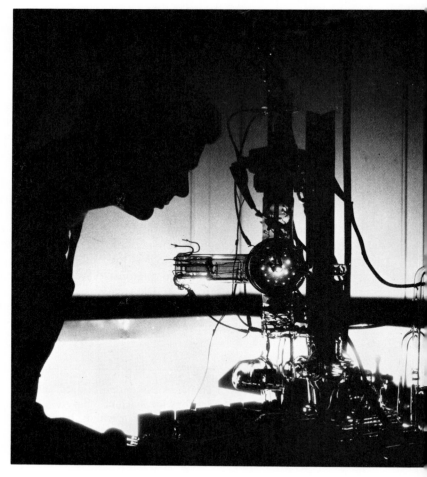

> Scientists study the world as it is, engineers create the world that never has been.
> —Theodore von Karman

The scientist

The primary objective of the scientist is to discover, to expand existing fields of knowledge, to correlate observations and experimental data into a formulation of laws, to learn new theories and to explore their meanings, and in general to broaden the horizons of science. The scientist is typically a theoretician who is concerned about *why* natural phenomena occur. As suggested above, his sole objective is to expand the world's reservoir of knowledge, and for this reason he seldom pauses to search for practical uses for any new-found truths that he may discover. There are, of course, *applied scientists* who work to find specific uses for new knowledge, such as devising a new instrument or synthesizing an improved medicine. For this reason it is difficult for the average person to tell the difference between an applied scientist and an engineer. In terms of schooling, the person educated as a scientist will generally have earned a doctorate (seven years of college minimum) in his chosen field of science.

The roles of the engineer and technologist

It is generally known that technological advances have made possible the many material things that now make our lives more enjoyable and provide extra time for recreation and study. It is also an accepted belief that engineers and technologists will continue to provide innovative and creative designs for the purpose of easing the burden of man's physical toil and to convert the materials and forces of Nature to the use of all mankind; however, when the time comes to make a choice of careers, each student finds himself groping for an answer to the question, "As an engineer, technologist, or technician, how would *I* fit into this picture."

First, we must realize that one does not become a graduate solely by studying a few courses. A technical education is more than knowing when to manipulate a set of formulas, where to search in an armful of reference books, or how to get accurate answers from a calculator or computer; it is also a state of mind. For example, through experience and training the engineer must be able to formulate problem statements and must conceive design solutions that many times involve novel ideas and creative thought processes. He must also exercise judgment and restraint, design with initiative and reliability, and be completely honest with himself and with others. These qualities should all mature as the student engineer advances from elementary to graduate-level studies.

What skills do engineers and technologists need?

The public expects engineers and technologists to be competent technically. Through the years the profession has built up a record of producing things that work. No one expects a company to produce television tubes that explode spontaneously, or a bridge that falls down, or an irrigation ditch that has the wrong slope. In fact, major technological failures are so rare that when they do occur they usually are front-page news.

The college engineering and technology curricula are designed to instill technical competence. The grading system generally used rewards acceptable solutions and penalizes inferior or unworkable solutions to practical problems. The subjects studied are not easy to master, and usually those persons who do not adapt themselves to the discipline of study and who do not accept the ideas of the exactness of Nature's laws will not complete a college program in engineering or technology. Such discipline has paid handsome dividends. The consistency of quality in graduates over many years has given the public a confidence that results in large measure from the rigorous but realistic accreditation standards of the Engineers Council for Professional Development.

It should also be realized, however, that the completion of a baccalaureate degree must not be the end of study. The pace of discovery is so rapid today that, even with constant study, one can barely keep abreast of technological improvements. If the graduate should resolve not to continue his technical study, he would be deficient within five years and probably would be completely out of step within ten years.

Engineers and technologists, then, must be capable of dealing with technological problems—not only those which they may have been trained to handle in college, but also new and unfamiliar situations that arise as a result of new discoveries. Of course, in preparation for the solution of real problems college courses can present problem areas only in general terms. Graduates are expected to provide solutions to new problem situations using a base of fundamental principles and an understanding of the most effective methods to use in problem solving.

Illustration 3-6
Many of today's complex engineering problems could not be solved without the team effort of technicians, technologists, and engineers.

After college what are the opportunities?

A question that arises frequently in the mind of high school graduates is, "What if I begin in engineering and decide later to change to another course of study? What would be the consequences?" Let us examine some of the possibilities. Normally an engineering student will follow engineering as a profession. However, many students change their mind during their college career or after graduation. Many authorities agree that engineering courses are excellent training for a great variety of careers, and records reveal that perhaps as many as 40 per cent of people on a management level have engineering educations. One of the basic and most valuable training concepts of an engineering education is that *engineering students are taught to think logically.* This means that, if a later career decision is made not to following engineering as a profession, the training and experience gained in studying engineering courses still will prepare a person for a wide variety of occupations.

What will the future hold for the serious student who proposes to make engineering his career? *First,* employment possibilities will be good. The rigor of the college courses usually removes those who are unable or unwilling to stick with a problem until they come up with a reasonable answer. Those who graduate in engineering usually are well qualified technically and, in addition, are well rounded in their knowledge of nonengineering courses.

Second, he will enjoy the profession. A sampling of questionnaires sent to engineers in large industrial concerns shows that those with several years experience almost unanimously enjoy their work. They like the opportunities for advancement, the challenges of new and exciting problems to be solved, the friendships gained in contacts with people with diverse backgrounds, and the possibilities of seeing their ideas develop into working realities. The engineer will find that, as a profession, the salary scale is among the higher groups, and his individual income usually is determined largely by the quality of his own efforts.

Third, engineering provides unlimited opportunities for creative design. As has been mentioned before, the pace of discovery is so great that the need for the application of discoveries provides countless places where the engineer with initiative and creative ability can spend his time in development. Also, in applications which are old and well known, the clever engineer can devise new, better, and more economical ways of providing the same services. For example, although roads have been built for centuries, the need for faster and more efficient methods of road building offers a continuing challenge to engineers.

Sometimes nontechnical people say that engineers are too dogmatic, that they think of things as being either positive or negative. To a certain extent this is true because of the engineer's training. He is educated to give realistic answers to real problems and to make the answers the most practical ones that he can produce. Within his knowledge of Nature's laws, the engineer usually can obtain a precise solution to a given problem, and he is always willing to defend his solution. To a nontechnical person accustomed to arriving at a solution by surmise, argument, and compromise, the positive approach of the engineer frequently is distressing. Part of the postgraduate training of an engineer is learning to convince nontechnical people of the worth of a design. The ability to reason, to explain by using simple applications, and to have patience in presenting ideas in simple terms that can be understood are essential qualities of the successful engineer.

The engineer does not claim to be a genius. However, by training, he is a leader. Because he has a responsibility to his profession and to his community to exercise that leadership, he should establish a set of technical and moral standards that will provide a wholesome influence upon all levels of his organization and upon his community as a whole.

Problems on the engineer's role

3-1. Interview an engineer and write a 500-word essay concerning his work.

3-2. Survey the job opportunities for engineers, scientists, and technicians. Discuss the differences in opportunity and salary.

3-3. Discuss the role of the engineer in government.

3-4. Frequently technical personnel in industry are given the title "engineer" in lieu of other benefits. Discuss the difficulties that arise as a result of this practice.

3-5. Write an essay on the differences between the work of the engineer, the scientist, and the technician.

3-6. Write an essay on the differences between the education of the engineer, the scientist, the technologist, and the technician.

3-7. Interview an engineering technician and write a 500-word essay concerning her work.

3-8. Discuss the role of the engineering technician in the aircraft industry.

3-9. Investigate the opportunities for employment of electronic technicians. Write a 500-word essay concerning your findings.

3-10. Investigate the differences in educational requirements of the engineer and the technician. Discuss your conclusions.

3-11. Classify the following items as to the most probable assignment in the occupational spectrum:

    a. Detail drawing of a small metal mounting bracket.

    b. Assisting an engineer in determining the pH of a solution.

    c. Boring a hole in an aluminum casting to fit a close tolerance pin.

    d. Determining the behavior of flow of a viscous fluid through a pipe elbow.

    e. Determining the percentage of carbon in a series of steel specimens to be used in fabricating cutter bits.

    f. Designing a device to permit the measurement of the temperature of molten zinc at a location approximately 150 feet from a vat.

    g. Preparation of a laboratory report concerning the results of a series of tests on an assortment of prospective heat-curing bonding adhesives.

    h. Preparing a work schedule for assigning manpower for a 2-day test of a small gasoline engine.

    i. Determining the effects of adding ammonia to the intake air of a gas turbine engine.

    j. Fabrication of 26 identical transistorized circuits using printed circuit boards.

    k. Calculation of the area of an irregular tract of land from a surveyor's field notes.

    l. Preparation of a proposal to study the effect of sunlight on anodized and unanodized aluminum surfaces.

    m. Design and fabrication of a device to indicate the rate of rainfall at a location several hundred feet from the sensing apparatus.

    n. Interpretation of the results of a test on a punch-tape-controlled milling machine.

    o. Preparing a computer program to determine the location of the center of gravity of an airplane from measured weight data.

3-12. A large office building is to be constructed, and tests of the load-bearing capacity of the underlying soil are to be made. Outline at least one way each member of the scientific team (scientists, engineers, technologists, technicians, and craftsmen) is involved in the work of determining suitability of the soil for supporting a building.

4

The spectrum of engineering work opportunities

During the years that he is in college, an engineering student will study courses in many subject areas. He will study language courses to prepare himself better in organizing and presenting ideas effectively, mathematics courses to learn the manipulation of symbols as an aid in problem solving, social science courses to help him better find his place in society as an informed citizen, and various technical courses to gain an understanding of natural laws. In his study of technical courses, he will become familiar with a store of factual information that will form the basis for his engineering decisions. The nature of these technical courses, in general, determines the major field of interest of the student. For example, he may decide to concentrate his major interest in some particular field such as civil, chemical, electrical, or mechanical engineering.

The college courses also provide training in learning facts and in developing powers of reasoning. Since it is impossible to predict what kind of work a practicing engineer will be doing after graduation, the objective of an engineering education is to provide a broad base of facts and skills upon which the engineer can practice his profession.

It usually is not sufficient to say that an engineer is working as a *civil engineer*. His work may vary over a wide spectrum. As a civil engineer, for example, he may be performing research on materials for surfacing highways, or he may be employed in government service and be responsible for the budget preparation of a missile launch project. In fact, there are many things that a practicing engineer will be called upon to do which are not described by his major course of study. The *type* of work that the engineer may do, as differentiated from his major field of specialization, can be called "engineering function." Some of these functions are research, development, design, production, construction, operations, sales, and management.

It has been found that in some engineering functions, such as in the management

> Research is an organized method of trying to find out what you are going to
> do after you find that you cannot do what you are doing now.
> —Charles F. Kettering

of a manufacturing plant, specialization is of lesser importance, whereas in other functions, such as research in transistor theory, specialization may be extremely important. In order to understand more fully the activities of a practicing engineer, let us examine some of these functions.

Research

In some respects research is one of the more glamorous functions of engineering. In this type of work the engineer delves into the nature of matter, exploring processes to use engineering materials and searching for reasons for the behavior of the things that make up our world. In many instances the work of the scientist and the engineer who are engaged in research will overlap. The work of scientists usually is closely allied

Illustration 4-1
Research is an important type of work performed by the engineer. In research he employs basic scientific principles in the discovery and application of new knowledge. This engineer is experimenting with a wind tunnel model of the city of Denver, Colorado, during a simulated atmospheric inversion.

with research. The objective of the research scientist is to *discover truths*. The objective of the research engineer, on the other hand, usually is directed toward the practical side of the problem: not only to discover but also *to find a use for the discovery*.

The research engineer must be especially perceptive and clever. He must be able to work patiently at tasks never before accomplished and must be able to recognize and identify phenomena previously unnoticed. As an aid to training an engineer to do research work, some colleges give courses in research techniques. However, the life of a research engineer can be quite disheartening. Since he is probing and exploring in new areas, much of his work is trial and error, and outstanding results of investigation usually occur only after long hours of painstaking and often discouraging work.

Until within the last few decades, almost all research was solo work by individuals. However, with the rapid expansion of the fields of knowledge of chemistry, physics, and biology, it became apparent that groups or "research teams" of scientists and engineers could accomplish better the aims of research by pooling their efforts and knowledge. Within the teams, the enthusiasm and competition provide added incentive to push the work forward, and since each person is able to contribute from his specialty, discovery is accelerated.

As has been indicated, a thorough training in the basic sciences and mathematics is essential for a research engineer. In addition, an inquiring mind and a great curiosity about the behavior of things is desirable. Most successful research engineers have a fertile and uninhibited imagination and a knack of observing and questioning phenomena that the majority of people overlook. For example, one successful research engineer has worked on such diverse projects as an automatic lawnmower, an electronic biological eye to replace natural eyes, and the use of small animals as electrical power sources.

Most research engineers secure advanced degrees because they need additional training in basic sciences and mathematics, and, in addition, this study usually gives them an opportunity to acquire useful skills in research procedures.

Development

After a basic discovery in natural phenomena is made, the next step in its utilization involves the development of processes or machines that employ the principles involved in the discovery. In the research and development fields, as in many other functions, the areas of activity overlap. In many organizations the functions of research and development are so interrelated that the department performing this work is designated simply as a research and development (R and D) department.

The engineering features of development are concerned principally with the actual construction, fabrication, assembly, layout, and testing of scale models, pilot models, and experimental models for pilot processes or procedures. Where the research engineer is concerned more with making a discovery that will have commercial or economic value, the development engineer will be interested primarily in producing a process, an assembly, or a system *that will work*.

The development engineer does not deal exclusively with new discoveries. Actually the major part of his work will involve using well-known principles and employing existing processes or machines to perform a new or unusual function. It is in this region that many patents are granted. In times past, the utilization of basic machines, such as a wheel and axle, and fundamental principles, including Ohm's Law and Lenz' Law, have eventually led to patentable articles, such as the electric dynamo. On the other hand, within a very short time after the announcement of the discovery of the laser in 1960, a number of patents were issued on devices employing this new principle. Thus the lag between the discovery of new knowledge and the use of that knowledge has been steadily decreasing through the years.

In most instances the tasks of the development engineer are dictated by immediate requirements. For example, a new type of device may be needed to determine at all times the position in space of an airplane. Let us suppose that the development engineer does not know of any existing device that can perform the task to the desired specifications. Should he immediately attempt to invent such a device? The answer, of course, is "usually not." First, he should explore the files of available literature for information pertaining to existing designs. Such information may come from two principal sources. The first source is library material on processes, principles, and methods of accomplishing the task or related tasks. The second source is manufacturers' literature. It has been said humorously that there is no need to reinvent the wheel. A literature search may reveal a device that can accomplish the task with little or no modification. If no device is available that will do the work, a system of existing subassemblies may be set up and joined to accomplish the desired result. Lacking these items, the development engineer must go further into basic literature, and, using results from experiments throughout the world, formulate plans to construct a model for testing. Previous research points a way to go, or perhaps a mathematical analysis will provide clues as to possible methods.

The development engineer usually works out his ideas on a trial or "breadboard" basis, whether it be a machine or a computer process. Having the parts or systems somewhat separated facilitates changes, modifications, and testing. In this process, improved methods may become apparent and can be incorporated. When the system or machine is in a workable state, the development engineer must then refine it and package it for use by others. Here again ingenuity and a knowledge of human nature are important. A device that works satisfactorily in a laboratory when manipulated by skilled technicians may be hopelessly complex and unsuited for field use. The development engineer is the important man behind every push button.

The training of an engineer for development work is similar to the training that the research engineer will expect to receive. However, creativity and innovation are perhaps of more importance, since the development engineer is standing between the scientist or the research engineer and the members of management who provide money for the research effort. He must be able to recognize the economic value of certain processes over others in achieving a desired result, and he must be able to convince others that his conclusions are the ones that should be accepted. A comprehensive knowledge of basic principles of science and an inherent cleverness in making things work are essential skills for the development engineer.

Design

In our modern way of life, mass production has given us cheaper products and has made more articles available than ever before in history. In the process of producing

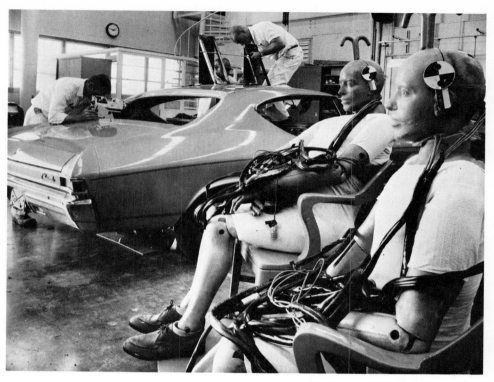

Illustration 4-2
Development engineers use the results of basic research and convert them into models and prototypes for full-scale testing and evaluation. In this picture, a team of engineers and technicians are shown preparing a test to evaluate the results of a head-on automobile crash into an immovable barrier.

these articles, the design engineer enters the scene just before the actual manufacturing process begins. After the development engineer has assembled and tested a device or a process and it has proved to be one that it is desirable to produce for a mass market, the final details of making it adaptable for production will be handled by a design engineer.

In his role of bridging the gap between the laboratory and the production line, the design engineer must be a versatile individual. He should be well grounded in basic engineering principles and in mathematics, and he must not only understand the capabilities of machines but also the temperament of the men who operate them. He must be conscious also of the relative costs of producing items, for it will be his design that will determine how long the product will survive in the open market. Not only must the device or process work, it must also be made in a style and at a price that will attract customers.

As an example, let us take a clock, a simple device widely used to indicate time. It includes a power source, a drive train, hands, and a face. Using these basic parts, engineers have designed spring-driven clocks, weight-driven clocks, and electrically driven clocks with all variations of drive trains. The basic hands and face have been modified in some models to give a digital display. The case has been made in many shapes and, perhaps in keeping with the slogan "time flies," it has even been streamlined! In the design of each modification the design engineer has determined the physical structure of the assembly, its aesthetic features, and the economics of producing it.

Of course the work of the design engineer is not limited solely to performing engineering on mass-produced items. Design engineers may work on items such as bridges or buildings in which only one of a kind is to be made. However, in such work he still is fulfilling the design process of adapting basic ideas to provide for making a completed product for the use of others. In this type of design the engineer must be able to use his training, in some cases almost intuitively, to arrive at a design solution which will provide for adequate safety without excessive redundancy. The more we learn about the behavior of structural materials, the better we can design without having to add additional materials to cover the "ignorance factor" area. Particularly in the aircraft industry, design engineers have attempted to use structural materials with minimum excess being allowable as a safety factor. Each part must perform without failure, and every ounce of weight must be saved. Of course to do this, fabricated parts of the design must be tested and retested for resistance to failure due either to static loads or to vibratory fatiguing loads. Also, since surface roughness has an important bearing on the fatigue life of parts which are subjected to high stress or repeated loads, much attention must be given to specifying in designs that surface finishes must meet certain requirements.

Since design work involves a production phase, the design engineer is always considering costs as a factor in our competitive economy. One way in which costs can be minimized in manufacture or construction is to use standard parts, and standard sizes and dimensions for raw material. For example, if a machine were designed using nonstandard bolt threads or a bridge designed using nonstandard steel I-beams, the design probably would be more expensive than needed to fulfill its function. Thus, the design engineer must be able to coordinate the parts of his design so that it functions acceptably and is produced at minimum cost.

The design engineer soon comes to realize also that there usually is more than one acceptable way to solve a design problem. Unlike an arithmetic problem with fixed numbers which give one answer, his problem can have many answers and many ways of obtaining a solution, *and all may be acceptable.* In such a case his decision becomes a matter of experience and judgment. At other times it may become just a matter of making a decision one way or the other. Regardless of the method used, his solution to a problem should be a conscious effort to provide the *best* method, considering fabrication, costs, and sales.

What are the qualifications of a design engineer? He, of course, must be creative. His every design will embody a departure from what has been done before. At the same time, he is constrained by the reality of the physical properties of materials and by economic factors. Therefore, he must be thoroughly knowledgeable in funda-mental engineering in a rather wide range of subjects. In addition, he must be familiar with basic principles of economics, both from the standpoint of employing people and using machines. As he progresses upward into supervisory and manage-ment duties, the employment of principles of psychology and economics becomes of even more importance. For this reason design engineers usually will have more use for management courses than will research or development engineers.

Illustration 4-3a and 4-3b
The design engineers pictured in 4-3a are evaluating a cost reduction proposal for the connector base plate assembly shown in 4-3b.

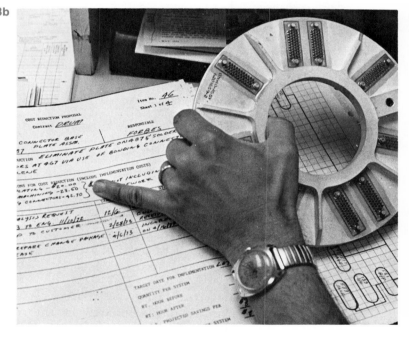

Production and construction

In the fields of production and construction, the engineer is more directly associated with the technician, mechanic, and laborer. The production or construction engineer must take the design engineer's drawings and supervise the assembly of the object as it was conceived and illustrated by drawings or models.

Illustration 4-4
The construction engineer is responsible for seeing that a project is carried out as designed. Usually he is working with large projects where weather and terrain are complicating factors.

Usually a production or construction engineer is associated closely with the process of estimating and bidding for competitive jobs. In this work he employs his knowledge of structural materials, fabricating processes, and general physical principles to estimate both time and cost to accomplish tasks. In construction work the method of competitive bidding is usually used to award contracts, and the ability to reduce an appropriate amount from an estimate by skilled engineering practices may mean the difference between a successful bid and one that is either too high or too low.

Once a bid has been awarded, it is usual practice to assign a "project engineer" as the person who assumes overall responsibility and supervision of the work from the standpoint of materials, labor, and money. He will have other production or construction engineers working under his direction who will be concerned with more specialized features of the work, such as civil, mechanical, electrical, or chemical engineering. Here the engineer must complete the details of the designers' plans. He must provide the engineering for employment of special tools needed for the work. He must also set up a schedule for production or construction, and he must be able to answer questions that technicians or workmen may raise concerning features of the design. He should be prepared to advise design engineers concerning desirable modifications that will aid in the construction or fabrication processes. In addition, he must be able to work effectively with construction or production crafts and labor unions.

Preparation of a schedule for production or construction is an important task of the engineer. In the case of an industrial plant, all planning for the procurement of raw materials and parts will be based upon this production schedule. An assembly line in a modern automobile manufacturing plant is one example which illustrates the necessity for scheduling the arrival of parts and subassemblies at a predetermined time. As another example, consider the construction of a modern multistory office building. The necessity for parts and materials to arrive at the right time is very important. If they arrive too soon, they probably will be in the way, and if they arrive too late, the building is delayed, which will cause an increase in costs to the builder.

Qualifications for a production or construction engineer include a thorough knowledge of basic engineering principles. In addition, he must have the ability to visualize the parts of an operation, whether it be the fabrication of a solid-state computer circuit or the building of a concrete bridge. From his understanding of the operations involved, he must be able to arrive at a realistic schedule of time, materials, and manpower. Therefore, emphasis should be placed upon courses in engineering design, economics, business law, and psychology.

Operations

In modern industrial plants, the number and complexity of machines, the equipment and buildings to be cared for, and the planning needed for expansion have brought out the need for specialized engineers to perform services in these areas. If a new manufacturing facility is to be constructed, or an addition made to an existing facility, it will be the duty of a plant engineer to perform the basic design, prepare the proposed layout of space and location of equipment, and to specify the fixed equipment such as illumination, communication, and air conditioning. In some cases, the work of construction will be contracted to outside firms, but it will be the general responsibility of the plant engineer to see that the construction is carried on as he has planned it.

After a building or facility has been built, the plant engineer and his staff are responsible for maintenance of the building, equipment, grounds, and utilities. This work varies from performing routine tasks to setting up and regulating the most complex and automated machinery in the plant.

The plant engineer must have a wide knowledge of several branches of engineering in order to perform these functions. For land acquisition and building construction, he will need courses in civil engineering; for equipment and machinery he will need mechanical training; for power he will need mechanical and electrical backgrounds; and for the specialized parts of the plant, his knowledge may need to be in such fields as chemical, metallurgical, nuclear, petroleum, or textile engineering.

In many plants, particularly in utility plants, the engineer also is concerned with operation of the plant. It is his duty to see that boilers, generators, turbines, and accessory equipment are operated at their best efficiency. He should be able to compare costs of operating under various conditions, and he attempts always to set schedules for machines so that best use will be made of them. In the case of chemical plants, he also will attempt to regulate the flows and temperatures at levels that will produce the greatest amount of desired product at the end of the line.

In his dual role as a plant and operations engineer, he will be constantly evaluating new equipment as it becomes available to see whether additional operating economies can be secured by retiring old equipment and installing new types. In this, he frequently must assume a salesman's role in order to convince management that it should discard equipment that, apparently, is operating perfectly and spend money for newer models. Here the ability to combine facts of engineering and economics is invaluable.

Plant engineering, of course, will be associated closely with production engineering processes. The production engineer will create needs for new machines, new facilities, and new locations. The plant engineer will correlate such things as the building layout, machine location, power supplies, and materials handling equipment so that they best will serve the needs of production.

The general qualifications of plant and operations engineers have already been mentioned. They must have basic knowledge of a wide variety of engineering fields such as civil, chemical, electrical, and mechanical, and also they must have specialized knowledge of areas peculiar to their plant and its operation. In addition, plant and operations engineers must be able to work with men and machines and to know what results to expect from them. In this part of their work, a knowledge of industrial

Illustration 4-5
The plant engineer is responsible for the maintenance of the building, equipment, grounds, and utilities.

engineering principles is valuable. In addition, it is desirable to have a basic understanding and knowledge of economics and business law. In this work, in general, training in detailed research procedures and abstract concepts is of lesser importance.

Sales

An important and sometimes unrecognized function in engineering is the realm of applications and sales. As is well known, the best designed and fabricated product is of little use unless a demand either exists or has been created for it. Since many new processes and products have been developed within the past few years, a field of work has opened up for engineers in presenting the use of new products to prospective customers.

Discoveries and their consequent application have occurred so rapidly that a product may be available about which even a recent graduate may not know. In this case, it will fall to the engineer in sales who has intimate knowledge of the principles involved to go out and educate possible users so that a demand can be created. In this work the engineer must assume the role of a teacher. In many instances he must present his product primarily from an engineering standpoint. If his audience is composed of engineers, he can "talk their language" and answer their technical questions, but if his audience includes nonengineers, he must present the features of the product in terms that they can comprehend.

In addition to his knowledge of the engineering features of his own product, the sales and application engineer must also be familiar with the operations of his customer's plant. This is important from two standpoints. First, he should be able to show how his product will fit into the plant, and also he must show the economics involved to convince the customer that he should buy it. At the same time, the engineer must point out the limitations of his product and the possible changes necessary to incorporate it into a new situation. For example, a new bonding material may be available, but in order for a customer to use it in an assembly of parts, a special refrigerator for storage may be necessary. Also the customer would need to have emphasized the necessity for proper cleaning and surface preparation of the parts to be bonded.

A second reason that the sales and application engineer must be familiar with a customer's plant operation is that many times new requirements are generated here. By finding an application area in which no apparatus is available to do the work, the sales engineer is able to report back to his company that a need exists and that a development operation should be undertaken to produce a device or process to meet the need.

Almost all equipment of any complexity will need to be accompanied by introductory instructions when it is placed in a customer's plant. Here the application engineer can create goodwill by conducting an instruction program outlining the capabilities and limitations of the equipment. Also, after the equipment is in service, maintenance and repair capabilities by competent technical personnel will serve to maintain the confidence of customers.

The sales and applications engineer should have a basic knowledge of engineering principles and should, of course, have detailed knowledge in the area of his own products. Here the ability to perform detailed work on abstract principles is of less

Illustration 4-6
*In sales, the engineer
must be able to describe a
technical product to
customers and show how
they will benefit from
using the latest
developments.*

importance than the ability to present one's ideas clearly. A genuine appreciation of
people and a friendly personality are desirable personal attributes. In addition to
basic technical subjects, courses in psychology, sociology, and human relations will
prove valuable to the sales and applications engineer.

Usually an engineer will spend several years in a plant learning the processes and
the details of his plant's operation and management policies before starting out to be
a member of the sales staff. Since the sales engineer represents his company in the
mind of the customer, he must present a pleasing appearance and give a feeling of
confidence in his engineering ability.

Management

Results of recent surveys show that the trend today is for corporate leaders in the
United States to have backgrounds in engineering and science. In a survey of some
600 large industrial firms, 20 technical and engineering colleges and universities have
four or more of their graduates serving as board chairmen, presidents, or senior
vice-presidents in these firms.

It has been predicted that within ten years, the *majority* of corporation executives
will be men who are trained in engineering and science as well as in business and the
humanities, and who can bridge the gap between these disciplines.

Since the trend is toward more engineering graduates moving into management
positions, let us examine the functions of an engineer in management.

The basic functions of the management of a company are generally similar whether the company objective is dredging for oyster shells or building diesel locomotives or digital computers. These basic functions involve using the capabilities of the company to the best advantage to produce a desirable product in a competitive economy. The use of the capabilities, of course, will vary widely depending upon the enterprise involved.

The executive of a company, large or small, has the equipment in the plant, the labor force, and the financial assets of the organization to use in conducting the plant's operations. In management, he must make decisions involving all three of these items.

In former years it was assumed that only persons trained and educated in business administration should aspire to management positions. However, now it has been recognized that the education and other abilities which make a good engineer also provide the background to make a good management executive. The training for correlating facts and evaluating courses of action in making engineering decisions can be carried over to management decisions on machinery, men, and money. In some cases, the engineer is technically strong but may be quite naïve in the realm of business practicability. Therefore, it is in the business side of an operation that the engineer usually must work harder to develop his skills.

The engineer in management is concerned more intimately with the long-range effects of policy decisions. Where the design engineer considers first the technical phases of a project, the engineer in management must consider how a particular

Illustration 4-7
The engineer in management must have the ability to reduce a large number of variables to the most significant factors and then move decisively to a plan of action.

decision will affect the men who work to produce a product and how the decision will affect the people who provide the financing of the operation. It is for this reason that the management engineer is concerned less with the technical aspects of his profession and relatively more with the financial, legal, and labor aspects.

This does not imply that engineering aspects should be minimized or deleted. Rather the growing need for engineers in management shows that the type and complexity of the machines and processes used in today's plants require a blending of technical and business training in order to carry forward effectively. Particularly is this trend noted in certain industries, such as aerospace and electronics, where the vast majority of executive managerial positions are occupied by engineers and scientists. As other industries become automated, a similar trend in those fields also will become apparent.

The education that an engineer in management receives should be identical to the basic engineering education received in other engineering functions. However, a young engineer usually can recognize early in his career whether or not he has an aptitude for working with men and directing their activities. If the young engineer has the ability to "sell his ideas" and to get others to work with him, probably he can channel his activities into managerial functions. He may start out as a research engineer, a design engineer, or a sales engineer, but the ability to influence others to his way of thinking, a genuine liking for people, and a consideration for their responses, will indicate that he probably has capabilities as a manager.

Of course, management positions are not always executive positions, but the ability to apply engineering principles in supervisory work involving large numbers of men and large amounts of money is a prerequisite in management engineering.

Other engineering functions

A number of other engineering functions can be considered that do not fall into the categories previously described. Some of these functions are testing, teaching, and consulting.

As in the other functions, there are no specific curricula leading directly toward these types of work. Rather a broad background of engineering fundamentals is the best guide to follow in preparing for work in these fields.

In testing, the work resembles design and development functions most closely. Most plants maintain a laboratory section that is responsible for conducting engineering tests of proposed products or for quality control on existing products. The test engineer must be qualified to follow the intricacies of a design and to build suitable test machinery to give an accelerated test of the product. For example, in the automotive industry, not only are the completed cars tested, but also individual components, such as engines, brakes, and tires, are tested on stands to provide data to improve their performance. The test engineer must be able also to set up quality control procedures for production lines to ensure that production meets certain standards. In this work, mathematics training in statistical theory is helpful.

A career in teaching is rewarding for many persons. A desire to help others in their learning processes, a concern for some of their personal problems, and a thorough grounding in engineering and mathematics are desirable for those considering teaching engineering subjects. In the teaching profession, the trend today is toward the more theoretical aspects of engineering, and a person will usually find that teaching is more closely allied with research and development functions than with others. Almost all colleges now require the faculty to obtain advanced degrees, and a person desiring to be an engineering teacher should consider seriously the desirability of obtaining a doctorate in his chosen field.

More and more engineers are going into consulting work. Work as an engineering consultant can be either part time or full time. Usually a consulting engineer is a person who possesses specific skills in addition to several years of experience. He may offer his services to advise and work on engineering projects either part time or full time.

Frequently two or more engineers will form an engineering consulting firm that employs other engineers, technicians, and draftsmen and will contract for full engineering services on a project. The firm may restrict engineering work to rather narrow categories, such as the design of irrigation projects, power plants, or aerospace facilities, or a staff may be available that is capable of working on a complete spectrum of engineering problems.

Illustration 4-8
Teaching is a rewarding activity of engineering. Frequently the engineering professor is the first person to introduce the student to the ethics and responsibilities of the profession.

On the other hand, a consulting engineer can operate alone. His firm may consist of a single individual with skills such that, in a minimum time, he is able to advise and direct an operation to overcome a given problem. For instance, he may be employed by an industrial plant. In this way the plant may be able to solve a given problem more economically, particularly if the required specialization is seldom needed by the plant.

As may be inferred, a consulting engineer must have *specific* skills to offer, and he must be able to use his creative ability to apply his skills to unfamiliar situations. Usually these skills and abilities are acquired only after several years of practice and postgraduate study.

Consulting work is an inviting part of the engineering profession for a person who desires self-employment, and is willing to accept its business risks to gain an opportunity for financial reward.

Engineering functions in general

As described in previous paragraphs, training and skills in all functions are basically the same, that is, fundamental scientific knowledge of physical principles and mathematics. However, it can be seen that research on one hand and management on the other require different educational preparations.

For work in research, emphasis is on theoretical principles and creativity, with little emphasis on economic and personnel considerations. On the other hand, in management, primary attention is given to financial and labor problems and relatively little to abstract scientific principles. Between these two extremes, we find the other functions with varying degrees of emphasis on research- or managerial oriented concepts.

Figure 4-1 shows an idealized image of this distribution. Bear in mind that this diagram merely depicts a trend and does not necessarily apply to specific instances.

To summarize the functions of the engineer, we can say that in all cases he is a problem identifier and solver. Whether it be a mathematical abstraction that may have an application to a nuclear process or a meeting with a bargaining group at a conference table, it is a problem that must be identified and reduced to its essentials and the alternatives explored to reach a solution. The engineer then must apply his knowledge and inventiveness to select a reasonable method to achieve a result, even in the face of vague and sometimes contradictory data. That the engineer has been able, in general, to accomplish this is proven by a long record of successful industrial management and productivity.

Problems on work of the engineer

4-1. Discuss an important scientific breakthrough of the past year that was brought about by an engineering research effort.

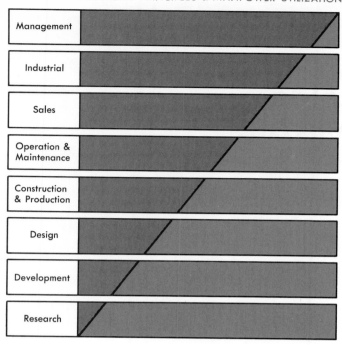

ABSTRACT SCIENTIFIC PRINCIPLES

Figure 4-1 Application of principles in various engineering functions.

4-2. Discuss the differences between engineering research and engineering development.

4-3. Interview an engineer and estimate the percentage of his work that is devoted to research, development, and design.

4-4. Discuss the importance of the engineer's design capability in modern industry.

4-5. Investigate the work functions of the engineer and write a brief essay describing the function that most appeals to you.

4-6. Discuss the importance of the sales engineer in the total engineering effort.

4-7. Interview an engineer in management. Discuss the reasons that many engineers rise to positions of leadership as managers.

4-8. Compare the engineering opportunities in teaching with those in industry.

4-9. Investigate the opportunities for employment in a consulting engineering firm. Discuss your findings.

4-10. Discuss the special capabilities required of the engineer in construction.

5

Career fields in engineering

Much of the change in our civilization in the past 100 years has been due to the work of the engineer. We hardly appreciate the changes that have occurred in our environment unless we attempt to picture the world of a few generations ago, without automobiles, telephones, radios, electronics, transportation systems, supersonic aircraft, automatic machine tools, electric lights, television, and all the modern appliances in our homes. In the growth of all these things the role of the engineer is obvious.

Development in the field of science and engineering is progressing so rapidly at present that within the last 10 years we have acquired materials and devices that are now considered commonplace but which were unknown to our parents. Through research, development, and mass production, directed by engineers, ideas are made into realities in an amazingly short time.

The engineer is concerned with more than research, development, design, construction, and the operation of technical industries, however, since many are engaged in businesses that are not concerned primarily with production. Formerly, executive positions were held almost exclusively by men whose primary training was in the field of law or business, but the tendency now is to utilize engineers more and more as administrators and executives.

No matter what kind of work the engineer may wish to do, he will find opportunities for employment not only in purely technical fields but also in other functions, such as general business, budgeting, rate analysis, purchasing, marketing, personnel, labor relations, and industrial management. Other opportunities also exist in such specialized fields of work as teaching, writing, patent practice, and work with the military establishment.

Although college engineering curricula contain many basic courses, there will be some specialized courses available that are either peculiar to a certain curriculum or

Illustration 5-1

Aeronautical engineers frequently work with models of their designs to confirm the validity of their calculations.

are electives. These specializations permit each student to acquire a particular proficiency in certain subjects so that, for example, he can be designated as an electrical, civil, chemical, mechanical, or industrial engineer.

Education in the application of certain subject matter to solve technological problems in a certain engineering field constitutes engineering specialization. Such training is not for manual skills as in trade schools, but rather is planned to provide preparation for research, design, operation, management, testing, maintenance of projects, and other engineering functions in any given specialty.

The principal engineering fields of specialization that are listed in college curricula and that are recognized in the engineering profession are described in the following sections.

Aerospace and astronautical engineering

The powered flight of man began in 1903 at Kitty Hawk, North Carolina. Perhaps no other single technological achievement has been so significant for mankind. Through faster transportation and improved communications almost every aspect of man's daily life has been affected. However, not all challenges are associated with space-flight. Problems associated with conventional aircraft, and the development of special vehicles such as hydrofoil ships, ground-effect machines, and deep-diving vessels for oceanographic research are all concerns of the industry.

Within the past few years many changes have taken place which have altered the work of the aeronautical engineer—not the least of which is man's successful conquest of space. Principal types of work vary from the design of guided missiles and spacecraft to analyses of aerodynamic studies dealing with the performance, stability, control, and design of various types of planes and other devices that fly. Most of such activity is concerned with the design, development, and performance testing of supersonic commercial transports and their propulsion systems.

Although aerospace engineering is one of the newer fields, it offers many possibilities for employment. Continued exploration and research in previously uncharted areas is needed in the fields of propulsion, materials, thermodynamics, cryogenics, navigation, cosmic radiation, and magnetohydrodynamics. It is predicted that within the near future the chemically fueled rocket engine, which has enabled man to explore lunar landscapes, will become obsolete as the need increases to cover greater and greater distances over extended periods of time.

The rapidly expanding network of airlines, both national and international, provides many openings for the engineering graduate. Since the demand for increasing numbers of aircraft of various types exists, there are opportunities for work in manufacturing plants and assembly plants and in the design, testing, and maintenance of aircraft and their component parts. The development of new types of aircraft, both civilian and military, requires the efforts of well-trained aeronautical engineers, and it is in this field that the majority of positions exists. Employment opportunities exist for specialists in the design and development of fuel systems using liquid oxygen propellants and solid propellants. Control of the newer fuels involves precision valving and flow sensing at very low and very high temperatures. Air traffic control is a problem that is becoming increasingly more complex, and trained people are needed here. The design of ground and airborne systems that will permit operation of aircraft under all kinds of weather conditions is also a part of the work of aeronautical engineers.

The aerospace engineer works on designs that are not only challenging and adventuresome but also play a major role in determining the course of present and future world events.

Agricultural engineering

Agricultural engineering is that discipline of engineering that spans the area between two fields of applied science—agriculture and engineering. It is directly concerned

Illustration 5-2
In this picture agricultural engineers at a research center test the safety features of a farm tractor. Agricultural engineers apply fundamental engineering principles of analysis and design to improve our methods of food production and land utilization.

with supplying the means whereby food and fiber are supplied in sufficient quantity to fill the basic needs of all mankind. In the next 30 years the world's population is expected to double. This factor, plus the increasing demands of people throughout the world for increased standards of living, provides unparalleled challenges to the agricultural engineer. Not only must the quantity of food and fiber be increased, but the efficiency of production also must be steadily improved in order that manpower may be released for other creative pursuits. Through applications of engineering principles, materials, energy, and machines may be used to multiply the effectiveness of man's effort. This is the agricultural engineer's domain.

In order that the agricultural engineer may understand the problems of agriculture and the application of engineering methods and principles to their solution, instruction is given in agricultural subjects and the biological sciences as well as in basic engineering. Agricultural research laboratories are maintained at schools for research and instruction using various types of farm equipment for study and testing. The young person who has an analytical mind and a willingness to work, together with an interest in the engineering aspects of agriculture, will find the course in agricultural engineering an interesting preparation for his life's work.

Many agricultural engineers are employed by companies that serve agriculture and some are employed by firms that serve other industries. Opportunities are particularly apparent in such areas as (1) research, design, development, and sale of mechanized farm equipment and machinery, (2) application of irrigation, drainage, erosion control, and land and water management practices, (3) application and use of electrical energy for agricultural production, and feed and crop processing, handling and grading, (4) research, design, sale, and construction of specialized structures for farm use, and (5) the processing and handling of food products.

Architectural engineering

The architectural engineer is interested primarily in the selection, analysis, design, and assembly of modern building materials into structures that are safe, efficient, economical, and attractive. The education received in college is designed to teach

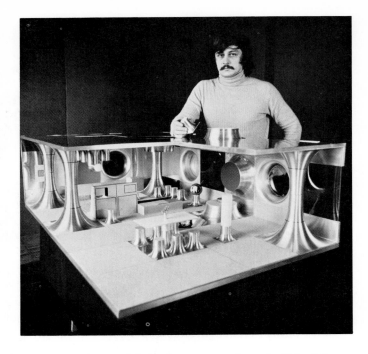

one how best to use modern structural materials in the construction of tall buildings, manufacturing plants, and public buildings.

The architectural engineer is trained in the sound principles of engineering and at the same time is given a background which enables him to appreciate the point of view of the architect. The architect is concerned with the space arrangements, proportions, and appearance of a building, whereas the architectural engineer is more nearly a structural engineer and is concerned with safety, economy, and sound construction methods.

Opportunities for employment will be found in established architectural firms, in consulting engineering offices, in aircraft companies, and in organizations specializing in building design and construction. Excellent opportunities await the graduate who may be able to associate himself with a contracting firm or who may form a partnership with an architectural designer. In the field of sales an interesting and profitable career is open to the individual who is able to present his ideas clearly and convincingly.

Bioengineering

Bioengineering encompasses all aspects of the application of engineering methods to the use and control of biological systems. It bridges the engineering, physical, and life sciences in identifying and solving medical and health-related problems. Bioengineers are team players in much the same way as many athletes. For example, engineers, physicists, chemists, and mathematicians routinely join with the biologist and physician in developing techniques, equipment, and materials.

The range of the bioengineers' interests is very broad. It would involve, for

Illustration 5-4
The work of bioengineers has made possible the development of many life-lengthening and life-enhancement systems.

example, the development of highly specialized medical instruments and devices—including artificial hearts and kidneys and the development of lasers for surgery and cardiac pacemakers that regulate the heartbeat. Other biomedical engineers may specialize more particularly in the adaptation of computers to medical science, such as in monitoring patients or processing electrocardiograph data. Some will design and build systems to modernize laboratory, hospital, and clinical procedures.

Those selecting a career in bioengineering should anticipate earning a graduate degree since advanced study beyond the bachelor's degree is acutely needed to attain a depth of knowledge from at least two diverse disciplines.

At present bioengineering is a small field because few engineers have attained the necessary depth of academic training and experience in the life sciences. Therefore, job opportunities for graduates are excellent. Here indeed is a promising new field for those so inclined.

Ceramic engineering

Today our technological world is amazingly dependent upon ceramics of all types. Unlike many other products they appear in every part of the spectrum of life, from

Illustration 5-5
Ceramic engineers hold in their hands a very important answer to the impending crisis in the shortage of metallic materials.

beautiful but commonplace table settings, to the protective coatings of electrical transducers or the refractories of space exploratory rocket nozzles, to the spark plugs of a farmer's tractor. Exactly what are ceramics? When did man first find a use for them?

Ceramics are nonmetallic, inorganic materials that require the use of high temperatures in their processing. In the earliest form, clay pottery of 10,000 B.C. has been found to be excellently preserved. The most common of ceramics, glass—an ancient discovery of the Phoenicians (about 4000 B.C.), is a miracle material in every sense. It may be made transparent, translucent, or opaque, weak and brittle or flexible and stronger than steel, hard or soft, water soluble or chemically inert. Truly it is one of the most versatile of engineering materials.

Although it is imperative today that all engineers have a fundamental understanding of the adaptive use of ceramics, it is ceramic engineers who are expert in the development and production of ceramic materials. Their activities cover a wide range of activities from the conception of the initial idea to the development, production, evaluation, application, and sale of the product. Therefore, ceramic engineers are employed by a variety of industries, from the specialized raw material and ceramic product manufacturers to the chemical, electrical and electronic, automotive, nuclear, and aerospace industries.

Chemical engineering

Chemical engineering is responsible for new and improved products and processes that affect every person. This includes materials that will resist extremities of heat and

Illustration 5-6
The chemical engineer is responsible for coordinating the work of many people to assure efficient plant operations.

cold, processes for life-support systems in other environments, new fuels for reactors, rockets, and booster propulsion, medicines, vaccines, serum, and plasma for mass distribution, and plastics and textiles to serve a multiplicity of human needs. Consequently, chemical engineers must be able to apply scientifically the principles of chemistry, physics, and engineering to the design and operation of plants for the production of materials that undergo chemical changes during their processing.

The courses in chemical engineering cover inorganic, analytical, physical, and organic chemistry in addition to the basic engineering subjects; and the work in the various courses is designed to be of a distinctly professional nature and to develop capacity for original thought. The industrial development of our country makes large demands on the chemical engineer. The increasing uses for plastics, synthetics, and building materials require that a chemical engineer be employed in the development and manufacture of these products. While well trained in chemistry, the chemical engineer is more than a chemist in that he applies the results of chemical research and discovery to the use of mankind by adapting laboratory processes to full-scale manufacturing plants.

The chemical engineer is instrumental in the development of the newer fuels for turbine and rocket engines. Test and evaluation of such fuels and means of achieving production of suitable fuels are part of the work of a chemical engineer. This testing must be carefully controlled to evaluate the performance of engines before the fuel is considered suitable to place on the market.

Opportunities for chemical engineers exist in a wide variety of fields of manufacture. Not only are they in demand in strictly chemical fields but also in nearly all types of manufacturing. The production of synthetic rubber, the uses of petroleum products, the recovery of useful materials from what was formerly considered waste products, and the better utilization of farm products are only a few of the tasks that will provide work for the chemical engineer. Although the first professional work of a chemical engineering graduate may be in production, other opportunities exist in the fields of engineering design, research and development, patents, and sales engineering.

Civil engineering

The civil engineer plans, designs, constructs, and operates physical works and facilities that are deemed essential to modern life. These include the broad categories of construction, soil mechanics and foundations, transportation systems, water resources, sanitation, city planning and municipal engineering, and surveying and mapping. Construction engineering is concerned with the design and supervision of construction of buildings, bridges, tunnels, and dams. The construction industry is America's largest industry today. Soil mechanics and foundation investigations are essential not only in civilized areas but also for successful conquest of new lands such as Antarctica

Illustration 5-7
The civil engineer pictured here is marking a structural member which will be used in the erection of a building that she has designed.

and the lunar surface. Transportation systems include the planning, design, and construction of necessary roads, streets, thoroughfares, and superhighways. Engineering studies in water resources are concerned with the improvement of water availability, harbor and river development, flood control, irrigation, and drainage. Pollution is an ever-increasing problem, particularly in urban areas. The sanitary engineer is concerned with the design and construction of water supply systems, sewerage systems, and systems for the reclamation and disposal of wastes. City planning and municipal engineers are concerned primarily with the planning of urban centers for the orderly, comfortable, and healthy growth and development of business and residential areas. Surveying and mapping are concerned with the measurements of distances over a surface (such as the earth or the moon) and the location of structures, rights of way, and property boundaries.

Civil engineers engage in technical, administrative, or commercial work with manufacturing companies, construction companies, transportation companies, and power companies. Other opportunities for employment exist in consulting engineering offices, in city and state engineering departments, and in the various bureaus of the federal government.

Electrical engineering

Electrical engineering is concerned, in general terms, with the utilization of electric energy. It is divided into broad fields, such as information systems, automatic control, and systems and devices. Electricity used in one form or another reaches nearly all our daily lives and is truly the servant of mankind.

The electrical engineer applies sound engineering principles, both mechanical and electrical, in the design and construction of computers. He must be familiar with the basic requirements of a computer so that he can design to provide for the necessary capabilities. In addition he must strive to build a machine that will furnish solutions of greater and greater problem complexity and at the same time have a means of introducing the problem into the machine in as simple a manner as possible.

Illustration 5-8
The electrical engineer works with many types of apparatus, both electrical and mechanical. In this picture, an electrical engineer adjusts an ultrasonic device to measure composite response to resonance. The superimposed resonance curve is a test result.

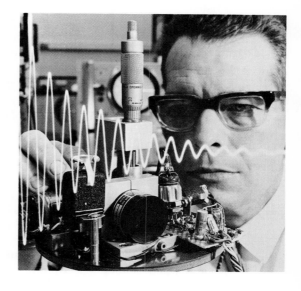

Although there are relatively few companies that build elaborate computing machines, employment possibilities in the design and construction part of the industry are not limited. Many industrial firms, colleges, and governmental branches have set up computers as part of their capital equipment, and opportunities exist for employment as computer applications engineers, who serve as liaison between computer programmers and engineers who wish their problems evaluated on the machines. Of course, in a field expanding as rapidly as computer design, increasing numbers of employment opportunities become available. More and more dependence will be placed on the use of computers in the future, and an engineer educated in this work will find ample opportunity for advancement.

The automatic control of machines and devices, such as autopilots for spacecraft and missiles, has become a commonplace requirement in today's technically conscious society. Automatic controlling of machine tools is an important part of modern machine shop operation. Tape systems are used to furnish signals to serve units on automatic lathes, milling machines, boring machines, and other types of machine tools so that they can be programmed to perform repeated operations. Not only can individual machines be controlled but also entire power plants can be operated on a program system. The design of these systems is performed usually by an electrical or mechanical engineer.

Energy conversion systems, where energy is converted from one form to another, also are a necessity in almost every walk of life. Power plants are constituted to convert heat energy from fuels into electrical energy for transmission to industry and homes. In addition to power systems, communication systems are a responsibility of the electrical engineer. Particularly in communications the application of modern electronics has been most evident. The electrical engineer who specializes in electronics will find that the majority of communication devices employs electronic circuits and components.

Other branches of electrical engineering that may include power or communication activities, or both, are illumination engineering, which deals with lighting using electric power; electronics, which has applications in both power and communications; and such diverse fields as x-ray, acoustics, and seismograph work.

Employment opportunities in electrical engineering are extremely varied. Electrical manufacturing companies use large numbers of engineers for design, testing, research, and sales. Electrical power companies and public utility companies require a staff of qualified electrical engineers, as do the companies which control the networks of telegraph and telephone lines and the radio systems. Other opportunities for employment exist with oil companies, railroads, food processing plants, lumbering enterprises, biological laboratories, chemical plants, and colleges and universities. The aircraft and missile industries use engineers who are familiar with circuit design and employment of flight data computers, servomechanisms, analog computers, vacuum tubes, transistors, and other solid-state devices. There is scarcely any industry of any size that does not employ one or more electrical engineers as members of its engineering staff.

Industrial engineering

Industrial engineers determine the most effective methods of using manpower, machines, and materials in a production environment. Whereas other branches of

Illustration 5-9
Industrial engineers are especially sought after to fill management roles in industry and government.

engineering tend to specialize in some particular phase of science, the realm of industrial engineering may include parts of all engineering fields. The industrial engineer then will be more concerned with the larger picture of management of industries and production of goods than with the detailed development of processes.

The work of the industrial engineer is rather wide in scope. His general work is with men and machines, and as a result he is educated in both personnel administration and in the relations of men and machines to production.

The advent of the electronic digital computer and other electronic support equipment has revolutionized the business world. Many of the resultant changes have been made as a result of industrial engineering designs. Systems analysis, operations research, statistics, queuing theory, information theory, symbolic logic, and linear programming are all mathematics-based disciplines that are used in industrial engineering work.

The industrial engineer must be capable of preparing plans for the arrangement of plants for best operation and then of organizing the workers so that their efforts will be coordinated to give a smoothly functioning unit. In such things as production lines, the various processes involved must be timed perfectly to ensure smooth operation and efficient use of the worker's efforts. In addition to coordination and automating of manufacturing activities, the industrial engineer is concerned with the development of data processing procedures and the use of computers to control production, the development of improved methods of handling materials, the design of plant facilities and statistical procedures to control quality, the use of mathematical models to stimulate production lines, and the measurement and improvement of work methods to reduce costs.

Opportunities for employment exist in almost every industrial plant and in many businesses not concerned directly with manufacturing or processing goods. In many cases the industrial engineer may be employed by department stores, insurance companies, consulting companies, and as engineers in cities. The industrial engineer is trained in fundamental engineering principles, and as a result may also be employed in positions which would fall in the realm of the civil, electrical, or mechanical engineer.

The courses prescribed for the student of industrial engineering follow the pattern of the other branches of engineering by starting with a thorough foundation in the engineering sciences. The engineering courses in the later semesters will be of a more general nature, and the curriculum will include such courses as economics, psychology, business law, personnel problems, and accounting principles.

Mechanical engineering

Mechanical engineering deals with power and the design of machines and processes used to generate power and apply it to useful purposes. These designs may be simple or complex, inexpensive or expensive, luxuries or essentials. Such items as the kitchen food mixer, the automobile, air-conditioning systems, nuclear power plants, and interplanetary space vehicles would not be available for man's use today were it not for the mechanical engineer. In general, the mechanical engineer works with systems, subsystems, and components that have motion. The range of work that may be classed as mechanical engineering is wider than that in any of the other branches of engineering, but it may be grouped generally under two heads: work that is concerned with power-generating machines, and work that deals with machines that transform or consume this power in accomplishing their particular tasks. The utilization of solar energy for domestic and industrial uses is one of the more important areas of the mechanical engineer's work at this time.

Some of the general subdivisions of mechanical engineering are as follows: Power or combustion engineers deal with the production of power from fuels. Design specialists may work with parts that vary in size from the microscopic part of the most delicate instrument to the massive part of heavy machinery. Included are the mass transit systems that are rapidly becoming a part of our nationwide transportation system. Automotive engineers work constantly to improve the vehicles and engines that we now have. Heating, ventilating, air-conditioning, and refrigeration engineers deal with the design of suitable systems for making our buildings more comfortable and for providing proper conditions in industry for good working conditions and efficient machine operation.

Employment may be secured by mechanical engineering graduates in almost every type of industry. Manufacturing plants, power-generating stations, public utility companies, transportation companies, airlines, and factories, to mention only a few, are examples of organizations that need mechanical engineers. Experienced engineers are needed in the missile and space industries in the design and development of such items as gas turbine compressors and power plants, air-cycle cooling turbines, electrically and hydraulically driven fans, and high-pressure refrigerants. Mechanical engineers are also needed in the testing of airborne and missile fuel systems, servovalves, and mechanical–electrical control systems. In addition, an

Illustration 5-10
*High-speed vehicle design is one area of specialization of the mechanical engineer.
Pictured here is a vehicle test in progress.*

engineer may be employed for research endeavor as a university professor, or in the governments of cities, states, and the nation.

Metallurgical engineering and materials science

In many respects the past 25 years may be said to be an "age of materials"—an age which has seen the maturing of space exploration, nuclear power, digital computer technology, and ocean conquest. None of these engineering triumphs could have been achieved without the contributions of the metallurgical engineer. Although his world is not confined to this planet, the unsolved problems of private industry are his menu. Metals are found in every part of the earth's crust, but rarely in immediately usable form. It is the metallurgical engineer's job to separate them from their ores and from other materials with which they exist in nature.

Metallurgical engineering may be divided into two branches. One branch deals with the location and evaluation of deposits of ore, the best way of mining and concentrating the ore, and the proper method of refining the ore into the basic metals. The other branch deals with the fabrication of the refined metal or metal alloy into various machines or metal products.

The metallurgist performs pure and applied research on vacuum melting, arc melting, and zone refining to produce metallic materials having unusual properties of strength and endurance. In addition the metallurgist in the aircraft and missile

Illustration 5-11

Almost every aspect of our life is affected by advances in metallurgy and materials science. For example, teeth can now be straightened because of the development of a special type of steel . . . rustproof, strong yet ductile, and hard yet smooth . . . unchanged through ice-cold sodas and red-hot pizzas.

industries is called upon to recommend the best materials to use for special applications, and is frequently called on to give an expert opinion on the results of fatigue tests of metal parts of machines.

The engineer who has specialized in materials science is in great demand today because of the urgent need for man-made composites—the joining of two or more different materials for the purpose of gaining advantageous or overcoming disadvantageous characteristics of each.

Mining and geological engineering

The mining and geological engineer of today who searches the earth for hidden minerals is necessarily a person of quite different stature than the traditional explorer of yesteryear. He must possess a combination of fundamental engineering and scientific education and field experience to enable him to unravel the story of the earth's crust. He must be expert in utilizing very sensitive instruments as he seeks to locate new mineral deposits and to anticipate the problems that might arise in getting them out and transporting them to civilization. For this reason it is not unusual to find a mining or geological engineer in a modern office building in New York one week, and the next in Arizona or Afghanistan—or commuting between an expedition campsite and technical laboratories.

Illustration 5-12
Geological engineers investigate new sources of essential mineral bodies.

The work of mining and geological engineers lies generally in three areas: finding the ore, extracting it, and preparing the resulting minerals for manufacturing industries to use. They design the mine layout, supervise the construction of mine shafts and tunnels in underground operations, and devise methods for transporting minerals to processing plants. Mining engineers are also responsible for mine safety and the efficient operation of the mine, including ventilation, water supply, power, communications, and equipment maintenance. Geological engineers are more directly concerned with locating and appraising mineral deposits.

An important part of the mining and geological engineer's work is to keep in mind inherent air- and water-pollution problems that might develop during the mining operation. This involves establishing efficient controls to prevent harmful side effects of mining and designing ways whereby the land will be restored for man to use after the mining operation terminates.

Naval architecture
and marine engineering

For many centuries the sea has played a dominant role in the lives of peoples of all cultures and geographical locations. For this reason in every era the designers of ships have been held in the highest regard for their knowledge and understanding of the sea's physical influences and for their artistry and ability in marine craftmanship. As our civilization increases in complexity, all peoples of future generations will depend to an even greater extent upon vessels of the sea to keep food, materials, and fuel flowing.

Ship design is a refined art as well as an exacting science since most ships are custom built—one at a time. Many large ships are virtually floating cities containing their own power sources, sanitary facilities, food preparation center, and recreational

Illustration 5-13
The marine engineer's role is significant in helping to solve the world's transportation problems.

and sleeping accommodations. Every service that would be provided to city dwellers must also be provided for the ship's crew. As with aircraft design, the ship's structural members and intricate networks of piping and electrical circuits must fit together harmoniously in the minimum space possible.

The marine engineer must have a broad-based engineering educational background. He will routinely establish specifications for the vessel, perform design calculations, and install and test ship machinery.

The basic design of seaworthy cargo ships has changed very little from 1900 to 1960. However, with the advent of nuclear power and sophisticated electronic computers a new era in ship design has begun to develop. Surface-effect or air-cushion-type vehicles, submarine tankers, and deep-submergence research vehicles have all emerged from the realm of science fiction to enter one of engineering reality. The application of these newer ideas for the shipbuilding industry awaits only a more positive commitment to the task by government and industry. With the shrinking world supply of food and energy, this commitment is certain to come.

Nuclear engineering

Nuclear engineering is one of the newest and most challenging branches of engineering. Although much work in the field of nucleonics at present falls within the realm of pure research, a growing demand for people educated to utilize recent discoveries for the benefit of mankind has led many colleges and universities to offer courses in nuclear engineering. The nuclear engineer is familiar with the basic principles involved in both fission and fusion reactions; and by applying fundamental

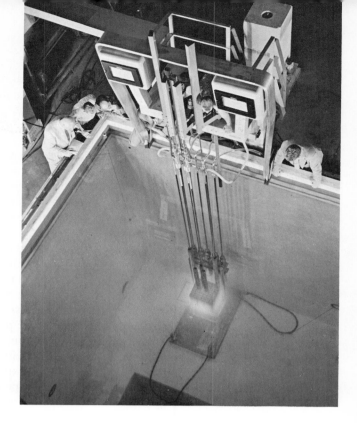

Illustration 5-14
The reactor pictured here is being used to study how radiation affects various materials. It also produces radioisotopes to use for industrial and research purposes.

engineering concepts, he is able to direct the enormous energies involved in a proper manner. Work involved in nuclear engineering includes the design and operation of plants to concentrate nuclear reactive materials, the design and operation of plants to utilize heat energy from reactions, and the solution of problems arising in connection with safety to persons from radiation, disposal of radioactive wastes, and decontamination of radioactive areas.

The wartime uses of nuclear reactions are well known, but of even more importance are the less spectacular peacetime uses of controlled reactions. These uses include such diverse applications as electrical power generation and medical applications. Other applications are in the use of isotopes in chemical, physical, and biological research, and in the changing of the physical and chemical properties of materials in unusual ways by subjecting them to radiation.

Recent advances in our knowledge of controlled nuclear reactions have enabled engineers to build power plants that use heat from reactions to drive machines. Submarine nuclear power plants, long a dream, are now a reality, and experiments are being conducted on smaller nuclear power plants that can be used for airborne or railway applications.

At present, ample opportunities for employment of nuclear engineers exist in both privately owned and government-operated plants, where separation, concentration, or processing of nuclear materials is performed. Nuclear engineers are also needed by companies that may use radioactive materials in research or processing involving agricultural, medical, metallurgical, and petroleum products.

Petroleum engineering

Throughout history the energy available to man beyond his own muscle power has been a measure of his hope for a more secure and improved material life. In early Greek and Roman civilizations wind and water provided much of man's energy needs. In early America, wood was the primary source of energy. Today the major source of energy is petroleum. It is the most widely used of all energy sources because of its mobility and flexibility in utilization. Approximately three fourths of the total energy needs of the United States are currently supplied by petroleum products, and this condition will likely continue for many years. Petroleum engineering is the practical application of the basic sciences (primarily chemistry, geology, and physics) and the engineering sciences to the development, recovery, and field processing of petroleum.

Petroleum engineering deals with all phases of the petroleum industry, from the location of petroleum in the ground to the ultimate delivery to the user. Petroleum products play an important part in many phases of our everyday life in providing our clothes, food, work, and entertainment. Because of the complex chemical structure of petroleum, we are able to make an almost endless number of different articles. Owing to the wide demand for petroleum products, the petroleum engineer strives to satisfy an ever-increasing demand for oil and gas from the ground.

The petroleum engineer is concerned first with finding deposits of oil and gas in quantities suitable for commercial use, in the extraction of these materials from the ground, and the storage and processing of the petroleum above ground. The petroleum engineer is concerned with the location of wells in accordance with the findings of geologists, the drilling of wells and the myriad problems associated with the drilling, and the installation of valves and piping when the wells are completed. In addition to the initial tapping of a field of oil, the petroleum engineer is concerned with practices that will provide the greatest recovery of the oil, considering all possible factors that may exist many thousand feet below the surface of the earth.

After the oil or gas has reached the surface, the petroleum engineer will provide the means of transporting it to suitable processing plants or to places where it will be used. Pipelines are providing an ever-increasing means of transporting both oil and gas from field to consumer.

Many challenges face the petroleum engineer. Some require pioneering efforts, such as with the rapidly developing Alaska field. Other opportunities lie closer at hand. For example, it is known that because of excessive costs in recovery less than one half of the oil already discovered in the United States *has yet to be brought to the surface of the earth*. It is estimated that even a 10 per cent increase in oil recovery would produce 3 billion barrels of additional oil, a worth of over 15 billion dollars.

Owing to the expanding uses for petroleum and its products, the opportunities for employment of petroleum engineers are widespread. Companies concerned with the drilling, producing, and transporting of oil and gas will provide employment for the majority of engineers. Because of the widespread search for oil, employment opportunities for the petroleum engineer exist all over the world; and for the young person wishing a job in a foreign land, oil companies have crews in almost every country over the globe. Other opportunities for employment exist in the field of technical sales, research, and as civil service employees of the national government.

The curriculum in petroleum engineering includes courses in drilling methods, engines, oil and gas recovery, storage and transportation, and geology.

Illustration 5-15
One of the tasks of the petroleum engineer is to locate oil deposits and to devise methods for oil recovery. As is the case in this photograph, design calculations frequently must be made at the drilling site.

Problems on career fields in engineering

5-1. Discuss the changing requirements for aerospace and astronautical engineers.

5-2. Investigate the opportunities for employment in agricultural engineering. Discuss your findings.

5-3. Write a short essay on the differences in the utilization and capability of the architectural engineer and the civil engineer who has specialized in structural analysis.

5-4. Interview a chemical engineer. Discuss the differences in his work and that of a chemist.

5-5. Assume that you are employed as an electrical engineer. Describe your work and comment particularly concerning the things that you most like and dislike about your job.

5-6. Explain why the demand for industrial engineers has increased significantly during the past ten years.

5-7. Write a 200-word essay describing the challenging job opportunities in engineering that might be particularly attractive for an engineering graduate.

5-8. Explain the importance of mechanical engineers in the electronics industry.

5-9. Describe the changes that might be brought about to benefit mankind by the development of new engineering materials.

5-10. Investigate the need for nuclear and petroleum engineers in your state and report on your findings.

Part Two

Preparation for a career in engineering

The engineer must be able
to communicate technical ideas quickly
and clearly.

6

Professional responsibilities of the engineer

The word "professional" is used in many ways and has many meanings. It can be used in the sense of the skill of a professional actor who receives pay for his efforts, as distinguished from an amateur who performs more for the joy of performing. It can be used in the sense of a type of work, as in describing a professional job of house painting done by an experienced painter. Also, it can be used merely to describe a degree of effort or line of conduct over a period of time, as used in the expression "a professional beggar." However, in the sense that engineers would employ the word "professional," it should be restricted to a particular and specialized group of people, identified by distinguishing characteristics that separate its members from nonprofessionals.

Within the last century, three groups have emerged with the title "learned professions." These professional groups are law, medicine, and theology. These groups came into being gradually over a long period of time and had certain characteristics in common, among which were higher levels of educational achievement and a sincere desire for performing a service for people. There is no formal naming of a person or group of persons to professional status, nor is there a schedule or procedure to follow to achieve recognition as a professional. Rather the group itself sets standards of training, skills, achievement, and service in order to call itself a professional group, and the public accepts the group's evaluation of itself.

Who is a professional? As generally used in the sense of the learned professions, a professional person is one who applies certain knowledge and skill, usually obtained by college education, for the service of people. In addition, a professional person observes an acceptable code of conduct, uses discretion and judgment in dealing with people, and respects their confidences. Also, professional persons usually have legal status, use professional titles, and associate together in groups. Although engineering has met most of these criteria for a long time, it has been only within the last few decades that legal status has been conferred upon the engineering profession.

The engineer as a professional person

Knowledge and skill above that of the average person is a characteristic of the professional man. Where a workman will have specific skills in operating a particular machine, a professional person is considered able to apply fundamental principles that are usually beyond the range of the average workman. The knowledge of these principles as well as the skills necessary to apply them distinguishes a professional man. The engineer, because of an education in the basic sciences, mathematics, and engineering sciences, is capable of applying basic principles for such diverse things as improving the construction features of buildings, developing processes that will provide new chemical compounds, or designing tunnels to bring water to aid areas.

An important concept in the minds of most persons is that a professional person will perform a service for people. This means that service must be considered ahead of any monetary reward that a professional person may receive. In this respect the professional person should, by himself, recognize a need for personal services and seek ways to provide a solution to these needs. Almost all engineering is performed to fill a need in some phase of our society. It may be to develop better appliances for the household, or to provide better transportation facilities, or to make possible a better life in regions of unfavorable climate.

Discretion and judgment also characterize a professional person. In most situations a choice of several methods to accomplish a given task will be available. The engineer must consider the facts available and the principles that apply and make decisions based upon these rather than upon expediency. Consideration must be

Illustration 6-1

As a professional person the engineer works in a confidential relationship with his client so that proprietary information is protected and trade secrets are not divulged.

given not only to the mechanical aspects of a solution but also to the effects that a particular decision will have upon the persons concerned.

A professional person is one in whom confidence can be placed. This confidence is not only in his skill and ability but also in that knowledge of his client's business or trade information or personal matters will not be divulged improperly. The engineer works in a relation of confidence to his client or employer not to divulge trade secrets or to take any advantage of his knowledge that may harm the client or employer. The public, in general, will have confidence that the engineer's design of buildings, bridges, or power systems will be adequate and safe to use. The engineer must not fail the public in this responsibility.

All professionals adhere to a code of ethical conduct. This code of ethics outlines the standards to which members of the group subscribe and gives an understanding of what the public can expect in its relationship with the profession. The code of ethics also serves as a guide to the members of the profession in their conduct and relations with each other. In engineering, general codes of ethics have been set up by the Engineers' Council for Professional Development and the National Society of Professional Engineers. The ECPD code is given in Appendix V.

Legal status usually is a characteristic of a professional. A medical doctor, for example, has certain rights and privileges afforded by law. Legal recognition of a professional group is afforded by a procedure of certification, licensing, or registration. In all states, a registration law is in effect which provides for legal registration of an engineer following submission of evidence of education and technical ability. Registration confers the legal title of "engineer" to the recipient, and he may use the initials "P.E." after his name to denote his registration as a "Professional Engineer."

Professionalism for the engineer

Professionalism is an individual state of mind. It is a way of thinking and living rather than the development of specific skills or the acquiring of certain knowledge. While the mere acquisition of knowledge may make a person more skilled as a clerk or laborer, knowledge alone does not often promote the desire within oneself to serve or be responsive to the needs of people. It is in this realm of service that the engineer joins with members of other learned professional groups in placing honesty and integrity of action above the legal or minimum level allowable.

Although knowledge and skill often exist apart from professionalism, professionalism can mature only where such competence creates a proper atmosphere. Where competence is an impersonal quality, professionalism, in contrast, is personal. In addition to a state of mind, it is a way of working and living—a way of adding something valuable to competence. For the engineer professionalism implies that he will make *maximum* use of his skill and knowledge, and that he will use his competence to its fullest extent:

□ With complete honesty and integrity.
□ With his best effort in spite of the fact that frequently neither client nor employer is able to evaluate that effort.
□ With avoidance of all possible conflicts of interest.

☐ With the consciousness that the profession of engineering is often judged by the performance of a single individual.

Professionalism for an engineer begins with good moral character, because he occupies a position of trust where he personally must set the standards. Consequently he is required to make decisions that sometimes differ from the preferences of his company or his client.

Professionalism for an engineer means:

☐ Striving to improve his work until it becomes a model for those in his field, as a minimum using the most up-to-date techniques and procedures.
☐ Proper credit for work done and ideas developed by subordinates.
☐ Loyalty to his employer or client, always with concern for the public safety in construction, product design, plant operation, and all other phases of engineering.
☐ Leadership of less experienced colleagues and subordinates toward personal development and an enthusiasm for the profession.
☐ Activity in technical societies in order to keep current in his field, and encouragement of those working under him to improve their technical competence the same way.
☐ Participation in professional societies, as well as technical societies, thereby demonstrating his interest in the profession and encouraging his coworkers to recognize the technical and the professional as of equal-ranking importance.
☐ Registration, not simply because it may be a legal requirement, but more particularly as a demonstration to his coworkers and the public that this is one important hallmark of a professional, a willingness to go beyond the minimum to help and encourage others to realize their full potential.

For engineers in various areas of work, professionalism will include special facets that are more particularly related to a particular field. For example, engineers in industry should be especially conscious of their responsibility in protecting "company proprietary" designs or processes. It also means the establishment of performance standards and safety criteria which protect the purchaser while maintaining a satisfactory return to the manufacturer. For the engineer in government or the engineer in private practice, professionalism may mean capitalizing on a special opportunity to project the profession to the public as a constructive force in society. For the engineer in education, professionalism means practicing at the frontier of knowledge in some field and pushing against that boundary, thus impressing on the students that boundaries need not be (and are rarely) static.

Professionalism for all engineers means an active participation in community life. Engineering cannot achieve general recognition as a profession unless engineers are publicly visible. It is in the realm of public and social service that professionalism shows up strongest. For this reason service to the public and the community and to those less fortunate is particularly significant.

Professionalism can be taught since it is an acquired condition and is not inherent in one's nature. It is most effectively taught by example by individuals whose lives are themselves models of integrity. The beginnings of a professional attitude for the engineering student should be established in the formative college years since, like character, it grows stronger with reinforcement. In laboratory work, for example, an honest reporting of facts and an intelligent evaluation of results are important ingredients in the development of the student's professional training. Design experiences in general involve many compromises—time, money, materials, and so on. Ethical consideration should necessarily become a part of each compromise (decision)

Illustration 6-2
Professionalism includes the establishment of performance standards and safety criteria.

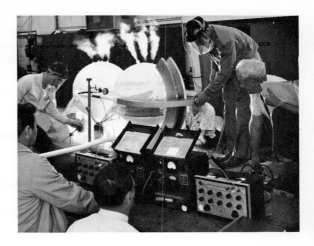

made by the young engineer. His professional career will, in fact, become one of compromise and he must prepare himself to face the realities of such a life.

Probably the student will not have achieved a mature professional attitude by the date of his graduation. However, responsibility of thought and decision should be firmly established by this time in order that entry into employment will be a continuation rather than the beginning of his professional advancement.

After graduation, opportunities for public service will present themselves. The engineer, as part of his professional responsibility, should seek and accept places of service in schools, community government, religious organizations, and charitable groups. Not only will he be able to contribute his talents to these causes, but also he will enhance his own outlook by contacts with both professional and nonprofessional persons. Each individual engineer should recognize within himself the need for a professional attitude and assume the ultimate responsibility for upholding this concept.

To sum up professionalism, engineering may be considered to be a profession insofar as it meets these characteristics of a learned professional group:

☐ Knowledge and skill in specialized fields above that of the general public.
☐ A desire for public service and a willingness to share discoveries for the benefit of others.

Illustration 6-3
The beginnings of a professional attitude for the engineering student should be established in the formative college years.

> Most men believe that it would benefit them if they could get a little from those who *have* more. How much more would it benefit them if they would learn a little from those who *know* more?
> —Wm. J. H. Boetcker

- ☐ Exercise of discretion and judgment.
- ☐ Establishment of a relation of confidence between the engineer and client or the engineer and employer.
- ☐ Acceptance of overall and specific codes of conduct.
- ☐ Formation of professional groups and participation in advancing professional ideals and knowledge.
- ☐ Recognition by law as an identifiable body of knowledge.

With these as objectives, the student should pursue his college studies and his training in his employment so as to meet these characteristics within their full meaning and take his or her place as a professional engineer in our society.

Technical societies

As suggested above, professionals band themselves together for the mutual exchange of ideas, to improve their knowledge, and to learn new skills and techniques. Meeting and discussing problems with others in the same field of endeavor affords an opportunity for the stimulation of thought to improve learning and skills. The National Society of Professional Engineers is concerned primarily with the *profes-*

Illustration 6-4

sional aspects of the whole field of engineering. In addition, engineers have organized a number of technical societies in various fields of specialization. For reference purposes, Table 6-1 lists the major engineering and scientific societies in the United States.

Problems on the professional role of the engineer

6-1. Discuss the factors that are common to all professions.

6-2. Investigate the laws of the state that pertain to serving as an expert engineering witness in court. What would you need to do to qualify as such a witness?

6-3. What engineering fields of specialization are recognized by the state registration board for licensing as professional engineers?

6-4. Using the Code of Ethics as a guide, discuss the procedures that you as a professional engineer in private practice may utilize to attract clients.

6-5. Discuss the value of humanities and social studies courses in relation to the work of the engineer in industry.

6-6. Investigate the need for graduate engineering education for the engineer in private practice.

6-7. List the reasons why it is important for engineers in education to become registered professional engineers.

6-8. Engineer Brown, P.E., is approached by Engineer Smith (nonregistered) who offers a fee of $100.00 to Brown if she will "check over" a set of engineering plans and affix her professional P.E. seal to them. Describe Brown's responsibilities and actions.

6-9. Discuss the reasons why a professional person such as a registered engineer, an attorney, or a physician will not bid competitively on the performance of a service.

6-10. In interviewing for permanent employment, a senior student in a California engineering school agreed to visit on two successive days a company in Chicago and a company in Detroit. Upon her return home both companies sent her checks to cover her expenses, including round-trip airfare. Discuss the appropriate actions that should have been taken by the student.

6-11. The majority of all engineering designs require some extension of the engineer's repertoire of scientific knowledge and analytical skills. How can the engineer determine whether or not this extension lies beyond the "areas of their competence" referred to in Section 2 of the Code of Ethics?

6-12. Engineer Jones, P.E., is the only registered engineer living in Smileyville. Two individuals, Green and Black, approach him with regard to employing his services in estimating the cost of constructing a small dam that would make it possible to reclaim 500 acres of swampland that is now owned by the city but will soon be offered for sale to the highest bidder. Jones learns that both individuals will be bidding against each other to purchase the acreage. Discuss Jones' responsibilities and actions.

6-13. Engineer White is approached by several of his neighbors to urge him to announce his candidacy for city mayor. In previous months the city administration has been accused of "selling rezoning authorizations" and of "enhancing personal fortunes" through the sale of privileged information pertaining to the location of the new proposed freeway. Discuss the course of action that you would recommend for White.

6-14. Engineer Williams and Contractor Smart have been good friends for several years. In March Smart is to begin construction on a multistory building that Williams designed. On Christmas a complete set of children's play equipment (swings, slides, gymnastic bars, etc.) is delivered to Williams' house—compliments of the Smart Construction Company. What course of action do you recommend for Williams?

Table 6-1

Code	Name	Address	Year organized	Total membership
AcSoc	Acoustical Society of America	335 East 45th St. New York, N.Y. 10017	1929	4,100
APCA	Air Pollution Control Association	4400 Fifth Ave. Pittsburgh, Pa. 15213	1907	5,012
AAAS	American Association for the Advancement of Science	1515 Massachusetts Ave., N.W. Washington, D.C. 20005	1848	125,000
AACE	American Association of Cost Engineers	308 Monongahela Bldg. Morgantown, W.V. 26505	1956	2,800
AAPM	American Association of Physicists in Medicine	335 East 45th St. New York, N.Y. 10017	1958	500
ACI	American Concrete Institute	22400 West 7 Mile Road Detroit, Mich. 48219	1905	14,988
ACM	Association for Computing Machinery	1133 Avenue of the Americas New York, N.Y. 10036	1947	30,000
ACS	American Ceramic Society, Inc.	4055 North High St. Columbus, Ohio 43214	1898	7,400
ACSM	American Congress on Surveying and Mapping	430 Woodward Bldg. 733 15th St. N.W. Washington, D.C. 20005	1941	6,000
AES	Audio Engineering Society, Inc.	60 East 42nd St., Room 428 New York, N.Y. 10017	1948	4,590
AGU	American Geophysical Union	2100 Pennsylvania Ave, N.W. Washington, D.C. 20037	1919	11,000
AIAA	American Institute of Aeronautics and Astronautics	1290 Sixth Ave. New York, N.Y. 10019	1932	40,614
AIA	American Institute of Architects	1735 New York Ave., N.W. Washington, D.C. 20006	1857	38,459
AIChE	American Institute of Chemical Engineers	345 East 47th St. New York, N.Y. 10017	1908	36,379
AICE	American Institute of Consulting Engineers	345 East 47th St. New York, N.Y. 10017	1910	420
AIIE	American Institute of Industrial Engineers	25 Technology Park/Atlanta Norcross, Ga. 30071	1948	20,500
AIME	American Institute of Mining, Metallurgical, and Petroleum Engineers, Inc.	345 East 47th St. New York, N.Y. 10017	1871	52,000
AIP	American Institute of Physics	335 East 45th St. New York, N.Y. 10017	1931	50,370
AIPE	American Institute of Plant Engineers	1021 Delta Ave. Cincinnati, Ohio 45208	1954	6,800
AMS	American Mathematical Society	P.O. Box 6248 Providence, R.I. 02904	1888	13,362
ANS	American Nuclear Society	244 East Ogden Ave. Hinsdale, Ill., 60521	1954	11,112
APS	American Physical Society	335 East 45th St. New York, N.Y. 10017	1899	26,000
APHA	American Public Health Association	1740 Broadway New York, N.Y. 10019	1872	21,791
ASAE	American Society of Agricultural Engineers	2950 Niles Ave. St. Joseph, Mich. 49085	1907	7,400
ASCE	American Society of Civil Engineers	345 East 47th St. New York, N.Y. 10017	1852	70,000
ASEE	American Society For Engineering Education	National Center for Higher Education One Dupont Circle Washington, D.C. 20046	1893	13,615
ASM	American Society for Metals	Metals Park, Ohio 44073	1913	36,390
ASNT	American Society for Nondestructive Testing	914 Chicago Ave. Evanston, Ill. 60202	1941	7,310
ASQC	American Society for Quality Control	161 West Wisconsin Ave. Milwaukee, Wisc. 53203	1946	22,800

Table 6-1 (*continued*)

Code	Name	Address	Year organized	Total member-ship
ASHRAE	American Society of Heating, Refrigerating and Air-Conditioning Engineers, Inc.	345 East 47th St. New York, N.Y. 10017	1894	30,000
ASLE	American Society of Lubrication Engineers	838 Busse Highway Park Ridge, Ill. 60068	1944	3,470
ASME	American Society of Mechanical Engineers	345 East 47th St. New York, N.Y. 10017	1880	74,000
ASNE	American Society of Naval Engineers, Inc.	1012 14th St. N.W. Suite 507 Washington, D.C. 20005	1888	4,009
ASSE	American Society of Safety Engineers	850 Busse Highway Park Ridge, Ill. 60068	1911	13,000
ASSE	American Society of Sanitary Engineering	228 Standard Building Cleveland, Ohio 44113	1906	2,342
ASTM	American Society for Testing and Materials	1916 Race St. Philadelphia, Pa. 19103	1898	16,500
AWRA	American Water Resources Association	St. Anthony Falls Hydraulic Laboratory Miss. River at 3rd Ave. S.E. Minneapolis, Minn. 55414	1964	1,935
AWWA	American Water Works Association, Inc.	6666 W. Quincy Ave. Denver, Colo. 80235	1881	25,000
CEC	Consulting Engineers Council of the United States of America	1155 15th St., N.W. Washington, D.C. 20005	1959	2,300
IES	Illuminating Engineering Society	345 East 47th St. New York, N.Y. 10017	1906	10,735
IEEE	Institute of Electrical & Electronics Engineers, Inc.	345 East 47th St. New York, N.Y. 10017	1884	174,000
IES	Institute of Environmental Sciences	940 East Northwest Highway Mt. Prospect, Ill. 60056	1959	1,750
ITE	Institute of Traffic Engineers	2029 K Street, N.W., 6th Floor Washington, D.C. 20006	1930	3,800
ISA	Instrument Society of America	400 Stanwix St. Pittsburgh, Pa. 15219	1945	20,400
NACE	National Association of Corrosion Engineers	2400 West Loop South Houston, Tex. 77027	1945	7,425
NAPE	National Association of Power Engineers, Inc.	174 West Adams St., Suite 1411 Chicago, Ill. 60603	1882	12,000
NICE	National Institute of Ceramic Engineers	4055 North High St. Columbus, Ohio 43214	1938	1,625
NSPE	National Society of Professional Engineers	2029 K Street, N.W. Washington, D.C. 20006	1934	68,000
ORSA	Operations Research Society of America	428 East Preston St. Baltimore, Md. 21202	1952	6,407
SAM	Society for Advancement of Management, Inc.	1472 Broadway New York, N.Y. 10036	1912	15,500
SESA	Society for Experimental Stress Analysis	21 Bridge Square Westport, Conn. 06880	1943	2,550
SIAM	Society for Industrial and Applied Mathematics	33 South 17th St. Philadelphia, Pa. 19103	1952	3,750
SAE	Society of Automotive Engineers	400 Commonwealth Dr. Warrendale, Pa. 15086	1905	30,220
SME	Society of Manufacturing Engineers	20501 Ford Rd. Dearborn, Mich. 48128	1932	42,152
SNAME	Society of Naval Architects and Marine Engineers	74 Trinity Place New York, N.Y. 10006	1893	10,200
SPE	Society of Plastics Engineers, Inc.	656 West Putnam Ave. Greenwich, Conn. 06830	1941	18,000
SWE	Society of Women Engineers	345 East 47th St. New York, N.Y. 10017	1952	3,000

7

Developing study habits

From high school to college

Students who have enrolled in a college or university for the first time often ask, "Is there a difference between a high school course and a college course?" and "Will I need to make any adjustments in my study habits, now that I have enrolled in college?"

The answer to both of these questions is probably *yes,* but let us examine some of the reasons why this may be so.

First, in high school you were competing against the *average* of high school students. However, of the total numbers graduating from high school in the United States each year, fewer than one third go to college. Thus, you are now competing with the average of *very good* high school students.

The study habits and learning process that you used in high school may not be adequate to cope with the increased requirements of college courses because of both the limited time available and the large quantity of material to be covered. A refinement of your study habits or perhaps a complete change in study habits may be necessary to enable you to keep up with the demands of new course material.

Many students, as they enter college, do not realize what will be expected of them. In general they are expected to bring basic skills in mathematical manipulation, in reading rapidly and comprehending, and in possessing a broad-based vocabulary. Engineering educators have observed that a high school graduate who has the ability to *read* and *add* also possesses the capability to succeed in a college engineering program. In high school, much time was taken in class to outline and drill on the daily

> Only the educated are free.
> —Epictetus, ca. A.D. 115

assignments. In college, relatively less time is taken in class, and much more study and preparation is expected from the student outside of class. The student is largely on his own, and his time can be used to a considerable extent as he sees fit. It can be used efficiently and profitably or it can be dissipated without plan and, in effect, be wasted.

Without parental urging or strong encouragement from teachers, the student must adopt personal methods of study that will produce desirable results. Specifically he must budget his time to permit adequate preparation for each course. There must be more than a casual desire to improve study habits. Positive steps must be taken to ensure effective study and learning conditions. It is for this reason that the following topics are included as suggestions to aid in improving the students effectiveness in study.

Preclass study

The object of study is to learn. Mere idle reading is not study. Particularly in scientific and technical courses, extreme attention to detail is necessary. With the learning process in mind, let us examine some basic principles.

1. The material must be organized into appropriate learning units. Random facts and concepts are more difficult to learn than facts which are related. For example, in learning the names of the bones of the body it is easier to remember the names if groups such as the arm or leg bones are studied as a unit.
2. Attempt to form the correct pattern of facts on the first try. This is necessary to eliminate the need for "unlearning" and relearning factual material. In the case of research or exploratory study, trial-and-error methods are necessary and frequently incorrect assumptions are made. However, by conscious effort to use reasoning and to incorporate other correct facts, false assumptions are minimized.
3. Correct errors immediately and reinforce correct learning responses. Experiments have shown that immediate confirmation of correct learning is more effective in remembering than when the confirmation is delayed. For example, if a mathematics problem is solved and its correctness verified immediately, the principle involved in the solution is retained better than if the verification is delayed.
4. Relate realistic experiences with the facts. Experiments have demonstrated that most people learn and retain information better if it is related in some way to their experiences. For example, an abstract idea such as "democracy" is difficult to present as a realistic picture unless the student has some related background of government upon which to draw a conclusion. On the other hand, a description of a new type of internal combustion engine may be simple to present to an experienced automobile mechanic because of his related experience with similar devices.
5. Give concise meanings to the facts. Particularly in scientific work the meanings of words may not always be clear. Frequently we misunderstand one another because we each may give different meanings to the same word. The use of dictionaries,

encyclopedias, and reference books is therefore necessary to gain a common understanding of new words.

6. Practice, review, and provide application for facts. Education specialists believe that facts are not actually learned until at least one perfect recitation or response is completed. After this has been accomplished, review and repeated use of the facts will greatly aid retention. Research also has shown that if the review is broken into spaced periods, retention and recall are increased (sometimes as much as doubled) over the retention when the reviewing is done all at one sitting. One should be alert to applications for the ideas being learned. This will help to relate them to previous experience and to place them into a pattern where they will become bricks in a wall of knowledge upon which other ideas can be added.

7. Evaluate the adequacy of the learning: A self-evaluation of the understanding of the new ideas which have been presented is one of the most valuable learning experiences in which a student can participate. Memorizing facts does not encourage self-evaluation. However, the ability to apply principles and *to use* facts is one important way in which a person can evaluate the adequacy of learning. For example, after studying a portion of text material in a physics book, are you able at once to apply the facts and principles discussed to the solution of related problems?

The realm of factual information available is so tremendous that a student at first should acquire only the essential and basic facts in a particular field of study. From this set of basic facts, the student then enlarges or details his information into more specialized subjects. For example, the electrical engineering student should begin his study of electricity with a consideration of basic principles, such as Ohm's Law, before beginning to consider the design of amplifiers.

Setting the stage

Provide a designated study area. It is desirable to find a place where you can concentrate and where other people will not bother you. Unfortunately distractions frequently abound in large study areas and interfere with study schedules. Other people in your home or your dormitory may have conflicting schedules and may not be concerned with respecting your own study periods. Radios, televisions, and "bull sessions" are always inviting diversions from study. For this reason many students find that libraries afford good study areas because of the absence of distractions and the ready availability of reference materials. Although a secluded spot is not always essential for study, for most people it does require a conscious effort to reject distracting sounds and backgrounds in order to concentrate.

The best place to study is usually at a desk or table, not on a bed. The effort of sitting helps to keep most people alert and in a mood for study. Good lighting is especially helpful and it is desirable for the whole study area to be illuminated, rather than a small portion of the area. Studies have shown that it is less fatiguing on the eyes if sharply defined regions of light and dark are excluded from the immediate study area. In addition, the work area should be large enough so that reference materials can be kept close at hand.

Prepare a schedule

Time is one of the most important factors to be considered in college study. In every course there is usually more material assigned than can be studied in detail in the time available. In addition outside activities will always compete for a student's time. Athletic events, social and educational programs, recreational activities, and unscheduled meetings with other people seem to disrupt the best laid plans. The student must realize that these contradictory conditions for study will always exist. Positive steps must be taken to ensure that the time for study is not taken away piecemeal by nonessentials.

In preparing a daily schedule, the question arises concerning the amount of time to allocate for study. Several rule-of-thumb principles are in common use, but the individual's capabilities in learning each specific subject will necessarily need to be the final guide. A recent survey of a cross section of students at a large university showed that the greatest number of students spent an average of 28 to 32 hours of study per week. Engineering students usually spend considerably more than this amount. Actually the number of hours of study is not always the most significant criterion to be used. Rather *how well* one studies is the factor that counts most. The results of study as shown by grades and by one's own personal satisfaction in doing a good job are usually the best indicators of the effective use of time.

A positive and direct approach to a schedule is necessary in the same way that a budget is necessary to manage the fiscal affairs of a business. No commercial enterprise can long exist that does not plan ahead for meeting expenses as they arise. In a similar way a student should prepare a budget of time for his school work and adhere to it, unless circumstances definitely indicate that it should be altered. Not only should the daily time be budgeted on a weekly basis, but also extra time must be allocated for major quizzes, term papers, and final examinations. It has been found, for instance, that the majority of the better students budget their time so that final examinations do not have to be prepared for on a frantic last minute rush. Study

Illustration 7-1
Because practice makes perfect, an unorganized student will in time become an unorganized adult.

skills, no matter how effective, will not be of much value if a student's time is not properly scheduled so that they can be employed.

Scheduling helps to allocate more time to the more difficult courses and to assign less time to the less demanding courses. It also helps to space the available study time so that it will be distributed in a manner to aid in better retention. (Refer to the sixth principle of learning on page 126.)

Studying to learn

Studying to learn is a skill that can be developed just as other skills are developed. All good golfers do not use exactly the same stance or swing, yet they all accomplish a reasonably consistent pattern of results. In like manner, good students may employ slightly different techniques of study and still accomplish an acceptable learning pattern. However, despite individual differences, in any activity a general set of principles can be found that will produce good results, whether it be golf or studying. Some of these principles for study are discussed below.

1. Remove or minimize distractions.
2. Arrange all necessary pencils, reference books, notebooks, note paper, and other supplies before beginning.
3. Put your full attention on the work at hand, and insist that your brain work accurately and rapidly. If the brain is not employed to its full capacity, it will tend to let other thoughts enter to distract it from the task at hand. Read with a purpose—to extract details from the printed page and to comprehend important ideas.
4. Practice reading as fast as feasible. This does not mean skipping from word to word randomly, but rather training the eyes and brain to group words, phrases, and even lines of reading material and to understand the thoughts therein. Many good books are available on how to improve your reading speed for comprehension. Time spent in learning this skill will aid immensely in the faster grasping of ideas.
5. Study as though you were going to teach someone else the subject matter. This will provide motivation for learning and also will encourage self-appraisal of the adequacy of learning as discussed in the seventh principle of learning on page 126.
6. Plan for review and repetition of the assignment. Principle of learning number six on page 126 points out the desirability of spaced periods of review.

The suggested methods and practices described above have been found to aid most students in their learning and the majority of good students follow the general outline of these practices.

Preparation for class recitation

A plan of study for each course is necessary to gain the greatest return in learning from your investment of study time. The plan will vary with the teacher, the textbook, the nature of the course, and the type of recitation and examination that is expected.

> Just as eating contrary to the inclination is injurious to the health, so study without desire spoils the memory, and it retains nothing that it takes in.
> —da Vinci, ca. A.D. 1500

However, before considering suggestions for specific types of courses, we should investigate study techniques that are applicable in general to all subjects.

Learning proceeds best from the general to the specific. It is therefore recommended at the beginning of a course to first skim the chapter and topic headings of the text without reading in detail any of the discussion material. This is done to get an overview of the whole organization of the book. Notice the order in which the topics appear and how the author has arranged the ideas to proceed from one to another. Next, read the lesson quickly to gain an insight into the nature of the material to be covered. Do not attempt to learn details nor to analyze any but the most emphasized points. If there is a summary, read it as part of your lesson survey to prepare you with background material that will be useful in understanding details.

Second, after a rapid survey of the material to be studied, start at the beginning of the assignment with the idea in mind that you will make notes during your study. Remember, you are going to learn as though you would have to teach the lesson content to someone else. Mere superficial reading here will not suffice. The notes can take various forms. They may include summaries of important facts, definitions of words, sample problems or examples, answers to questions, sketches, diagrams, and graphs. A better mental picture is formed and retained if the hand and eye work together on an idea and if you are forced to participate more actively and completely in the learning process.

These notes should be made in semipermanent form, not on random scraps of paper. Bound notebooks or loose-leaf notebooks can be used, but it is important that the notes be organized and retrievable. In addition to separate notes, it is helpful to underline key words and phrases in the text. Don't worry about the appearance of the book. However, do not overdo the underlining; it is better to note the crucial words and phrases so that they are more obvious for review than to underline whole sentences and paragraphs. Usually from three to eight words per paragraph will point out the central idea that has been presented.

Third, reread and review the lesson assignment and prepare your own questions and answers to the topics. At first this may seem to be an unnecessary step but it will pay dividends. Attempt, if possible, to foresee the questions the instructor may ask later concerning the material. Your notes on these predictions can be invaluable at examination time. Check to see that you have noted all the important details and related facts that bring out the main ideas. Particularly in the technological courses, you can do this easily, since much of the material is completely factual.

Fourth, if given an opportunity, plan to participate in the classroom recitation. Force yourself, if necessary, to volunteer to recite. Recitation is a form of learning and it aids in acquiring ideas from others. If the class does not afford an opportunity for recitation, recite the lesson in your room. Review and recitation are the best methods for making a final check of your retention of information. Tests show that you begin to forget even while learning, but if you participate in some form of recitation as soon after study as possible, the retention of facts may be increased by as much as 50 per cent.

Recitation is an effective way of self-appraisal of learning. Just reading a book is not enough to convince anyone—yourself included—that you have learned what you should. When you study, break the topics into groups, and upon completion of each group of topics, as a summary close the book and see if you can recite the important facts either mentally or in writing. When you repeat them satisfactorily then continue; if you cannot repeat them, for further study, go back and pick out the ones that you have missed. It is particularly important to recite if the subject matter consists of somewhat disconnected material such as names, dates, formulas, rules, laws, or items. If the material to be studied is more narrative in style and well organized, the recitation time can occupy a small part of the study period, but it should never be left out altogether.

The general principles above apply to all courses, but certain study plans will apply better to one course than to another. We shall examine study plans for several types of courses in more detail.

Technical courses

In this type of course your study plan should be to direct your study toward understanding the meanings of words and toward grasping the laws and principles involved. In order to understand the words, a dictionary, encyclopedia, and reference books are necessary. The first step is to write down definitions of unfamiliar terms. Remember also that a word does not always have the same meaning in different courses. For example, the word "work" as used in economics has a meaning quite different from the word "work" as used in physics.

When the definitions of words are obtained, study for complete understanding. Texts in technical courses tend to be concise and extremely factual. A technique of reading must be adopted here for reading each word and fitting it into its place in the basic idea. Except for the initial survey reading of the lesson, do not skim rapidly through the explanations, but rather read to locate the particular ideas in each paragraph. If example problems are given, try working them yourself without reference to the author's solution.

After definitions and basic ideas are studied, apply the principles to the solution of

Illustration 7-2
Technical jargon includes not only spoken and written words, but symbols and graphical notation as well. Many lectures will frequently involve the use of all the various media available to the teacher.

problems. It has been said by students that it is impractical to study for examinations where problems are to be solved because you are unable to predict the problem questions. This statement is not correct, for you can predict the principles which will be used in solving the problems. For example, in chemistry a vast number of compounds can be used in equation-balancing problems. However, a very few basic principles are involved. If the principle of balancing is learned, all problems, regardless of the chemical material used, are solved the same way. The objective of this part of study then is to determine the few principles involved and the few problem patterns that can be used. After this, all problems, regardless of their number arrangement and descriptive material, can be classified into one of the problem patterns for which a general method of solution is available. For instance, a problem in physics may involve an electrical circuit in which both current and voltage are known and an unknown resistance is to be determined. Another problem may suggest a circuit containing a certain resistance and with a given current in it. In this case a voltage is to be found. The problems are worded differently, but a general principle involving Ohm's Law applies to each situation. The same problem structure is used in each case. The only difference appears in where the unknown quantity lies in the problem pattern.

Do not become discouraged if you have difficulty in classifying problems. One of the best ways to aid in learning to classify problems is to work an abundance of problems. It is then likely that any examination problem will be similar to a problem that you have solved before.

Learn to analyze each problem in steps. Examine the problem first for any operations that may simplify it. Sometimes a change in units of measure will aid in pointing toward a solution. Try rewriting the problem in a different form. Frequently in mathematical problems, this is a useful approach. Write down each step as the solution proceeds. This approach is particularly helpful if the solution will involve a number of different principles. If a certain approach is not productive, go back and reexamine the application of the principles to the data. For problems which have definite answers, these techniques usually will provide a means for obtaining a solution.

It usually is better in studying technological courses to divide the study periods into several short sessions, rather than one continuous and long study period. For most people a period of incubation (where the idea is allowed to soak into the subconscious) is helpful in grasping the new ideas presented. After returning to do subsequent study, make a quick review of the material previously studied, and look at the notes you have prepared to provide continuity for your thinking.

Literature courses

Most writings classed as literature are written to be interesting and to entertain. For this reason, not as much attention to detail and to individual words is needed as is required in technological books. Usually the ideas are presented descriptively and are readily distinguished. However, the interpretation of the ideas may vary from person to person, and it is with this in mind that the following suggestions are given.

Examine the ideas not only from your point of view but also from the point of view of your teacher. Try to find out from his discussions and examinations the pattern of thought toward which he is directing you, and study the things *in which he is interested.*

Consciously look for these items while you read prose: the setting, central characters (note the realism or symbolism of each), the theme, the point of view of the author (first person, omniscent, etc.), the author's style of writing, the tone, and the type of the writing. For poetry, the ideas may be more obscure, but certain things may be noted. For example, the authors' style, the type of verse, the rhyme scheme, the theme, the symbols, allusions, images, similes, metaphors, personifications, apostrophes, and alliterations are all basic and important parts of the study of poetry.

Social science courses

These courses can be interesting and satisfying or dull and dry, depending upon the student's attitude and interest. Most texts use a narrative style in presenting the material and, as a consequence, the assignment should be surveyed quickly for content and then in more detail for particular ideas. Here the use of notes and underlining is invaluable, and summaries are very helpful in remembering the various facts.

If the course is history, government, sociology, psychology, or a related subject, consider that it contains information that is necessary to help you as a citizen. A knowledge of these subjects will aid you in dealing with other people, and it will give you background information to aid in the evaluation of material that has been specifically designed to influence and control people's thinking. Study the course for basic ideas and information and, unless the instructor indicates otherwise, do not exaggerate the importance of detail and descriptive information.

These principles apply also to courses in economics, statistics, and related courses except that they frequently are treated on a more mathematical basis. Here a combination of techniques described above together with problem-solving procedures can be helpful. Again, since the volume of words usually is quite large, it is necessary to use notes and summaries to keep the ideas in a space to be handled easily.

Language courses

Many techniques have been developed to aid in learning foreign languages. In the absence of specific study guides from your instructor, the following procedures have been found to be helpful.

Learn a vocabulary first. Study new foreign words and form a mental image of them with a conscious effort to think in the new language. As you study, practice putting words together, and, if the course includes conversation, say the words aloud. Space your vocabulary study and review constantly, always trying to picture objects and actions in the language rather than in English.

Rules of grammar are to be learned as any rule or principle: first as statements and then by application. Reading and writing seem to be the best ways of aiding retention of grammar rules. Read a passage repeatedly until it seems natural to see or hear the idea in that form. Write a summary in the language, preferably in a form that will employ the rules of grammar which you are studying. Unfortunately, there is no way to learn a new language without considerable effort on your part. Even English, our native tongue, when studied as a subject, gives some students trouble. However, many students have said that they really understand basic English much better after having taken a foreign language.

Classroom learning

The discussion so far has been concerned with learning by study. An equally effective and more widely used method of learning is by listening. From earliest childhood you have learned by listening and imitating. Do not stop now but rather use the classroom to supplement your home study. You will find that things are covered in classroom work that you do not find in your texts. The interchange of ideas with others stimulates your thinking and retention processes. The classroom can also be a place to practice and to demonstrate your learning and problem-solving skills before the examination periods.

The skill of listening seldom is used to the fullest extent: If your attention is only partly on the lecture, the part missed may make a major difference in your grade. Use the time in class to evaluate your instructor, find out what he will expect of you, watch for clues for examination questions, and make notes to be used for later study. If you plan to make the classroom time profitable, you will find it also will be enjoyable.

Come to class with a knowledge of the assignment to be discussed. The instructor then can fill in your knowledge pattern rather than present entirely new material. This also saves time in taking notes because the notes will be needed only for amplification rather than as semiverbatim recording. If a point arises at variance with your knowledge from study, you have an opportunity to question it. Prior study also permits you to predict what the instructor will say next. This serves as a valuable psychological device to hold your attention throughout the class period.

Note taking during class is a skill that can be learned. The inexperienced will try to take notes verbatim and thus get so involved in writing that they cannot listen for ideas. Usually, they cannot write fast enough to copy all the words anyway. Rather than take notes verbatim, practice your listening skills and evaluate the critical points in the lecture. A few critical points will be amplified with descriptive materials. Practice taking down these critical points in your own words. Such note taking serves to keep your attention focused and to encourage a better understanding of the principles being discussed. If you do not understand the points completely, make notes for later study or for questioning the instructor.

If the course is such that you can, recite during the class period. Push your shyness aside, and place your ideas and information before the class. It stimulates your thinking and retention, it will help to clear up obscure points, and will give you much needed practice in hearing your voice in the presence of others.

> I am a great believer in luck. The harder I work the more of it I seem to have.
> —Coleman Cox

Attention and listening during a class period together with participation, either in recitation or in anticipating what the instructor will cover next, will save you hours of study time outside the classroom.

Preparation for themes, papers, and reports

The purpose of writing is to transmit information. For an engineer this is a valuable means of communicating his ideas to others and the practicing engineer takes pride in the conciseness and adequacy of his reports. No matter how good your ideas may be, if you cannot communicate them to others, they are of little value. Since part of the work of an engineer is writing reports and papers, the opportunity to learn and to practice this skill in school should be exploited. There are many good books on composition and manuals on writing that will help you. For this reason the suggestions given here are to be considered as supplementary helps in the preparation of written work.

In general, good writing involves good grammar, correct spelling, and an orderly organization of ideas. The basic rules of grammar should be followed and a logical system of punctuation used. If in doubt as to the application of a somewhat obscure rule in grammar, either look it up or reword your idea in a more conventional manner. Punctuation is used to separate ideas and should follow, in general, the pauses you would use if you were reading the material aloud. Usually, reports are written in an impersonal manner; seldom is the first person used in formal writing.

Little needs to be said about spelling except to say, spell correctly. There is so little room for choice in spelling that there is no excuse for a technical student to misspell words. If you don't know how to spell a word, look it up in a dictionary or reference book and remember how to spell it correctly thereafter.

The last characteristic of good writing is a clear orderly organization of ideas. They may take several forms depending upon the type of writing. For themes, usually a narrative or story form is used in which a situation is set up, possibly with characters, and a story is told or a condition is described. Engineering papers and reports generally are concerned with technical subjects. Therefore, they describe the behavior of objects or processes, or they provide details of events in technical fields. Frequently, the first paragraph summarizes the thoughts in the whole report in order to give the reader a quick survey without his having to skim through the manuscript first. Following paragraphs outline the contents in more detail. They frequently end with graphs, drawings, charts, and diagrams to support the conclusions reached.

In preparation of written work, some research usually is needed. In order to aid in keeping the notes for the material in usable form, it is helpful to record abbreviated notes from research works on cards in order that the arrangement of the writing of the paper can be made in a logical order. Usually, in compiling information you do not know how much will be used so the notes on cards provide a flexibility of choice that is a great aid in the final organization of the paper. The cards can be 3 by 5 in. or 4 by 6 in., with the latter usually being the better choice because of more available space.

An outline of the material to be discussed or described is necessary, even for brief reports. An outline ensures a more logical arrangement of ideas and helps to make the writing follow from concept to concept more smoothly.

Write a draft copy first and plan on making alterations. Write first to get your ideas down on paper, and then go over the copy to improve the rough places in grammar, spelling, punctuation, and wording. If possible, wait a short time before taking these corrective steps to get a more detached and objective approach to the suggested changes. When the rework of the draft has been made, copy it over neatly, still maintaining a critical attitude on the mechanics of the writing. It is helpful if the final draft can be typed, but if not, you will find that good handwriting or lettering frequently makes a favorable impression upon the person who grades your paper.

Preparation for examinations

Have you ever felt after taking an examination (for which you studied) that everything you studied was inappropriate? Perhaps you used the wrong techniques in studying for the examination. Certain rules have been found to be very useful in preparation for tests of any kind. The type of preparation you make is often more important in the final grading than how long you spend in preparation. Let us discuss some of these rules that have been found to be effective.

Start preparing for tests the day the course starts. You know that they will be assigned, so do not close your eyes to this fact. From the very first, start studying two things: (1) the big overall ideas of the course, and (2) the instructor. Keeping these two things in mind will help you to learn while in class and while studying. This will also mean shorter reviews before tests.

Illustration 7-3
Close attention in class frequently will give major clues pertaining to the nature of future examination questions.

Keep writing reminders of important points—as notes in the margins of your books, as flagged notes in class, or as short statements while studying. Use nontext material as it becomes available, such as outlines, old tests, and information from students who have taken the course.

It is not unethical to study the instructor. After all, he is a person qualified to present the subject matter, and, because of his training and experience, frequently you will learn much more from him than you will from any text. From a study of the instructor, you can follow his pattern of lesson organization and find out what he wants from you. In your preparation, attempt to think and study along these lines. Close attention in class frequently will give major clues pertaining to the nature of future examination questions.

Make a final review of the subject matter. This should be a planned review and not a "last ditch" cramming session. Schedule it in several short sessions and do these things: First, review the general organization of the material before the final class periods of the semester in order to take advantage of the instructor's summaries and reviews. Second, set up the major topics or ideas and associate them with specific facts or examples. In the case of problem courses, this is the time to work out and review sample problems that will illustrate laws or principles. Third, study for more detailed information and to complete the areas of uncertainty. If the first two phases of the review have been adequate, this last phase should take relatively little time.

Predict the type of test that will be given. If the test is to be an essay type, practice outlining key subject matter, summarizing important concepts, comparing or contrasting trends, and listing factual data. This may seem to be an excessive amount of work, but if you have noted the points the instructor has stressed during the course, you can narrow the field considerably. However, a word of caution here: Do not try to outguess the instructor; it usually will not pay. Study the topics you honestly feel are important, and avoid unjustified evaluation of different concepts in hope that there will not be questions on them. Remember that on this type of test not only are the ideas to be recalled but also the organization and sequence of the ideas is important.

For objective or short-answer type tests, follow the three-step program of study given above, giving more attention to relating key ideas to specific items of information. Here, short periods of highly concentrated study usually are to be preferred. Think about each idea long enough to form mental pictures and precise answers, but guard against merely memorizing words. Frequently sketches and diagrams will aid in retaining a mental picture of a concept.

If the test is to be a problem-solving test, first review the principles that may be encountered in a problem. Work at least one sample problem that will illustrate the principle. Ask yourself, "If there is a change in the quantity to be solved for, can I still place this problem in the correct problem-solving pattern?" This is important because, although the variation in problem statements is infinite, the applicable principles and consequently the problem patterns are relatively few.

What if you have more than one examination on the same day? Most students have found it better to do the last review on the subject that comes first. A quick check of notes then can be made before the next examination begins. An old but useful maxim states that the best preparation for taking tests is to practice the things you will need to do on the test.

Taking examinations

We shall assume that a primary objective in taking an examination is to make a high grade on it. Your grade will be based on what you put down on the test paper—not on what you know. It is crucially important then to get the correct sampling of your knowledge on record. Sometimes students fail, not because of lack of knowledge, but rather because of lack of skill in proving on the test paper that they do understand the material.

In taking any examination, be prepared. The ability to think clearly is of most importance, but the ability to recall facts is also very important. Enter the examination room with a feeling of confidence that you have mastered the subject and, while waiting for the questions to be distributed, be formulating plans for taking the test so that your mind will not be blocking itself with worry.

If the test is an objective type, turn quickly through the pages and note the kinds of questions; true–false, multiple choice, matching, completion. Make a rapid budget of time, and read the directions for answering the questions. If there is no penalty for guessing, answer every question; otherwise plan to omit answers to questions on which you are not reasonably sure. Be certain you understand the ground rules for marking and scoring so that you will not lose points on technicalities.

A basic principle of taking any examination is: *Answer the easy questions first.* If time runs out, at least you have had an opportunity to consider the questions you could answer readily; and usually an answer to an easy question counts as much as an answer to a difficult question. Do not carry over thoughts from one question to another. Concentrate on one question at a time and do not worry about a previous answer until you return to it on the next trial. It helps to relax for a moment between questions and get a fresh breath, and to help dismiss one set of thoughts before concentrating on another. Look for key words that may point to whether a statement is true or false. Usually statements are worded so that a key word or phrase tips the balance one way or the other. In case of doubt, try substituting a similar word into the statement and see whether it may aid in identification of truth or falsity. When you go back over the examination, do not change your answers unless you have obviously misread the questions or you are reasonably certain your original answer is incorrect. Tests have shown that your first response to a question on which you have some doubt is more likely to be correct than not.

If the test is an essay type, again read quickly through the questions, budget your time, and answer the easy questions first. It is helpful to plan to put an answer to each question on a separate sheet of paper unless the answer obviously will be short. The one-answer-per-page system permits easy addition of material after your initial trial. Watch your time schedule, since it is easy to write so much on one question that you are forced to slight others. A help on answering lengthy questions is to jot down a hasty outline of points to include so they will not be overlooked in the process of composition.

After the questions have been answered, take a final critical look at your paper to correct misspelling, grammatical errors, punctuation, and indistinct writing. If time permits, add sketches, examples, or diagrams that may come to mind. Sometimes a period of quiet contemplation, mentally reviewing your notes, will help recall needed additional facts.

If the test has mathematical problems to solve, again read through the questions and budget your time. Plan to answer the easy questions first. Determine the "ground rules," such as whether points will be given for correct procedure regardless of the correctness of the arithmetic. Unless the problem solution is obviously short, plan to work only one problem per page. If a mistake is detected you can more easily and more quickly line out the mistake than attempt erasure. One answer per page also permits room for computations and makes checking your work easier. Usually it is better to do all the work on that page and avoid scratch paper.

If the test is an open book test, use the reference books only for tabular or formula data that you reasonably could not remember. If you try to look up things you should already know, you will surely run out of time.

Let each problem stand by itself. First, analyze it from the standpoint of a pattern into which it can be fitted. Consider then what steps will need to be used in the solution, and finally determine how these solution steps will be presented. When the analysis is complete, solve the problem in the framework of the analysis.

Usually it is better to go ahead and work through all problems and then come back and check for arithmetic mistakes and incorrect algebraic signs. This is the place also to take an objective look at the answer and ask whether it seems reasonable. A questioning attitude here may reveal mistakes that can be corrected.

Analysis of results of tests

Finally, when the test is ended and you get an opportunity to see your graded paper, analyze it and yourself critically. Assuming that you knew the material but that your grade did not reflect your knowledge, find out why the grade was not as good as it should have been. Blame only yourself for any deficiencies. Look for clues such as the ones given below that will help you not to make the same mistake again.

If it was an essay test, was your trouble poor handwriting, incorrect grammar, incorrect spelling, or incorrect punctuation? Correction of these faults is a matter of the mechanics of learning the rules and making a conscious effort to improve on your shortcomings.

Was your trouble failure to follow instructions, lack of organization of ideas, or lack of examples? Look for clues such as marks on your paper by graders stating "not clear," "not in sequence," "why?," "explain," "?," "trace," "compare," "contrast," and so forth, which indicate a failure on your part to follow instructions. The remedies are twofold. Look for key words in instructions on tests, and practice beforehand the listing, contrasting, or comparing of factual data.

When the grader's marks include words such as "incomplete," "hard to follow," "meaning not clear," or "rambling," these comments indicate that your ideas need to have better organization. A remedy is to consider carefully what is being asked for in the question. Make a brief outline before you start writing. This affords a means of placing ideas in the most effective sequence and also helps to avoid omitting good points.

The grader's marks may be "for example," "explain," "be more specific," or "illustrate." These marks usually indicate a need for illustrations and examples. Your answers may show that you know something about the subject, but they may not convey precise information. Examples will convince the grader that you know the material covered.

For objective tests, evaluate the patterns of the questions missed. Did you misread the questions? Were you tripped up by double negatives in true–false questions? Did you fail to look for key words in multiple-choice questions? Did you realize immediately after the examination that you had answered incorrectly? Some aids in improving grades on objective tests follow.

For true–false questions, did you give each question undivided attention, and were you careful not to read something into the question that was not there. If you missed several questions in sequence, you probably were thinking about more than the question at hand. Try rewording questions that have double negatives next time if you show a pattern of missing them.

For multiple-choice questions, determine whether you concentrated on each question alone and determined, if possible, what the answer was before looking at the set of multiple-choice answers. You should have eliminated as quickly as possible answers that obviously did not fit the question and concentrated on key words that would have provided clues to select from the remainder.

If the test consisted of problems to be solved, check for mistakes in two things: analysis and arithmetic. If your paper shows false starts on a problem, if you worked partway and could go no further, or if the solution process was incorrect from the beginning, your principal trouble probably is lack of skill in problem analysis. The remedy, of course, is to work more problems illustrating the principles so that the test situations will be more familiar. If you use no scratch paper on tests, and keep all parts of your solution on the page, checking to ascertain your mistakes should be easy.

If the processes are correct but the answers are incorrect, look for careless mistakes: in arithmetic, in employing algebraic signs, in mixing systems of units such as feet and centimeters, in copying the problem or in copying from one step to the next, or in making numbers so indistinct that they are misread. The remedy for these mistakes is to go over the solution carefully checking for these things. If time permits, one independent solution will help. Finally, look at the answer—does it seem reasonable?

The employment of the techniques discussed above should help you to achieve grades based on your knowledge of a subject without a handicap in the skill of presenting the knowledge on an examination. No more should you have to say, "I knew it but I couldn't put it down on paper."

8

Spoken and written communication

Compared to Europeans and even to other nationalities who speak English, we Americans have a peculiar uneasiness about our language. Some of us are reluctant to speak out because we are shy about our "grammar." Others confuse language ability with the ability to spell. Many Americans think English is difficult in some obscure way. It is a common misconception that English is extraordinarily difficult for foreigners to learn.

Some linguists attribute our attitude toward language to the American self-image: we think of ourselves as men of action, tight-lipped, monosyllabic, like Gary Cooper with his "yup" and "nope." But what really sets us apart from much of the world is that we have no language establishment. France has a venerated Academy that sets standards, Great Britain a university-bred establishment, and in the totalitarian states the ruling party controls the language along with education. In contrast, we have no "official" American English.

To further add to our confusion, linguistic researchers have revealed some new discoveries about language that are going to demand some rethinking on our part. Just as today's anthropologists are learning that some societies of the past that have been shrugged off as "primitive" were in many respects just as complex as ours, linguists are finding that no language is really primitive. For example, the Australian Aborigine speaks with a grammar every bit as complicated as ours. Every society seems to develop a language that wholly satisfies its need.

In some cultures, linguistic skill is lavishly rewarded; in others it is regarded with suspicion. The American attitude lies somewhere between these extremes. In general, we admire the plainspoken man, but also chuckle at Li'l Abner's crude English. We respond to the power of Winston Churchill's oratory, but anyone who oversteps such very narrow bounds will be labeled a windbag politician.

Skills in communication are important for the engineering student and for the

A bolt is a thing like a stick of hard metal such as iron with a square bunch on one end and a lot of scratching wound around the other end.

A nut is similar to a bolt only just the opposite being a hole in a little chunk of iron sawed off short with wrinkles around the inside of the hole.
—An 8th-grade student's definition and description of two mechanical objects.

engineer in practice. If an engineer cannot express clearly his ideas and the results of his endeavor to others, even though he may have the intellect of a genius and the capability of performing the most creative work, the benefits of his intellect and creative abilities will be of little use to others. A surprisingly large amount of an engineer's time and effort is devoted to communicating—principally writing and speaking. While in college, most engineering students don't recognize this, and they are not easily convinced that the ability to write and speak effectively has a *significant* effect on their professional competence.

What are the skills that are needed in communications? For the engineer they generally are classed as verbal, graphical, and mathematical. In this respect we shall consider that *verbal* means language communication, either oral or written; that *graphical* constitutes all pictorial language such as engineering drawings, charts, diagrams, graphs, and pictures; and that *mathematical* includes all symbolic language in which concepts and logic processes are presented by use of a system of prearranged symbols.

A question may arise as to whether models and demonstrations constitute communication. In the truest sense, they do, but usually they are inadequate within themselves to convey all concepts. Since usually words, pictures, or symbols are used as a supplement to explain such devices, these methods should not be considered to be a separate means of communication.

Illustration 8-1
The engineer must be able to convey his ideas to others by means of spoken and written communications. Conciseness without loss of clarity is most desirable.

Verbal communication

We begin practicing oral communication at an early age. Unfortunately, the basics of English composition, spelling, and grammar and an understanding of what the language is and how it works *are not* being effectively mastered in the public schools by an increasingly greater percentage of students. For this reason, the student who realizes the need and has the desire to improve his communication skills should either avail himself of college courses in speech and English composition or undertake a self-study program.

As we progress through school, we add to our vocabulary and pick up experience in presenting ideas in writing. By the time a student is a freshman in college, he is expected to have a working vocabulary of several thousand words, to be able to organize ideas into a coherent pattern, and to present these ideas either orally or in writing. How does the objective of verbal communication for an engineer differ from this objective for other people?

Primarily, the engineer communicates to present ideas and to gain ideas. For some people, talking or writing is purely an entertainment, or an outlet for creative feelings. On the other hand, for the engineer, verbal communication is a part of his professional life. He communicates on a technological level with other engineers and with scientists and technicians and on a layman's level with nontechnically trained people.

Although public awareness and understanding of engineering, science, and technology are increasing each year—the world is becoming more technologically complex—many people are uneasy or uncomfortable about "what technology is doing." In their minds, the "what" is frequently assumed to be bad or, at least, constitutes a veiled threat because of its mysteriousness. The mystery, of course, is a result of their lack of understanding and knowledge in these areas (the United States' conversion to the metric system is too often met with fear, or at least discomfort, by large numbers of people).

The point of all this for the engineer, of course, is that he has an important job to do in selling his competence, knowledge, and judgment. His profession and its activities must be understood and appreciated by the public—not condemned through ignorance. In general, his works bring about much public good, but he must let this be known broadly.

His skill in communicating to the nontechnically trained public is not only a professional responsibility, it is his personal responsibility. Ideally, an engineer should be able to take a most complex topic—nuclear fission, a jet turbine, holography, or systems analysis—and explain it in such a way that his brother in the fifth grade or his Aunt Bessie will have a basic, simple understanding of it.

For example, suppose you have been working on a device for inclusion in the design of an autopilot control for an airplane. This device includes a potentiometer and a gyroscope. Consider how different your description of the device would be to an engineer and to an accountant. The ability to "speak their language" is an important skill which the engineer should possess in dealing with diverse people.

Not only should the engineer's verbal skill be descriptive, but also it should be persuasive. Frequently, good ideas are considered by the uninformed to be too impractical or too revolutionary. The engineer should have not only an adequate vocabulary, but also skills in presenting his ideas in a way that others will be led to accept them. In situations like this, practice in idea organization, a knowledge of psychology, and training in debate are all helpful.

In college, many opportunities are available for participation in group discussions, for the presentation of concepts both written and oral, and for gaining vocabulary skills. This is the time for the engineering student to learn by trial and error his best ways to communicate. After graduation, trial-and-error methods may be economically impractical. A conscious effort while in college to improve one's ability to communicate verbally will make the transition to work as a practicing engineer after graduation much easier.

Graphical communication

How often have you heard someone exclaim, after a futile attempt to describe an object to another person, "Here, let me draw you a picture of it!" The old adage of a picture being worth a thousand words is still true. The ability to present ideas by such means as pictures, diagrams, and charts is a valuable asset. In general, the engineer is expected by nontechnical people to be able to sketch and draw better than the average person. Today, most engineering curricula include some work in engineering graphics, and although the engineer may not be a professional draftsman, he should be able to attain and maintain an acceptable level of performance in engineering drawing and lettering.

Since ideas in research and development are frequently somewhat abstract, diagrams and graphs not only help to present ideas to others but also help the engineer to crystalize his own thought processes.

For the design engineer, the ability to present ideas graphically is a necessity. In almost every case, instructions prepared by a design engineer for use by technicians or workmen in building or fabricating articles are transmitted in the form of drawings. In the case of machined parts, for example, usually the workman has only

Illustration 8-2
The engineer must be able to understand graphical as well as oral and written instructions.

the vaguest idea of the application of the part, so the drawing prepared by the design engineer must tell the complete story to the machinist.

The engineer engaged in sales will need to make frequent use of graphic aids. It usually is easier and faster to project ideas and application by graphic means than by verbal communication alone. Pictures fill the gap between verbal description and actual observance of an operating device. So effective are these techniques that considerable experimentation is now being done in teaching by means of television.

Mathematical communication

Mathematics involves the use of symbols to represent concepts and their manipulation in logic processes. It has been stated humorously, but nevertheless somewhat truthfully, that mathematics is a form of shorthand used to describe science and that higher mathematics is shorter shorthand. In his study of mathematics, the engineer learns the meaning of symbols such as π, $+$, and $\int$, the rules for manipulating mathematical quantities, and the logic processes involved.

The question sometimes arises as to why an engineer needs so much mathematical training. The answer in simple terms is that it is such a valuable and powerful tool that the engineer cannot afford to ignore its use. By using mathematics, not only is space conserved in the presentation of ideas but also the task of carrying the ideas through logic processes is simplified. Since many engineering science operations follow elementary mathematical laws, it is much easier to transform ideas into symbols, and manipulate the symbols according to prearranged mathematical procedures, and finally to come up with a set of symbols which can be reconverted into ideas.

The engineer's way of thinking is so consistently geared to mathematical processes that it becomes almost impossible for him to think otherwise. For example, if you are asked to find the area of a circle whose radius is known, you may immediately visualize $A = \pi r^2$. Now try to think of finding the area of a circle without using such a mathematical formula—you will probably find such thought to be difficult and unnatural.

Since your mathematical training has given you a skill in communication, as an engineer you should make full use of it. As has been pointed out, the logic processes enable one to predict mathematically the behavior of many engineering science operations. In addition, the mathematical presentation of the ideas enables others familiar with mathematical rules to envision the practical application of the concepts.

For example, if the effects of gravitational forces, air loads, centrifugal forces, temperatures, and humidity are expressed properly in a mathematical formulation, the path of a missile over the earth's surface can be predicted with surprising accuracy. It is not actually necessary to perform the flight and to measure the trajectory if the parameters involved are known accurately.

Of course, mathematics is not restricted to an application of the known behavior of objects. By mathematical extrapolation, fundamentals of natural laws have been determined even before it has become known that such behavior is possible. For example, the principles of atomic fission were predicted mathematically many years before it was possible to verify them experimentally.

The use of mathematics by engineers permits more time to be given to creative thought, since ideas can be explored symbolically without having to make physical determinations. Of course, the advent of high-speed automatic computing machines also has aided both in accelerating exploratory research and in executing routine mathematical operations.

Technical reporting

One of the first things a young graduate must learn is that writing or speaking as an engineer on the job is not at all like writing or speaking as a student in college. For example, engineering students usually write for professors who know more about the subject than they do. On the job, however, the engineering graduate will write for management, for technicians, for skilled and semiskilled workers, and only occasionally for other engineers. Therefore, in college the young engineer writes to impress the reader with his knowledge and diligence; on the job, he writes to inform, to instruct, and not infrequently to persuade. However, for both audiences his goals are clarity, logic, and economy of words.

Much of the engineer's communication is executed by reports. These reports may be oral presentations in the form of technical talks or they may be written presentations as technical reports. In either case, information must be presented in a form so that the desired meaning can be understood.

Since the objective of a report is to present information, it must be prepared with the reader in mind. Clarity is therefore a prerequisite for a good report. A report that uses rare words or uncommon foreign phrases may serve to point up the brilliance of the author, but it may also discourage readers from attempting to unravel the

Illustration 8-3
Technical talks should be supplemented with appropriate charts, slides, graphs, models, and so on, to add clarity and interest.

> Short words are best and the old words, when short, are best of all.
> —Sir Winston Spencer Churchill

meaning. A report should be prepared using words and phrases with which the reader will be familiar.

In addition to clarity, a report should state clearly and honestly the results obtained. In the case of reporting on tests, frequently data are taken and the test assembly is dismantled before the test results are available. Therefore, the tests cannot always be rerun, and the data are usually used as recorded. If the results should turn out to be less than desirable, as an engineer you are obligated to report the facts completely and honestly. Even though reporting the true facts may be distressing to the writer, an honest statement will instill a feeling of confidence in the reader that the results are trustworthy.

In preparing reports, in general, only factual material should be covered. There is often a temptation to include irrelevant subjects or personal opinions as a part of the factual material. In some cases, it may be desirable to give a personal opinion, but such opinion should be identified clearly as a matter of judgment and not as factual data.

Written reports

There may have been a time when engineers were not required to write reports, that is, when most of them worked with small groups of people and they could let their ideas be known by word of mouth or by circulating an occasional sketch or drawing. These times are gone for all but a very few engineers. Most engineers today work in large organizations. They cannot be "heard" unless their ideas are written down in proposals and their findings are recorded in reports. This does not mean that the spoken word and the drawing or sketch have lost their importance, but rather that they must be supplemented by the written word. Therefore, it is important for the engineer to know how best to communicate his ideas to a reader; how to put his best foot forward with a client whom he may never see, or with the company vice-president to whom he will report.

Although the engineer's writing is often directed toward other engineers, occasionally he may be called upon to write for an audience unfamiliar with technical terms. Then he must be able to express his thoughts in terminology that can be understood by an intelligent layman. During the years to come, when engineers must solve the problems of society, such as the urban crisis, air and water pollution, energy, food production, and so on, the need for cooperation between technical and non-technical persons becomes increasingly important. The engineer must learn to communicate with people of all types of background, and he must be able to state his views clearly and concisely.

In whatever he writes, an engineer needs to be aware of a few commonsense rules of technical communication. He should not try to show off his knowledge, but should keep his vocabulary as simple as possible without "talking down" to the reader. It should be kept in mind that the reader is not a specialist (most engineers direct their

> Anyone who wishes to become a good writer should endeavor, before he allows himself to be tempted by the more showy qualities, to be direct, simple, brief, vigorous, and lucid.
> —H. W. Fowler

writing to management) and that this same reader is probably in a hurry. He does not want to be held in a prolonged state of suspense awaiting some thrilling outcome. Rather, what the reader wants is information as quickly and as painlessly as possible, and he wants it to connect in some way with his own experience.

Keeping the reader in mind, the successful engineer/writer will organize his reports in reverse order to the way that he accomplishes his work. An engineer proceeds in his design work from problem recognition to solution (see page 456), bringing order out of chaos, isolating bits of data, selecting and interpreting, and, after trial and error, arriving at a specific finding. This finding might be a particular choice of structural material, an optimal shape, or even the decision to terminate a project.

When his design or analysis work is complete he is ready to write his report—but in reverse order. He must avoid the temptation "to write a detective story"—the temptation to lead the reader from the beginning over the same arduous route that he himself followed in his quest for truth. (All detective fiction is synthetically organized this way.) Instead, he organizes his paper to reveal all major conclusions in the first few paragraphs. Now the reader immediately knows what the paper is about. If he chooses, he may read further to find out *why* the engineer's conclusions are true and *how* he arrived at his decisions.

So many books have been written about the art of writing, about grammar, syntax, and style, that it would be presumptuous to try here to summarize them in a few words. However, we would like to quote a few phrases from a small but exceedingly valuable book on style.[1] These authors recommend some 21 rules, among which are the following:

- ☐ Place yourself in the background. Write in a way that draws the reader's attention to the sense and substance of the writing.
- ☐ Write in a way that comes naturally. But do not assume that because you have acted naturally, your product is without flaws.
- ☐ Write with nouns and verbs, not with adjectives and adverbs.
- ☐ Revise and rewrite.
- ☐ Do not overstate, because it causes your reader to lose confidence in your judgment.

If we can assume that we know how to write, how to express our thoughts in words and sentences that are clear to the reader, we still need to know how to organize our ideas. Organization is important to the writer so that his ideas will be presented in a logical sequence and to assure that the important things are included in his writing. Organization is important to the reader so that he can follow the presentation and conclusions of the author easily, so that he need not jump back and forth in his thoughts (a tiring exercise for any reader).

Over the years certain minimum conventions (standards) have been established

[1] William Strunk, Jr., and E. B. White, *The Elements of Style*, second edition, Macmillan, New York (1972).

concerning the writing of engineering reports and proposals, conventions that are not binding but have proven to be useful guidelines for technical writers. Let us look at the typical organization of such a report.

Organization of a technical report

Abstract
Table of Contents
Table of Figures
Acknowledgments
Nomenclature

Introduction

Body of report
Analysis
Design
Experiments
Test Results

Discussion

Summary and Conclusions
References
Appendixes

The essential features of the technical report are shown in italics. The other items *may* appear in the report if appropriate.

Although it appears first in the report, the *abstract* is usually the last item to be written. It is a summary of the summary, containing in less than a page a statement of the problem, the way in which it was solved, and the results and conclusions that were drawn from the work. One may well ask, "Why repeat the contents of the report first in the summary and then again in the abstract?" One reason is the variation in the interest of the readers. One man may have only a general interest in the report and is satisfied with a well-written abstract; the second, wanting to go somewhat deeper, may wish to read the summary and conclusions; and only a few (those particularly interested in the subject) may be sufficiently interested to read the entire report. Yet, it is important that all these readers obtain a clear picture of why it was done and what was accomplished. Another reason is the need for some repetition in communication. This attitude is exemplified in the philosophy of the successful southern preacher who, when asked why his sermons were so successful, answered, "Well, first Ah tells 'em what Ah's going' to tell 'em—then Ah tells 'em—then Ah tells 'em what Ah done told 'em."

The *introduction* tells what the problem is and why it was studied. It will discuss the *background* for the study, the literature that pertains to the subject, the solutions that have been tried before, and why these are not adequate for the present investigation. It is here that the majority of the literature references are mentioned. If there are three or less, it may be adequate to list them in footnotes. However, when there are more than three references, it is common practice to list them together in a reference section at the end of the report.

The *body of the report* may have any of a number of titles and may, in fact, consist of several chapters with different titles. The author has considerable latitude here and he should make use of the titles that appear to him to be appropriate. For example, if the work was essentially analytical in nature, he may wish to entitle the section "Analysis," or he may wish to be more specific and to discuss first the assumptions that were made, then the construction of the model, the pertinent equations, and finally the solution of the equations. If the report contains information on experiments, the writer may wish to discuss the experimental apparatus, the construction of the test model, and the performance and organization of the test. He may then follow it up with a chapter discussing the test results.

The preparation of the body of the report requires considerable judgment. The engineer must provide enough information to give the reader a very clear picture of what was done and to allow him to arrive at the *conclusions* of the report. On the other hand, it is essential that the reader not be bored by unnecessary detail. Many writers find it appropriate to give only the major outline of their work in the body of the report and to relegate all important but minor details to appendixes at the end of the report. This technique gives a report a highly desirable conciseness.

Young engineers often feel impelled to write their reports in the same chronological sequence that the work was accomplished. This is both unnecessary and undesirable, for very rarely does one proceed in a straight line from the beginning to the finish of his work. Rather one detours down side roads and retraces one's steps. If the report follows the same path of procedures, it will be very difficult to follow. It is much more important to present the data in the sequence that the engineer would have used in his work if he had been knowledgeable of all the difficulties and errors in the beginning. The actual chronological sequence of the study is of little interest to anyone but the author.

It is usually expedient to illustrate the body of the report with tables, charts, graphs, sketches, drawings, and photographs. The old adage that "one picture is worth a thousand words" is often true, but care should be taken to avoid unnecessary illustrations.

Another important feature of functional technical writing is the use of headings and subheadings. These allow the reader to skip over parts he already knows, or parts that are too specialized for him to follow, or parts that simply don't interest him.

Since it is customary to limit the body of the report to facts, the *discussion* section permits a review of the author's opinion. It is as if he were able to stand back and look at the work and say why this or that was done, to speculate on why the results are the way they are, and what they might have been if the experiment had been done differently. The discussion sections should anticipate the type of questions the listener would ask and attempt to answer them as forthrightly and honestly as possible.

The *summary and conclusions* is, as the name implies, a concise statement of the work done—including goals, background, analysis, experiment, and a review of the work accomplished. The concise statement of the conclusions reached is most important. For the reader, the conclusions should be the "pot of gold" at the end of the rainbow, the information that will be directly useful to him. Therefore, the development of meaningful conclusions, well stated, is one of the most important parts in writing a report. They should include all that is new and important, and yet they should be so stated that they leave no question in the reader's mind as to what is incontrovertible fact and what is opinion. Wherever possible the writer should make estimates of the accuracy and repeatability of his results. It is often useful to number the conclusions, much as a patent attorney will number the claims in a patent application.

After the conclusions have been written, the author should write the abstract as if it were a summary of the "summary and conclusions" just finished. Only the most important conclusions need be included in the abstract.

A note on the convention for *references*. In most engineering reports, it is now customary to list the last name of the senior author first, followed by his initials, and followed by initials and name of coauthors. The names are then followed by the title of the report and this by the name of the journal in which it was published, or the publisher and year, in case it is a book. Typical references are as follows:

Smith, A. B., and T. D. Jones, "Air Pollution at the North Pole," *J. Arctic Society,* Vol. 15, No. 6 (1964), pp. 317–320.
Beakley, G. C., and E. G. Chilton, *Introduction to Engineering Design.* New York: Macmillan Publishing Co., Inc., 1974.

The formal report of a feasibility study often may be in the form of a proposal which suggests how the problem should be pursued. There are many similarities between an engineering report and a proposal, but their purposes are quite different. The report exists to present the results of a study and to present them so clearly and completely that other engineers can use them as stepping stones in the further development of engineering knowledge and use. The proposal, on the other hand, proposes to sell an idea—tries to convince a client or a superior to make funds available for the preliminary design. Thus, while the report is written for a general audience, not necessarily all engineers, the proposal is always written for just one person or organization. A proposal is an attempt to sell an idea. Therefore, what is good advice for the salesman is also good advice for the proposal writer: *try to put yourself in the position of the client.* Find out what his needs and wants are and see to what extent your idea meets these needs. Find out who else competes for the funds which might be used to further your idea and emphasize those special points that make your idea or talents superior to that of others.

There is no general format for the organization of a proposal, but the following order is used frequently.

Typical organization of a technical proposal

Technical Part
Introduction
Objectives
Background
Method of Approach
Qualifications
Management Part
Statement of Work
Schedule and Reporting
Cost Estimate
[Other special paragraphs, i.e.,
   Rights to inventions
   Security provisions
   Time at which work can begin
   Time limit on proposal acceptance]

The proposal is often split into a technical part which discusses the technical aspects and a management part which considers the financial and legal aspects. The first part of the technical portion introduces the reason for the proposed work and clarifies why its solution should be of importance to the potential client. The introduction is followed by a brief statement of the objectives, that is, what the author hopes to be able to achieve by performing the work. This may be followed, if appropriate, by a study of background information, such as the literature surveyed, to indicate that the author is well informed on the subject. Following this, a plan or method of approach is suggested which shows the client that the author has a well-thought-out plan of how he is going to proceed with the work. The method of approach should indicate not only what the author wishes to do, but also what results he expects to obtain from the various portions of his program, what he is going to do if the outcome of the results is as expected, and what if it is not. In conclusion, the technical part of the proposal should include the qualifications of the author or his organization to perform the work.

The management part of the proposal is a precise specification of the work—what it costs and what it entails. This is followed by the schedule for the work, including the time and type of reports to be presented, and by a cost estimate.

Although many young engineers may believe that fancy cover designs and big words can sell proposals, it is a fact that the most successful proposals are those that convince the reader of the sincerity and expertise of the writer and his ability to accomplish the objective.

Oral communication

Although most engineers like to talk, too few enjoy speaking. Talking is casual, random, and unrehearsed, but speaking requires a plan, an organization, and practice. Public speaking, like writing, is an art of increasing necessity for the successful engineer; an art that he must perfect if he is to succeed in his profession and his society. In his professional life the engineer will be called upon to present his ideas clearly and concisely to his peers, his supervisors, or the board of directors of the company for which he is working. If he has conducted research or development, he may wish to present the results at a meeting of his professional society. As an effective member of civic, social, and religious organizations, he will want to express his opinions clearly and convincingly.

Many persons make the mistake of assuming that the written report can be "talked" exactly as it is written; a common but deadly mistake is made when the speaker reads his paper—usually word for word—in a dull, monotonous voice. When this happens, even the listener who has a natural interest in or curiosity about the subject will usually lose it; and the listener who is not too interested, or who might differ basically with the speaker's views or recommendations, remains either antagonistic, indifferent, or unconcerned.

There is an appreciable difference between effective written and spoken words. The reader can proceed as quickly or as slowly as he wishes, or retrace his steps, and in this way absorb difficult and complicated thoughts. The listener, on the other hand, cannot control the speed of the speaker nor can he retrace his steps if he has lost the thread of the conversation. It is important, therefore, for the speaker to retain the interest of his listeners by the conviction of his presentation and by the presentation of a forthright, orderly, and logical sequence of thoughts. The most effective speakers

> There are three things to aim at in public speaking: first to get into your
> subject, then to get your subject into yourself and lastly to get your subject into
> your hearers.
> —Gregg

do not try to present more than two or three important ideas in one speech, and they
get these ideas across by using clear logic, simple illustrations, and by similes or
analogies, knowing that different listeners have different ways of seeing things.

In preparing a speech or oral presentation, first make an outline of the principal
ideas that you wish to project. Place them in a logical sequence and prepare your
illustrations and similes, but do not attempt to write every word of your speech. Few
things are more likely to put an audience to sleep than a speaker who reads his
speech. If you tend to be nervous, memorize the first sentence or two, which will get
you started, and then use notes only as reminders for the sequence of your talk and to
make sure that you have said everything that you wanted to say. Since an audience
can best follow simple ideas, it is rarely advisable to present mathematical developments in a speech unless it is to an audience of mathematicians. Nor is it often useful
or desirable to delve into the circuitous routes that were used during the development
of the idea or the research that is being presented. *The audience is interested in the
results and in the usefulness of the results for their own purposes.* All of us are interested
primarily in our own life and work, and the better a speaker can convince us that his
findings are useful to us, the more successful we believe him to be. Therefore, in
preparing a speech, first, find out to whom you will be speaking, and then ask yourself
what it is that you can give to the audience that is useful to them. What will they
remember after you have stopped speaking?

Your speeches will be successful if you fulfill these four goals of effective public
speaking:

1. *Command attention.* (This can be done if you achieve the other three goals.)

Illustration 8-4
*Remember, you should not
expect that all listeners
will be interested in
hearing your observations.*

152

2. *Be confident.* (Make a few strong points.)
3. *Know the material.* (Repeat, restate.)
4. *Be enthusiastic.* (An audience will forgive a speaker many things but will not forgive being bored.)

Many things a speaker does or doesn't do may create distraction, causing the audience to miss the main points the speaker is attempting to make. Here are a few hints to remember when making a speech.

Voice: It should be loud enough; vary the pitch, volume, and speed; speak slowly and clearly—but don't drag. Take a tip from the TV commercial—speak loudly the first few sentences. Inexperienced speakers have a tendency to speak too rapidly. They are afraid of pauses, which seem much longer to them than they do to the audience. Speakers are often afraid that if there are moments of silence the audience thinks the speaker has forgotten what he wants to say, or that he doesn't know the material.

Nervous habits: (These are personal mannerisms that are largely subconscious reactions—stage fright.) Don't play with notes, pencil, or pointer. Don't jingle the change in your pockets. Avoid excessive movements of your hands and feet. You may move around a little or change positions and posture, but don't move constantly—or stand statuelike. Many nervous habits express themselves in your language. You've listened to speakers who constantly clear their throats, and you've counted the uh's and aah's, "you know," or some other of the speaker's "favorite" words, which he uses to excess.

Poor grammar: Technical talks require, for the most part, the use of formal English

> If you make people think they're thinking, they'll love you; but if you *really* make them think, they'll hate you.
> —Don Marquis

grammar and pronunciation. Avoid slang, technical jargon, and long, unfamiliar words when short, simple terms will convey the desired information. A few expressions such as "we was," "you people setting in the audience," and "ya know" will cause many in the audience to feel you are less than literate.

Eye contact: Look at the audience—look at individuals in the audience—look at their foreheads. Don't look at only one section of the audience or talk to the wall, ceiling, or blackboard.

Gestures: "Elocution," a person's manner of speaking or reading in public, in the early part of this century was a form of highly stylized gestures, head and eye movements, and changes in posture. It was artificially eloquent and theatrical—full of bombast sometimes—but would be considered humorous and old-fashioned today. Be natural! A few gestures are effective, but be careful not to use too many. You may move your body forward for emphasis or sideways for transitions or change of thought. Don't attempt to be "professorial" and hang from or drape yourself over the lectern or podium. Be casual—but not too casual. Remember, all speakers and public performers are somewhat nervous—"excited" may be a more descriptive word, perhaps; just like athletes, they must get "up" for the event. Eventually practice and experience will help you to transmit your excitement to the audience in the proper form—not of nervousness, but as enthusiasm.

The successful speech, like a successful athletic contest, requires practice and rehearsal. In practicing, use a "sparring partner"—a person not afraid to criticize or interrupt and ask questions when something is not clear. Go over a speech with your "sparring partner" again and again, until you are sure that you could present it even if you lost all your notes.

Part Three

Preparation for problem solving

The advent of electronic
hand calculators has
dramatically increased
the computational skills
of the engineer.

Presentation of technical calculations

Format

In problem solving, both in school and in industry, considerable importance is attached to a proper analysis of the problem, to a logical recording of the problem solution, and to the overall professional appearance of the finished calculations. Neatness and clarity of presentation are distinguishing marks of the engineer's work. Students should strive always to practice professional habits of problem analysis and to make a conscious effort to improve the appearance of each paper, whether it is submitted for grading or is included in a notebook.

The computation paper used for most calculations is $8\frac{1}{2}$ by 11 in. in size, with lines ruled both vertically and horizontally on the sheet. Usually these lines divide the paper into five squares per inch, and the paper is commonly known as cross-section paper or engineering calculation paper. Many schools use paper that has the lines ruled on the reverse side of the paper so that erasures will not remove them. A fundamental principle to be followed is that the problem work shown on the paper should not be crowded and that all steps of the solution should be included.

Engineers use slant or vertical lettering (see Figure 9-1); either is acceptable as long as there is no mixing of the two forms. The student should not be discouraged if he finds that he cannot letter with great speed and dexterity at first. Skills in making good letters improve with hours of patient practice. Use a well-sharpened H or 2H pencil and follow the sequence of strokes recommended in Figure 9-1.

Several styles of model problem sheets are shown in Figures 9-2 to 9-5. Notice in each sample that an orderly sequence is followed in which the known data are given first. The data are followed by a brief statement of the requirements, and then the engineer's solution.

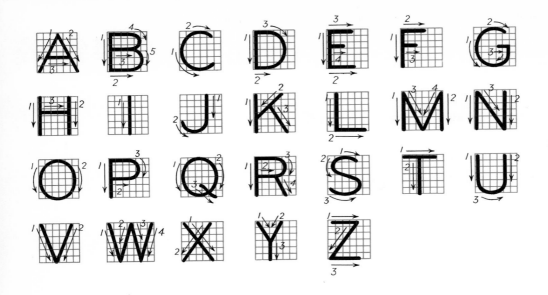

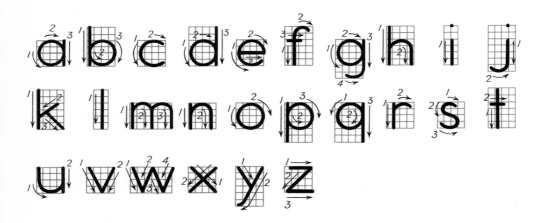

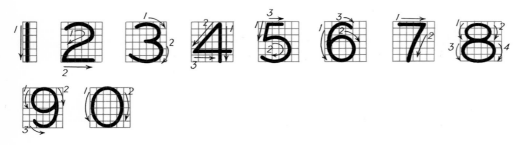

Figure 9-1 Vertical lettering.

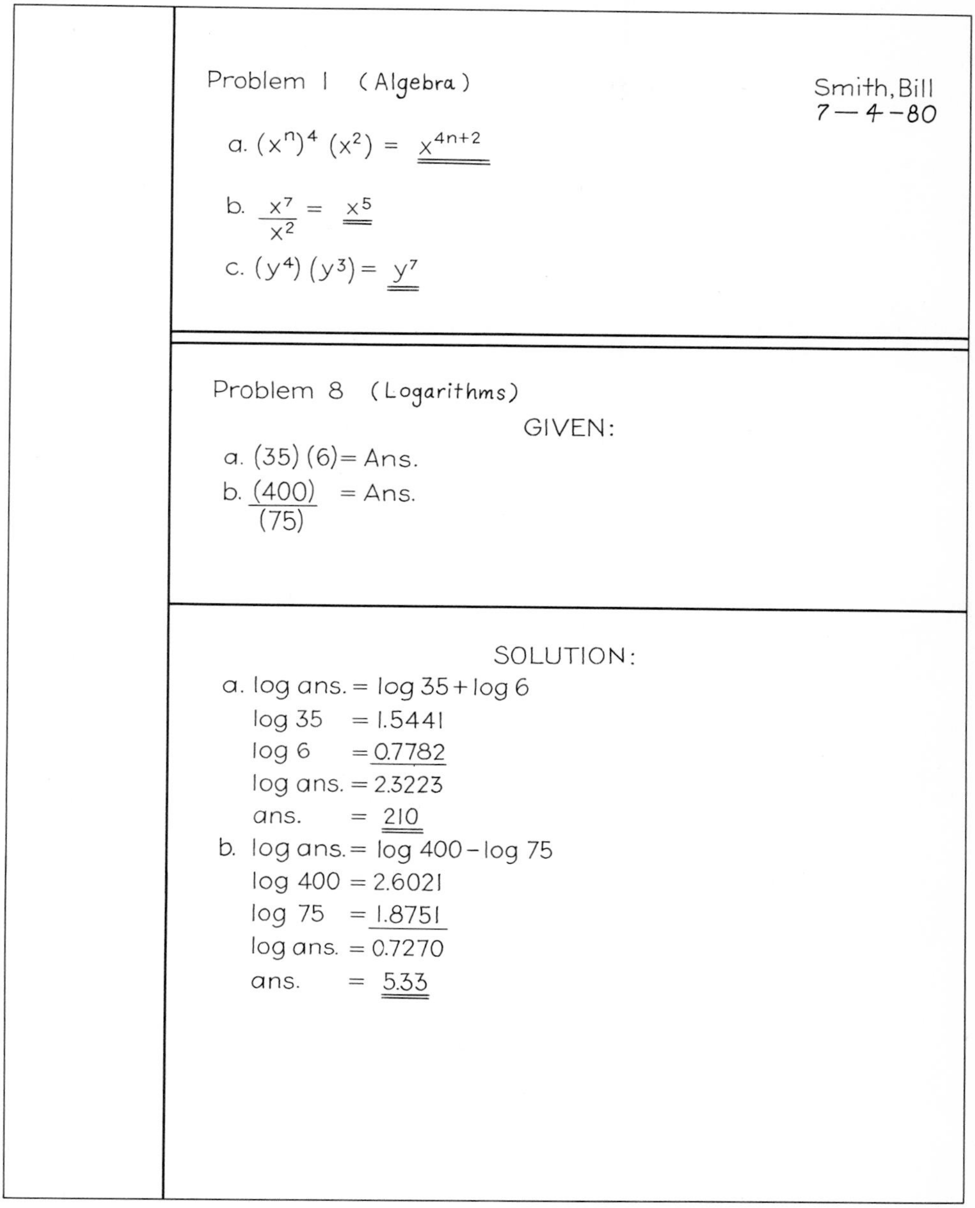

Problem I   (Algebra)

Smith, Bill
7 — 4 —80

a. $(x^n)^4 (x^2) = \underline{\underline{x^{4n+2}}}$

b. $\dfrac{x^7}{x^2} = \underline{\underline{x^5}}$

c. $(y^4)(y^3) = \underline{\underline{y^7}}$

Problem 8   (Logarithms)

GIVEN:

a. $(35)(6) = $ Ans.

b. $\dfrac{(400)}{(75)} = $ Ans.

SOLUTION:

a. log ans. $= \log 35 + \log 6$
   $\log 35 \quad = 1.5441$
   $\log 6 \quad\ = \underline{0.7782}$
   log ans. $= 2.3223$
   ans. $\quad = \underline{\underline{210}}$

b. log ans. $= \log 400 - \log 75$
   $\log 400 = 2.6021$
   $\log 75 \ = \underline{1.8751}$
   log ans. $= 0.7270$
   ans. $\quad = \underline{\underline{5.33}}$

Figure 9-2   Model problem sheet, style A. This style shows a method of presenting short, simple exercises.

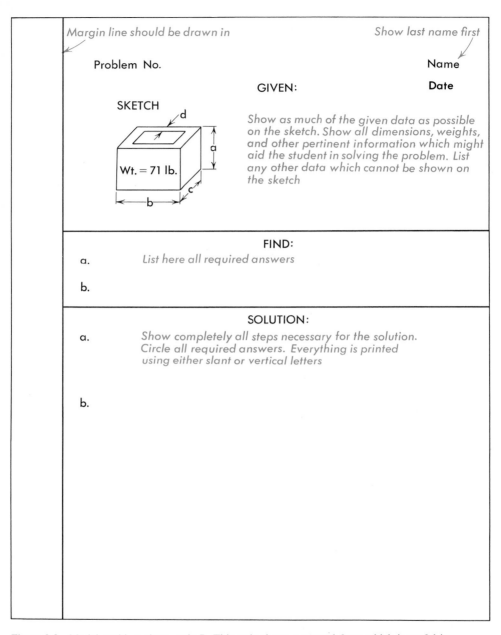

Figure 9-3 Model problem sheet, style B. This style shows a general form which is useful in presenting the solution of mensuration problems.

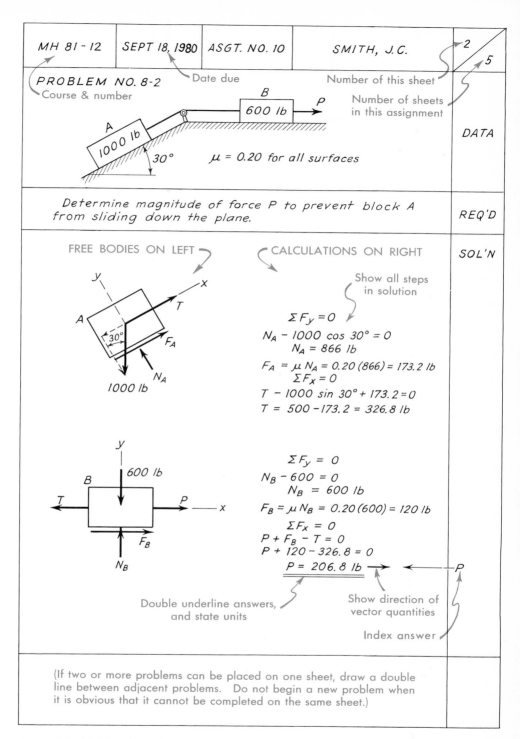

| MH 81-12 | SEPT 18, 1980 | ASGT. NO. 10 | SMITH, J.C. | 2 / 5 |

PROBLEM NO. 8-2 — Date due   Number of this sheet
— Course & number   — Number of sheets in this assignment

DATA

$\mu = 0.20$ for all surfaces

Determine magnitude of force P to prevent block A from sliding down the plane.

REQ'D

FREE BODIES ON LEFT   CALCULATIONS ON RIGHT   SOL'N

Show all steps in solution

$\Sigma F_y = 0$
$N_A - 1000 \cos 30° = 0$
$N_A = 866 \; lb$
$F_A = \mu N_A = 0.20 \,(866) = 173.2 \; lb$
$\Sigma F_x = 0$
$T - 1000 \sin 30° + 173.2 = 0$
$T = 500 - 173.2 = 326.8 \; lb$

$\Sigma F_y = 0$
$N_B - 600 = 0$
$N_B = 600 \; lb$
$F_B = \mu N_B = 0.20 \,(600) = 120 \; lb$
$\Sigma F_x = 0$
$P + F_B - T = 0$
$P + 120 - 326.8 = 0$
$\underline{\underline{P = 206.8 \; lb}}$

Double underline answers, and state units

Show direction of vector quantities

Index answer

(If two or more problems can be placed on one sheet, draw a double line between adjacent problems. Do not begin a new problem when it is obvious that it cannot be completed on the same sheet.)

Figure 9-4   Model problem sheet, style C. This style shows a method of presenting stated problems. Notice that all calculations are shown on the sheet and that no scratch calculations on other sheets are used.

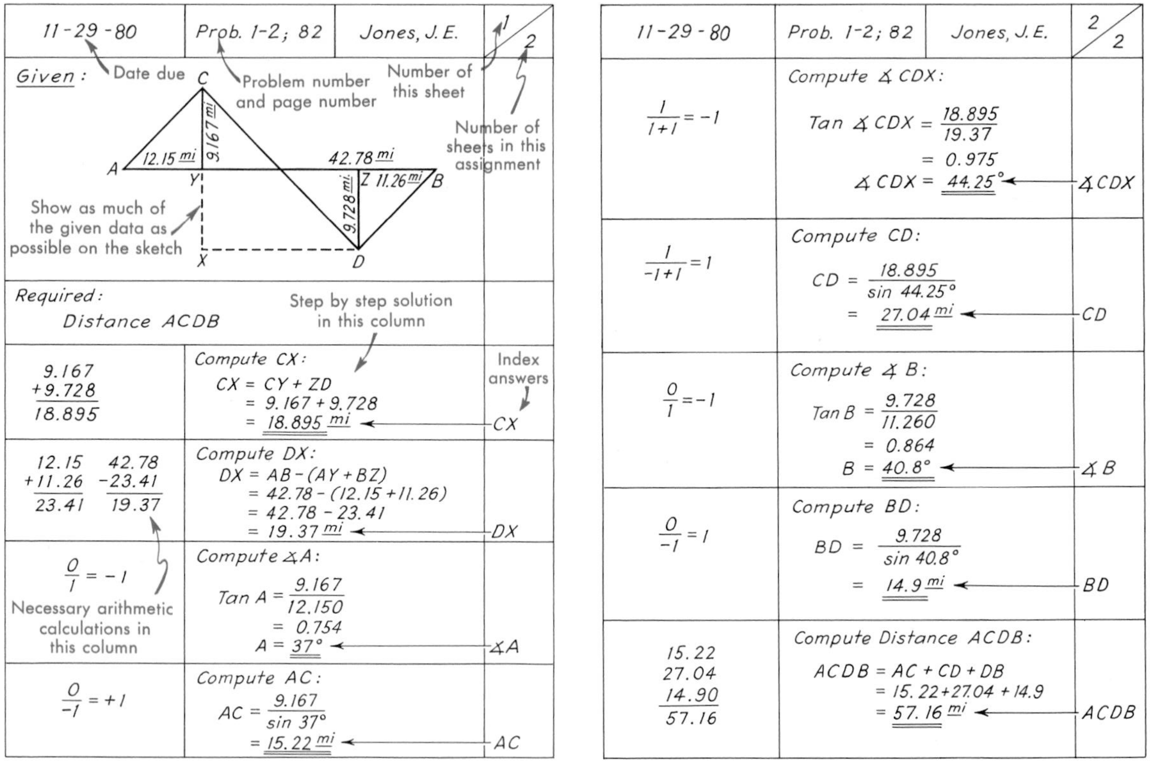

Figure 9-5 Model problem sheet, style D. This style employs a sheet with heading and margin lines preprinted. Notice that all calculations are shown on the solution sheet.

Sheet 1:

| 11-29-80 | Prob. 1-2; 82 | Jones, J.E. | 1 → 2 |

Given:
Date due — Problem number and page number — Number of this sheet — Number of sheets in this assignment

Show as much of the given data as possible on the sketch

Sketch: triangle with points C, A, B, D, X, Y; 12.15 mi, 9.167 mi, 42.78 mi, 9.728 mi, 11.26 mi

Required:
  Distance ACDB

Step by step solution in this column — Index answers

Compute CX:
$$CX = CY + ZD$$
$$= 9.167 + 9.728$$
$$= 18.895 \ mi \rightarrow CX$$

9.167
+9.728
18.895

Compute DX:
$$DX = AB - (AY + BZ)$$
$$= 42.78 - (12.15 + 11.26)$$
$$= 42.78 - 23.41$$
$$= 19.37 \ mi \rightarrow DX$$

12.15   42.78
+11.26   −23.41
23.41   19.37

Necessary arithmetic calculations in this column

Compute ∡A:
$$Tan \ A = \frac{9.167}{12.150}$$
$$= 0.754$$
$$A = 37° \rightarrow ∡A$$

$$\frac{0}{1} = -1$$

Compute AC:
$$AC = \frac{9.167}{sin \ 37°}$$
$$= 15.22 \ mi \rightarrow AC$$

$$\frac{0}{-1} = +1$$

Sheet 2:

| 11-29-80 | Prob. 1-2; 82 | Jones, J.E. | 2 / 2 |

Compute ∡CDX:
$$Tan \ ∡CDX = \frac{18.895}{19.37}$$
$$= 0.975$$
$$∡CDX = 44.25° \rightarrow ∡CDX$$

$$\frac{1}{1+1} = -1$$

Compute CD:
$$CD = \frac{18.895}{sin \ 44.25°}$$
$$= 27.04 \ mi \rightarrow CD$$

$$\frac{1}{-1+1} = 1$$

Compute ∡B:
$$Tan \ B = \frac{9.728}{11.260}$$
$$= 0.864$$
$$B = 40.8° \rightarrow ∡B$$

$$\frac{0}{1} = -1$$

Compute BD:
$$BD = \frac{9.728}{sin \ 40.8°}$$
$$= 14.9 \ mi \rightarrow BD$$

$$\frac{0}{-1} = 1$$

Compute Distance ACDB:
$$ACDB = AC + CD + DB$$
$$= 15.22 + 27.04 + 14.9$$
$$= 57.16 \ mi \rightarrow ACDB$$

15.22
27.04
14.90
57.16

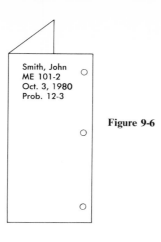

Smith, John
ME 101-2
Oct. 3, 1980
Prob. 12-3

Figure 9-6

When the problem solution is finished, the paper may be folded and endorsed on the outside or may be submitted flat in a folder. Items that appear on the endorsement should include the student's name, and the course, section, date, problem numbers, and any other prescribed information. An example of a paper that has been folded and endorsed is shown in Figure 9-6.

Scientific presentation of measured data

Since measured data inherently are not exact, it is necessary that methods of manipulating data be examined so that information derived therefrom can be evaluated properly. It should be obvious that the diameter of a saucepan and the diameter of a diesel engine piston, although each may measure about 6 in., usually will be measured with different accuracies. Also a measurement of the area of a large ranch which is valued at $50 per acre would not be made as accurately as a measurement of a piece of commercial property that is valued at $1000 per square foot. In order to describe the accuracy of a single measurement, it can be given in terms of a set of significant figures.

Significant figures

A significant figure in a number can be defined as a figure that may be considered reliable as a result of measurements or of mathematical computations. In making measurements, it is customary to read and record all figures from the graduations on the measuring device and to include one estimated figure which is a fractional part of the smallest graduation. Any instrument can be assumed to be accurate *only* to one half of the smallest scale division that has been marked by the manufacturer. All figures read are considered to be significant figures. For example, if we examine the sketch of the thermometer in Figure 9-7, we see that the mercury column, represented in the sketch by a vertical line, lies between 71° and 72°. Since the smallest graduation is 1°, we should record 71° and include an estimated 0.5°. The reading would then be recorded as 71.5° and would contain three significant figures.

As another example, suppose that it is necessary to record the voltmeter reading shown in Figure 9-8. The needle obviously rests between the graduations of 20 and 30 volts. A closer inspection shows that its location can be more closely determined as being between 25 and 26 volts. However, this is the extent of the aid which we can get

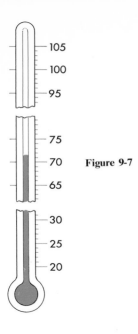

Figure 9-7

from the individual graduations. Any further refinement must be accomplished by eye.[1] Since the scale of the voltmeter is calibrated to the nearest volt, we can estimate the reading to the nearest half-volt—in this case 25.5 volts. An attempt to obtain a more precise reading (such as 25.6 or 25.7) would result only in false accuracy, as discussed below.

The designated digits, together with one doubtful digit, are said to be "significant figures." In reading values previously recorded, assume that only one doubtful digit has been recorded. This usually will be the last digit retained in any recorded measurement.

False accuracy

In analysis of engineering problems one must prevent false accuracy from appearing in the calculations. False accuracy occurs when data are manipulated without regard to their degree of precision. For example, it may be desirable to find the sum of three lengths, each having been measured with a different type of instrument. These

---

[1] In most cases, estimation by eye (beyond the precision obtainable from the graduations) is acceptable. It should be recognized that this final subdivision (by eye) will give doubtful results.

Figure 9-8

lengths might have been recorded in tabular form (rows and columns) as:

Columns

		a b c d e f g	
First measurement:	Row A	1 5 7 . 3 9	±0.02 ft
Second measurement:	Row B	1 8 . 0 2 5	±0.001 ft
Third measurement:	Row C	8 5 3 .	±2 ft
		1 0 2 8 . 4 1 5	(by regular addition)

Although the sum of the columns would be 1028.415, it would not be proper to use this value in other calculations. Since the last measurement (Row C) could vary from 851 to 855 (maximum variation in Column d), it would be trivial to include the decimal numbers in Rows A and B in the sum. The final answer should be expressed as 1028 ± 2, or merely 1028. In this case the last digit (8) is of doubtful accuracy.

In the tabulation of data (readings from meters, dials, gages, verniers, scales, etc.), only one doubtful digit may be retained for any measurement. In the preceding example, the doubtful digits are 9 (Row A), 5 (Row B), and 3 (Row C). The example also shows that, when numbers are added, the sum should not be written to more digits than the digit under the first column which has a doubtful number.

Scientific notation

The decimal point has nothing to do with how many significant figures there are in a number, and therefore it is impossible to tell the number of significant figures if written as 176,000., 96000., or 1000. This doubt can be removed by the following procedure:

1. Move the decimal point to the left or right until a number between 1 and 10 remains. The number resulting from this process should contain *only* significant figures.

Illustration 9-1
The accurate collection of data is essential to successful experimentation.

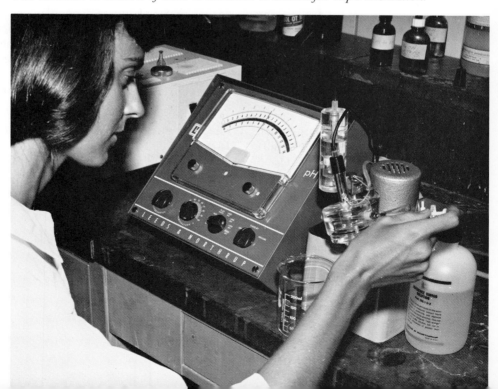

2. This remaining number must now be multiplied by a power of ten, $(10)^{\text{number of decimal moves}}$. If the decimal is moved to the left, the power of 10 is positive.

Example Express the number 1756000 to five significant figures:

1 756000.   (Move the decimal point to the left
6 5 4 3 2 1   to get a number between 1 and 10.)

Answer **(1.7560)(10)^6** (The power of 10 is the number of decimal moves.)

Note Only the five significant figures remain to be multiplied by the power of 10.

Example Express the number 0.016900 to three significant figures:

0 . 0 1 6 9   (Move the decimal point to the right
1 2   to get a number between 1 and 10.)

Answer **(1.69)(10)$^{-2}$** (The power of 10 is the number of decimal moves and is negative in sign.)

Note The three significant figures remain to be multiplied by the power of 10.

Examples of significant figures:

385.1	four significant figures
38.51	four significant figures
0.03851	four significant figures
3.851×10^7	four significant figures
7.04×10^{-4}	three significant figures
25.5	three significant figures
0.051	two significant figures
0.00005	one significant figure
27,855	five significant figures
8.91×10^4	three significant figures
2200	May have two, three, or four significant figures depending on the accuracy of the measurement that obtained the number. Where such doubt may exist, it is better to write the number as 2.2×10^3 to show two significant figures; or as 2.20×10^3 to show three significant figures.
55	two significant figures
55.0	Three significant figures. The zero is significant in this case, since it is not otherwise needed to show proper location of the decimal point.

In engineering computations it is necessary to use standard computed constants, such as π (3.14159265 . . .) and ϵ (2.71828 . . .). It is feasible to simplify these values to fewer significant figures. Usually three or four significant figures are sufficient, but this may vary somewhat with the nature of the problem. Since we do not need a large number of significant figures, let us examine some rules concerning "rounding off" the excess figures which need not be used in a given calculation.

Retention of significant figures

1. In recording measured data, only one doubtful digit is retained, and it is considered to be a significant figure.

2. In dropping figures which are not significant, the last figure retained should be increased by 1 if the first figure dropped is 5 or greater.

3. In addition and subtraction, do not carry the result beyond the first column which contains a doubtful figure.

4. In multiplication and division, carry the result to the same number of significant figures that there are in the quantity entering into the calculation which has the least number of significant figures.

9-1. Determine the proper value of X for each problem.

a. $0.785 = 7.85 = 7.85(10)^x$

b. $0.005066 = 6.066(10)^x$

c. $6.45 = 64.5(10)^x$

d. $10.764 = 10764(10)^x$

e. $1973 = 0.01973(10)^x$

f. $0.3937 = 3937000(10)^x$

g. $30.48 = 0.03048(10)^x$

h. $2.54 = 254(10)^x$

i. $1000 = 10(10)^x$

j. $0.001 = 1(10)^x$

k. $44.2 = 0.442(10)^x$

l. $0.737 = 73.7(10)^x$

m. $1.093 = 10930(10)^x$

n. $4961 = 0.4961(10)^x$

Addition of laboratory data

9-2. Add and then express the answer to the proper number of significant figures.

a.	b.	c.	d.
11.565	858.7	1.39395	757.1
4.900	404.3	8.7755	54.540
226.55	54.42	10.6050	11.5
82.824	19.8	49.201	1.0375
17.668	8.775	88.870	378.64
108.77	12.04	108.887	4372.1

e.	f.	g.	h.
16.59	0.32	6282.6	38.808
0.0531	6171.0	545.81	11.955
11.72	255.5	122.55	35.306
285.5	80.60	334.75	67.332
4.41	715.55	98.88	105.65
0.0748	3707.	28.77	575.75
		1.059	

i.	j.	k.	l.
0.005754	17.306	61.309	1.0585
0.006434	1.6535	1.9792	18.08
0.018466	0.0762	0.005531	675.5
0.085405	653.22	122.88	70.08
0.131876	29.969	52.8	111.0
0.97574	0.02202	37.075	828.

Subtraction of laboratory data

9-3. Subtract and then express the answer to the proper number of significant figures.

a.	b.	c.	d.
6508.	8.104	0.04642	731.16
3379.	7.891	0.0199	189.28

e.	f.	g.	h.
7.114	10276.	118.72	0.016
16.075	61581.	366.	0.1513

i. 766. −516.16	j. 0.8280 −0.023	k. −933.0 77.12	l. −156.2 0.0663
m. −610.01 −355.66	n. −1.9767 −113.54		

Multiplication of laboratory data

9-4. Multiply and then express the answer to the proper number of significant figures.

a. 5167. 238.	b. 32105. 5.28	c. 535.58 0.2759	d. 84.636 30869.
e. 1.03975 54682.	f. 0.0548 0.00376	g. 14.7410 0.7868	h. 47.738 0.065
i. 15903. 0.00469	j. −9757 0.05478	k. 7.5427 −542.16	l. −0.0989 −11.6507
m. 17.66 0.0307	n. 558.0 80.08	o. 141.8 0.37	p. 0.0051 1.06

Division of laboratory data

9-5. Divide and then express the answer to the proper number of significant figures.

a. $\dfrac{3928.}{5636.}$ b. $\dfrac{216.75}{53.83}$ c. $\dfrac{7.549}{3.069}$

d. $\dfrac{539.77}{1.6303}$ e. $\dfrac{0.5322}{0.343}$ f. $\dfrac{8831.}{128.75}$

g. $\dfrac{73.65}{127.1}$ h. $\dfrac{4.91}{1598.}$ i. $\dfrac{0.2816}{5383.}$

j. $\dfrac{-0.005295}{1728.}$ k. $\dfrac{0.07737}{-0.1293}$ l. $\dfrac{-0.3343}{-52.1}$

m. $\dfrac{3.58}{100}$ n. $\dfrac{13.550}{120}$ o. $\dfrac{4.001}{2.5}$

p. $\dfrac{0.0507}{350.1}$ q. $\dfrac{1.8}{0.006075}$

Calculation of error

The word "error" is used in engineering work to express the *uncertainty* in a measured quantity. When used with a measurement, it shows the probable reliability of the quantity involved. *Error,* as used here, does not mean the same as the word "mistake," and care should be exercised to call operations or results which are mathematically incorrect "mistakes" and not "errors."

Errors are inherent in making any measurement and as such cannot be eradicated by any practical means. Errors can be made smaller by care in making measurements, by employing more precise measuring instruments, and by performing repeated measurements to afford statistical accuracy. Statistical accuracy defines a region in which the true value probably will fall.

Since the reliability of engineering data is of extreme importance, familiarity with

> If you can measure that of which you speak, and can express it by a number, you know something of your subject; but if you cannot measure it, your knowledge is meager and unsatisfactory.
> —Lord Kelvin

methods of computing probable error is essential. As the student has more opportunity to collect his own data, the need for means of expressing the reliability or uncertainty involved in measured quantities will become even more apparent. Although a detailed study of theory of errors is beyond the scope of this book, a general discussion of some of the basic computations of errors is desirable.

Measurement and error

Experimentation in the laboratory is necessary to verify the engineer's design analysis and to predict results in processes of manufacture. For certain tests the laboratory technician will attempt to secure data to prove the analytical results as predicted by the engineer. At other times, emphasis will be directed to routine testing of items for acceptance. In any case, the results obtained in the laboratory will only approximate the true values, and the data tabulated will not be exact. Rather, every measurement taken and every gage reading or scale deflection noted will reflect the accuracy with which the individual measuring instruments were designed and manufactured—as well as the human errors that may have appeared in the readings.

For example, it is convenient and many times expedient to estimate distances by eye when under other circumstances an unknown distance could be more accurately measured by using a surveyor's tape or perhaps a graduated scale. In a similar manner we may lift a given object and, from experience, estimate its weight. A more accurate procedure would be to weigh it on some type of balance. In general, the more precise the measuring device, the more accurate the measurement obtained.

As we know from practical experience, length, weight, or time can be measured to various degrees of precision, depending upon the accuracy that has been designed into the measuring instrument being used. The engineer must therefore have some method whereby he can evaluate the degree of accuracy obtained in any given measurement. Where a numerical error of plus or minus ($\pm$) 1 in. would not ordinarily make too much difference in a measured distance of 100 mi, the same numerical error (of 1 in.) would cause considerable concern if it occurred in a measured distance of 2 in. For this reason the engineer will frequently express the maximum error present in a measurement as "per cent error" instead of "numerical error."

By "per cent error" is meant how many parts out of each 100 parts that a number is in error. For example, if a yardstick is too long by 0.02 yd, the numerical error is 0.02 yd, the relative error is 0.02 yd in 1.00 yd, and the "per cent error" is therefore 2 per cent. In other words:

$$\text{per cent error} = \frac{(\text{numerical error})(100 \text{ per cent})}{\text{measured value}}$$

$$= \frac{(1.02 - 1.00)(100 \text{ per cent})}{1.00} = \mathbf{2 \text{ per cent}}$$

In any measured quantity, the true value is never known. The measured value is usually expressed to the number of digits corresponding to the precision of measure-

Illustration 9-2
While a difference of a few degrees Celsius might not be important to many people, there are some who would be severely discomforted by any significant departure from the prescribed temperature.

ment followed by a number showing the maximum probable error of the measurement. For example, if we measure the length of a desk to be 5.712 ft and we have estimated the last digit, 2, because of our inability to read our measuring device closely, we would need to know what the probable variation in this last digit could be. Assuming that we can estimate to the nearest 0.001 ft, we could show this measurement with its error as

$$5.712 \pm 0.001 \text{ ft}$$

In order to compute the per cent error of our measurement, we proceed as follows:

$$\text{per cent error} = \frac{\text{numerical error} \times 100 \text{ per cent}}{\text{measured value}}$$

$$= \frac{0.001 \times 100 \text{ per cent}}{5.712}$$

$$= \mathbf{0.02 \text{ per cent}}$$

The error in measurement could be less than 0.02 per cent, but this shows the maximum probable error in the measurement.

As another example, a measurement can be shown as a number, and a per cent error as

$$7.64 \text{ lb} \pm 0.2 \text{ per cent}$$

To express this measurement as a number and a numerical error, the procedure is as follows:

$$\text{numerical error} = (\text{measured value})\frac{\text{per cent error}}{100 \text{ per cent}}$$

$$= (7.64)\frac{0.2 \text{ per cent}}{100 \text{ per cent}}$$

$$= 0.02 \text{ lb}$$

Expressing the measurement as a number,

7.64 ± 0.02 lb

Problems

(Note that the proper number of significant figures may not be given in the reading.)

9-6. Compute the per cent error:
    a. Reading of 9.306 ± 0.003 *b.* Reading of $19165 \pm 2.$
    c. Reading of 756.3 ± 0.7 *d.* Reading of 2.596 ± 0.006
    e. Reading of 13.750 ± 0.009 *f.* Reading of 0.0036 ± 0.0006
    g. Reading of 0.7515 ± 0.02 *h.* Reading of $12,835 \pm 20$
    i. Reading of 382.5 ± 5 *j.* Reading of 0.03 ± 0.03

9-7. Compute the numerical error:
    a. Reading of 35.219 ± 0.03 per cent
    b. Reading of 651.79 ± 0.01 per cent
    c. Reading of 11.391 ± 0.05 per cent
    d. Reading of 0.00365 ± 2 per cent
    e. Reading of 0.03917 ± 0.6 per cent
    f. Reading of 152 ± 4.0 per cent
    g. Reading of 0.0575 ± 10 per cent
    h. Reading of $7.65(10)^7 \pm 7$ per cent
    i. Reading of $3.080(10)^{-4} \pm 2.5$ per cent
    j. Reading of $32.5(10)^{-2} \pm 30$ per cent

9-8. A surveyor measures a property line and records it as being 3207.7 ft long. The distance is probably correct to the nearest 0.3 ft. What is the per cent error in the distance?

9-9. The thickness of a spur gear is specified as 0.875 in., with an allowable variation of 0.3 per cent. Several gears that have been received in an inspection room are gaged, and the thickness measurements are as follows: 0.877, 0.881, 0.874, 0.871, 0.880. Which ones should be rejected as not meeting dimensional specifications?

9-10. A rectangular aluminum pattern is laid out using a steel scale which is thought to be exactly 3 ft long. The pattern was laid out to be 7.42 ft by 1.88 ft, but it was subsequently found that the scale was incorrect and was actually 3.02 ft long. What were the actual pattern dimensions and by what per cent were they in error?

9-11. A resident of a city feels that his bill for water is considerably too high, probably because of a defective water meter. He proposes to check the meter on a do-it-yourself basis by using a gallon milk bottle to measure a volume of water. He believes that the volume of the bottle is substantially correct and that the error of filling should not exceed plus or minus 2 tablespoons.

a. What would be the probable maximum error in gallons per 1000 gallons of water using this measurement?

b. Using the milk bottle, he draws ten full bottles of water and observes that the meter indicates a usage of 1.345 ft^3 of water. If the average rate for water is $1.05 per 1000 ft^3, by how much could his water bill be too high?

10

Using electronic hand calculators

Since mathematics began many centuries ago, there has been a continuing need for some type of calculational aid. In the earliest period of history notches cut in a stick or a collection of beads or chips may have been used to tally animals, events, or whatever. As the need for more complex mathematical operations arose, however, additional calculation aids were designed. Today the engineer–scientist–technician, or for that matter any person needing a calculational aid, has available more computing power and versatility than was thought possible just a few decades ago. In fact, calculators may be purchased today whose capability compares favorably with the first electronic computer built. The wide range of calculator types available have so many features that it is difficult for the average buyer to make a judicious choice to best fit his needs. For example, each of the different kinds of machine logic available has both advantages and disadvantages, yet the novice often has inadequate insight as to which type would be best for him. In this chapter many features of calculators will be examined and, hopefully, some of the confusion will be eliminated.

One of the first forms of calculational aids to be developed was the *abacus*. The word *abacus* is derived from the Phoenician word "abak" meaning *sand spread on a surface for the purpose of writing*. It was in this configuration that the ancient Babylonians used the abacus. Later, counters of glass or metal discs were placed loosely on a waxed board. Still later wires were used to constrain the counters and the abacus eventually assumed the general shape that it has today. Herodotus, about 450 B.C., made reference to the Egyptians using the abacus. In the Far East the Chinese version of the abacus appeared about the sixth century B.C. In the sixteenth century, the Japanese developed a slightly altered form of the Chinese abacus called the "suan-pan," or computing tray. In the Orient, Russia, and Western Europe, it became the standard manual calculating tool (Figure 10-1). In many places the abacus can still be found in use, although it has now been replaced by improved calculational aids in most technologies.

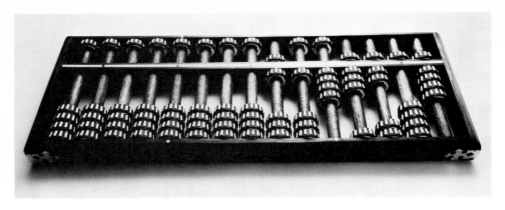

Illustration 10-1
The abacus was one of man's first "calculation" machines.

Although the abacus can perform simple mathematical operations, because of its simplicity it is limited in its usefulness. The invention of logarithms by John Napier in 1614 simplified some of the more difficult mathematical operations. In 1620, Edmund Gunter plotted logarithm values on a long line and then performed multiplication and division by the simple process of using a pair of dividers to subtract line segments

Figure 10-1   For centuries the abacus, in its various forms, served the major cultures of the world as the only form of manual calculation.

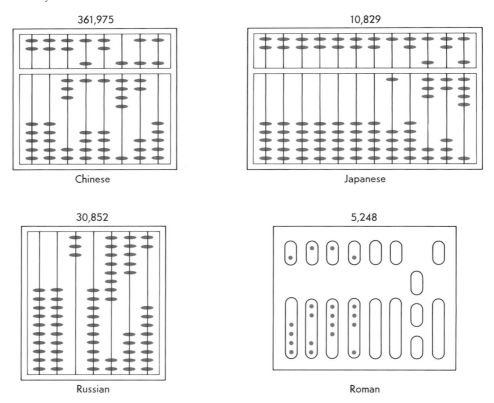

> One evening I was sitting in the rooms of the Analytical Society at Cambridge . . . with a table of logarithms lying open before me. Another member coming into the room, and seeing me half asleep called out, "Well, Babbage, what are you dreaming about?" to which I replied, "I am thinking that all these tables might be calculated by machinery. . . .
> —Charles Babbage

of the logarithm scale. This technique gave rise eventually to the development of the slide rule. William Oughtred in 1633 is said to have constructed two lines, similar to the ones made by Gunter, and moving one alongside the other eliminated the need for dividers.

As far as is now known the first practical mechanical calculator was invented by Gabriel Pascal in 1642. His design was similar in principle to that which had been recorded in sketches made over two centuries previously by Leonardo da Vinci—although Pascal was unaware of Leonardo's visionary work. Pascal's calculator consisted of a set of wheels, each rotated from 0 to 9 using a stylus, transmitting through gears to a set of result wheels. The positions of each wheel could be observed and sums read through windows in the wheel covers. Subtraction was executed by adding complements and multiplication by repeated addition. Up to six figures could be handled, with a fixed decimal point.[1] Later Wilhelm Leibniz improved Pascal's basic design to include division and the extraction of square roots. Design principles utilized in his machines are still used in modern mechanical calculating machines.

Charles Babbage is perhaps the undisputed father of the modern automatic computer. In 1822 he developed a "difference machine," which was used to produce the calculations for simple tables. However, due to the unavailability of suitable materials and advanced engineering technology his more elaborate "analytical engine" was never constructed.

The modern type of slide rule with a hairline on a cursor was designed originally in 1859 by Amedee Mannheim (a Prussian officer in the French artillery) to serve as a quick calculational aid for figuring the trajectories of cannon balls. Since the Mannheim slide rule was designed, there have been a number of improvements in its capabilities. Several different scales were placed on both sides of the slide rule body, and all scales were related to the C and D scales. Later, convenience scales were added that increased the rule's flexibility for problem solving.

For many years the slide rule served adequately as a computational tool for those in science, engineering, and technology. For $20 to $40 high-quality rules could be purchased that would perform a number of quite complicated tasks. However, the slide rule has several inherent limitations. In most calculations it does not consistently provide an accuracy beyond three significant figures. Some scales can be read to four significant figures, but frequently such apparent preciseness is meaningless. Also the slide rule cannot automatically locate the decimal point of a calculation. Still, this instrument has served several generations as an adequate and convenient calculation tool.

The abacus, Pascal's mechanical calculator, the Mannheim slide rule, and the other calculational tools mentioned above are all mechanical in nature. As such they suffer a number of limitations, such as versatility, speed, and size. To overcome these

---

[1] Maurice Trask, *The Store of Cybernetics*, Studio Vista, Dutton Pictureback, London (1971).

Illustration 10-2

For many years the slide rule served adequately as a computational tool for those in science, engineering, and technology.

limitations designers began to employ electronic components rather than mechanical components. The characteristics of electronic calculational tools depend on the type of components they contain. Thus a brief examination of each of the major types of components would be in order. First the vacuum tube, then the transistor, and then the integrated circuit will be discussed.

Following the discovery by Edison that electric charges could be transferred from a heated element in an evacuated space to another element in that space, DeForrest developed a device that could amplify electrical currents. The essential parts are shown in Figure 10-2.

In this simplified diagram, a cell or battery at *A* heats a filament of tungsten that frequently is coated with material such as cesium or thorium. The heat "boils off" electrons from the filament surface and produces a cloud of negatively charged particles around the filament. If the plate is made electrically positive with respect to the filament by the battery *B*, it is possible for the charges to flow through the

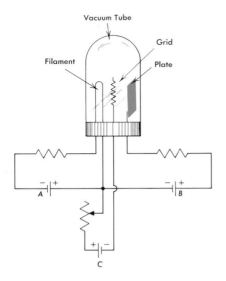

Figure 10-2 The essential parts of a three-element (triode) vacuum tube.

evacuated space from the filament area to the plate and constitute a current flow. If the positive voltage of the plate is below a certain level or if the polarity is reversed to make the plate negative, no current will flow.

These two elements in an evacuated space constitute a diode and can be used to rectify alternating currents—that is, change the alternating current to pulsating direct current.

If a third electrode is introduced between the filament and plate and is connected so it is negative with respect to the filament, the voltage of this third element, called a *grid,* can block the current flow, even though the positively charged plate is attempting to attract electrons. In fact, because the grid is near the filament, a very small change in its voltage will make a large change in the filament-to-plate current flow. This constitutes the amplifying capability of the vacuum tube.

If a person needed more speed or capability than available on his slide rule or desk calculator, he utilized an electronic computer containing vacuum tubes to solve his problem. Such computers were physically large and needed substantial amounts of power to operate their vacuum tubes. Initially electronic computers had not been made smaller because of the physical and electrical limitations of their components. Although the heritage of calculating tools had advanced considerably since the days of the original abacus, the breakthrough to a new generation of calculational aids had to await the development of solid-state electronics. If the barriers of size and power requirements could be overcome, then the development of even more powerful calculating tools could be undertaken.

Shortly after the close of World War II, an announcement was made of the discovery of a solid-state device, requiring no heated filament, that could be used as an amplifier. The discovery has in only a few years revolutionized the electronics industry. Although the solid-state diode as a rectifier had been in use for many years, the introduction of another element to permit amplification provided a tremendous opportunity for miniaturizing electronic components. This new device was called a *transistor* and, as it made its appearance almost at the same time that the computer was being developed, it was incorporated into almost all modern computers.

The theory of the transistor is fairly complex, but its action depends essentially on the presence of minute quantities of an "impurity" material (such as arsenic) in a crystal of pure material (such as germanium) permitting current to flow in one direction but not in the other. A proper assembly of three sections of negative carrier and positive carrier material permits a small voltage to control a much larger current flow in a manner similar to the way a vacuum tube behaves in a circuit.

The major advantages of the transistor as used in electronic circuits are light weight, small size, low power consumption, and long life. The modern integrated circuit is made possible only by the use of semiconductor techniques, and permits a tiny chip of material to perform the same functions as a vacuum-tube type of amplifier which would be thousands of times larger.

When the transistor was first invented, no one realized the eventual impact that this device would have upon society or how broad would be the extent and range of

> Machines made by the hand of God are incomparably better ordered and have in themselves more admirable movements than any that can be invented by the mind of man.
> —Rene Descartes

Illustration 10-3

The first transistor was a crude device by today's manufacturing standards, but it worked . . . and it opened a whole new world of solid-state electronics.

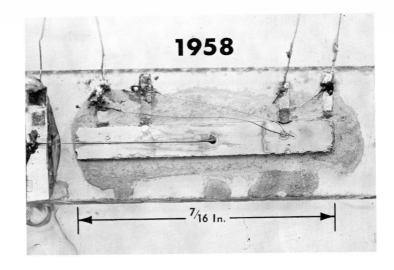

its areas of application (Illustration 10-3). However, few inventions of the last several hundred years have had more influence than the tiny transistor. In fact, the transistor and its numerous forms have found their way into most fields of human endeavor. Great advances in the areas of communications, science and technology, medicine, travel, and even entertainment have been made. In the field of calculating tools, the effect has been nothing short of revolutionary.

If one combines the desirable features of transistors (such as small size, no heat requirements, reliability, low heat production, little space requirement, lower voltage and power requirement, and shock durability), it is easy to see why they have found such widespread application. Also, mass production has reduced the price per unit to such a low level that the economic comparison between the vacuum tube and transistor now favors the transistor. The transistor does have some disadvantages, but its advantages are so numerous that it is still preferred over the vacuum tube for most applications.

The operation of the transistor is not really important to this discussion but we are interested in its fabrication. Understanding how the device is made will give us a clue as to how integrated circuits (abbreviated as IC's) are made. In general, the way in which IC's are made is not appreciably different than the way a transistor is made.

In making a transistor we may begin with silicon or, less commonly, germanium. Silicon is the second most abundant element in the earth's crust but, unlike gold, because of its strong chemical activity it is seldom found as an isolated element. It is normally combined with oxygen and usually also with one or more other elements, such as iron, aluminum, potassium, calcium, sodium, or magnesium. In order to get pure silicon, one must first refine an ore that contains silicon and then separate out the silicon. The separated, nearly pure silicon is then placed in a furnace and melted. A crystal of pure silicon, used as a *seed,* is placed in the melt and slowly withdrawn and rotated. The molten silicon attaches itself to the seed to form a single crystal that ultimately may be several centimeters across and 30 to 40 or more centimeters long. From this long single crystal is cut a number of thin wafers of the pure silicon, which are polished and prepared for the manufacture of transistors or integrated circuits.

Through an intricate and delicate set of operations involving metal etch resistant chemicals, ultraviolet light, microphotographic circuit plates, and furnaces, certain impurities are intentionally injected into the crystal structure. The depth, amount, and surface area of this injection are all carefully controlled in order to assure that the result will perform according to expectations. This operation is repeated as many times as necessary to make the final device.

On a single silicon wafer, several hundred (or several thousand) devices can be made at the same time. The wafer is then scribed and broken carefully and the different devices are separated from each other. Each device may be quite small, such as 0.25 mm², or smaller. With the aid of a microscope, these small chips are mounted onto some type of a base and wired to connecting pins, then capped or covered. The completed device finally proceeds to other processors for assembly into radios, television sets, or other applications.

Early in the 1960's a logical question was asked about the manufacture of transistors. Since the device was tiny, and since there was a lot of extra room on each chip that could be used for other devices, why not manufacture several devices onto the same chip and thereby save still more space? When this was done, the integrated circuit was born. Aside from the advantage of saving space by using the same chip to accommodate several devices, there was the additional advantage of saving processing and assembly time, because the devices could be connected internally by utilizing the same processes that were used to make the transistors themselves. Also, other devices such as diodes and resistors could be added to the chip at the same time. Thus in this manner whole circuits could be fabricated using the same processes as those used to develop a single transistor. Some devices could not conveniently be placed in any numbers on the chip (such as inductors or large capacitors), so some integrated circuits still had to have some additional electronic elements added outside the package. However, as circuits were modified to eliminate these outside elements, the IC package and its appendages became smaller.

At the same time, researchers found that still more devices could be added onto a chip by alternating the separate devices between insulating layers of glass. Connections between separate layers could be made at specified points. Thus the IC became a three-dimensional structure and the number of devices that could be located on an IC chip grew. Medium-scale integration (abbreviated as MSI) gave way to large-scale integration (abbreviated as LSI). With LSI the number of components in a single package has exceeded several thousand. In this manner, fairly complicated circuits could be placed into very small packages. Whole circuits such as a stereo preamplifier, power amplifier, logic circuits, and computer networks could now be produced in packages not much larger than the size of a fingernail. Thus one of the major barriers to a truly revolutionary calculation aid, that of size, has been eliminated.

Circuits that performed a variety of mathematical operations soon became available, and hand-sized calculators began to appear on the market. Initially the price was quite high because of the cost of the engineering development. Also the first calculators could perform only limited operations. The first electronic hand calculators that could add, subtract, multiply, and divide cost several hundred dollars. However, with the decrease in the cost of producing LSI circuitry, the field was opened, and it was only a matter of time before the calculator costs diminished and capability increased. As development costs were amortized, the rise in popularity began.

Several developments have been responsible for the production of the hand calculator. The invention of the transistor with all its advantages permitted design in a small space with low power consumption. The IC permitted the manufacture of circuits of high complexity in a small space. The technology for production of complex IC's in great numbers existed, so the cost of each unit IC was very low.

Illustration 10-4

There are 64 complete electronic memory circuits on this chip of silicon, shown for size comparison on the nib of a pen. The circuits can transmit electronic signals in as little as 3-billionths of a second. Because of this characteristic an instruction fed into a computer can be processed in 54-billionths of a second . . . or the time for light to move only 16 meters.

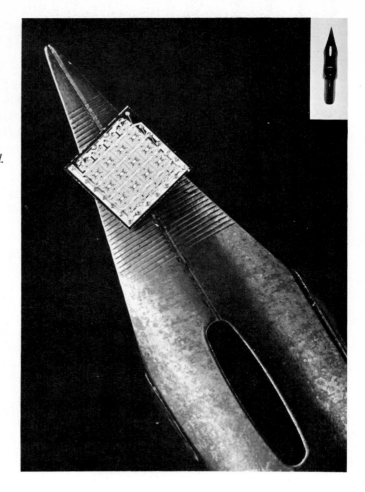

However, one other development has also been important to calculator production. A form of a readout or light display was necessary so that the user could identify a response from the machine. Mechanical printers were not small enough, so the answer had to be a type of optical display. Gas tubes could be used, and indeed are now being used in several units, but the power drain is rather large. Some diodes, when a voltage is applied across them, give off light (usually red). The invention of the light-emitting diode (called an LED) was the popular answer. The LED's are small and consume little power, so they are well suited for use in hand calculator design.

Even though all the individual components for the hand calculator now existed, a number of practical questions had to be considered in their design. Which features ought to be included for a particular calculator? How much would be the retail cost? What kind of a market would be anticipated for the product? What kind of sequence operations should be designed into the calculator? How durable should it be? How important was "human engineering" (i.e., such factors as size, weight, color, etc.)? The actual size of the calculator is dictated by the user, not the electronic contents. The keys, for example, must be set far enough apart so that the user will not inadvertently press an incorrect key. The readout must be of an adequate size and brightness or the user (with normal eyesight) cannot see the numbers in daylight. The "feel" of the calculator in the user's hand is an important consideration since the unit must be easy to hold. Weight and outside dimensions must be considered because the unit may be carried in a coat or shirt pocket. The calculator should be attractive in appearance with good color coordination. The arrangement of keys and functions must be logical for easy use by the operator. These and other considerations are all a part of human engineering.

Some of these questions are not easily answered. For example, the features of a calculator will depend on a particular consumer. The price will be a function of the number of features offered. So, some measure of balance must be established for each type of consumer. Overall guidelines seemed to be: offer as many desirable features as possible for the price range of the potential customer.

The specific mathematical functions that are designed into a calculator depend on the needs of the user. The next several pages will discuss the various features available on some of the more popular calculators. Sample problems for practice are also included.

One of the first questions that a potential buyer has to answer about calculators is "which type of *logic* should the calculator have?" Logic is defined as the sequence of operations that the user must follow to obtain the answer in the simplest manner possible. There are several types of logic from which to choose and each has some advantage. *Straight algebraic* logic has appeal for some because the keystroke sequence to be followed in the simpler problems is identical to the sequence that the operator will read when the problem is read from left to right. For some this kind of keyboard logic is the easiest to understand and learn. A second type of logic permits *nested parentheses,* a feature which allows the solution of more complicated problems than will the straight algebraic logic. The third type of logic was developed by a Polish mathematician named Jan Lukasiewicz. For simplicity, this type of logic is called *reverse Polish notation* (abbreviated as RPN). RPN is the most efficient way of handling more complicated mathematical expressions. Although the novice may find this latter format somewhat unfamiliar by comparison to previous thought processes, ease of working problems comes very quickly and the method poses no difficulty to the rapid solution of problems. When RPN is combined with a "stack register," a feature to be explained later, it can be a potent tool in calculation.

Since the type of logic chosen varies with the different types of calculators available for selection, it is recommended that one study in some detail the way that the several types of logic are used in the keystroke sequence. For the calculators that use algebraic logic, the equal $\boxed{=}$ key is used as a final step to obtain the answer. On the RPN calculators, the $\boxed{\text{ENTER}}$ key is used during the problem solution. Let us illustrate this difference with a simple problem.

Example $\qquad\qquad\qquad 4 \times 5 = ?$

With algebraic logic, the sequence is

$$\boxed{4}\ \boxed{\times}\ \boxed{5}\ \boxed{=}$$

The four steps give 20 as the answer that is displayed. With RPN the keystroke sequence is

$$\boxed{4}\ \boxed{\text{ENTER}}\ \boxed{5}\ \boxed{\times}$$

RPN also takes four steps. From this simple problem, the reader might jump to the conclusion that the two types of logic use the same number of steps for any problem solution. Another problem illustrates that this is not the case.

Ilustration 10-5
*The HP-21 is a
popular RPN-type
calculator used by
students of engineering
and technology.*

Example $(4 + 5) \times (6 + 7) = ?$

With algebraic logic the solution is

$$\boxed{4}\ \boxed{+}\ \boxed{5}\ \boxed{=}\ \boxed{STO}\ \boxed{6}\ \boxed{+}\ \boxed{7}\ \boxed{=}\ \boxed{\times}\ \boxed{RCL}\ \boxed{=}$$

and the answer is displayed as 117. A total of 12 steps were used. Notice that the keystrokes $\boxed{STO}$ and $\boxed{RCL}$ were used in the sequence. It was necessary to store the first part of the problem in a memory register while the second part of the problem was being solved. When the second part was finished, the answer to the first part was recalled and used as a product to obtain the final answer. With problems such as this, it is necessary to have a storage register or "empty reservoir" available for use. With RPN, the solution is

$$\boxed{4}\ \boxed{ENTER}\ \boxed{5}\ \boxed{+}\ \boxed{6}\ \boxed{ENTER}\ \boxed{7}\ \boxed{+}\ \boxed{\times}$$

for a total of 9 steps. Also, note that in this case no storage of the answer to the first part of the problem was required. One characteristic of RPN logic calculators is the availability of stack registers. These are automatic storage registers that can be used during the calculation for storing (remembering) intermediate answers. In the above example, after the first four steps in the RPN keystroke sequence, the partial answer was stored in one of these stack registers.

Perhaps the operation of the stack register can best be understood by thinking of the stack as being arranged vertically. The X register is on the bottom, topped by the Y, the Z, and finally the T register. The result of the first four keystrokes was stored in the Y register while the next number, 6, was entered. When the 7 was keyed in, the contents of the Y register were shifted to the Z register, the contents of the X register were lifted to the Y register, and the 7 was displayed in the X register. The $\boxed{+}$ keystroke combined the X and Y contents and brought the Z contents down to the Y register. The final $\boxed{\times}$ operation recalled the contents from the Y register and multiplied them with the X contents to obtain the final answer. These stack registers are separate from any other memory registers available and are only used for storing partial answers during a calculation.

Another type of logic uses the parentheses as a calculational aid. To illustrate, let us use the same problem demonstrated above. The keystroke sequence is

$$\boxed{4}\ \boxed{+}\ \boxed{5}\ \boxed{\times}\ \boxed{(}\ \boxed{6}\ \boxed{+}\ \boxed{7}\ \boxed{)}\ \boxed{=}$$

The logic for this solution requires ten steps. The final parenthesis could have been left off and the answer would still have been the same. The inclusion of the final parenthesis, however, permits the intermediate answer to be determined if it is desired.

The important thing to remember about calculator logic is that, no matter which type has been designed into a calculator, the operator soon becomes familiar with it and will feel little hindrance in the solution of a problem. Practice with a specific calculator in the solution of a number of problems gives one the familiarity needed for accuracy and speed. For this reason, the type of logic is not as important as the other features of the calculator, such as functions, price, or other considerations.

Once the question of the choice of logic has been settled, the buyer is faced with a number of other decisions that must be made. Some of the physical considerations should be considered, for example, size and weight. The factor of portability may be of importance to the buyer, especially if the calculator is to be used as a mobile instrument and not affixed to some specific location. If it is to be used primarily at a desk, the size and weight may not be of particular concern. The size of the readout may also be a consideration of some importance. A small readout with light-emitting diodes (LED's) may not be desirable because of the difficulty of seeing the numbers in bright sunlight or high-glare lights. Also, it is far more comfortable for the user to see a bright large display if he is to use the calculator for an extended period of time. However, large displays are also power consumers. The batteries for the larger units may be quite large and heavy, making the unit even less portable. In some instances a paper-tape machine will be desirable because of the necessity to record intermediate calculations for later review. Again, however, a printer of this type requires power, space, and has considerable weight. Usually, students are satisfied with a small hand-sized unit that has LED readout and is portable for carrying in pocket or briefcase.

How many digits in the readout are necessary? Some calculators only have five, some have eight, and some have ten. In addition to these digits, many have two additional digits for use in scientific notation (page 165). Scientific notation is necessary and desirable in calculators using numbers which are larger than 100,000,000 or smaller than 0.00000001 in size. For balancing checkbooks or figuring the monthly budget, the use of scientific notation is not important since few family budgets exceed this range. However, for calculations in science and engineering, often the range limitation becomes impossibly restrictive. Therefore, the buyer who plans to use the calculator in school or laboratory would be well advised to consider purchasing a model which features scientific notation.

Perhaps the most important of decisions to be made is to determine which mathematical functions are to be included on the calculator. A partial list of desirable functions found on a variety of calculators is given below. No calculator has all the features listed, but some have the majority. Of course, those that have an enlarged capability are more expensive and may offer more versatility than you need. Therefore, pick out the features that you desire, and then select a calculator which has those features.

Four functions

Addition, subtraction, multiplication, and division are considered to be standard functions for all calculators.

Trigonometric functions

The trigonometric functions usually include the sin, cos, tan, Arcsin, Arccos, and Arctan. Those entering fields of science or engineering will find these functions a necessity. However, if one is using the calculator primarily for business applications, they will not be as useful. On some calculators there are a number of keys that manipulate angular measure. For example, degrees can be converted to radian measure, or vice versa. Some calculators can be set to compute in either decimal degrees, decimal radians, or the relatively new decimal grads mode—the latter being a 100th part of a right angle in the centesimal system of measuring angles. While all these conversions are convenient to have on occasion, they are not all necessary since conversion factors can be used to convert from one to another. It is also most desirable to be able to convert interchangeably between rectangular and polar coordinates, since the conversion between these two types of coordinate systems is used frequently in numerous areas of science and engineering.

Exponential functions

Those calculators that have trigonometric functions likely will also have available a number of exponential functions. These include both the common and natural logarithms of a number, the base of natural logs, ϵ, raised to a power, 10 raised to a power, and a number raised to a power. For convenience, a number of calculators have the capability of solving directly for both the square root and square of numbers. While convenient, these latter functions are not as necessary as the more general exponential functions.

Conversion functions

With emphasis nationally now being placed on conversion to the SI metric system, a number of calculators have several preprogrammed conversion keys. For example, by depressing one or two keys, it is possible to convert between degrees Fahrenheit and degrees Celsius, inches and centimeters, gallons and liters, etc. For some fields, the ability to easily convert from one unit system to another is an advantage. Several calculators have the conversion between hours–minutes–seconds and hours–decimal parts of hours. If one is working in fields such as astronomy, this feature can be convenient. The ability to add or subtract times or angles expressed in hours–minutes–seconds is included as a special feature on several calculators. However, as was the case with the trigonometric unit conversions above, conversion factors can be obtained from a number of sources, so the availability of these functions should not normally be a deciding factor in selecting a calculator.

Hyperbolic functions

These special functions are used for certain applications by several different fields, such as electrical engineering and electronics. If the calculator does not have these functions, it is a rather simple procedure to calculate the hyperbolic functions using standard keys. As is the case with a number of other special functions, one should not consider these keys as essential unless the calculator is to be used often in an application where these functions are necessary. If you are in doubt, ask a teacher in science or engineering, or call someone working in your selected field.

Alternate-function keys

Because a number of calculators have more function capacity than they have keys available on the keyboard, it is necessary for several of the keys to serve multiple purposes. This is accomplished using *alternate-function keys*. The *alternate-function* (designated by a separate color or letter designation such as F or AF) key is pressed before the desired function key is pressed. This informs the calculator that when a regular key is depressed it is to be regarded in a different manner. For example, a number of calculator models have a SIN key and another key to be used for the arcsin (most frequently the same key). If one wants to find the sin of an angle, the SIN key is depressed. If he wants the arcsin, he first presses an alternate-function key (specially designated by the symbol AF, F, ARC, INV, a color, or by some other symbol), then the SIN key. The arcsin of the value will be displayed. Several calculator models will have two or three alternate-function keys. A little practice will permit the user to gain speed in using alternate-function keys. The time lost in pressing more than one key in a calculation is more than offset by the convenience of having the added functions available.

Statistical keys

Functions for statistical calculations are included in a number of calculators. For instance, a *per cent* key and a *per cent change* key are usually included. Like a number of the functions discussed above, these are to be regarded as a convenience and not a necessity. Two other useful keys are the *mean* and *standard deviation* keys. These functions permit the characteristics of a series of numbers to be examined (i.e., the mean value and standard deviation). Their presence is useful for some applications but again they are secondary in importance to the more general functions discussed earlier. The *linear regression* function combines a number of coordinate points that may represent a crude straight line and gives the slope (or angle of inclination relative to the horizontal) and the point of interception with the vertical axis. This function is important for analyzing data from some types of experiments.

Decimal-point placement

Considering the rules for handling significant figures (page 166), most calculators give far more accuracy for an answer than is really needed. A desirable feature for calculators is the ability to designate a specific number of decimal places to be displayed. The FIX, or FIX PT key (or some other key with equivalent notation) is used to blank out the digits in the display that appear after a designated number of significant figures. Generally this cutting off of the display in no way affects the calculations recorded inside the calculator logic since the operations will always be executed to the limit of accuracy of the unit. If the operator wishes to obtain a specific number accuracy, he needs only to limit the desired number of significant figures by first depressing the FIX key and then the number key designating the desired number of significant figures. The readout will adjust and automatically round off the display to the new length.

Other keys

Seemingly with each new calculator that appears on the market new functions have been added that make the calculator more desirable for certain users. The design of the new calculators is strongly dictated by the potential market, so companies making calculators will continue to include those functions that will appeal to the customer. As more and more functions are added, the complexity of the keyboard increases. This increased complexity may not always be an advantage to the user, however.

Examples of other keys will be mentioned here but will not be explained in detail, since they are special keys whose application may be somewhat limited. Such keys are a *factorial* key, "LAST X," *random number, permutations,* and *summation plus and minus.*

Usually a key is included for the constant π. Generally *clear display* and *clear everything* keys are standard. *Change sign* and *enter exponent* keys are also found on most calculators. The Hewlett-Packard calculators have a *stack roll-down* key as well as an *x–y interchange* key. Other companies have their own set of individualized keys.

Memory

A handy feature of a number of calculators is their array of additional storage reservoirs of "registers" to retain constants designated by the user. In a particular operation an operator may wish to store an answer or some other number. For some calculators the M key is utilized. Others use a STO–RCL pair of keys. STO refers to storage of the displayed number in a memory spot, and RCL refers to the ability to recall that stored number. The tendency has been to increase the number of available memories offered. Originally, there was one, but later the number went to three, eight, nine, and now even twenty. Some of these storage registers might be used for calculator functions such as summation plus and minus or mean-standard deviation functions, but there are usually an adequate number available for the operator to store whatever he wishes. A handy feature is the ability to perform operations on the numbers stored in the registers. For this purpose a set of keys might be found such as $\boxed{M+}$, $\boxed{M-}$, and $\boxed{M\times}$, $\boxed{M\div}$, or some equivalent operational set. It is convenient to have a number of registers available since some calculations are rather involved. To interrupt a calculation in order to write down a number is both annoying and time consuming. Such a procedure can also introduce errors since the operator can miscopy or miskey the number out or into the calculator.

Programmability

With the introduction of the programmable calculator, a big leap forward has been made in calculator capability. Now the calculator is able to operate in a number of respects as if it were a pocket computer. Although the programmable calculator has a long way to go before it reaches the capability of modern computers, its usefulness is great in the scientific and business communities. A computer simply uses data fed into the program, automatically executes a series of steps, and displays or prints the answer. To a much lesser degree, the programmable calculator can accomplish the same result. It can perform a series of steps automatically, make elementary-type decisions and perform tests, branch to other parts of a miniprogram, and provide an answer. The implication and application of such capability should not be overlooked.

However, as with the computer, one must be careful to use the programmable calculator with some measure of common sense. There are a number of programs written for the computer that could most economically be solved by hand calculator because the time required to access the computer, write the program, and obtain an answer is longer than the time that would be spent by the operator using a hand calculator. Therefore, the proper instrument should be used for each job. Programmable hand calculators fill a gap that has existed between the hand calculator and the computer.

The programmable hand calculator is most useful to solve the type of problem for which a number of the same kind of keystroke sequences will be used repeatedly. In this case, the operator puts the calculator into program mode, keys in the sequence of keystrokes that he would normally use in solving the problem, changes back into run mode, and then runs the program. The data can be fed into the program at various times by having the calculator stop at predetermined points and wait for the operator to feed in pertinent values. After the data are input, the user presses a *proceed* key, and the program continues executing the programmed instructions. Another method of data entry is to first store the data in one or more storage registers, and then instruct the calculator to recall the data from the appropriate register as needed. The answers can be either displayed or stored in other registers for later retrieval.

Illustration 10-6
Where algebraic entry and programmability are desirable, the SR-52 is a popular model.

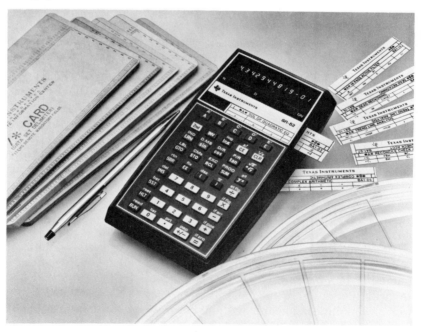

An example of an appropriate problem for a programmable calculator is the following. Suppose you had been asked to plot the parabola $y = 2x^2 + 3x + 4$. Normally the way one plots a curve is to calculate a number of data pairs. A value for x is picked and the resulting value for y is calculated. If a programmable calculator is available, the work of plotting is much simpler. On one such calculator, the mode switch is moved to "PRGM," which means the calculator is ready to accept the program. This program is nothing more than a sequence of keystrokes that will solve for y given a value for x. Using an RPN programmable calculator, the sequence entered into the program memory is

$$\boxed{\text{ENTER}}\ \boxed{g}\ \boxed{x^2}\ \boxed{2}\ \boxed{x}\ \boxed{x \geqslant y}\ \boxed{3}\ \boxed{x}\ \boxed{+}\ \boxed{4}\ \boxed{+}$$

Then the calculator is switched to "RUN" mode and the program pointer is set to 0, which means that the calculator is ready to execute the program from the beginning. A value for x is keyed in, and the R/S or RUN/STOP key is pressed. In a few seconds the calculator calculates the value for y. In this manner, the operator needs only key in a value for x and press the R/S key. In a short time enough data pairs can be obtained to plot a smooth curve.

Calculator future

It would take a "crystal ball" to be able to predict calculator capability in the future. The progress of the last several years indicates that this field is expanding rapidly. By virtue of price reduction or expanded features, the models that are available and most desirable today may be obsolete tomorrow.

Programmable calculators will likely become more common as versatility increases and price decreases. More specialized calculators also may be available. Tape cassette units may become common attachments to calculators for increased program length.

Which calculator should you buy? As with automobiles the particular brand is a personal preference, and the most desirable features are determined by the field you plan to enter. Remember, however, that the cost, although seemingly expensive at first, will be amortized over a number of years of usage. Thus the cost per year is quite low.

Regardless of the calculator that is purchased, it is necessary to be able to use it with both speed and accuracy. Problem answers are only as accurate as the operator using the unit. As marvelous as these calculators are, they still cannot perform miracles in correcting operator errors. The adage of "garbage in–garbage out" for computers is also true for calculators.

It is wise to practice with your new calculator. When you consistently get the correct answer, you can feel reasonably confident that you can use it accurately in problem solving. For practice, a number of problems are included below. In each case, the instructions are simple. Solve as many as necessary until the answers obtained are *consistently correct*. Then, make certain that you do not get sloppy with technique or become careless with your fingers.

As with the slide rule, the majority of problems solved with the hand calculator will involve multiplication and division. For this reason a substantial number of this type of practice problems is included.

Multiplication practice problems

10-1. $(23.8)(31.6) = (7.5208)(10)^2$

10-2. $(105.6)(4.09) = (4.3190)(10)^2$

10-3. $(286,000)(0.311) = (8.8946)(10)^4$

10-4. $(0.0886)(196.2) = (1.7383)(10)^1$

10-5. $(0.769)(47.2) = (3.6297)(10)^1$

10-6. $(60.7)(17.44) = (1.0386)(10)^3$

10-7. $(9.16)(115.7) = (1.0598)(10)^3$

10-8. $(592.)(80.1) = (4.7419)(10)^4$

10-9. $(7.69 \times 10^3)(0.722 \times 10^{-6}) = (5.5522)(10)^{-3}$

10-10. $(37.5 \times 10^{-1})(0.0974 \times 10^{-3}) = (3.6525)(10)^{-4}$

10-11. $(23.9)(0.715)(106.2) = (1.8148)(10)^3$

10-12. $(60.7)(1059)(237,000) = (1.5235)(10)^{10}$

10-13. $(988)(8180)(0.206) = (1.6649)(10)^6$

10-14. $(11.14)(0.0556)(76.3 \times 10^{-6}) = (4.7259)(10)^{-5}$

10-15. $(72.1)(\pi)(66.1) = (1.4972)(10)^4$

10-16. $(0.0519)(16.21)(1.085) = (9.1281)(10)^{-1}$

10-17. $(0.001093)(27.6)(56,700) = (1.7105)(10)^3$

10-18. $(0.379)(0.00507)(0.414) = (7.9551)(10)^{-4}$

10-19. $(16.05)(23.9)(0.821) = (3.1493)(10)^2$

10-20. $(1009)(0.226)(774) = (1.7650)(10)^5$

10-21. $(316)(825)(67,600) = (1.7623)(10)^{10}$

10-22. $(21,000)(0.822)(16.92) = (2.9207)(10)^5$

10-23. $(0.707)(80.6)(0.451) = (2.5700)(10)^1$

10-24. $(1.555 \times 10^3)(27.9 \times 10^5)(0.902 \times 10^{-7}) = (3.9133)(10)^2$

10-25. $(0.729)(10)^3(22,500)(33.2) = (5.4456)(10)^8$

10-26. $(18.97)(0.216)(899)(\pi)(91.2) = (1.0554)(10)^6$

10-27. $(7160)(0.000333)(26)(19.6)(5.01) = (6.0873)(10)^3$

10-28. $(1.712)(89,400)(19.5)(10^{-5})(82.1) = (2.4503)(10)^3$

10-29. $(62.7)(0.537)(0.1137)(0.806)(15.09) = (4.6561)(10)^1$

10-30. $(10)^6(159.2)(144)(7,920,000)(\pi) = (5.7040)(10)^{17}$

10-31. $(0.0771)(19.66)(219)(0.993)(7.05) = (2.3239)(10)^3$

10-32. $(15.06)(\pi)(625)(0.0963)(43.4) = (1.2359)(10)^5$

10-33. $(2160)(1802)(\pi)(292)(0.0443) = (1.5818)(10)^8$

10-34. $(437)(1.075)(0.881)(43,300)(17.22) = (3.0859)(10)^8$

10-35. $(\pi)(91.6)(555)(0.673)(0.00315)(27.7) = (9.3787)(10)^3$

10-36. $(18.01)(22.3)(1.066)(19.36)(10)^{-5} = (8.2886)(10)^{-2}$

10-37. $(84.2)(15.62)(921)(0.662)(0.1509) = (1.2100)(10)^5$

10-38. $(66,000)(25.9)(10.62)(28.4)(77.6) = (4.0008)(10)^{10}$

10-39. $(55.1)(7.33 \times 10^{-8})(76.3)(10)^5(0.00905) = (2.7889)(10)^{-1}$

10-40. $(18.91)(0.257)(0.0811)(92,500)(\pi) = (1.1453)(10)^5$

10-41. $(264)(564)(522) = (7.7724)(10)^7$

10-42. $(387)(7.32)(176) = (4.9858)(10)^5$

10-43. $(0.461)(4.79)(1140) = (2.5173)(10)^3$

10-44. $(6.69)(1548)(92,000) = (9.5276)(10)^8$

10-45. $(561)(3.30)(1.94) = (3.5915)(10)^3$

10-46. $(1456)(0.351)(0.835) = (4.2673)(10)^2$

10-47. $(1262)(0.405)(65,100) = (3.3273)(10)^7$

10-48. $(0.1871)(5.04)(53,000) = (4.9978)(10)^4$

10-49. $(7.28 \times 10^{-5})(4.16)(14.10) = (4.2702)(10)^{-3}$

10-50. $(10.70)(19,400)(0.0914) = (1.8973)(10)^4$

10-51. $(4.56)(47.4)(87.1) = (1.8826)(10)^2$

10-52. $(0.510)(68.9)(3.370) = (1.1842)(10)^2$

10-53. $(2,030)(14.72)(129.7) = (3.8756)(10)^6$

10-54. $(1824)(29.1)(21,800) = \mathbf{(1.1571)(10)^9}$
10-55. $(0.0255)(0.0932)(0.867) = \mathbf{(2.0605)(10)^{-3}}$
10-56. $(93.6)(3.99)(5,680) = \mathbf{(2.1213)(10)^6}$
10-57. $(4.48)(103.5)(0.198) = \mathbf{(9.1809)(10)^1}$
10-58. $(0.580)(43,700)(40.3) = \mathbf{(1.0214)(10)^6}$
10-59. $(7.05)(62.0)(34.9) = \mathbf{(1.5255)(10)^4}$
10-60. $(74.8)(8.)(483,000) = \mathbf{(2.8903)(10)^8}$

Multiplication problems

10-61. $(46.8)(11.97)$
10-62. $(479.)(11.07)$
10-63. $(9.35)(77.8)$
10-64. $(10.09)(843,000.)$
10-65. $(77,900)(0.467)$
10-66. $(123.9)(0.00556)$
10-67. $(214.9)(66.06)$
10-68. $(112.2)(0.953)$
10-69. $(87.0)(1.006)$
10-70. $(1,097,000)(1.984)$
10-71. $(43.8)(0.000779)$
10-72. $(10.68)(21.87)$
10-73. $(88,900.)(54.7)$
10-74. $(113,900.)(48.1)$
10-75. $(95,500.)(0.000479)$

10-76. $(0.0956)(147.2)(0.0778)$
10-77. $(15.47)(82.5)(975,000.)$
10-78. $(37.8)(22,490,000.)(0.15)$
10-79. $(1.048)(0.753)(0.933)$
10-80. $(31.05)(134.9)$
10-81. $(117.9)(98.9)$
10-82. $(55.6)(68.1)$
10-83. $(1.055)(85.3)$
10-84. $(33,050.)(16,900.)$
10-85. $(6.089)(44.87)$
10-86. $(34.8)(89.7)$
10-87. $(43,900.)(19.07)$
10-88. $(41.3)(87.9)$
10-89. $(99.7)(434,000.)$
10-90. $(0.0969)(0.1034)(0.1111)(0.1066)$

10-91. $(1.084 \times 10^{-5})(0.1758 \times 10^{13})(66.4)(0.901)$
10-92. $(234.5)(10)^4(21.21)(0.874)(0.0100)$
10-93. $(\pi)(26.88)(0.1682)(0.1463)(45.2)(1.007)$
10-94. $(75.8)(0.1044 \times 10^8)(10)^{-2}(54,000)(0.769)$
10-95. $(34.5)(31.09)(10)^{-6}(54.7)(0.677)(0.1003)$
10-96. $(6.08)(5.77)(46.8)(89.9)(3.02)(0.443)(\pi)$
10-97. $(1.055)(6.91)(31.9)(11.21)(\pi)(35.9)(4.09)$
10-98. $(1.856)(10)^3(21.98)$
10-99. $(57.7)(46.8)(3.08)$
10-100. $(0.045)(0.512)(115.4)$
10-101. $(0.307)(46.3)(7.94)$
10-102. $(2.229)(86.05)(16,090.)(\pi)$
10-103. $(44,090.)(38.9)(667.)(55.9)$
10-104. $(568.)(46.07)(3.41)(67.9)$
10-105. $(75.88)(0.0743)(0.1185)(0.429)$
10-106. $(10)^{-7}(69.8)(11.03)(0.901)$
10-107. $(46.3)(0.865)(10)^{-9}(0.953)(\pi)$
10-108. $(665.)(35,090)(0.1196)(0.469)$
10-109. $(888.)(35.9)(77.9)(0.652)$
10-110. $(43.4)(0.898)(70.09)(0.113)(\pi)$

Division practice problems

10-111. $(1.532) \div (72.6) = \mathbf{(2.1102)(10)^{-2}}$
10-112. $(0.1153) \div (70.3) = \mathbf{(1.6401)(10)^{-3}}$
10-113. $(89.3) \div (115.6) = \mathbf{(7.7249)(10)^{-1}}$
10-114. $(0.1052) \div (33.6) = \mathbf{(3.1310)(10)^{-3}}$
10-115. $(40.2) \div (50.8) = \mathbf{(7.9134)(10)^{-1}}$
10-116. $(0.661) \div (70,500) = \mathbf{(9.3759)(10)^{-6}}$

10-117. $(182.9) \div (0.00552) = (3.3134)(10)^4$

10-118. $(0.714) \div (98,200) = (7.2709)(10)^{-6}$

10-119. $(4.36) \div (80,300) = (5.4296)(10)^{-5}$

10-120. $(1.339) \div (22.6 \times 10^4) = (5.9248)(10)^{-6}$

10-121. $(17.03) \div (76.3) = (2.2320)(10)^{-1}$

10-122. $(0.511) \div (0.281) = (1.8185)(10)^0$

10-123. $(67.7) \div (91,300) = (7.4151)(10)^{-4}$

10-124. $(5.04) \div (29,800) = (1.6913)(10)^{-4}$

10-125. $(18.35) \div (0.921) = (1.9924)(10)^1$

10-126. $(29.6 \times 10^5) \div (0.905) = (3.2707)(10)^6$

10-127. $(0.1037) \div (92.5 \times 10^5) = (1.1211)(10)^{-8}$

10-128. $(537) \div (15.63 \times 10^{-7}) = (3.4357)(10)^8$

10-129. $(26,300) \div (84.3 \times 10^5) = (3.1198)(10)^{-3}$

10-130. $(6.370) \div (0.733) = (8.6903)(10)^0$

10-131. $(1.066) \div (7.51 \times 10^3) = (1.4194)(10)^{-4}$

10-132. $(29.6 \times 10^4) \div (0.973) = (3.0421)(10)^5$

10-133. $(0.912) \div (10.31 \times 10^5) = (8.8458)(10)^3$

10-134. $(17.37 \times 10^{-4}) \div (0.662) = (2.6239)(10)^{-3}$

10-135. $(0.693 \times 10^5) \div (1.008 \times 10^{-6}) = (6.8750)(10)^{10}$

10-136. $(89.1 \times 10^3) \div (189.3 \times 10^4) = (4.7068)(10)^{-2}$

10-137. $(0.617) \div (29,600) = (2.0845)(10)^{-5}$

10-138. $(18.06 \times 10^7) \div (15.29) = (1.1812)(10)^7$

10-139. $(56.8)(10)^4 \div (29.6)(10)^{-3} = (1.9189)(10)^7$

10-140. $(183,600) \div (76.3 \times 10^{-3}) = (2.4063)(10)^6$

10-141. $(75.9) \div (0.000813) = (9.3358)(10)^4$

10-142. $(43.6) \div (0.0837) = (5.2091)(10)^2$

10-143. $(156.8 \times 10^3) \div (0.715) = (2.1930)(10)^5$

10-144. $(216 \times 10^{-3}) \div (1557) = (1.3873)(10)^{-4}$

10-145. $(88.3 \times 10^{-1}) \div (29.1 \times 10^{-4}) = (3.0344)(10)^3$

10-146. $(1.034 \times 10^3) \div (0.706 \times 10^{-8}) = (1.4646)(10)^{11}$

10-147. $(55.2)(10)^3 \div (0.1556 \times 10^3) = (3.5476)(10)^2$

10-148. $(0.01339) \div (1896 \times 10^5) = (7.0622)(10)^{-11}$

10-149. $(4030 \times 10^{-7}) \div (75.3 \times 10^{-9}) = (5.3519)(10)^3$

Problems in division

10-150. $\dfrac{89.9}{45.}$

10-151. $\dfrac{147.}{22.}$

10-152. $\dfrac{9.06}{7.1}$

10-153. $\dfrac{1,985.}{78.55}$

10-154. $\dfrac{19,230.}{64.88}$

10-155. $\dfrac{87,600.}{43.8}$

10-156. $\dfrac{54.8}{9.10}$

10-157. $\dfrac{0.877}{33.07}$

10-158. $\dfrac{11.44}{24.9}$

10-159. $\dfrac{187,900.}{71.45}$

10-160. $\dfrac{0.00882}{87.04}$

10-161. $\dfrac{0.675}{54.8}$

10-162. $\dfrac{87.9}{45.7}$

10-163. $\dfrac{164,800.}{3.88}$

10-164. $\dfrac{7.09 \times 10^3}{18.45}$

10-165. $\dfrac{0.001755}{6.175}$

10-166. $\dfrac{0.0000559}{0.00659}$

10-167. $\dfrac{5.065}{0.0003375}$

10-168. $\dfrac{469,000}{793}$

10-169. $\dfrac{5,100,000}{933 \times 10^5}$

10-170. $\dfrac{3765 \times 10^3}{760.3}$

10-171. $\dfrac{4917}{0.391}$

10-172. $\dfrac{5516}{1.65}$

10-173. $\dfrac{0.0916}{0.331}$

10-174. $\dfrac{193.7}{5.06}$ 10-181. $\dfrac{6607}{1.91 \times 10^5}$ 10-188. $\dfrac{0.000497}{38.9 \times 10^{-5}}$

10-175. $\dfrac{113.05}{72.35}$ 10-182. $\dfrac{1.993 \times 10^{-8}}{72.31 \times 10^{-6}}$ 10-189. $\dfrac{48.6 \times 10^{-9}}{1.977 \times 10^5}$

10-176. $\dfrac{32.33}{46.77}$ 10-183. $\dfrac{461 \times 10^3}{0.003617}$ 10-190. $\dfrac{69,990. \times 10^{18}}{43.9 \times 10^{-2}}$

10-177. $\dfrac{3.17}{3.1416}$ 10-184. $\dfrac{9903 \times 10^{-5}}{47.31 \times 10^3}$ 10-191. $\dfrac{5.06 \times 10^{-7}}{0.001853 \times 10^9}$

10-178. $\dfrac{0.221}{56.91}$ 10-185. $\dfrac{0.711}{11,980.}$ 10-192. $\dfrac{1.097 \times 10^{-6}}{458. \times 10^{-1}}$

10-179. $\dfrac{233.17}{5506}$ 10-186. $\dfrac{0.01253}{66.8}$ 10-193. $\dfrac{89.99 \times 10^{-3}}{40.7 \times 10^{-6}}$

10-180. $\dfrac{72.13}{52.03}$ 10-187. $\dfrac{0.974}{1.058}$

Combined multiplication and division practice problems

10-194. $\dfrac{(29.6)(18.01)}{937} = (5.6894)(10)^{-1}$

10-195. $\dfrac{(625,000)(0.0337)}{48.2} = (4.3698)(10)^2$

10-196. $\dfrac{(0.887)(1,109)}{5.22} = (1.8845)(10)^2$

10-197. $\dfrac{(0.1058)(937,000)}{0.218} = (4.5475)(10)^5$

10-198. $\dfrac{(43,800)(0.0661)}{87.2 \times 10^5} = (3.3202)(10)^{-4}$

10-199. $\dfrac{(114.3)(0.567)}{66,400} = (9.7603)(10)^{-4}$

10-200. $\dfrac{76.5 \times 10^4}{(0.733)(49.7 \times 10^{-6})} = (2.0999)(10)^{10}$

10-201. $\dfrac{11.03}{(20,100)(8.72 \times 10^3)} = (6.2931)(10)^{-8}$

10-202. $\dfrac{0.226}{(87.3 \times 10^4)(0.717)} = (3.6106)(10)^{-7}$

10-203. $\dfrac{43.2}{(9.09)(0.000652)} = (7.2891)(10)^3$

10-204. $\dfrac{94.9 \times 10^{-9}}{(33,800)(0.609)} = (4.6103)(10)^{-12}$

10-205. $\dfrac{737,000}{(0.1556)(61.9 \times 10^3)} = (7.6519)(10)^1$

10-206. $\dfrac{(17.01)(0.0336)}{(52,600)(0.01061)} = (1.0241)(10)^{-3}$

10-207. $\dfrac{(66.6)(0.937)}{(7.05 \times 10^2)(184,300)} = (4.8029)(10)^{-7}$

10-208. $\dfrac{(2.96)(1000)(62.1)}{(0.911)(432,000)} = (4.6707)(10)^{-1}$

10-209. $\dfrac{(45.8)(10.33)}{(29,200)(0.702)} = \mathbf{(2.3081)(10)^{-2}}$

10-210. $\dfrac{(0.604)(9,270)}{(0.817 \times 10^4)(1.372)} = \mathbf{(4.9951)(10)^{-1}}$

10-211. $\dfrac{(176,300)(42.8 \times 10^3)}{(68.3)(15.01)} = \mathbf{(7.3603)(10)^6}$

10-212. $\dfrac{(39,200)(89.3 \times 10^{-7})}{(20.4 \times 10^{-6})(155.5)} = \mathbf{(1.1035)(10)^2}$

10-213. $\dfrac{(0.763 \times 10^{-4})(0.01004)}{(44.3)(7,150,000)} = \mathbf{(2.4185)(10)^{-15}}$

10-214. $\dfrac{(152,300)(88,100)}{(0.00339)(60.4)} = \mathbf{(6.5530)(10)^{10}}$

10-215. $\dfrac{(90,400)(2.05 \times 10^6)}{(24.3 \times 10^{-2})(0.0227)} = \mathbf{(3.3596)(10)^{13}}$

10-216. $\dfrac{(14.36 \times 10^2)(0.907)}{(51.6 \times 10^2)(0.00001118)} = \mathbf{(2.2577)(10)^4}$

10-217. $\dfrac{(991,000)(60.3 \times 10^4)}{(23.3 \times 10^{-1})(0.1996)} = \mathbf{(1.2849)(10)^{12}}$

10-218. $\dfrac{(8.40)(10)^3(29.6 \times 10^{-5})}{(0.369)(10.02 \times 10^9)} = \mathbf{(6.7248)(10)^{-10}}$

10-219. $\dfrac{(54.9)(26.8)(0.331)}{(21.6)(11.03)(54.6)} = \mathbf{(3.7438)(10)^{-2}}$

10-220. $\dfrac{(17,630)(0.1775)(92.3)}{(0.433)(0.0061)(57.3)} = \mathbf{(1.9084)(10)^6}$

10-221. $\dfrac{(0.821)(0.221)(0.811)}{(0.0907)(10.72)(66,300)} = \mathbf{(2.2827)(10)^{-6}}$

10-222. $\dfrac{(0.00552)(89.6)(0.705)}{(19.52 \times 10^3)(18.03)(22.4)} = \mathbf{(4.4230)(10)^{-8}}$

10-223. $\dfrac{(30,600)(29.9)(0.00777)}{(485)(19.32)(62.6)} = \mathbf{(1.2120)(10)^{-2}}$

10-224. $\dfrac{(54.1)(0.393)(16,070)}{(49.3 \times 10^3)(11.21)(61.6)} = \mathbf{(1.0036)(10)^{-2}}$

10-225. $\dfrac{(44.2)(100.7)(62,400)}{(90.3)(75,100)(0.01066)} = \mathbf{(3.8419)(10)^3}$

10-226. $\dfrac{(78.4)(15.59)(0.01669)}{(33.6)(88,100)(0.432)} = \mathbf{(1.5952)(10)^{-5}}$

10-227. $\dfrac{(994,000)(21,300)(0.1761)}{(44.4)(71.2)(32.1 \times 10^4)} = \mathbf{(3.6742)(10)^0}$

10-228. $\dfrac{(16.21)(678,000)(56.6)}{(0.01073)(4980)(30.3)} = \mathbf{(3.8420)(10)^5}$

10-229. $\dfrac{(61.3 \times 10^3)(0.1718)(0.893)}{(21.6)(0.902)(0.01155)} = \mathbf{(4.1792)(10)^4}$

10-230. $\dfrac{(20,900)(16.22 \times 10^4)(0.1061)}{(877)(20.1 \times 10^{-4})(5.03)} = \mathbf{(4.0565)(10)^7}$

10-231. $\dfrac{(999,000)(17.33)(0.1562)}{(0.802)(0.0443)(29.3 \times 10^{-1})} = \mathbf{(2.5978)(10)^7}$

10-232. $\dfrac{(16.21)(0.0339)(151.6)(0.211)}{(0.00361)(0.785)(93.2)(406)} = (1.6393)(10)^{-1}$

10-233. $\dfrac{(84.3)(0.916)(0.1133)(21.3)}{(66.2)(0.407)(55.3)(462)} = (2.7072)(10)^{-4}$

Problems

10-234. $\dfrac{0.916}{(90.5)(13.06)}$

10-235. $\dfrac{0.00908}{(22.3)(33.2)}$

10-236. $\dfrac{(24.5)(43)}{36}$

10-237. $\dfrac{(82)(9.3)}{56.5}$

10-238. $\dfrac{(167)(842)}{0.976}$

10-239. $\dfrac{(5.72)(3690)}{95.7}$

10-240. $\dfrac{(925)(76.9)}{37.6}$

10-241. $\dfrac{9.87}{(1.76)(89)}$

10-242. $\dfrac{85.4}{(26.3)(213)}$

10-243. $\dfrac{1525}{(73.6)(0.007)}$

10-244. $\dfrac{84,500}{(126)(37.3)}$

10-245. $\dfrac{(76)(23.7)}{(13.5)(373)}$

10-246. $\dfrac{(6.23)(2.14)}{0.00531}$

10-247. $\dfrac{(21.3)(370)}{(10.9)(758)}$

10-248. $\dfrac{(0.00215)(2520)}{(7.57)(118)}$

10-249. $\dfrac{(755)(1.15)}{(51.4)(0.093)}$

10-250. $\dfrac{(916)(0.752)}{5.16}$

10-251. $\dfrac{(23.1)(1.506)}{6.27}$

10-252. $\dfrac{(42.6)(1.935)}{750.3}$

10-253. $\dfrac{(77.1)(10.53)}{(331.0)(73)}$

10-254. $\dfrac{(56.7)(0.00336)}{(15.06)(8.23)}$

10-255. $\dfrac{(14.5)(10)^3(6.22)}{(53.3)(0.00103)}$

10-256. $\dfrac{(42)(1000)}{(5.23)(0.00771)}$

10-257. $\dfrac{1.331}{(916)(506)}$

10-258. $\dfrac{(4320)(0.7854)}{(134)(0.9)}$

10-259. $\dfrac{(0.00713)(329)}{(0.0105)(1000)}$

10-260. $\dfrac{(103.4)(0.028)}{0.0798}$

10-261. $\dfrac{(1573)(4618)}{(3935)(97)}$

10-262. $\dfrac{(47.2)(0.0973)}{(85)(37.6)}$

10-263. $\dfrac{(0.0445)(0.0972)}{(0.218)(0.318)}$

10-264. $\dfrac{(39.1)(680,000)(3.52)(1.1 \times 10^6)}{(0.0316)(9.6 \times 10^6)(26.3)}$

10-265. $\dfrac{(7.69)(76,000)(5.63)(0.00314)}{(0.00365)(10 \times 10^6)}$

10-266. $\dfrac{(3.97)(6.71 \times 10^{-3})(0.067)}{(63.1)(3 \times 10^7)(7.61)(80,175)}$

10-267. $\dfrac{(697)(0.000713)(68.1)}{(234)(9.68)(5.1 \times 10^4)}$

10-268. $\dfrac{(43,400)(9.16)(8.1 \times 10^{-6})}{(0.00613)(67,000)(0.416)}$

10-269. $\dfrac{(691.6)(7.191)(3 \times 10^7)}{(410,000)(6.39)(0.0876)}$

10-270. $\dfrac{(37.615)(81.4)(9.687)(0.0017)}{(13.13)(0.076)(43)}$

10-271. $\dfrac{(51.2 \times 10^{-6})(3.41 \times 10^5)(36.1)}{(96.69)(7 \times 10^{-2})(0.134}$

10-272. $\dfrac{(6.716)(3.2 \times 10^3)(0.0173)(413)}{(0.0000787)(6.6 \times 10^4)}$

10-273. $\dfrac{(1.061 \times 10^{-1})(96,000)(3.717)}{(7.34 \times 10^{-6})(3.9 \times 10^4)(13.5)}$

10-274. $\dfrac{(361)(482)(5.816)(38.91)(0.00616)}{(0.07181)(3 \times 10^3)(39.36)}$

10-275. $\dfrac{(0.019 \times 10^8)(111.15)(0.0168)}{(7.96)(58.6)(0.0987)(3000)}$

10-276. $\dfrac{(21.4)(0.82)(39.6 \times 10^{-1})}{(10.86)(6.7 \times 10^{-2})(37,613)}$

10-277. $\dfrac{(63,761)(43,890)(0.00761)}{(8 \times 10^6)(0.0781)(67.17)}$

10-278. $\dfrac{(516.7)(212 \times 10^3)(0.967)(34)}{(76,516)(2 \times 10^{-6})(618)}$

10-279. $\dfrac{(5.1 \times 10^8)(370)(8.71)(3698)}{(0.00176)(36,170)}$

10-280. $\dfrac{(59.71 \times 10^{-6})(0.00916)(0.1695)(55.61)}{(17.33 \times 10^5)(0.3165)(10.56)(1.105)}$

Cubes and cube roots, or for that matter any other power or root, can be solved using the exponential operation key found on most scientific calculators. Since any root can be expressed as an exponential fraction (i.e., the cube root can be considered as the $\frac{1}{3}$ power), the $\boxed{y^x}$ key can handle any root within the range of capability of the calculator. If the exponent is negative, the calculator can still solve the problem, but an additional step may be needed.

Example $\qquad\qquad (3)^{-4.2} = ?$

Use the following keystroke sequence (using RPN logic):

$$\boxed{3}\ \boxed{\text{ENTER}}\ \boxed{4}\ \boxed{.}\ \boxed{2}\ \boxed{\text{CHS}}\ \boxed{y^x}$$

The answer is displayed as 0.00991 (to three significant figures). Since the negative exponent indicates that a reciprocal is involved, another procedure would be to first calculate $(3)^{4.2}$ and then press the reciprocal $\boxed{1/x}$ key. On calculators with algebraic logic, the keystroke sequence for this problem would be

$$\boxed{3}\ \boxed{y^x}\ \boxed{4}\ \boxed{.}\ \boxed{2}\ \boxed{+/-}\ \boxed{=}$$

In the case of RPN calculators, the $\boxed{\text{CHS}}$ key changes the sign of the number displayed, and on the algebraic calculators the $\boxed{+/-}$ key accomplishes the same result. If it is desired to find the root of a number such as

$$\sqrt[4.2]{3.5} = ?$$

the keystroke sequence would be (using RPN logic)

$$\boxed{3}\ \boxed{.}\ \boxed{5}\ \boxed{\text{ENTER}}\ \boxed{4}\ \boxed{.}\ \boxed{2}\ \boxed{1/x}\ \boxed{y^x}$$

For algebraic logic, it would be

$$\boxed{3}\ \boxed{.}\ \boxed{5}\ \boxed{y^x}\ \boxed{4}\ \boxed{.}\ \boxed{2}\ \boxed{1/x}\ \boxed{=}$$

In either case the answer to five significant figures is 1.3475.

Squares and square roots practice problems

10-281. $(408)^2 = (1.6646)(10)^5$
10-282. $(8.35)^2 = (6.9723)(10)^1$
10-283. $(3,980)^2 = (1.5840)(10)^7$
10-284. $(0.941)^2 = (8.8548)(10)^{-1}$
10-285. $(57.4)^2 = (3.2948)(10)^3$
10-286. $(0.207)^2 = (4.2849)(10)^{-2}$
10-287. $(784)^2 = (6.1466)(10)^5$
10-288. $(296,000)^2 = (8.7616)(10)^{10}$

10-289. $(1037)^2 = (1.0754)(10)^6$
10-290. $(8.93)^2 = (7.9745)(10)^1$
10-291. $(30.9)^2 = (9.5481)(10)^2$
10-292. $(43,300)^2 = (1.8749)(10)^9$
10-293. $(0.00609)^2 = (3.7088)(10)^{-5}$
10-294. $(0.846)^2 = (7.1572)(10)^{-1}$
10-295. $(55.2 \times 10^3)^2 = (3.0470)(10)^9$
10-296. $(0.0707)^2 = (4.9985)(10)^{-3}$

10-297. $(11.92 \times 10^{-4})^2 = (1.4209)(10)^{-6}$
10-298. $(0.291 \times 10^{-5})^2 = (8.4681)(10)^{-12}$
10-299. $(449,000)^2 = (2.0160)(10)^{11}$
10-300. $(0.000977)^2 = (9.5453)(10)^{-7}$
10-301. $(33.5 \times 10^{-6})^2 = (1.1223)(10)^{-9}$
10-302. $(8810)^2 = (7.7616)(10)^7$
10-303. $(50.9 \times 10^6)^2 = (2.5908)(10)^{15}$
10-304. $(99,300)^2 = (9.8605)(10)^9$
10-305. $(0.0714 \times 10^{-6})^2 = (5.0980)(10)^{-15}$
10-306. $\sqrt{96,100} = (3.1000)(10)^2$
10-307. $\sqrt{0.912} = (9.5499)(10)^{-1}$
10-308. $\sqrt{24.9} = (4.9900)(10)^0$
10-309. $\sqrt{0.01124} = (1.0602)(10)^{-1}$
10-310. $\sqrt{5256} = (7.2498)(10)^1$
10-311. $\sqrt{0.3764} = (6.1351)(10)^{-1}$
10-312. $\sqrt{43,800,000} = (6.6182)(10)^3$
10-313. $\sqrt{0.01369} = (1.1700)(10)^{-1}$
10-314. $\sqrt{73.6} = (8.5790)(10)^0$

10-315. $\sqrt{1.1025} = (1.0500)(10)^0$
10-316. $\sqrt{487,000} = (6.9785)(10)^2$
10-317. $\sqrt{580.8} = (2.4100)(10)^1$
10-318. $\sqrt{0.00002767} = (5.2602)(10)^{-3}$
10-319. $\sqrt{0.1399} = (3.7403)(10)^{-1}$
10-320. $\sqrt{6368} = (7.9800)(10)^1$
10-321. $\sqrt{1.142 \times 10^{-3}} = (3.3793)(10)^{-2}$
10-322. $\sqrt{6.496 \times 10^1} = (8.0598)(10)^0$
10-323. $\sqrt{190,970} = (4.3700)(10)^2$
10-324. $\sqrt{3,204,000} = (1.7900)(10)^3$
10-325. $\sqrt{0.003807} = (6.1701)(10)^{-2}$
10-326. $\sqrt{0.08352} = (2.8900)(10)^{-1}$
10-327. $\sqrt{3069} = (5.5399)(10)^1$
10-328. $\sqrt{61.78 \times 10^{-4}} = (7.8600)(10)^{-2}$
10-329. $\sqrt{3.648 \times 10^{-8}} = (1.9100)(10)^{-4}$
10-330. $\sqrt{9.92 \times 10^5} = (9.9599)(10)^2$

Problems

10-331. $(1468.)^2$
10-332. $(0.886)^2$
10-333. $(67.4)^2$
10-334. $(11.96)^2$
10-335. $(0.00448)^2$
10-336. $(0.000551)^2$
10-337. $(9.22)^2$
10-338. $(64,800.)^2$
10-339. $(0.0668)^2$
10-340. $(16.85)^2$
10-341. $(1.802 \times 10^9)^2$
10-342. $(0.00358)^2$
10-343. $(5089)^2$
10-344. $(44,900.)^2$
10-345. $(64.88)^2$
10-346. $\sqrt{11.81}$
10-347. $\sqrt{4567.}$
10-348. $\sqrt{0.01844}$
10-349. $\sqrt{0.9953}$
10-350. $\sqrt{1395.}$
10-351. $\sqrt{0.1148}$
10-352. $\sqrt{0.2776}$

10-353. $\sqrt{9.31}$
10-354. $\sqrt{73,800.}$
10-355. $\sqrt{13.38}$
10-356. $\sqrt{93.07}$
10-357. $\sqrt{0.0001288}$
10-358. $\sqrt{1.082 \times 10^2}$
10-359. $\sqrt{75.9}$
10-360. $\sqrt{\pi}$
10-361. $(0.774)^2(11.47)^{1/2}$
10-362. $(0.1442)^{1/2}(33.89)^{1/2}$
10-363. $(54.23)^2(88,900)^{1/2}$
10-364. $\sqrt{234.5}\,\sqrt{55,900.}$
10-365. $\sqrt{16.38}\,\sqrt{45.6}\,\sqrt{0.9}$
10-366. $\sqrt{415.}\,\sqrt{\pi}\,\sqrt{86.4}$
10-367. $\sqrt{15.66}\,\sqrt{0.1904}\,\sqrt{\pi}$
10-368. $(34.77)^2(54.8)^2(0.772)^{1/2}$
10-369. $\sqrt{7.90}\,\sqrt{7.02}\,\sqrt{11.54}$
10-370. $\sqrt{31.19}\,\sqrt{56.7}\,\sqrt{54.8}$

Cubes and cube roots practice problems

10-371. $(206)^3 = (8.7418)(10)^6$
10-372. $(7.68)^3 = (4.5298)(10)^2$
10-373. $(0.00519)^3 = (1.3980)(10)^{-7}$

10-374. $(33.5)^3 = (3.7595)(10)^4$
10-375. $(0.229)^3 = (1.2009)(10)^{-2}$
10-376. $(1090)^3 = (1.2950)(10)^9$

10-377. $(0.0579)^3 = (1.9410)(10)^{-4}$
10-378. $(9.89)^3 = (9.6736)(10)^2$
10-379. $(419)^3 = (7.3560)(10)^7$
10-380. $(52.4)^3 = (1.4388)(10)^5$
10-381. $(0.0249)^3 = (1.5438)(10)^{-5}$
10-382. $(14.9)^3 = (3.3079)(10)^3$
10-383. $(2.96)^3 = (2.5934)(10)^1$
10-384. $(397)^3 = (6.2571)(10)^7$
10-385. $(63.4)^3 = (2.5484)(10)^5$
10-386. $(9040)^3 = (7.3876)(10)^{11}$
10-387. $(0.0783)^3 = (4.8005)(10)^{-4}$
10-388. $(0.844)^3 = (6.0121)(10)^{-1}$
10-389. $(5.41)^3 = (1.5834)(10)^2$
10-390. $(35.5)^3 = (4.4739)(10)^4$
10-391. $(0.1270)^3 = (2.0484)(10)^{-3}$
10-392. $(20.7)^3 = (8.8697)(10)^3$
10-393. $(691)^3 = (3.2994)(10)^8$
10-394. $(0.719)^3 = (3.7169)(10)^{-1}$
10-395. $(4.34)^3 = (8.1747)(10)^1$

10-396. $\sqrt[3]{30,960,000} = (3.1400)(10)^2$
10-397. $\sqrt[3]{0.001728} = (1.2000)(10)^{-1}$
10-398. $\sqrt[3]{491} = (7.8891)(10)^0$
10-399. $\sqrt[3]{9.91 \times 10^{11}} = (9.9699)(10)^3$
10-400. $\sqrt[3]{0.272} = (6.4792)(10)^{-1}$

10-401. $\sqrt[3]{118,400} = (4.9104)(10)^1$
10-402. $\sqrt[3]{22.91} = (2.8402)(10)^0$
10-403. $\sqrt[3]{527,500} = (8.0799)(10)^1$
10-404. $\sqrt[3]{1.295} = (1.0900)(10)^0$
10-405. $\sqrt[3]{0.0001804} = (5.6504)(10)^{-2}$
10-406. $\sqrt[3]{460,100,000} = (7.7200)(10)^2$
10-407. $\sqrt[3]{261,000} = (6.3907)(10)^1$
10-408. $\sqrt[3]{0.11620} = (4.8798)(10)^{-1}$
10-409. $\sqrt[3]{0.0030486} = (1,4500)(10)^{-1}$
10-410. $\sqrt[3]{0.03096} = (3.1400)(10)^{-1}$
10-411. $\sqrt[3]{504.4} = (7.9602)(10)^0$
10-412. $\sqrt[3]{8,869,000} = (2.0699)(10)^2$
10-413. $\sqrt[3]{174,700,000} = (5.5902)(10)^2$
10-414. $\sqrt[3]{5.886 \times 10^{10}} = (3.8899)(10)^3$
10-415. $\sqrt[3]{5.885 \times 10^{-1}} = (8.3801)(10)^{-1}$
10-416. $\sqrt[3]{76.105 \times 10^{-5}} = (9.1300)(10)^{-2}$
10-417. $\sqrt[3]{327.1} = (6.8901)(10)^0$
10-418. $\sqrt[3]{0.02567} = (2.9499)(10)^{-1}$
10-419. $\sqrt[3]{0.0004118} = (7.4398)(10)^{-2}$
10-420. $\sqrt[3]{68,420} = (4.0900)(10)^1$

Problems

10-421. $(86)^3$
10-422. $(148)^3$
10-423. $(395,000)^3$
10-424. $(47.6)^3$
10-425. $(1.074)^3$
10-426. $(76.9)^3$
10-427. $(220.8)^3$
10-428. $(9.72)^3$
10-429. $(110.7)^3$
10-430. $(91.3)^3$
10-431. $(1.757 \times 10^4)^3$
10-432. $(3.06 \times 10^{-7})^3$
10-433. $(44.8 \times 10^{-1})^3$
10-434. $(0.933 \times 10^{-2})^3$
10-435. $(0.1184 \times 10^8)^3$
10-436. $(51.5 \times 10^2)^3$

10-437. $\sqrt[3]{118}$
10-438. $\sqrt[3]{2197}$
10-439. $\sqrt[3]{9}$
10-440. $\sqrt[4]{0.0689}$
10-441. $\sqrt[3]{0.001338}$
10-442. $\sqrt[3]{0.1794}$

10-443. $\sqrt[3]{0.0891}$
10-444. $\sqrt[3]{34,690.}$
10-445. $\sqrt[3]{0.3329}$
10-446. $\sqrt[3]{1,258,000}$
10-447. $\sqrt[3]{0.1853}$
10-448. $\sqrt[3]{12.88}$
10-449. $\sqrt[3]{4.98 \times 10^7}$
10-450. $\sqrt[3]{1.844 \times 10^{-5}}$
10-451. $\sqrt[3]{3.86 \times 10^{-1}}$

10-452. $(9.94)(0.886)^{1/3}$
10-453. $(284.)(11.98)^{1/3}$
10-454. $(0.117)(0.0964)^{1/3}$
10-455. $(\pi)^3(44.89)^3$
10-456. $(6.88)^3(0.00799)^3$
10-457. $(0.915)^{1/3}(0.366)^{1/3}\sqrt[3]{11,250}(36.12)^{1/3}$
10-458. $(2.34)^3(3.34)^3(4.56)^3(5.67)^3$
10-459. $(8.26)^{1/3}(8.26)^3(1000)^{1/3}(10)^3$
10-460. $\sqrt[3]{2670}\sqrt[3]{3165}\sqrt[3]{1065}\sqrt[3]{7776}$
10-461. $\sqrt[3]{206}\sqrt[3]{0.791}(12.35)^3(26.3)^3$

Powers of numbers: practice problems

10-462. $(53.2)^{0.84} = (2.8169)(10)^1$

10-463. $(4.65)^{3.68} = (2.8591)(10)^2$

10-464. $(0.836)^{0.47} = (9.1926)(10)^{-1}$

10-465. $(1.0042)^{217} = (2.4831)(10)^0$

10-466. $(0.427)^4 = (3.3244)(10)^{-2}$

10-467. $(0.3156)^4 = (9.9208)(10)^{-3}$

10-468. $(0.159)^{0.67} = (2.9170)(10)^{-1}$

10-469. $(1.0565)^{49.5} = (1.5189)(10)^1$

10-470. $(32.5)^{0.065} = (1.2539)(10)^0$

10-471. $(3.45)^{4.65} = (3.1685)(10)^2$

10-472. $(0.759)^5 = (2.5189)(10)^{-1}$

10-473. $(2.127)^4 = (2.0468)(10)^1$

10-474. $(2.03)^{-5} = (2.9008)(10)^{-2}$

10-475. $(4.00)^{0.0157} = (1.0220)(10)^0$

10-476. $(0.0818)^{-0.777} = (6.9950)(10)^0$

10-477. $(1.382)^{21.3} = (9.8361)(10)^2$

10-478. $(0.071)^{-0.46} = (3.3761)(10)^0$

10-479. $(0.232)^{0.0904} = (8.7627)(10)^{-1}$

10-480. $(2.718)^{0.405} = (1.4992)(10)^0$

10-481. $(0.916)^{0.724} = (9.3845)(10)^{-1}$

10-482. $(1.1106)^{1.72} = (1.1892)(10)^0$

10-483. $(59.2)^{-0.43} = (1.7294)(10)^{-1}$

10-484. $(883)^{0.964} = (6.9168)(10)^2$

10-485. $(7676)^{0.001102} = (1.0099)(10)^0$

10-486. $(4.30)^{0.521} = (2.1381)(10)^0$

Roots of numbers practice problems

10-487. $\sqrt[7.81]{5.85} = (1.2538)(10)^0$

10-488. $\sqrt[6]{0.0835} = (6.6112)(10)^{-1}$

10-489. $\sqrt[5]{0.0763} = (5.9773)(10)^{-1}$

10-490. $\sqrt[194]{460.} = (1.0321)(10)^0$

10-491. $\sqrt[6]{0.0001} = (2.1544)(10)^{-1}$

10-492. $\sqrt[1.65]{8.26} = (3.5954)(10)^0$

10-493. $\sqrt[0.34]{0.862} = (6.4612)(10)^{-1}$

10-494. $\sqrt[2.3]{85.9} = (6.9321)(10)^0$

10-495. $\sqrt[60]{45.} = (1.0655)(10)^0$

10-496. $\sqrt[21.5]{1.606} = (1.0223)(10)^0$

10-497. $\sqrt[1.91]{92.5} = (1.0700)(10)^1$

10-498. $\sqrt[50]{0.05} = (9.4184)(10)^{-1}$

10-499. $\sqrt[7]{0.0108} = (5.2367)(10)^{-1}$

10-500. $\sqrt[0.006]{0.9762} = (1.8050)(10)^{-2}$

10-501. $\sqrt[5.21]{2000} = (4.3031)(10)^0$

10-502. $\sqrt[0.04]{0.9792} = (5.9127)(10)^{-1}$

10-503. $\sqrt[2.7]{81} = (5.0915)(10)^0$

10-504. $\sqrt[2.81]{1.218} = (1.0727)(10)^0$

10-505. $\sqrt[2.15]{52.5} = (6.3107)(10)^0$

10-506. $\sqrt[400]{100} = (1.0116)(10)^0$

10-507. $\sqrt[0.75]{2.37} = (3.1598)(10)^0$

10-508. $\sqrt[0.073]{1.060} = (2.2215)(10)^0$

10-509. $\sqrt[1.51]{6.50} = (3.4542)(10)^0$

10-510. $\sqrt[5.6]{0.0018} = (3.2350)(10)^{-1}$

10-511. $\sqrt[0.67]{0.954} = (9.3213)(10)^{-1}$

Problems

10-512. $(2.89)^6$

10-513. $(4.11)^{5.2}$

10-514. $(19.01)^{1.6}$

10-515. $(1.185)^{2.7}$

10-516. $(1.033)^{5.8}$

10-517. $(1.0134)^{25}$

10-518. $(3.95)^{0.65}$

10-519. $(8.46)^{0.134}$

10-520. $(81.2)^{0.118}$

10-521. $(7850.)^{0.0775}$

10-522. $(1.399)^{0.883}$

10-523. $(10.06)^{0.0621}$

10-524. $(0.569)^4$

10-525. $(0.157)^8$

10-526. $(0.985)^{1.568}$

10-527. $(0.318)^{4.65}$

10-528. $(0.078)^{0.458}$

10-529. $(17.91)^{0.012}$

10-530. $(4780.)^{0.913}$

10-531. $(253.)^{0.269}$

10-532. $(0.428)^{0.559}$

10-533. $(4.08)^{24}$

10-534. $(3.91)^{20}$

10-535. $(8.45)^{16}$

10-536. $(7.77)^{42}$

10-537. $(16.89)^{1.402}$

10-538. $(87.8)^8$

10-539. $(0.1164)^{0.33}$

10-540. $(0.779)^{0.43}$

10-541. $(867.)^6$

10-542. $(91.05)^{14}$

10-543. $(0.775)^{0.0259}$

10-544. $\sqrt[6]{8.69}$

10-545. $\sqrt[3]{1.094}$

10-546. $\sqrt[1.3]{8.74}$

10-547. $\sqrt[0.6]{19.77}$

10-548. $\sqrt[18]{54.8}$

10-549. $\sqrt[7]{1.004}$

10-550. $\sqrt[1.95]{0.642}$

10-551. $\sqrt[14]{0.1438}$

10-552. $\sqrt[3.6]{0.952}$

10-553. $\sqrt[2.4]{0.469}$

10-554. $\sqrt[1.7]{0.1975}$

10-555. $\sqrt[0.55]{0.2218}$

10-556. $\sqrt[0.46]{16,430}$

10-557. $\sqrt[0.133]{507.}$

10-558. $\sqrt[0.57]{0.964}$

10-559. $\sqrt[5.09]{6.49}$

10-560. $\sqrt[13.6]{0.1574}$

10-561. $\sqrt[2.09]{0.1268}$

Solve for X.

10-562. $X = (43.8)^{6.4}$

10-563. $X = (1.853)^{0.447}$

10-564. $(31.77)^x = 1.164$

10-565. $(2.388)^{3x} = 3.066$

10-566. $(X)^{5.8} = 8.57$

10-567. $(4.92)^{0.66x} = 24.1$

10-568. $(0.899)^{4.7x} = (1.552)(10)^{-8}$

10-569. $(0.1135)^{0.77x} = 0.775$

10-570. $(11.774)^{8.31x} = 12.88$

10-571. $(18.73)^{6.4x} = 8688.$

10-572. $(34.86)^{1.117x} = 9.44$

10-573. $(0.631)^{0.64x} = 0.318$

10-574. $(0.1299)^{0.68x} = 0.443$

10-575. $(15.84)^x = 4.87$

10-576. $(0.679)^x = 0.337$

10-577. $(1.461)^{19.66x} = 9.07$

10-578. $(0.766)^{5.8x} = 0.239$

10-579. $(X)^{7.99} = 0.775$

10-580. $(X)^{0.175} = 8.53$

10-581. $(X)^{3.33} = 1.055$

10-582. $(X)^{0.871} = 0.1557$

10-583. $(X)^{4.77} = 1.088$

10-584. $(X)^{0.771} = 0.0521$

10-585. $(4.51)^{0.199} = \dfrac{X}{3}$

Trigonometric operations are simple to execute with the hand calculator. Although the keys for cotangent, secant, and cosecant are not found on the keyboard, the $\boxed{1/x}$ key can be easily employed. These functions then can be found with the following table of keystroke sequences:

$$\text{Cot (cotangent)} = \boxed{\text{TAN}}\ \boxed{1/x}$$
$$\text{Sec (secant)} = \boxed{\text{COS}}\ \boxed{1/x}$$
$$\text{Csc (cosecant)} = \boxed{\text{SIN}}\ \boxed{1/x}$$

Remember also that an alternative way to write the arc function is with an exponent -1 such that $\arcsin 0.5 = \sin^{-1} 0.5$.

Sines practice problems

10-586. $\sin 26° = (4.3837)(10)^{-1}$

10-587. $\sin 81° = (9.8768)(10)^{-1}$

10-588. $\sin 16° = (2.7564)(10)^{-1}$

10-589. $\sin 15.5° = (2.6724)(10)^{-1}$

10-590. $\sin 42.6° = (6.7688)(10)^{-1}$

10-591. $\sin 3.33° = (5.8087)(10)^{-2}$

10-592. $\sin 10.17° = (1.7657)(10)^{-1}$

10-593. $\sin 63.2° = (8.9259)(10)^{-1}$

10-594. $\sin 70.83° = (9.4455)(10)^{-1}$

10-595. $\sin 26.67° = (4.4885)(10)^{-1}$

10-596. $\sin 7.33° = (1.2758)(10)^{-1}$

10-597. $\sin 2.83° = (4.9373)(10)^{-2}$

10-598. $\sin 51.5° = (7.8261)(10)^{-1}$

10-599. $\sin 5.17° = (9.0111)(10)^{-2}$

10-600. $\sin 33.8° = (5.5630)(10)^{-1}$

10-601. $\sin 20.3° = (3.4694)(10)^{-1}$

10-602. $\sin 68.2° = (9.2849)(10)^{-1}$

10-603. $\arcsin 0.557 = (3.3849)(10)^{1}$

10-604. $\sin^{-1} 0.032 = (1.8338)(10)^{0}$

10-605. $\sin^{-1} 0.242 = (14.005)(10)^{1}$

10-606. $\arcsin 0.709 = (4.5154)(10)^{1}$

10-607. $\sin^{-1} 0.581 = (3.5521)(10)^{1}$

10-608. $\arcsin 0.999 = (8.7437)(10)^{1}$

10-609. $\sin^{-1} 0.569 = (3.4681)(10)^{1}$

10-610. $\sin^{-1} 0.401 = (2.3641)(10)^{1}$

Cosines practice problems

10-611. $\cos 18.8° = (9.4665)(10)^{-1}$

10-612. $\cos 33.17° = (8.3705)(10)^{-1}$

10-613. $\cos 71° = (3.1730)(10)^{-1}$

10-614. $\cos 45° = (7.0711)(10)^{-1}$

10-615. $\cos 68.3° = (3.6975)(10)^{-1}$

10-616. $\cos 26.9° = (8.9180)(10)^{-1}$

10-617. $\cos 55.7° = (5.6353)(10)^{-1}$

10-618. $\cos 5.5° = (9.9540)(10)^{-1}$

10-619. $\cos 81.3° = (1.5126)(10)^{-1}$

10-620. $\cos 8.9° = (9.8796)(10)^{-1}$

10-621. $\cos 77.6° = (2.1474)(10)^{-1}$

10-622. $\cos 39.1° = (7.7605)(10)^{-1}$

10-623. $\cos 50.7° = (6.3338)(10)^{-1}$

10-624. $\cos 11.5° = (9.7992)(10)^{-1}$

10-625. $\cos 49.2° = (6.5342)(10)^{-1}$

10-626. $\arccos 0.901 = (2.5710)(10)^{1}$

10-627. $\cos^{-1} 0.727 = (4.3365)(10)^1$
10-628. $\cos^{-1} 0.0814 = (8.5331)(10)^1$
10-629. $\arccos 0.284 = (7.3501)(10)^1$
10-630. $\cos^{-1} 0.585 = (5.4197)(10)^1$
10-631. $\cos^{-1} 0.658 = (4.8852)(10)^1$

10-632. $\cos^{-1} 0.1190 = (8.3166)(10)^{-1}$
10-633. $\arccos 0.303 = (7.2362)(10)^1$
10-634. $\cos^{-1} 0.505 = (5.9669)(10)^1$
10-635. $\cos^{-1} 0.693 = (4.6132)(10)^1$

Tangents practice problems

10-636. $\tan 29.6° = (5.6808)(10)^{-1}$
10-637. $\tan 48.2° = (1.1184)(10)^0$
10-638. $\tan 11.5° = (2.0345)(10)^{-1}$
10-639. $\tan 71.9° = (3.0595)(10)^0$
10-640. $\tan 5.7° = (9.9813)(10)^{-2}$
10-641. $\tan 61.4° = (1.8341)(10)^0$
10-642. $\tan 33.3° = (6.5688)(10)^{-1}$
10-643. $\tan 69.2° = (2.6325)(10)^0$
10-644. $\tan 40.6° = (8.5710)(10)^{-1}$
10-645. $\tan 8.7° = (1.5302)(10)^{-1}$
10-646. $\tan 17.5° = (3.1530)(10)^{-1}$
10-647. $\tan 85.1° = (1.1664)(10)^1$
10-648. $\tan 58.6° = (1.6383)(10)^0$

10-649. $\tan 39.3° = (8.1849)(10)^{-1}$
10-650. $\tan 20.9° = (3.8186)(10)^{-1}$
10-651. $\tan 42.1° = (9.0357)(10)^{-1}$
10-652. $\arctan 0.362 = (1.9900)(10)^1$
10-653. $\arctan 0.841 = (4.0064)(10)^1$
10-654. $\tan^{-1} 0.119 = (6.7863)(10)^0$
10-655. $\tan^{-1} 0.0721 = (4.1239)(10)^0$
10-656. $\tan^{-1} 1.732 = (5.9999)(10)^1$
10-657. $\arctan 21.6 = (8.7349)(10)^1$
10-658. $\tan^{-1} 0.776 = (3.7811)(10)^1$
10-659. $\arctan 89.3 = (8.9358)(10)^1$
10-660. $\tan^{-1} 0.661 = (3.3465)(10)^1$

Trigonometric functions: problems

10-661. $\sin 35°$
10-662. $\sin 14°$
10-663. $\sin 78°$
10-664. $\sin 3.7°$
10-665. $\sin 88.3°$
10-666. $\sin 55.3°$
10-667. $\cos 35°$
10-668. $\cos 66°$
10-669. $\cos 21.3°$
10-670. $\cos 11.1°$
10-671. $\cos 7.9°$
10-672. $\cos 43.8°$
10-673. $\tan 33.8°$
10-674. $\tan 9.4°$
10-675. $\tan 37.7°$
10-676. $\tan 22.5°$
10-677. $\tan 86.1°$
10-678. $\tan 54.4°$
10-679. $\tan 70.3°$
10-680. $\tan 29.7°$
10-681. $\tan 36.5°$
10-682. $\tan 13.3°$
10-683. $\tan 45.8°$
10-684. $\cot 14.7°$
10-685. $\cot 81.8°$
10-686. $\cot 36.9°$
10-687. $\cot 61.2°$
10-688. $\cot 54.3°$
10-689. $\cot 18.7°$

10-690. $\cot 3.77°$
10-691. $\cot 66.4°$
10-692. $\csc 38.1°$
10-693. $\csc 75.2°$
10-694. $\csc 88.3°$
10-695. $\csc 12.8°$
10-696. $\csc 46.4°$
10-697. $\csc 81.1°$
10-698. $\csc 32.6°$
10-699. $\csc 9.03°$
10-700. $\sec 6.14°$
10-701. $\sec 59.2°$
10-702. $\sec 79.4°$
10-703. $\sec 19.5°$
10-704. $\sec 2.77°$
10-705. $\sec 45.9°$
10-706. $\arcsin 0.771$
10-707. $\arccos 0.119$
10-708. $\arctan 34.8$
10-709. $\text{arcsec } 7.18$
10-710. $\text{arccsc } 1.05$
10-711. $\cos 33.4°$
10-712. $\cos 3.6°$
10-713. $\arccos 0.992$
10-714. $\cos 24.67°$
10-715. $\cos^{-1} 0.496$
10-716. $\cos 36.6°$
10-717. $\arccos 0.238$
10-718. $\cos 0.75°$

10-719. $\cos 36.6°$
10-720. $\tan 32.6°$
10-721. $\tan 16.34°$
10-722. $\tan 88.30°$
10-723. $\arctan 0.62$
10-724. $\tan^{-1} 0.75$
10-725. $\arctan 0.392$
10-726. $\tan^{-1} 1.53$
10-727. $\tan 37.24°$
10-728. $\arctan 0.567$
10-729. $\tan^{-1} 0.0321$
10-730. $\cot 19.33°$
10-731. $\sec 46.46°$
10-732. $\csc 32.12°$
10-733. $\sin 37°$
10-734. $\sin 51.50°$
10-735. $\sin 68.37°$
10-736. $\sin 75.10°$
10-737. $\arcsin 0.622$
10-738. $\sin 13.6°$
10-739. $\sin^{-1} 0.068$
10-740. $\sin 14.6°$
10-741. $\arcsin 0.169$
10-742. $\sin 34.67°$
10-743. $\cos 26.26°$
10-744. $\csc 20.20°$
10-745. $(\csc 20°)(\sin 46°)$
10-746. $(\cos 32°)(\tan 43°)$

10-747. $\dfrac{\sin 13.9°}{\cot 13.9°}$

10-748. $\dfrac{\cot 33.22°}{\sec 4.53°}$

10-749. $\dfrac{\cos 33.15°}{\cot 46.19°}$

10-750. $\dfrac{(\sec 10°)(\cot 10°)}{(\sin 10°)(\csc 10°)}$

10-751. $\dfrac{(\sin 35°)(\tan 22°)}{\sqrt[3]{\sin 5.96°}}$

10-752. $\dfrac{(\sec 11°)(\tan 4°)}{\cot 49°}$

10-753. $\dfrac{(\sin 8°)(\tan 9°)}{\cot 82°}$

10-754. $\dfrac{(\sin 1.36°)(\cot 26°)}{\sqrt[3]{0.00916}}$

10-755. $\dfrac{\cot \sin^{-1} 0.916}{(1.32)(5.061)}$

10-756. $\dfrac{(77.19)(\sec 46°)}{\tan 3.91°}$

10-757. $\dfrac{(\sqrt[3]{\tan 25.9°})(\sin \cos^{-1} 0.5)}{(\sin 5.16°)(\tan 22°)}$

10-758. $\dfrac{(0.0311)(\sec 69°)\sqrt[3]{9.0}}{(\sin 9°)(\cos 9°)}$

10-759. $\dfrac{(1.916)(\sqrt[3]{1.916})(\sqrt[3]{\sin 20°})}{(\sqrt[3]{\sec 40°})(\tan 10.22°)}$

10-760. $\dfrac{(6.17)(\tan 6.17°)(\sqrt[3]{6.17})}{(6.17)^2(\sin 61.7°)(\cos 6.17°)}$

If the calculator does not have the hyperbolic functions, they can be solved with simple formulas. They are

$$\text{hyperbolic sine } x \text{ (sinh } x) = \frac{e^x - e^{-x}}{2}$$

$$\text{hyperbolic cosine } x \text{ (cosh } x) = \frac{e^x + e^{-x}}{2}$$

$$\text{hyperbolic tangent } x \text{ (tanh } x) = \frac{e^{2x} - 1}{e^{2x} + 1}$$

$$\sinh^{-1}(x) = \ln[x + (x^2 + 1)^{1/2}]$$

$$\cosh^{-1}(x) = \ln[x + (x^2 - 1)^{1/2}]x \geq 1$$

$$\tanh^{-1}(x) = \frac{1}{2}\ln\left(\frac{1 + x}{1 - x}\right)x^2 < 1$$

Problems on hyperbolic functions

10-761. Find the values of sinh x for the following values of x: (a) 0.12, (b) 1.07, (c) 1.91, (d) 2.30, (e) 3.11, (f) 4.26, (g) 5.00

10-762. Find the values of x for the following values of sinh x: (a) 0.1304, (b) 0.956, (c) 1.62, (d) 4.10, (e) 8.70, (f) 19.42, (g) 41.96

10-763. Find the values of cosh x for the following values of x: (a) 0.28, (b) 1.03, (c) 1.98, (d) 2.37, (e) 3.56, (f) 4.04, (g) 5.00

10-764. Find the values of x for the following values of cosh x: (a) 1.204, (b) 1.374, (c) 2.31, (d) 5.29, (e) 8.50, (f) 21.7, (g) 52.3

10-765. Find the values of tanh x for the following values of x: (a) 0.16, (b) 0.55, (c) 1.14, (d) 1.94, (e) 2.34, (f) 2.74, (g) 5.00

10-766. Find the values of x corresponding to the following values of tanh x: (a) 0.1781, (b) 0.354, (c) 0.585, (d) 0.811, (e) 0.881, (f) 0.980, (g) 0.990

Conversion between rectangular and polar notation on some calculators is accomplished using a single keystroke. Many calculators, however, still require a conversion routine such as that discussed below.

Conversion to polar form from the rectangular form $x + jy$: Here the j serves to separate the vertical (or Y component) from the X (or horizontal) component. The conversion formulas are

$$R \text{ (or hypotenuse)} = (x^2 + y^2)^{1/2}$$

$$\theta \text{ (or angle)} = \arctan\left(\frac{y}{x}\right)$$

Example Solve for R and θ if the rectangular form is $3 + j4$. The keystroke sequence would be (RPN logic)

for R: ③ $\boxed{x^2}$ ④ $\boxed{x^2}$ $\boxed{\div}$ $\boxed{\sqrt{x}}$, which equals 5

for θ: ④ $\boxed{\text{ENTER}}$ ③ $\boxed{\div}$ $\boxed{\text{ARC}}$ $\boxed{\text{TAN}}$, which equals 53.13°

For algebraic logic the sequence would be

for R: ③ $\boxed{x^2}$ $\boxed{+}$ ④ $\boxed{x^2}$ $\boxed{=}$ $\boxed{\sqrt{x}}$

For θ: ④ $\boxed{\div}$ ③ $\boxed{=}$ $\boxed{\text{ARC}}$ $\boxed{\text{TAN}}$

Conversion to rectangular form from polar form $R\underline{/\theta}$ the conversion formulas are

$$x = R \cos \theta$$
$$y = R \sin \theta$$

Example Solve for x and y if the polar form is $10\underline{/30°}$. The keystroke sequence for RPN would be

for x: ① ⓪ $\boxed{\text{ENTER}}$ ③ ⓪ $\boxed{\text{COS}}$ $\boxed{x}$, which equals 8.66

for y: ① ⓪ $\boxed{\text{ENTER}}$ ③ ⓪ $\boxed{\text{SIN}}$ $\boxed{x}$, which equals 5

For algebraic logic the sequence would be

for x: ① ⓪ $\boxed{x}$ ③ ⓪ $\boxed{\text{COS}}$ $\boxed{=}$

for y: ① ⓪ $\boxed{x}$ ③ ⓪ $\boxed{\text{SIN}}$ $\boxed{=}$

Some calculators have a key marked $\boxed{\to R}$ and $\boxed{\to P}$, or some equivalent. This key permits the direct and immediate conversion from one system to another with the pressing of a single key (or possibly two if an alternate-function key is needed).

Problems on complex numbers

10-767. Express in polar form: (a) $8 + j3$, (b) $2 + j6$, (c) $1 + j4$, (d) $5 + j5$

10-768. Express in rectangular form: (a) $6.2\underline{/39°}$, (b) $3.6\underline{/48°}$, (c) $9.2\underline{/21.4°}$, (d) $2.7\underline{/71°}$

10-769. Express in polar form: (a) $-8.9 + j4.2$, (b) $-16.8 + j9.3$, (c) $-5.3 + j2.1$, (d) $-18.4 + j3.3$

10-770. Express in rectangular form: (a) $9.7\underline{/118°}$, (b) $115\underline{/137°}$, (c) $2.09\underline{/160°}$, (d) $5.72\underline{/110°}$

10-771. Express in polar form: (a) $-7.3 - j6.1$, (b) $-4.4 - j8.2$, (c) $-8.8 - j2.5$, (d) $-1.053 - j5.13$

10-772. Express in rectangular form: (a) $81.3\underline{/200°}$, (b) $62.1\underline{/253°}$, (c) $1059\underline{/197°}$, (d) $0.912\underline{/231°}$

10-773. Express in polar form: (a) $160.5 - j147$, (b) $89.3 - j46.2$, (c) $0.0062 - j0.0051$, (d) $3.07 - j1.954$

10-774. Express in rectangular form: (a) $557\underline{/297°}$, (b) $6.03\underline{/327°}$, (c) $0.9772\underline{/344°}$, (d) $19,750\underline{/300°}$

10-775. Express in polar form: (a) $15.61 + j7.09$, (b) $-14.9 - j61.7$, (c) $0.617 - j0.992$, (d) $-41.2 + j75.3$

10-776. Express in rectangluar form: (a) $1.075\underline{/29.1°}$, (b) $10.75\underline{/136°}$, (c) $107.5\underline{/253°}$, (d) $1075\underline{/322°}$

Review problems

10-777. $(51)(9)$

10-778. $(426)(51)$

10-779. $(6.03)(5.16)$

10-780. $(561)(4956)$

10-781. $(43.2)(0.617)$

10-782. $(6617)(0.00155)$

10-783. $(99.043)(3.091)$

10-784. $(0.0617)(0.4417)$

10-785. $(1.035)(2.31 \times 10^5)$

10-786. $(79.81 \times 10^{-4})(0.617)$

10-787. $(516 \times 10^{-8})(0.391 \times 10^{-2})$

10-788. $(51)(97)(32)$

10-789. $(52.3)(759.3)$

10-790. $(716.5)(0.03166)$

10-791. $(11.65)(-0.9213)$

10-792. $(76.2)(-31.45)$

10-793. $(-0.6175)(-12,391)$

10-794. $\dfrac{-759.6}{0.6175}$

10-795. $\dfrac{-19.96}{3346}$

10-796. $\dfrac{-1.0366}{29.31}$

10-797. $\dfrac{7575}{695.2}$

10-798. $\dfrac{-516.6}{0.06052}$

10-799. $(116.5)(4619)(0.317)$

10-800. $(210.9)(151.3)(7716)$

10-801. $(706.5)(1.695 \times 10^{-6})(0.006695)$

10-802. $(1033)(7.339 \times 10^{-6})(0.0317 \times 10^{-3})$

10-803. $(4.017 \times 10^{-8})(0.0991)(0.1756)$

10-804. $(5.576)(0.0917)(1.669 \times 10^4)$

10-805. $(6.991)(0.75)(0.993)(4.217)$

10-806. $(56.88)(0.971 \times 10^{-5})$

10-807. $(59.17)(0.3617)(0.5916)(0.00552)$

10-808. $(5.691)(0.3316)(0.991)(0.00554)(0.1712)$

10-809. $(6.523)(71.22)(4.091)(591)(600)(0.1332)$

10-810. $(43.06)(0.2361)(0.905 \times 10^{-4})(3.617 \times 10^{-3})$

10-811. $(0.3177)^{2.06}$

10-812. $\sqrt[5]{26.31}$

10-813. $\sqrt[3]{0.03175}$

10-814. $(1917)^{2.16}$

10-815. $(4.216)^{1.517}$

10-816. $(2.571)^{2.91}$

10-817. $\sqrt{116.75}$

10-818. $\sqrt[3]{0.6177}$

10-819. $\sqrt{3167}$

10-820. $(179 \times 10^3)(0.3165)$

10-821. $(5033 \times 10^{-4})(0.9116)$

10-822. $(0.06105)(77.165)$

10-823. $(\sqrt{216})(34)(\pi)^2$

10-824. $(\sqrt{819})(107)(\sqrt{\pi})$

10-825. $\dfrac{(\sqrt{616})(6.767)}{\sqrt{39.6}}$

10-826. $\dfrac{Y}{28} = \dfrac{3.2}{4/118}$

10-827. $\dfrac{Y}{42} = \dfrac{39.1}{1/45}$

10-828. $(37.3)(X)(46.6) = (175)(\pi)$

10-829. $(\sqrt{256})(3) = (X)(197.6)$

10-830. $\dfrac{(54.6)(\tan 10.6°)}{(\sqrt{0.0967})(8.1 \times 10^3)}$

10-831. $\dfrac{\sqrt[3]{(15.1)^2}(31.4)^2}{\sin \arccos 0.617}$

10-832. $\dfrac{(0.954)(0.06 \times 10^3)}{(\tan 59°)^{1/2}(6.5)^2}$

10-833. $\dfrac{\sqrt[3]{(15.6)^2}(0.9618)}{(0.08173)(61,508)(2\pi)}$

10-834. $\dfrac{(68)(765)(391)(0.0093 \times 10^3)}{(571)^2(\sqrt[3]{64})}$

10-835. $\dfrac{(\cos 11.5°)(\sqrt{6.87})}{(0.00081)(7.7 \times 10^4)}$

10-836. $\dfrac{\sqrt[4]{(1.71)^5}(6.87)}{(\tan 53°)(5.1)^2}$

10-837. $\dfrac{(0.000817)(\tan 81°)}{(0.00763)(\tan 81°)}$

10-838. $\dfrac{(\sin \arctan 3.17)(71.7)}{(\sqrt{89.6})(\sqrt[4]{(76.5)^2})}$

10-839. $\dfrac{(\sqrt{(16)^3})(\log_{10} 100)}{(6.71 \times 10^{-1})(3.71)^3}$

10-840. $\dfrac{1045}{X} = \dfrac{0.0278}{0.0798}$

10-841. $\dfrac{1.486}{33} = \dfrac{(0.37)(X)}{467}$

10-842. $816 = \dfrac{(244)(2\pi)}{(0.049)(X)}$

10-843. $(0.0036)(\sin 49.8°)$

10-844. $\dfrac{(20.5)^2(7.49)(\sin 49°)}{(30.5)(0.0987)}$

10-845. $\sqrt{\dfrac{(38)^2(6.71)^2}{\pi}}$

10-846. $(7.61)(\sqrt[3]{7.61})(\pi)$

10-847. $\dfrac{96.5}{3.9} = \dfrac{X}{\sin 46.6°} = \dfrac{(Y)^2}{3.14 \times 10^{-2}}$

10-848. $\dfrac{(X)^2}{Y} = \dfrac{(67.3)^2(Y)}{96.61} = \dfrac{497.1}{\tan 75°}$

10-849. $\dfrac{(3.7)(4.9)}{X} = \dfrac{46.7}{564}$

10-850. $\dfrac{(13.1)(\sin 3.12°)}{\tan 41.9°}$

10-851. $\dfrac{2}{3} = \dfrac{(X)(\pi)}{8.37}$

10-852. $\dfrac{9616}{X} = \dfrac{3.1416}{0.0142}$

10-853. $(\sqrt[3]{64.9})(2.1 \times 10^3)$

10-854. $(4 \times 10^6)(0.007) = (X)(10{,}980)$

10-855. $Y = \left(\dfrac{1}{4}\right)\left(\dfrac{16}{6}\right)\left(\dfrac{1}{17}\right)$

10-856. $\dfrac{X}{\pi} = \dfrac{(\sqrt{46.2})(3.14)^2}{\sin 3.7°}$

10-857. $\dfrac{(3.98)(X)}{(1.07)(38)} = \dfrac{3 \times 10^6}{17{,}680}$

10-858. $\dfrac{\sqrt[3]{986}}{X} = \dfrac{14}{1/116}$

10-859. $\dfrac{(X)^2}{9.2} = \dfrac{(18.17)(3.4)}{166}$

10-860. $\dfrac{3.6}{(X)^2} = \dfrac{9.6 \times 10^2}{67.4} = \dfrac{(Y)^{1/2}}{64}$

10-861. $\dfrac{(X)^{1/2}}{31.1} = \dfrac{(\sqrt{196})(189.1)}{4/76}$

11

Statistics and graphical analysis

Certain statistical operations are commonly encountered in engineering work, particularly when data are acquired and evaluated. We might even say that science is based on statistics and that the scientific "laws" that we use relate not to how nature will behave with certainty, but rather to how nature has behaved within limits and to how nature is likely to continue to behave under similar conditions. It is within the realm of statistics to determine what these limits are and to attempt to determine the probability of recurrence of any given set of events based on the frequency and regularity of their occurrence in the past.

If we consider some natural phenomena, such as the sun's rising in the east rather than in the west, we cannot say with certainty that the same thing will happen tomorrow. We can say, however, that as far as any records show the sun has always risen in the east, and so far as we know no changes in conditions have occurred to alter the probability of its rising in the east, so we conclude that it is highly probable that tomorrow's sunrise will be in the east.

In a like manner, we can make a general statement that at some given city in the United States, it will be colder on New Year's Day than it will be on the first day of July. Statistically, we can show that for many years it has always been colder on New Year's Day than on the first of July, but it is within the realm of possibility, knowing how local weather conditions can vary, that a set of weather circumstances can occur that could make, for a certain year, a colder July 1st than New Year's Day.

The probability of such an occurrence is slight but it definitely is greater than that of the sun's sudden appearance in the morning on the western horizon. Thus, we see that there not only is an uncertainty even in well-ordered natural phenomena, but also there is a degree of uncertainty of future happenings which is based on the variability in the past of certain occurrences.

Statistics then is the science of making decisions based on observation, collection, analysis, and interpretation of data that are affected by chance causes. The importance of the use of statistics has been emphasized in recent years by the national effort to place a man on the moon and effect his safe return to earth.

The best statistical methods are useless unless the data obtained have been collected so that the methods are applicable. Accuracy in tabulation, in calculation, and in thought are essential ingredients to all statistical work, since the data used are themselves subject to chance errors. Care and neatness in preparing all calculations are important aids to accuracy. However, it is no less important that the engineer develop a natural skepticism and inquisitive attitude toward all data collected and toward their methods of collection, their analyses, and their interpretation.

Variables

From a statistical point of view, the chance variations which occur in measured data are a major problem in any evaluation. Fortunately, in the physical sciences and in engineering work the variables usually are easier to control and are better known than they are in some other fields, such as psychology, where animal and human behavior is being studied. For example, if we should desire to determine the relation between current and voltage in an electrical circuit, we can establish a test setup of a power source, a set of conductors, and a power receiver together with appropriate meters to measure the electrical quantities. However, we recognize that variables inherent in the test setup must be controlled, held constant, or evaluated for effects on the meter readings. Of all the possible variables, temperature is most likely to change the circuit resistance and consequently alter the data secured. Fortunately, it usually is fairly easy to maintain constant temperature conditions and then to obtain a relation between a voltage change and the corresponding current change in order to establish a relationship describing their behavior within the limits of error of the experiment. Again, if we should try other circuit conditions of power and resistance, we probably would arrive at substantially the same results; therefore, we can reasonably conclude that the ratio between the voltage change and the corresponding current change in a circuit is a constant.

On the other hand, if a psychological test were to be made to determine the effect of loss of sleep on ability to perform simple arithmetical operations, we would find quite a wide variance between subjects and even between the ability of the same subject at various times. We conclude then that the relationship of loss of sleep to arithmetical accomplishments involves variables of many sorts, most of which are hard to control or evaluate.

Normal probability law

If a large glass jar were filled to the top with marbles and placed in view of a large class of students and each student was asked to write down his estimate of the number

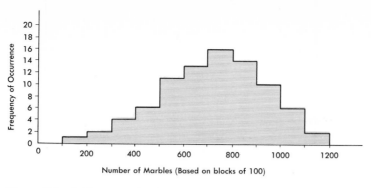

Figure 11-1 A histogram of estimates of marbles in a jar.

of marbles in the jar, it is extremely unlikely that every student would estimate the same number and that this number would be the exact number of marbles in the jar. Rather it is likely that, if the answers were compiled, a pattern of distribution of estimates would focus upon a certain estimated number of marbles.

If, for simplicity in plotting, the estimates are grouped into blocks to the nearest 100 marbles, a graph of this distribution might look like Figure 11-1. This figure is plotted so that the width of a column is equal to the interval, in this case 100 marbles, and the height is equal to the frequency, which is the number of persons making any given block of estimates.

If the number of persons making estimates of the marbles were doubled and the blocks within which the estimates fall were made smaller, the histogram probably would take on an appearance similar to Figure 11-2.

If this process were to be continued, we would see that the appearance of the graph would begin to assume the shape of a smooth curve. Although the proof of this statement is beyond the scope of this book, we can show that, for a large number of types of observations, the pattern becomes similar to the graph in Figure 11-3.

This graph shows the usual frequency distributions of a large number of observations and is typical of the distribution of any set of chance events. In practice, it can be taller or shorter, fatter or thinner, but it is usually symmetrical and bell shaped.

If a person should take ten coins and toss them on a table many times and keep a tally of the number of heads that show up each time, he would find that the occurrence of ten heads is extremely rare, that the occurrences of zero heads is extremely rare, and that the greatest number of occurrences is for five heads to show

Figure 11-2 A histogram of a large number of estimates.

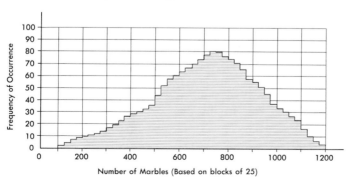

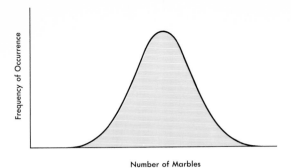

Frequency of Occurrence

Number of Marbles

Figure 11-3   Normal probability curve.

up. If the frequencies of occurrences are plotted against the number of heads, we would find that the bell-shaped curve described above will result.

This graph, which pictures the distribution of frequencies of certain chance events, is called the *normal probability curve*. It is of great use in many forms of testing in engineering and science.

The horizontal axis (abscissa) of the graph represents the values of the measurements made (X_1, X_2, X_3, etc.) and the vertical axis (ordinate) represents the fraction of the total number of observations made corresponding to each value of X.

The general mathematical expression for the probability curve is a log function of the form $y = Ce^{Kx^2}$. From an inspection of the probability curve determined either by trial or by derivation, several principles can be observed:

1. Small errors occur more frequently than large ones.
2. Errors of any given size are as likely to be positive as they are to be negative.
3. Very large random errors seldom occur.

Some fundamental statistical measures

The terms "arithmetic mean," "median," and "mode" are used extensively in statistical work. These terms will be defined here in order that the student can gain a better appreciation of their use.

Arithmetic mean   This term is used to denote the point about which the data tend to cluster. It is often referred to as the *average* or *central tendency*. It is calculated by obtaining the sum of the individual measurements and dividing this quantity by the number of measurements made. This process may be represented mathematically by

$$\bar{X} = \sum_{i=1}^{i=n} \frac{X_i}{n} = \frac{X_1 + X_2 + X_3 + \cdots + X_n}{n}$$

where $\bar{X}$ = arithmetic mean
X_i = individual measurements
n = total number of measurements

> Chance favors the prepared mind.
> —Louis Pasteur

Example   Find the arithmetic mean of the following data:

$$7, 4, 9, 5, 6, 8, 3$$

$$\bar{X} = \frac{7 + 4 + 9 + 5 + 6 + 8 + 3}{7} = \frac{42}{7} = 6$$

Median   This term is similar to the mean in that it is also a measure of the tendency of the data to collect about a central point. The median is the midpoint (not average) of a group of data. When the total number of data is odd, the median is the middle number of the set of numbers. If the total number of data is even, the median is the arithmetic mean of the two middlemost numbers in the set.

Examples

a. 4, 6, 9, 10, 11, 12, 15   Median = 10
b. 5, 7, 7, 8, 10, 11, 15, 19   Median = 9
c. 2, 5, 7, 9, 9, 11, 15, 16   Median = 9

Mode   As in the other cases, the mode is also a measure of the "central tendency" of the data. The mode is that value which occurs with the greatest frequency in the set of data. It is the most common value, and for this reason it may exist in some sets of data. In other cases there may be more than one mode.

Examples

a. 2, 4, 5, 5, 5, 3, 2, 6   Mode = 5
b. 2, 3, 4, 6, 7, 8   There is no mode
c. 2, 3, 4, 4, 5, 5, 6, 7   There are two modes—4 and 5
                               (This is called bimodal)

Deviations from the normal curve

The most common deviation from the normal probability curve is a condition known as "skewness." In this condition the curve is distorted, and the high part of the curve corresponding to the greatest frequency is nearer to one end, rather than being in the middle. One of the most common causes of this nonnormality is that the distribution may be restricted from going beyond a certain point. This situation would exist, for example, if the measurement has a physical limit of zero. Such a graph could also be formed if the scores on an examination given to students were plotted, and the test had been much too easy or much too difficult. In another case, if the lengths of a group of parts made by an improperly adjusted machine are measured with a steel scale, the plot of the measurements could be distorted or skewed.

Another abnormal condition is produced when the group being sampled is not homogeneous. The curve produced could have two peaks and would be known as a "bimodal" distribution. Such a plot could be obtained if an examination were given to a group of students some of whom were rather dull and the remainder of whom were very apt and intelligent. As another example, if a box of similar-type resistors is measured to determine the distribution of resistance values, and the box contains resistors from two different machines set to produce slightly different values of resistance, it is likely that a "two-humped" graph would result, showing that two somewhat independent groups are present in the test sequence.

If deviations from the normal curve are excessive, accurate results cannot be obtained from the statistical tools described in the following topics. Usually it is necessary to examine the method of measurement to see whether systematic errors are present or to examine the group being measured to determine whether a proper sample is taken, so that results can be made to approximate the normal curve.

Theory of errors

As suggested above, the normal curve may be considered to be the frequency distribution of the infinite number of possible measurements of the quantity being observed. When practical, *all possible* measurements in a given situation should be tabulated. When this can be accomplished it is said to be a study of the *total population* or *universe*. In many situations such measurement is not possible. For example, if someone wanted to obtain the heights of all the men in the world, he could not do so. In such cases, it is necessary to examine a small part of the total population, called a *sample*. If the *sample* is representative of the *total population*, certain important conclusions can be drawn about the nature of the *total population*. The size of the *sample* chosen will depend upon how close it is desired to approximate the *total population*.

Standard deviation   Since, in any group of measured quantities the true value is never known, it is desirable to have a means of estimating the uncertainty, and consequently the accuracy, of a measurement. In order to do this, we must make use of several statistical tools, one of which is known as the "standard deviation." The standard deviation may be calculated for a *total population* (usually designated as σ)[1] and for the *sample*. For the sample, the standard deviation is given by the equation

$$s = \sqrt{\frac{\Sigma (X_i - \bar{X})^2}{n - 1}}$$

where σ and s represent the standard deviations for the respective situations, Σ is the Greek capital letter *sigma*, which represents the sum, $(X_i - \bar{X})$ represents the deviation of a single observation from the mean, and n is the number of observations. For example, if we weigh a block of wood on ten different scales and record the weight from each weighting in a tabular form, the deviations from the mean can be obtained readily by subtracting any single reading from the mean of the values. The standard deviation for the *sample* can then be calculated as shown in Table 11-1.

From this table the value of $\Sigma (X_i - \bar{X})^2 = 274$, which, if substituted in the expression for standard deviation, gives

$$s = \sqrt{\frac{274}{10 - 1}} = \sqrt{\frac{274}{9}} = \sqrt{30.44} = 5.52 \text{ g}$$

The use of the standard deviation will be discussed in more detail later in this chapter.

Population dispersion

If the plot of a series of measurements is made to produce a histogram and the tops of the rectangles are connected by a smooth curve, the bell-shaped curve is a typical probability curve.

---

[1] Greek letters are usually used to represent descriptive quantities about the population whereas Arabic letters are used to represent descriptive quantities about a sample.

Table 11-1

Trial	Weight, grams	$X_i - \bar{X}$	$(X_i - \bar{X})^2$
1	522	+10	100
2	506	− 6	36
3	513	+ 1	1
4	510	− 2	4
5	519	+ 7	49
6	508	− 4	16
7	512	0	0
8	504	− 8	64
9	512	0	0
10	514	+ 2	4
	$\Sigma = 5120$	$\Sigma = 0$	$\Sigma = 274$

$\bar{X} = \text{mean} = 5120/10 = 512$

After a normal probability curve is obtained from a histogram, if the values of σ (assuming a *total* population study) are plotted on the abscissa, it will be found that about 68.27 per cent of the measurements will fall within the $\pm 1\sigma$ range of the mean (that is, one standard deviation on either side of the mean). This means that there is a 68.27 per cent chance that the value of any single observation will fall between $+\sigma$ and $-\sigma$ of the mean, $\bar{X}$.

Referring to Figure 11-4, we can show experimentally that 68.27 per cent of the values will be plotted between $-\sigma$ and $+\sigma$. However, this percentage can be shown best where a large number of observations, 100 or more for example, are made.

If an abscissa value of $\pm 2\sigma$ is plotted on a probability curve, it can be shown that about 95.45 per cent of the observations will fall between -2σ and $+2\sigma$ of $\bar{X}$, Figure 11-5.

Use of probability curve

A probability curve does not give the true value of a quantity being measured. If we assume that the mean, or arithmetical average, of a number of observations is acceptable as a value to use to present a measured quantity, then the standard deviation gives an indication of the reliability of any single observation.

Figure 11-4 Normal probability curve.

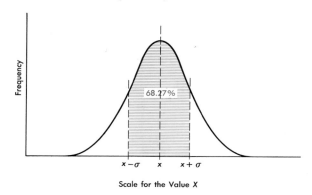

Scale for the Value X

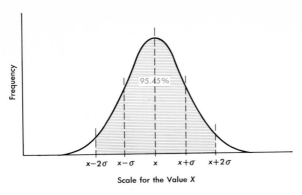

Scale for the Value X

Figure 11-5 Sigma error on a normal probability curve.

Standard error

It usually is desirable to evaluate the uncertainty of the arithmetic mean. We know that the uncertainty of the mean, $\overline{X}$, is considerably less than the uncertainty of any single observation, X_i. The uncertainty of the mean can be expressed in the following form:

$$\sigma_m = \frac{\sigma}{\sqrt{n}}$$

and this is usually approximated by

$$s_m = \frac{s}{\sqrt{n}}$$

where σ_m is the standard error of the mean, S_m is the standard deviation of the mean, s is the standard deviation of a sample, and n is the number of observations in the sample.

Example The mean of 25 measurements of an angle gives a value of 32° 17.1′; s is 1.2′. What is the probable range of the true value?

Solution
   1. The mean of 32° 17.1′ is the most likely true value.

   2. $s_m = \dfrac{1.2'}{5} = 0.24' \cong \pm 0.2'$

There is a 68.27 per cent certainty that the true value lies between 32° 16.9′ and 32° 17.3′ ($\pm s_m$).
   3. There is a 95.45 per cent certainty that the true value lies between 32° 16.7′ and 32° 17.5′ ($\pm 2s_m$).

Since the true value is never known, an estimate based on mathematical processes can be made as to the confidence that can be placed in the mean as an assumed true value.

Problems

11-1. A series of weighings of a sample of metal powder are made with the following results:

Weight of a sample, grams

2.020	2.021	2.021	2.019	2.019
2.018	2.021	2.018	2.021	2.017
2.017	2.020	2.016	2.019	2.020

Compute the mean, s, and s_m values for the weighings. What is the probable weight of the sample?

11-2. A series of measurements of the length of a concrete runway is made using a steel tape. The results (in metres) are tabulated below:

1363.7	1364.5	1364.0	1363.8	1364.0
1364.1	1363.9	1364.1	1363.9	

Compute the mean and give the s_m limits for the measurements.

11-3. A series of readings was taken, using an electronic interval timer, for one complete swing of a pendulum to occur. The data are tabulated as follows:

Time, in seconds	Number of occurrences	Time, seconds	Number of occurrences
1.851	1	1.859	18
1.852	3	1.860	15
1.853	6	1.861	12
1.854	9	1.862	10
1.855	12	1.863	5
1.856	14	1.864	4
1.857	18	1.865	2
1.858	19	1.866	1

What is the mean time of a swing, and what would be the standard error of the mean?

11-4. The test scores on an intelligence test given to a class of elementary students are tabulated as follows:

35	58	46	67	47	53
55	38	50	47	50	53
46	54	45	52	62	48
45	51	48	42	48	65
51	55	60	53	55	
56	43	47	58	34	
42	55	46	59	68	
60	52	61	39	31	
52	44	42	39	70	

Plot a histogram of the scores and sketch in a probability curve. Compute the mean, median, and mode. Is there any tendency to skewness or bimodality? Does the mean value of the scores have a significance comparable to the mean of, for example, a series of measurements of the length of a steel block?

11-5. Take ten coins and toss them at least 25 times, keeping count of the number of heads and tails for each toss. Plot a probability curve and determine whether s does represent 68.27 per cent of the total observations.

11-6. The distribution of ages of a group of recruits at an Army camp is given in the accompanying table. Plot a histogram and sketch a probability curve for the ages. Show the s and $2s$ locations. Does this graph show any unusual departures from a standard probability curve? Compute the mean, median, and mode.

Age, years– months	Number of persons	Age, years– months	Number of persons
18–1	1	19–7	9
18–2	0	19–8	5
18–3	1	19–9	3
18–4	3	19–10	3
18–5	8	19–11	0
18–6	5	19–12	2
18–7	8	20–1	5
18–8	10	20–2	1
18–9	14	20–3	0
18–10	7	20–4	2
18–11	12	20–5	6
18–12	11	20–6	0
19–1	11	20–7	1
19–2	6	20–8	0
19–3	10	20–9	2
19–4	8	20–10	0
19–5	7	20–11	0
19–6	6	20–12	0

11-7. Measurements were made of the lengths of a number of steel rods which were supposed to be cut to a length of 6.80 in. The measurements are as follows:

6.81	6.80	6.79	6.80
6.82	6.80	6.78	6.80
6.81	6.83	6.79	6.77
6.82	6.80	6.78	6.80
6.81	6.81	6.79	6.87

What is the average length of the rods, and what maximum tolerance can be set up if 95.45 per cent of the rods is to be acceptable?

Graphical analysis

Graphs are a valuable aid in presenting many types of information where facts must be readily grasped. They aid in the analysis of engineering data and facilitate the presentation of statistical information. Graphs generally can be classified as those used for technical purposes and those used for general presentation of information. To be of greatest value, graphs should be prepared in accord with the best current practice.

A graphical display of information may take any of several forms, depending upon the type of information to be presented and the use to be made of the information. For rapid dissemination of information, pictographs are convenient. Where more exact representation is desired, bar graphs or circle graphs may be employed. Most engineering data are displayed in line graphs. Such information usually is more exact and offers opportunity to interpolate values, to extrapolate values, and to draw conclusions as to the behavior of the variable quantities involved. Examples of several types of graphs are shown in Figures 11-6 to 11-10.

Since line graphs offer the best opportunity to present engineering data, the

Illustration 11-1
In this picture an engineer identifies himself by telephone to a "graphics" computer. The computer compares the characteristics of his pronounciations with reference characteristics from a previously collected sample stored in the computer's memory. The bright wave form on the computer display screen represents characteristics of the speaker's voice.

discussion here will be concerned chiefly with the preparation and use of line graphs. The general form of the graph sheet illustrated by Figure 11-11 is the form used by the majority of engineering schools and is the style widely used in industry.

Notes on the preparation of graphs

1. Graphs usually are prepared in pencil on printed coordinate graph paper. Carbon paper backing should be used where sharpness of reproduction is a factor. For more permanent work or for display purposes, India ink should be used.
2. Arrange the data in tabular form for convenience in plotting, and determine

Figure 11-6 Pictograph (data comparative in nature).

North Carolina

Georgia

South Carolina

California

Virginia

Wyoming

Leading States in
the Quarrying of Granite
Each symbol represents
1,000,000 tons

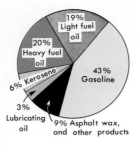

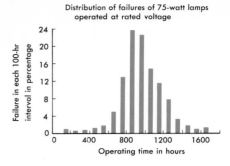

Distribution of failures of 75-watt lamps
operated at rated voltage

Figure 11-7 Circle graph (data expressed as parts of a whole).

Figure 11-8 Vertical bar graph (a family of individual sets of data).

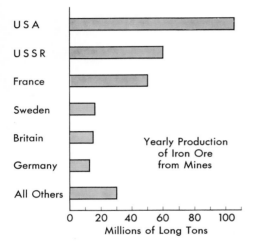

Yearly Production
of Iron Ore
from Mines

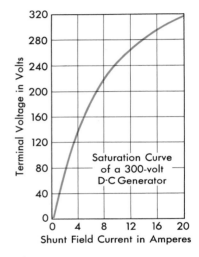

Saturation Curve
of a 300-volt
D-C Generator

Figure 11-9 Horizontal bar graph (numerical data).

Figure 11-10 Line graph used for display purposes.

the type of scales that most logically portray the functional relationship between the variables (see Figure 11-11).

3. Graphs usually are designated by naming ordinate values first, then abscissa quantities. It is customary to plot the dependent variable along the ordinate and the independent variable along the abscissa.

4. Make a trial computation to select the scale on each axis:

$$\text{scale} = \frac{\text{range in the variable}}{\text{scale length available}}$$

5. The scale must be suitable for the paper used. For graph paper having 20 divisions per inch, scale divisions of $1, 2, 5, 10$, or a multiple of these numbers are desirable for ease of plotting and reading. Do not use a scale that will require awkward fractions in the smallest calibration on the paper. The scale should be consistent with the precision of the data. If the numbers are very large or very small, they may be written as a number times 10 to a power, for example, $(3.22)(10)^{-5}$, or $(7.50)(10)^{6}$.

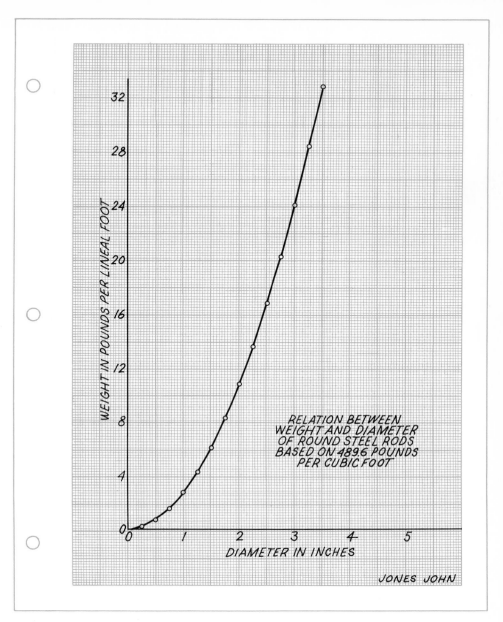

Figure 11-11

6. It is desirable to show zero as the beginning of the ordinate and abscissa quantities unless this would compress the curve unnecessarily. The origin is usually placed in the lower left corner except in cases where both positive and negative values of a function are to be plotted. In such cases the origin should be located so all desired values can be shown.

7. Printed rectangular coordinate paper is not normally available with sufficient margins to accommodate the axes and the description of the quantities plotted. Therefore, the axes should be set in far enough from the edge of the paper to allow for lettering. The sheet may be turned so that the abscissa is along either the short or

long side of the paper. If the graph is prepared for a report, the holes in the paper should be either to the left or at the top of the sheet.

8. Lettering usually is three squares, or approximately $\frac{3}{20}$ in. high. Either vertical or slant lettering may be used.

9. The ordinate and abscissa variables together with their respective units of measure should be labeled. For example: WEIGHT IN POUNDS.

10. The plotted points are fine, tiny dots in pencil. After the points are located, draw a circle, not more than $\frac{1}{16}$ in. in diameter, around each point. Where multiple curves are plotted on one sheet, the points for each curve may be identified by using distinctive identification symbols such as squares, triangles, diamonds, or other simple geometric figures. Distinctive line work such as solid line, dashed line, or long dash–short dash also may be used to aid identification.

11. Graphs may be drawn for theoretical relationships, empirical relationships, or measured relationships. Curves of theoretical relationships will not normally have point designations. Empirical relationships should form smooth curves or straight lines, depending upon the form of the mathematical expression used. Datum points in measured relationships, not supported by mathematical theory or empirical relationships, should be connected by straight lines drawn from point to point. Otherwise, the data obtained from measured relationships will be drawn to average the plotted points. For this reason curves showing measured data do not necessarily go from center to center of the points.

12. Much experimentally determined data when plotted will show a dispersion of the points about an average position due to the many variable factors entering into the measurement. For this condition draw a smooth curve or a straight line, as the data indicate, which as nearly as possible will average the plotted points. A light pencil freehand line will aid in locating the average, but the final line should be mechanically drawn. The example of Figure 11-12 is taken from an actual test to show the dispersion that may occur.

13. In drawing the final curve do not draw the line through the symbols that enclose the plotted points, but rather stop at the perimeter.

14. The title of the graph should include the names of the plotted quantities and should include other descriptive information such as sizes, weights, names of equipment, date that the data were obtained, where data were obtained, serial numbers of apparatus, name of manufacturer of apparatus, and any other information that would help describe the graph.

Figure 11-12 An example of a graph displaying data which were subject to considerable variation. Obviously the curves can be only approximately located. Such curves are sometimes referred to as "paintbrush" curves.

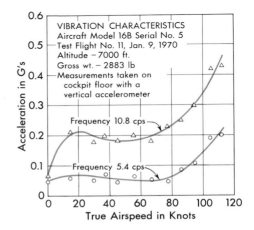

15. The title should be placed on the sheet where it will not interfere with the curve. The title section of display graphs is usually placed across the top of the sheet. Simple graphs that comprise parts of reports frequently have the title in either the lower right quadrant or the upper left quadrant.

16. The name of the person preparing the graph and the date the graph is plotted should be placed in the lower right hand corner of the sheet.

Problems on graphs

11-8. Plot a graph showing the relation of weight to diameter for round steel rods. Plot values for every quarter-inch to and including $3\frac{1}{2}$ in. in diameter. (See model, Figure 11-11.)

Weight of round steel rods in pounds per lineal foot (based on 489.6 lb/ft³)

Size, in.	Weight, lb/ft	Size, in.	Weight, lb/ft
$\frac{1}{4}$	0.167	2	10.66
$\frac{1}{2}$	0.668	$2\frac{1}{4}$	13.50
$\frac{3}{4}$	1.50	$2\frac{1}{2}$	16.64
1	2.68	$2\frac{3}{4}$	20.20
$1\frac{1}{4}$	4.17	3	24.00
$1\frac{1}{2}$	6.00	$3\frac{1}{4}$	28.30
$1\frac{3}{4}$	8.18	$3\frac{1}{2}$	32.70

11-9. Plot a graph showing the relation of normal barometric pressure of air to altitude. Plot values up to and including 15,000 ft.

Altitude, feet above sea level	Normal barometric pressure, inches of mercury	Altitude, feet above sea level	Normal barometric pressure, inches of mercury
0	29.95	5,000	24.9
500	29.39	6,000	24.0
1000	28.86	7,000	23.1
1500	28.34	8,000	22.2
2000	27.82	9,000	21.4
2500	27.32	10,000	20.6
3000	26.82	15,000	16.9
4000	25.84		

11-10. Plot a graph showing the relation between horsepower transmitted by cold-drawn steel shafting and diameter for a speed of 72 rpm based on the formula

$$\text{hp} = \frac{D^3 R}{50}$$

where hp = horsepower
    D = diameter of shaft, in.
    R = rpm of shaft

Calculate and plot values for every inch diameter up to and including 8 in.

11-11. Plot a graph for the following experimental data showing the relation between the period in seconds and the mass of a vibrating spiral spring.

Period, seconds	Mass, grams	Period, seconds	Mass, grams
0.246	10	0.650	70
0.348	20	0.740	90
0.430	30	0.810	110
0.495	40	0.900	130
0.570	50	0.950	150

11-12. Using data in Problem 11-11 plot a graph between period squared and mass on a vibrating spring.

11-13. Plot a graph of the variation of the boiling point of water with pressure.

Boiling point, °C	Pressure, cm of mercury	Boiling point, °C	Pressure, cm of mercury
33	3.8	98	72.9
44	5.3	102	85.8
63	17.2	105	93.7
79	34.0	107	102.2
87	48.1	110	113.5
94	69.1		

11-14. *a.* Plot a graph showing the variation of the following measured values of sliding force with the normal force for a wood block on a horizontal wood surface.

Sliding force, grams	Normal force, grams
100	359
130	462
155	555
185	659
210	765
240	859

b. Determine the slope of the line plotted and compare with the average value of the coefficient of sliding friction obtained from individual readings of normal force and sliding force.

Slope = tan θ (where θ is the angle that the line makes with the abscissa axis)
$$\tan \theta = \frac{y_2 - y_1}{x_2 - x_1}$$

11-15. Plot the variation of pressure with volume, using data as obtained from a Boyle's law apparatus.

Pressure, cm of mercury	Volume, cm^3	Pressure, cm of mercury	Volume, cm^3
50.3	23.2	76.8	15.1
52.5	22.4	79.7	14.7
54.5	21.5	82.7	14.1
56.9	20.9	84.2	13.6
59.4	19.6	87.9	13.2
63.0	18.5	90.6	12.8
65.3	17.8	93.5	12.5
67.2	17.3	95.7	12.3
72.6	16.1	101.9	11.4
74.5	15.6		

Illustration 11-2
*Performance diagram
of a transistor.*

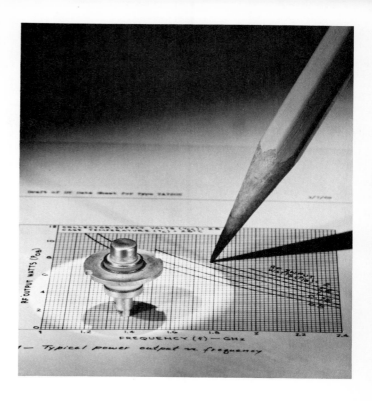

11-16. Using data in Problem 11-15, plot a graph of the relation between the pressure and the reciprocal of the volume.

11-17. Plot the relation between magnetic flux density in kilolines per square centimeter (B) and magnetizing force in gilberts per centimeter (H) for a specimen of tool steel. This graph will form what is customarily called a $B–H$ curve.

B, kilolines/ cm^2	H, gilberts/ cm	B, kilolines/ cm^2	H, gilberts/ cm
9.00	27.1	14.66	189.7
11.80	54.2	14.86	216.8
13.02	81.3	14.98	243.9
13.75	108.4	15.23	271.0
14.09	135.5	15.35	298.1
14.22	162.6	15.57	352.2

11-18. The formula for converting temperatures in degrees Fahrenheit to the equivalent reading in degrees Celsius is

$$C° = \frac{5}{9}(F° - 32°)$$

Plot a graph so that by taking any given Fahrenheit reading between 0 and 220° and using the graph, the corresponding Celsius reading can be determined.

11-19. Plot a graph showing the relation between drill speed and size of drill for carbon steel drills in brass.

Diameter of drill, in.	Drill speed, rpm	Diameter of drill, in	Drill speed, rpm
$1/16$	6112	$5/8$	612
$1/8$	3056	$11/16$	555
$3/16$	2036	$3/4$	508
$1/4$	1528	$13/16$	474
$5/16$	1222	$7/8$	438
$3/8$	1018	$15/16$	407
$7/16$	874	1	382
$1/2$	764	$1 1/16$	359
$9/16$	679	$1 1/8$	340

11-20. Plot a graph showing the variation of temperature with electric current through a heating coil, using the following data, which were taken in the laboratory.

Current, amp	Temperature change, °C
0.0	0.0
0.46	0.5
1.05	1.2
1.50	2.0
2.06	5.1
2.20	7.7
2.35	8.8

11-21. The following data were taken in the laboratory for a 16-cp, carbon-filament electric light bulb. Plot a resistance-voltage curve.

Voltage, volts	Resistance, ohms	Voltage, volts	Resistance, ohms
10	169.5	70	114.5
20	140.0	80	113.2
30	129.0	90	112.5
40	121.5	100	111.8
50	117.0	110	111.2
60	113.2		

11-22. The following data were taken in the laboratory for a 60-w, gas-filled, tungsten-filament light bulb. Plot a resistance-voltage curve.

Voltage, volts	Resistance, ohms	Voltage, volts	Resistance, ohms
10	47.5	70	160.2
20	77.5	80	170.0
30	100.3	90	178.3
40	119.0	100	189.0
50	132.6	110	200.1
60	144.2		

11-23. The equation which expresses the variations of electric current with time in an inductive circuit is

$$i = I_0 \epsilon^{(-Rt)/L}$$

where i = current, amperes

I_0 = original steady-state value of current and is a constant

ϵ = base of the natural system of logarithms and is approximately 2.7183

R = resistance, ohms, in the circuit and is constant

t = time, seconds, measured as the current i varies

L = inductance, henries, and is constant

Let

$$I_0 = 0.16 \text{ amp}$$
$$R = 1.2 \text{ ohms}$$
$$L = 0.5 \text{ henry}$$

Calculate and plot values of i as t varies from 0 to 0.5 sec.

11-24. Plot the variations of efficiency with load for a $\frac{1}{4}$-hp, 110-v, direct-current electric motor, using the following data taken in the laboratory.

Load output, horsepower	Efficiency, per cent
0	0
0.019	24.0
0.050	42.0
0.084	44.9
0.135	50.7
0.175	56.5
0.195	58.0
0.248	59.1
0.306	58.0
0.326	56.2

Plotting on semilogarithmic graph paper

The preceding discussion has concerned the graphing of data on rectangular coordinate paper. There are cases where the variation of the data is such that it may be desirable to compress the larger values of a variable. To do this, semilogarithmic graph paper may be used. Semilog paper, as it is usually called, is graph paper which has one coordinate ruled in equal increments and the other coordinate ruled in increments which are logarithmically expressed. When plotting on this type of paper, it can be turned so that either the horizontal coordinate or the vertical coordinate will have the logarithmic divisions. Semilog paper is available in either one-cycle, two-cycle, three-cycle, four-cycle, or five-cycle ruling.

A semilog grid is especially useful in the derivation of relationships where it is difficult to analyze the rate of change or trend as depicted on rectangular coordinate paper. Data that will plot as a curve on rectangular coordinate paper may plot as a straight line on semilog paper. In many instances this is desirable because the trends are more easily detected. Where straight lines do not occur on a semilog grid, the rate of change is varying.

The same rules apply for plotting on semilog paper as were given for rectangular coordinate paper, except that the numbering of the logarithmic divisions cannot begin with zero. Each cycle on the paper represents a multiple of 10 in value, and the graduations may begin with any power of 10. When reading from a logarithmic graph, interpolations should be made logarithmically rather than arithmetically. An example of data plotted on semilog paper is shown in Figure 11-13.

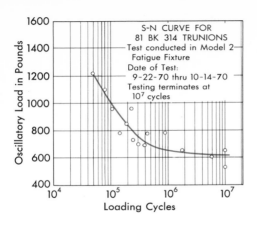

Figure 11-13

Plotting on log-log graph paper

Log-log graph paper, as its name indicates, has both coordinate divisions expressed as logarithmic functions. This subdivision of the sheet serves to compress the larger values of the plotted data. In addition, data that plotted as a curve on rectangular coordinate paper may plot as a straight line on log-log paper. For example, the graphs of algebraic equations representing multiplication, division, powers, and roots may be straight lines on log-log paper.

As an example, the plot of the algebraic expression

$$X = Y^2$$

on rectangular coordinate paper is a parabola. However, if its values are plotted on log-log paper, it is equivalent to taking the logarithm of the expression

$$\log X = 2(\log Y)$$

This expression has the form of a linear equation having a slope of 2. Thus, a relationship of variable quantities that may be expressed as $X = Y^2$ when plotted on log-log paper will be a straight line with a slope of 2.

Log-log paper may be secured in $8\frac{1}{2}$ by 11 in. or larger sheets that have one or more cycles for each coordinate direction. The axis lines are drawn on the sheet in a manner similar to the procedure described for plotting on rectangular coordinate paper. However, the beginning values for the axes will never be zero but always will be a power of 10.

An example of data plotted on log-log paper is shown in Figure 11-14.

Plotting on polar graph paper

Polar graphs are sometimes used where a variable quantity is to be examined with respect to various angular positions. The same general principles of plotting apply as were outlined for rectangular plots except that the outer border is marked off in degrees for the independent variable, and either the horizontal or vertical radial line is marked off for the dependent variable.

Polar graphs frequently are used to display the light output of luminous sources, the response of microphone pickups, and the behavior of rotating objects at various angular positions. An example of a graph plotted on polar coordinate paper is shown in Figure 11-15.

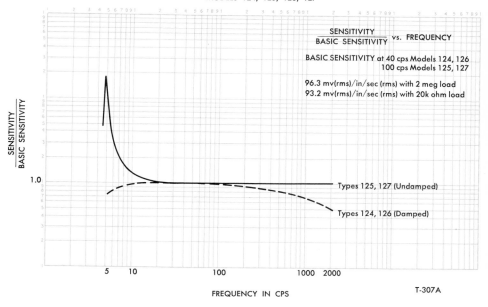

$\dfrac{\text{SENSITIVITY}}{\text{BASIC SENSITIVITY}}$ vs. FREQUENCY

BASIC SENSITIVITY at 40 cps Models 124, 126
100 cps Models 125, 127

96.3 mv(rms)/in/sec (rms) with 2 meg load
93.2 mv(rms)/in/sec (rms) with 20k ohm load

$\dfrac{\text{SENSITIVITY}}{\text{BASIC SENSITIVITY}}$

Types 125, 127 (Undamped)

Types 124, 126 (Damped)

FREQUENCY IN CPS

T-307A

Figure 11-14

Illustration 11-3

The engineer's work is greatly diminished by the use of plotting instruments that are directly connected to the electronic computer.

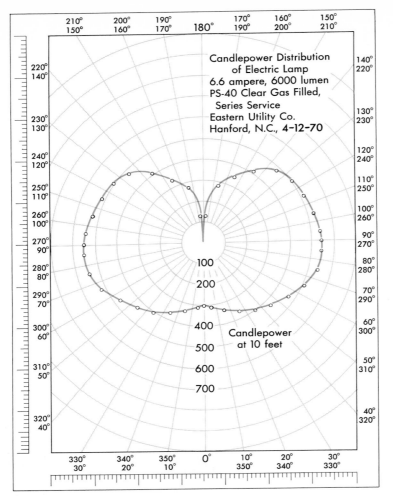

Candlepower Distribution
of Electric Lamp
6.6 ampere, 6000 lumen
PS-40 Clear Gas Filled,
Series Service
Eastern Utility Co.
Hanford, N.C., 4-12-70

Candlepower
at 10 feet

Figure 11-15

Determining empirical equations from curves

Experimentally determined data when plotted usually will approximate a straight line or a simple curve. By plotting experimentally determined data, frequently it is possible to obtain a mathematical equation that closely expresses the relations of the variables.

Many equations encountered in engineering work have the form

$$y - k = m(x - h)^n$$

where n may have either positive or negative values. If the exponent n is 1, the equation reduces to the familiar straight-line slope-intercept form. If the value of n is positive, the equation is a parabolic type, but if the value of n is negative, the equation is a hyperbolic type. This expression affords a means of securing empirical equations from experimental data.

If experimental data are to be plotted and an empirical equation is to be determined, it is advisable first to plot the test data on rectangular coordinate paper in order to gain some idea of the shape of the graph. If the locus approximates a straight

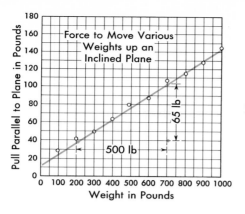

Figure 11-16

line, the general equation $y = mx + b$ may be assumed. The Y intercept b and the slope m may be measured by taking a straight line drawn so as to average the plotted points.

Figure 11-16 is a plot of data taken in the laboratory for a test involving the magnitude of frictional forces. A straight line is drawn to average the plotted points, and the slope of the line is found by taking any two points along the line and determining the X component and the Y component between the two points according to the plotted scales. In this example the slope is approximately 65/500, or 0.13. If the line is projected to the Y axis, corresponding to a value of $x = 0$, the Y intercept is seen to correspond approximately to 14 lb. An approximate equation of these data would be $y = 0.13x + 14$.

If a plot of experimental data on rectangular coordinate paper should appear to be approximately parabolic in shape, an empirical equation may be obtained by plotting the datum points on log-log paper. The slope of the line determines the exponent of the independent variable, and the y intercept, when $x = 1$, defines the coefficient of the independent variable. For example, a plot of data taken in the laboratory is shown in Figure 11-17.

A straight line is drawn to average the plotted points. Using a linear scale, measure the X-component and Y-component values for two points on the plotted line. The slope of the graph in Figure 11-17 is 2.2/2.0, or 1.1, and the Y intercept is 23.3. Substituting these values in the basic equation of a parabola gives $y = 23.3(x^{1.1})$ for the approximate equation.

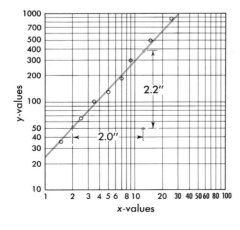

Figure 11-17

In case the plotted points on log-log paper curve upward as x increases, the expression may approximate the form $y = ax^n + k$. To straighten the curve, try subtracting a constant from the y values. By trial and error, a value of k may be found that will cause the plot to follow a straight line. If this is done, the approximate equation may be determined.

If log-log paper is not available, it is still possible to use rectangular coordinate paper to plot a curve as a straight line. If the data indicate the equation may be of the form $y = ax^n$, we can take the logarithm of the equation and plot logarithmic values for the datum points. For example, if we express $y = ax^n$ in logarithmic form, it will be $\log y = \log a + n \log x$. Let $v = \log y$; $C = \log a$; and $u = \log x$. The straight-line equation will then be

$$v = nu + C$$

Plot the logarithm of the data values on rectangular coordinate paper. Measure the slope and the Y intercept. Assume that the slope is measured to be 1.8, using the scales of the plot, and the Y intercept is 0.755. The straight line equation is

$$v = 1.8u + 0.755$$

or $\qquad \log y = 1.8 \log x + 0.755$

Since $\qquad C = \log a = 0.755$

then $\qquad a = 5.69$

The equation then is $\qquad y = 5.69(x^{1.8})$

There are other methods of determining empirical equations, such as the method of least squares, but a complete discussion of such techniques is beyond the scope of this book. Also, data that plot into curves following harmonic laws or exponential laws are not discussed here.

Nomographs

Nomographs are a pictorial method of solving problems which involve equations of various types. Nomographs consist of scales graduated so that distances are proportional to the variables involved. A simple example would be a single line having graduations corresponding to inches on one side and graduations corresponding to centimeters on the other (see Figure 11-18).

The layout of nomographs is beyond the scope of this book, but since the solution to problems involving repeated readings of process or laboratory data may be obtained readily by use of nomographs, a brief discussion of the types and uses of the charts is presented.

Figure 11-18 is an example of a *functional chart*. Charts of this type are frequently used when two variables are related by a constant coefficient.

An *alignment chart* is another example of a nomograph. A simple form consists of

Figure 11-18

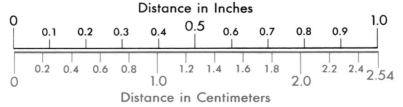

Distance in Inches

Distance in Centimeters

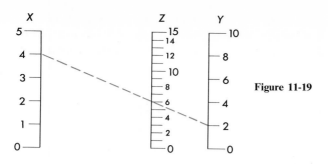

three parallel lines graduated so that a straight line passing through points on two of the graduated lines will intersect the third graduated line at a point that will satisfy the relations between the variables (see Figure 11-19).

Example Given an alignment chart for the equation $x + y = z$. Solve for the value of z when $x = 4$ and $y = 2$.

Solution Lay out a straight line connecting the point on the x scale corresponding to 4 and the point on the y scale corresponding to 2. The intersection of this line with the z scale at 6 is a solution to this problem. Repeating this procedure with other values will enable one to locate the position of the z scale with regard to the x and y scales.

Another form of alignment chart that is of considerable use is the Z chart, so named because the center graduated line runs diagonally. It may be set up to provide a solution to equations of the form $x = (y)(z)$. Other alignment charts may provide solutions to problems having three or four variable quantities by employing multiple interior graduated lines.

Alignment charts are all used in the same manner; that is, a straight line connects two points on the graduated lines and intersects another graduated line, thereby providing a solution to a given problem.

As an example, the nomograph given in Figure 11-20 permits an evaluation of

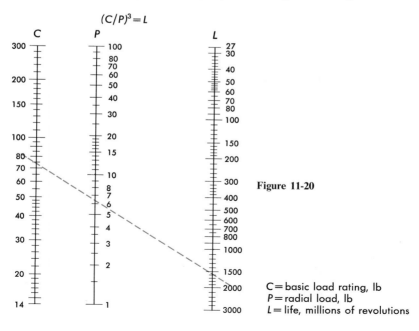

Figure 11-20

C = basic load rating, lb
P = radial load, lb
L = life, millions of revolutions

factors concerned with the life of a ball bearing. The straight line drawn across the chart shows that, for a basic load rating of 100 lb and a radial load of 10 lb, the expected service life of the ball bearing should be a billion revolutions.

Problems

11-25. Plot the values given in Problem 11-8 on semilog paper.
11-26. Plot the values given in Problem 11-11 on semilog paper.
11-27. Plot the values given in Problem 11-15 on semilog paper.
11-28. Plot the values given in Problem 11-23 on semilog paper.
11-29. Plot the values given in Problem 11-8 on log-log paper.
11-30. Plot the values given in Problem 11-10 on log-log paper.
11-31. Plot the values given in Problem 11-15 on log-log paper.

Determine the slope of the line and give the approximate form of the equation shown by the plot.

11-32. The following data were taken from an acoustical and electrical calibration curve for a Type 1126 microphone. The test was run with an incident sound level of 85 db perpendicular to the face of the microphone.

Frequency, cps	Relative response, db
20	−40
50	−29
100	−19
400	−5
1,000	+1
2,000	+1
3,000	0
6,000	−4
10,000	−11

Plot a graph on semilog paper showing the decibel response with frequency.

11-33. The electrical frequency response of a Type X501 microphone is given below.

Frequency, cps	Relative response, db
20	−40
40	−33
80	−22
100	−18
200	−11
400	−5
600	−2
1,000	+1
2,000	+2
4,000	−1
6,000	−4
10,000	−10

Plot a graph on semilog paper showing the decibel response with frequency.

11-34. According to recommendations of the Thrust Bearing Engineers Committee, bearing loads for bearings lubricated with oil having a viscosity range of 115 to 165 Saybolt sec at 100°F should fall between specified values. Plot graphs of bearing loads against speeds to show the range of acceptable operating speeds.

Speed, rps	Bearing load, lb
10	400
100	170
1,000	74
10,000	32
40,000	20
10	1,700
100	650
1,000	275
10,000	123
40,000	70

11-35. The variation of sensitivity of a Model 932 vibration sensing unit with frequency is given below. The basic sensitivity is taken as 96.3 mv (rms) with a 2-MΩ load. Plot sensitivity frequency on three-cycle log-log paper.

Frequency, cps	Ratio of sensitivity at various frequencies to basic sensitivity
4.0	4.6
4.8	19.0
5.0	11.0
6.0	2.7
7.0	1.9
8.0	1.6
10	1.3
20	1.05
40	1.00
80	0.98
100	0.97
300	0.85
600	0.76
1000	0.66
2000	0.46

11-36. A series of test specimens of a crank arm, part No. 466-1, was tested for the number of cycles needed to produce fatigue failure at various loadings. The results of the tests are shown below.

Specimen number	Oscillatory load, lb	Operating cycles to produce failure
1	960	1.1×10^5
2	960	2.2×10^5
3	850	1.5×10^5
4	850	2.4×10^5
5	800	4.2×10^5
6	800	6.0×10^5
7	700	2.4×10^5
8	700	3.1×10^5
9	700	5.1×10^5
10	650	1.8×10^6
11	650	2.6×10^6
12	600	7.7×10^6
13	550	1.0×10^7

Plot a graph of load against operating cycles (*S–N* curve) on semilog paper for the tests on page 231.

11-37. A Weather Bureau report gives the following data on the temperature over a 24-hr period for October 12.

Midnight	47°	2 pm	73°
2 am	46°	4 pm	75°
4 am	44°	6 am	63°
6 am	43°	8 pm	58°
8 am	49°	10 pm	57°
10 am	55°	Midnight	57°
Noon	68°		

Plot the data.

11-38. A test on an acorn-type street lighting unit shows the mean vertical candlepower distribution to be as shown below.

Midzone angle, degrees	Candlepower at 10 ft	Midzone angle, degrees	Candlepower at 10 ft
180	0	85	156
175	0	75	1110
165	0	65	1050
155	1.5	55	710
145	3.5	45	575
135	5.5	35	500
125	8.5	25	520
115	13.5	15	470
105	22.0	5	370
95	40.0	0	370

Plot the data. (While data for only half the plot are given, the other half of the plot can be made from symmetry of the light pattern.)

11-39. The candlepower distribution of a 400-w, Type J-H1 fluorescent lamp used for street light service was measured with a photometer, and the following data were obtained:

Midzone angle, degrees	Candlepower at 10 ft	Midzone angle, degrees	Candlepower at 10 ft
180	0	75	7700
165	0	72	8600
145	0	65	7100
135	3	55	5300
125	20	45	4300
115	100	35	3500
105	700	25	2700
95	1200	15	2300
85	3000	5	2100
		0	2000

Plot the data. (While data for only half the plot are given, the other half of the plot can be made from symmetry of the light pattern.)

11-40. From data determined by the student, draw a circle chart (pie graph) to show one of the following.

a. Consumption of sulfur by various industries in the United States.

b. Budget allocation of the tax dollar in your state.

c. Chemical composition of bituminous coal.

d. Production of aluminum ingots by various countries.

11-41. Make a bar chart showing the number of men students registered in your school for each of the past 10 years.

11-42. Plot the following data and determine an empirical equation for the plotted points:

X:	100	200	300	400	500	600	700	800	900
Y:	0.25	0.38	0.53	0.66	0.79	0.90	1.06	1.17	1.30

11-43. Determine the empirical equation, using the following data which were taken in the laboratory for a test involving accelerated motion:

t:	5	10	20	40	60	80	100
s:	0.93	5.6	32	175	490	989	17,600

11-44. Laboratory data taken on an adjustable time-delay relay show the following values:

Dial index settings D:		2	4	6	8	10
Seconds delay time T:	0.124	0.084	0.063	0.026	0.014	

Find an empirical equation to express the data.

11-45. The following data were recorded during a laboratory test of a system of gears. Find an empirical equation to express the data.

Applied force F:	11.0	13.0	21.5	26.0	34.0	39.0	41.0	49.0	50.5
Weight lifted W:	135	180	210	345	275	310	340	370	400

11-46. Data taken on a laboratory test involving pressure–volume relations of a gas are as follows:

P:	14.6	17.5	20.9	25.0	29.0	33.6	39.0	45.5
V:	26.4	22.3	19.1	16.3	14.1	12.2	10.5	9.2

Determine an empirical equation for the data.

11-47. Determine an empirical equation to express the data given in Problem 11-9.

11-48. Determine an empirical equation that will express the data given in Problem 11-11.

11-49. Plot a graph on rectangular coordinate paper of $N = (1.296)^x$ for values of x from -9.0 to $+9.0$ in 0.5 increments.

11-50. Plot a graph on rectangular coordinate paper of the equation $N = (0.813)^x$ for values of x from -9.0 to $+9.0$ in 0.5 increments.

11-51. Using the nomograph of Figure 11-20, what will be the allowable radial load on a ball bearing if the basic load rating is 22 lb and the expected life of the bearing is to be $1.4(10)^8$ revolutions?

11-52. Construct a functional scale about 6 in. long that will relate temperatures in degrees Fahrenheit and degrees Celsius for the range -40 to $100°C$.

11-53. Construct a functional scale about 10 in. long that will show the relation between the diameter and circumference of a circle for values of diameter from 2 to 9 in.

11-54. Prepare a line chart that will permit converting readings from grams to ounces up to 64 oz and a line chart that will convert readings from pounds to kilograms up to 10 lb.

12

The metric (SI) and other unit systems

Man interprets the universe in which he lives by evaluating those things that he perceives. Through experience he has learned that there are certain physical quantities that are unique and *fundamental* and that can be used to describe all other physical relationships. Among the fundamental dimensions most commonly recognized are *length, force,* and *time,* which are used extensively by peoples of all cultures, economic classes, and educational levels. Engineering and scientific calculations make use of measurements of all types and, therefore, use not only these, but other fundamental dimensions as well. Fundamental dimensions may be combined in numerous ways to form *derived dimensions;* it is by this means that man is able to portray accurately the physical laws of nature that he observes.

Some measurements are made with precise instruments, while others are the result of crude approximations. Regardless of the accuracy of the measurements or of the particular type of measuring instrument used, the measurements are themselves merely representative of certain comparisons previously agreed upon.

The length of a metal cylinder, for example, can be determined by laying it alongside a calibrated scale or ruler. The 12-in. ruler is known to represent one third of a yard, and a yard is recognized as being equivalent to 36.00/39.37 metre[1]—which used to be the distance between two marks on a platinum–iridium bar kept in a vault in Sèvres, France, but is now defined in terms of the wavelength of a particularly uniform monochromatic light. All these methods of measurements are comparisons. Other similar standards exist for the measurement of temperature, time, and force.

Physical quantities to be measured may be of two types: those concerned with *fundamental dimensions* of length (L), time (T), force (F), mass (M), electrical charge

---

[1] The SI spelling "metre" is used instead of the more traditional spelling, "meter," which has been used in the past in this country. Both spellings are acceptable.

(Q), luminous intensity (I), and temperature (θ); and those concerned with *derived dimensions*, such as area, volume, pressure, or density. *Fundamental dimensions* may be subdivided into various-sized parts, called *units*. The dimension *time* (T), for example, can be expressed in the units of seconds, hours, days, and so forth, depending upon the application to be made or the magnitude of the measurement. *Derived dimensions* are categorical descriptions of some specific physical characteristic or quality of an entity, and they are brought into being by combining *fundamental dimensions*. Area, therefore, is expressed dimensionally as length times length, or length squared (L^2), pressure as force per unit area (F/L^2, or FL^{-2}), and acceleration as length per time squared (L/T^2, or LT^{-2}).

Most measured quantities must be expressed in both magnitude and units. To state that an area was 146 would have no meaning. For example, an area could be tabulated as 146 mi² or 146 cm²; a pressure could be recorded as 0.0015 N/cm² or 0.0015 lb$_f$/in.²; an acceleration could be indicated as 159 in./sec² or 159 ft/sec², and so forth. However, some values used in engineering computations are dimensionless (without dimensions). These should be ignored in the unit balancing of an equation. *Radians, π, coefficient of friction, ratios,* and *per cent error* are examples of dimensionless quantities.

Equations involving measured quantities must be balanced dimensionally as well as numerically. Both dimensions and units can be multiplied and divided or raised to powers just like ordinary algebraic quantities. When all of the dimensions (or units) in an equation balance, the equation is said to be *dimensionally homogeneous*.

Example An alloy has a specific weight of 400 lb$_f$/ft³. What is the weight of 2 ft³ of the alloy? Show the numerical and dimensional solutions to the problem.

$$W = V\rho \qquad \text{[Algebraic equation]}$$

or

$$(\text{Weight}^2 \text{ of metal}) = (\text{volume of metal})(\text{specific weight of metal})$$

Fundamental Dimensions: $F = (L^3)\left(\dfrac{F}{L^3}\right)$ [Dimensional equation]

Units: $F = (2 \text{ ft}^3)\left(400\dfrac{\text{lb}_f}{\text{ft}^3}\right) = \mathbf{800\ lb}_f$ [Unit equation]

Check: $\text{lb}_f = \text{lb}_f$

Frequently it will be necessary to change unit systems, that is, feet to inches, hours to seconds, pounds to grams, and so on. This process can be accomplished by the use of unity conversion factors that are multiplied by the expression to be changed. Refer to Appendix III for a listing of commonly used conversion factors.

Example Change a speed of 3000 miles per hour (mi/hr) to feet per second (ft/sec).

Fundamental dimensions: $\dfrac{L}{T} = \dfrac{L}{T}$

Units: $V = \left(3000\dfrac{\text{mi}}{\text{hr}}\right)\left(\dfrac{5280 \text{ ft}}{1 \text{ mi}}\right)\left(\dfrac{1 \text{ hr}}{3600 \text{ sec}}\right) = \mathbf{4400}\ \dfrac{\text{ft}}{\text{sec}}$

[2]Weight is expressed in the dimensions of force.

The two conversion factors, (5280 ft/1 mi) and (1 hr/3600 sec), are each equivalent to unity, since the numerator of each fraction is equal to its denominator (5280 ft = 1 mi, and 1 hr = 3600 sec).

Note that the word *per* means *divided by*. To avoid misunderstandings in computations, the units should be expressed in fractional form.

Example

a. $(X \text{ per } Y) \text{ per } Z = (X \div Y) \div Z = [(X/Y)/Z] = \dfrac{X/Y}{Z} = \dfrac{X}{YZ}$

b. Acceleration = 156 ft per sec per min = 156 ft/sec/min

$$= 156 \frac{\text{ft}}{(\text{sec})(\text{min})}$$

c. Pressure = 65.4 newtons per square centimetre = $65.4 \dfrac{\text{N}}{\text{cm}^2}$

Example Solve for the fundamental dimensions of Q and P in the following dimensionally homogeneous equation if C is a velocity and B is an area.

$$Q = C(B - P)$$

Fundamental Dimensions: $\qquad Q = \dfrac{L}{T}(L^2 - P)$

Since the equation is dimensionally homogeneous, P must also be length squared (L^2) in order that the subtraction can be carried out. If this is true, the units of Q are[3]

$$Q = \frac{L}{T}(L^2 - L^2) = \frac{L}{T}(L^2) = \frac{L^3}{T}$$

Example Solve for the conversion factor k:

a. $\dfrac{L^2 T^3 \theta}{F^4} = k\left(\dfrac{L^5 T F^2}{Q^2}\right)$

Solving for k: $\qquad\qquad\qquad k = \dfrac{T^2 \theta Q^2}{F^6 L^3}$

and $\qquad \dfrac{L^2 T^3 \theta}{F^4} = \left(\dfrac{T^2 \theta Q^2}{F^6 L^3}\right)\left(\dfrac{L^5 T F^2}{Q^2}\right)$

Check: $\qquad \dfrac{L_2 T^3 \theta}{F^4} = \dfrac{L^2 T^3 \theta}{F^4} = L^2 T^3 \theta F^{-4}$

b. $\left(\dfrac{F^3 T^2 \theta}{L^2 Q}\right) k = \dfrac{M F^2 Q^3}{T^2 L^3}$

$\qquad k = \dfrac{M Q^4}{F T^4 L \theta}$

Check: $\qquad \left(\dfrac{F^3 T^2 \theta}{L^2 Q}\right)\left(\dfrac{M Q^4}{F T^4 L \theta}\right) = \dfrac{M F^2 Q^3}{T^2 L^3} \quad$ or $\quad M F^2 Q^3 T^{-2} L^{-3}$

---

[3] Remember that the terms L^2 represent a particular length squared in each instance. Thus the remainder (depending on the numerical magnitude of each term) will also be length squared or will be zero for the special case of the original lengths being equal.

Problems

Solve for the conversion factor k.

12-1. $k = \dfrac{L^3 \theta T Q^5}{FM} = \dfrac{M^3 \theta Q}{L^2}$

12-2. $\dfrac{FTL^2}{\theta M^3} = k \dfrac{\theta^5 M}{T^2}$

12-3. $k \left(\dfrac{QM}{TF^2} \right) = \sqrt{L^4 I \theta Q^8}$

12-4. $\theta^2 \sqrt{LM^5} = k \left(\dfrac{FT^2}{M^3} \right)$

12-5. $k(F\theta^2 TL^{-2}M^{-3}) = M^5 L\theta F^{-3}$

12-6. $M^2 FT^{-5}L^{-2} = k \sqrt{MT\theta}$

12-7. $\sqrt{LT^3 F^{-2}M} = k \sqrt{TF^3 M^6}$

12-8. $k \dfrac{\sqrt{T^3 Q}}{L^2 F^{-2}} = MTLF$

12-9. $k(F^2 T \sqrt{L\theta^{-2}}) = \theta^{-3} T^{-2}$

12-10. $FL^3 Q^{-1}M^{-3} = k \sqrt{L^2 Q^{-1}}$

12-11. Convert 76 newtons to dynes and lb$_f$.

12-12. Convert 2.67 in. to angstroms and miles.

12-13. Convert 26 knots to feet per second and metres per hour.

12-14. Convert $8.07(10)^3$ tons to grams and ounces.

12-15. Convert 1.075 atmospheres to dynes per cm^2 and inches of mercury.

12-16. Convert 596 Btu to foot-pounds and Joules.

12-17. Convert 26,059 watts to horsepower and ergs per second.

12-18. Convert 92.7 coulombs to faradays.

12-19. Convert 75 angstroms to feet.

12-20. Convert 0.344 henry to abhenries.

12-21. Express 2903 ft^3 of sulfuric acid in gallons and cubic metres.

12-22. Change a Btu to horsepower-seconds.

12-23. A car is traveling 49 mi/hr. What is the speed in feet per second and metres per second?

12-24. A river has a flow of $3(10)^6$ gallons per 24-hr day. Compute the flow in cubic feet per minute.

12-25. Convert 579 qt/sec to cubic feet per hour and cubic metres per second.

12-26. A copper wire is 0.0809 cm in diameter. What is the weight of 1000 m of the wire?

12-27. A cylindrical tank 2.96 ft high has a volume of 136 gallons. What is its diameter?

12-28. A round iron rod is 0.125 in. in diameter. How long will a piece have to be to weigh 1 lb?

12-29. Find the weight of a common brick that is 2.6 in. by 4 in. by 8.75 in.

12-30. Convert 1 yd^2 to acres and square metres.

12-31. A white pine board is 14 ft long and 2 in. by 8 in. in cross section. How much will the board weigh? At $200.00 per 1000 f.b.m., what is its value?

12-32. A container is 12 in. high, 10 in. in diameter at the top, and 6 in. in diameter at the bottom. What is the volume of this container in cubic inches? What is the weight of mercury that would fill this container?

12-33. How many gallons of water will be contained in a horizontal pipe 10 in. in internal diameter and 15 ft long, if the water is 6 in. deep in the pipe?

12-34. A hemispherical container 3 ft in diameter has half of its volume filled with lubricating oil. Neglecting the weight of the container, how much would the contents weigh if kerosene were added to fill the container to the brim?

12-35. What is the cross-sectional area of a railroad rail 33 m long that weighs 94 lb/yd?

12-36. A piece of cast iron has a very irregular shape and its volume is to be determined. It is submerged in water in a cylindrical tank having a diameter of 16 in. The water level is raised 3.4 in. above its original level. How many cubic feet are in the piece of cast iron? How much does it weigh?

12-37. A cylindrical tank is 22 ft in diameter and 8 ft high. How long will it take to fill the tank with water from a pipe which is flowing at 33.3 gallons/min?

12-38. Two objects are made of the same material and have the same weights and diameters. One of the objects is a sphere 2 m in diameter. If the other object is a right cylinder, what is its length?

12-39. A hemisphere and cone are carved out of the same material and their weights are equal. The height of the cone is 3 ft. $10\frac{1}{2}$ in., while the radius of the hemisphere is 13 in. If a flat circular cover were to be made for the cone base, what would be its area in square inches?

12-40. An eight-sided wrought iron bar weighs 3.83 lb per linear foot. What will be its dimension across diagonally opposite corners?

12-41. Is the equation $a = (2S/t^2) - (2V_1/t)$ dimensionally homogeneous if a is an acceleration, V_1 is a velocity, t is a time, and S is a distance? Prove your answer by writing the equation with fundamental dimensions.

12-42. Is the equation $V_2{}^2 = V_1{}^2 + 2as$ dimensionally correct if V_1 and V_2 are velocities, a is an acceleration, and s is a distance? Prove your answer by rewriting the equation in fundamental dimensions.

12-43. In the homogeneous equation $R = B + \frac{1}{2}CX$, what are the fundamental dimensions of R and B if C is an acceleration and X is a time?

12-44. Determine the fundamental dimensions of the expression $B/g \sqrt{D - m^2}$, where B is a force, m is a length, D is an area, and g is the acceleration of gravity at a particular location.

12-45. The relationship $M = \sigma I/c$ pertains to the bending moment for a beam under compressive stress. σ is a stress in F/L^2, C is a length L, and I is a moment of inertia L^4. What are the fundamental dimensions of M?

12-46. The expression $V/K = (B - \frac{7}{3}A)A^{5/3}$ is dimensionally homogeneous. A is a length and V is a volume of flow per unit of time. Solve for the fundamental dimensions of K and B.

12-47. Is the expression $S = 0.031 \ V^2/fB$ dimensionally homogeneous if S is a distance, V is a velocity, f is the coefficient of friction, and B is a ratio of two weights? Is it possible that the numerical value 0.031 has fundamental dimensions? Prove your solution.

12-48. If the following heat transfer equation is dimensionally homogeneous, what are the units of k?

$$Q = \frac{-kA(T_1 - T_2)}{L}$$

A is a cross-sectional area in square feet, L is a length in feet, T_1 and T_2 are temperatures (°F), and Q is the amount of heat (energy) conducted in Btu per unit of time.

Fundamental and derived units

The most commonly used fundamental and derived units in engineering calculations are the following:

Units of length The concept of *length* as a measure of space in one direction is easily understood. People in every country use this concept because the position of any point in our universe may be described in relation to any other point by specifying three lengths. The world standard of length is the metre (m), defined now in terms of the wavelength of a particularly uniform monochromatic light. It is quite close to being equal to the distance from the earth's equator to the North Pole divided by 10 million, which was its original definition. This unit of length is commonly used by engineers and scientists in most countries for the usual engineering problems as well as in the field of space mechanics.

In the United States, the most common units of length that are used in engineering calculations are the inch (in.), the foot (ft), and the mile (mi). Less common are the yard and the nautical mile. They are defined as

$$1 \text{ in.} = 2.54(10)^{-2} \text{ m (exactly by definition)}$$
$$1 \text{ ft} = 12 \text{ in.}$$
$$1 \text{ m} = 39.37 \text{ in.}$$
$$1 \text{ yd} = 3 \text{ ft}$$
$$1 \text{ mi} = 5280 \text{ ft}$$
$$1 \text{ mi} = 1609.34 \text{ m}$$
$$1 \text{ nautical mi} = 6080.27 \text{ ft approximately}$$
$$1 \text{ nautical mi} = 1852 \text{ m}$$

Often feet and inches or feet and miles are used in the same problem. The foot is sometimes decimalized and sometimes the last fractional foot is given in inches and fractions of an inch. Sometimes the inch is decimalized and sometimes it is frac-

Illustration 12-1
For the next decade, engineers and technicians must be equally proficient in using metric and English units.

tionalized; sometimes it is both decimalized and fractionalized in the same problem. One of the disadvantages of the English system of measurement is the tendency to use a mixture of methods of showing a measured quantity.

Units of force Force is most commonly thought of as a "push" or a "pull" and represents the action of one body on another. The action may be exerted by direct contact between the bodies or at a distance, as in the case of magnetic and gravitational forces.

A common unit of force used by American engineers is the pound (lb_f). This is the force that is required to accelerate a pound mass with the mean acceleration of gravity, or $g = 32.174$ ft/sec². Its value is

$$1 \ lb_f = 4.448 \ \text{newtons}$$

The newton (N) is derived from the kilogram by means of Newton's law, $F = Ma$. Thus,

$$1 \ N = 1 \ kg_m \times 1 \ \text{m/sec}^2$$
$$1 \ N = 0.2248 \ lb_f$$

and it is the force required to accelerate a 1-kg mass 1 m/sec². This unit of force (N) is the standard unit in the SI (International System of Units) unit system.

Units of time Time cannot be defined in simple terms, but in general it is a measure of the interval separating the occurrence of two events. The mean solar day is the standard unit of time in all systems of units used at present. The hour (hr), the minute (min), and the second (sec) are all derived from the mean solar day, but not decimally. Since all four of these units are used, sometimes even in the same problem, it is easy to see how mistakes can be made and considerable extra work required. This is also the case with the International System of Units (SI), since the unit of time is

Illustration 12-2
The contrast of energy sources is depicted in this picture, which shows the relative energy conversion from coal and nuclear fuels. In working with such energy sources, the engineer must be able to convert from one unit system to another.

the same in all systems of units. The most common unit used in engineering calculations and the standard unit in the SI unit system is the second, defined as

$$1 \text{ sec} = \frac{1 \text{ mean solar day}}{86,400}$$

Then, 1 min = 60 sec and 1 hr = 3600 sec.

Units of mass While length, force, and time are readily understood concepts, *mass* is somewhat more difficult to perceive. The universe is filled with matter—the accumulation of electrons, protons, and neutrons. *Mass* is a measure of the quantity of these subatomic particles that a particular object possesses and to which it owes its inertia. Although a quantity of matter can change form—for example, when a block of ice is melted to water and then vaporized to steam—its "quantity" does not change.

In contrast to length, force, and time, there is no direct measure for mass. Its quantity may be measured only through an examination of its properties, such as the amount of force that must be provided to give it a certain acceleration. The world standard of mass is the kilogram (kg), defined originally as being one thousandth of 1 m^3 of water at a temperature of 4°C and standard atmospheric pressure, but now defined as the mass of a block of platinum kept at the French Bureau of Standards. This unit has been used by American electrical and space engineers and is the standard unit of mass in the SI unit system.

The pound mass (lb_m) is the unit that the average American engineer thinks he is using most of the time. In most instances this is incorrect. Generally the pound that he uses is the pound force (lb_f), from which he derives units of mass by means of Newton's law, $F = Ma$. Thus,

$$1 \text{ lb}_m = 0.4535924277 \text{ kg (by definition)}$$

$$1 \frac{\text{lb}_f \text{ sec}^2}{\text{ft}} \text{ (also called a slug)} = 32.174 \text{ lb}_m$$

$$1 \frac{\text{lb}_f \text{ sec}^2}{\text{in.}} = 386.088 \text{ lb}_m$$

$$1 \text{ kg} = 2.205 \text{ lb}_m = 6.85 \times 10^{-2} \text{ slug}$$

Units of temperature Temperature is an arbitrary measure which is proportional to the average kinetic energy of the molecules of an ideal gas. Four temperature scales are used by American engineers. The degree Celsius (°C), formerly called centigrade, reads zero at the freezing point of water and 100°C at the boiling point of water under standard conditions of pressure. However, the kelvin is the standard unit of temperature in the SI system. The temperature in kelvins (K) is derived from the Celsius temperature reading by the following equation:

$$\text{Temperature in K} = \text{temperature in °C} + 273.15$$

The degree Fahrenheit (°F) reads 32°F at the freezing point of water and 212°F at the boiling point of water under standard conditions of pressure.[4] The degree Rankine (°R) is derived from the Fahrenheit scale by the equation

$$\text{Temperature in °R} = \text{temperature in °F} + 459.69$$

---

[4] G. D. Fahrenheit thought that 0°F was the lowest possible temperature that could be obtained and that 100°F was the uniformly standard temperature of human blood.

When a temperature is measured in either K or in °R, it is said to be the *absolute* temperature, because these scales read zero for the condition where the kinetic energy of the molecules of an ideal gas is presumed to be zero.

Units of area and volume   Units of area and volume are derived from the units of length for the most part. However, in the United States the gallon is a commonly used measure of volume that is in no way related to units of length. The units of area and volume in the SI system are, respectively,

$$area = square\ metre\ (m^2)$$
$$volume = cubic\ metre\ (m^3)$$

Units of velocity and acceleration   These units are all derived from the fundamental units of length and time. The units of area and volume in the SI system are, respectively,

$$velocity = metre\ per\ second\ (m/s)$$
$$acceleration = metre\ per\ second\ per\ second\ (m/s^2)$$

Units of work and energy   Work is the product of a force and a distance through which that force acts. Energy is the ability or capacity for doing work. Although the two quantities are conceptually different, they are measured by the same units. Several different units are used. For example, foot-pounds, inch-pounds, and horse-power-hours are all used for work and mechanical energy; both the joule and the kilowatt-hour are used for electrical energy; and both the calorie (two types) and the British thermal unit (Btu) are used for heat energy. In some problems all these units occur, which frequently makes the task of unit conversion the most formidable part of the solution. However, in the SI system both work and energy are

$$work = joule\ (J) = N\text{-}m$$
$$energy = joule\ (J) = N\text{-}m$$

Units of power   Power is the time rate of accomplishing work. The average power is the work performed divided by the time required for the performance. Since power units are derived units, they also involve the various work and energy units described above. In addition, the ton of refrigeration (3517 watts) is sometimes used in air-conditioning design calculations. In the SI system power is expressed as

$$power = watt\ (W) = J/s$$

Units of pressure   Pressure is the result of a force distributed over an area. In general, the units of pressure have been derived from conventional units of force and area. However, other measures are also used. For example, the standard atmospheric pressure is commonly used as a unit, and for fractional atmospheres the millimeter of mercury, the inch of mercury, and the inch of water are used. In the SI system pressure is expressed as

$$pressure = pascal\ (P) = N/m^2$$

Units of density   Density is a measure of the mass that a body of uniform substance possesses per unit volume. The units of density are derived units that are made up by dividing the chosen unit of *mass* by the unit of volume, for example, grams per cubic centimeter. In the SI system density is expressed as

$$\text{density} = \text{kilogram per cubic metre (kg/m}^3)$$

Units of specific weight Specific weight is a measure of the *weight* of a substance per unit volume. Many people confuse density with specific weight. Remember that they are not the same, but rather that they are related to each other by the relationship

$$(\text{specific weight}) = (\text{density})(\text{acceleration of gravity})$$

The units most commonly used for specific weight are lb/ft^3 and lb/in.^{3}, where the lb is a unit of force, lb$_f$. They represent the attraction, in lb$_f$, of the earth on either 1 ft^3 or 1 in.^{3} of the material. To convert specific weights to density for use in a gravitational system, one must divide these quantities by the acceleration of gravity, g. If one divides the first unit by $g = 32.174$ ft/sec^2, one obtains the unit of density lb$_f$-sec^2/ft^4, or slugs/ft^3; and, if one divides the second unit by $g = 386.088$ in./sec^2, one obtains the unity of density lb$_f$-sec^2/in.^{4}.

Unit systems

The need for unit systems was first evident when it became necessary to explain the fundamental relationships between force, mass, and acceleration. Sir Isaac Newton (1642–1727) expressed several basic laws that he believed to govern the motion of particles. Only recently has it become evident that in studying the motion of atoms and certain planets Einstein's theory of relativity must supplant Newton's concepts. However, Newton's "second law" still serves as a basis for much of today's engineering mechanics. Briefly this law may be stated as follows:

When an external unbalanced force F acts on a rigid particle of mass, the motion of the particle will be changed. The particle will be accelerated. Its rate of change in motion will be in the direction of the unbalanced force and will be proportional to it.

Stated mathematically, $F_1/a_1 = F_2/a_2 = F_3/a_3 = F_n/a_n = $ a constant, where F_1, F_2, F_3, and so forth, are external unbalanced forces acting on a particle, and a_1, a_2, a_3, and so forth, are consequential accelerations of the particle.

The quotient of (F/a) is a quantity which is invariant. The units of this term depend upon the units arbitrarily chosen to define F and a. This constant has been called the *mass* of the particle under consideration. It is properly designated by the symbol M, or in some cases by the product of the two symbols, km. In this latter case, k could be a value of 1, or it could be some other dimensional expression whose resultant value is unity. The mass M of a particular body is independent of the location of the body in the universe. Thus

$$F = Ma$$
or
$$= kma$$

Unit systems are of two general types—absolute and gravitational. The absolute systems are independent of gravitational effects on the earth or other planets and have most generally been used for scientific calculations. In absolute systems the dimensions of force are derived in terms of the fundamental units of *time, length,* and *mass.* There are three absolute systems in use today, as shown in Table 12-1. How-

Table 12-1   Unit systems

	Absolute			Gravitational		
	(1) SI (Modified MKS)	(2) CGS	(3) FPS	(4) FPS	(5) MKS	(6) American engineering
Fundamental dimensions						
Force (F)	—	—	—	lb_f	kg_f	lb_f
Length (L)	m	cm	ft	ft	m	ft or in.
Time (T)	sec*	sec	sec	sec	sec	sec
Mass (M)	kg_m	g	lb_m	—	—	lb_m
Derived dimensions						
Force (F)	$\dfrac{kg_m\text{-m}}{\text{sec}^2}$ (called a newton†)	$\dfrac{\text{g-cm}}{\text{sec}^2}$ (called a dyne)	$\dfrac{lb_m\text{-ft}}{\text{sec}^2}$ (called a poundal)	—	—	—
Mass (M)	—	—	—	$\dfrac{lb_f\text{-sec}^2}{\text{ft}}$ (called a slug)	$\dfrac{kg_f\text{-sec}^2}{\text{m}}$	—
Energy (LF)	N-m	cm-dyne (called an erg)	ft-poundal	$\text{ft-}lb_f$	$\text{m-}kg_f$	$\text{ft-}lb_f$
Power $\left(\dfrac{LF}{T}\right)$	$\dfrac{\text{N-m}}{\text{sec}}$	$\dfrac{\text{erg}}{\text{sec}}$	$\dfrac{\text{ft-poundal}}{\text{sec}}$	$\dfrac{\text{ft-}lb_f}{\text{sec}}$	$\dfrac{\text{m-}kg_f}{\text{sec}}$	$\dfrac{\text{ft-}lb_f}{\text{sec}}$
Velocity $\left(\dfrac{L}{T}\right)$	$\dfrac{\text{m}}{\text{sec}}$	$\dfrac{\text{cm}}{\text{sec}}$	$\dfrac{\text{ft}}{\text{sec}}$	$\dfrac{\text{ft}}{\text{sec}}$	$\dfrac{\text{m}}{\text{sec}}$	$\dfrac{\text{ft}}{\text{sec}}$
Acceleration $\left(\dfrac{L}{T^2}\right)$	$\dfrac{\text{m}}{\text{sec}^2}$	$\dfrac{\text{cm}}{\text{sec}^2}$	$\dfrac{\text{ft}}{\text{sec}^2}$	$\dfrac{\text{ft}}{\text{sec}^2}$	$\dfrac{\text{m}}{\text{sec}^2}$	$\dfrac{\text{ft}}{\text{sec}^2}$
Area (L^2)	m^2	cm^2	ft^2	ft^2	m^2	ft^2
Volume (L^3)	m^3	cm^3	ft^3	ft^3	m^3	ft^3
Density $\left(\dfrac{M}{L^3}\right)$	$\dfrac{kg_m}{\text{m}^3}$	$\dfrac{\text{g}}{\text{cm}^3}$	$\dfrac{lb_m}{\text{ft}^3}$	$\dfrac{lb_f\text{-sec}^2}{\text{ft}^4}$	$\dfrac{kg_f\text{-sec}^2}{\text{m}^4}$	$\dfrac{lb_m}{\text{ft}^3}$
Pressure $\left(\dfrac{F}{L^2}\right)$	$\dfrac{\text{N}}{\text{m}^2}$ (called a pascal)	$\dfrac{\text{dyne}}{\text{cm}^2}$	$\dfrac{\text{poundal}}{\text{cm}^2}$	$\dfrac{lb_f}{\text{ft}^2}$	$\dfrac{kg_f}{\text{m}^2}$	$\dfrac{lb_f}{\text{ft}^2}$

*The abbreviation for seconds in the SI international system of units is s rather than sec. However, in this text both abbreviations will be used interchangeably.
† A newton (N) is the force required to accelerate a 1-kg mass at 1 m/sec². The acceleration of gravity at sea level and 45° latitude has the measured value of 9.807 m/sec². A force of 1 N equals 0.2248 lb_f.

ever, of these the SI system (International System of Units) has in recent years been adopted by all the major countries of the world, except the United States. In this country the SI system [a modified version of the previously used metre–kilogram–second (MKS) system] is rapidly being accepted into every area of industrial life, and there is no longer any doubt that this system will soon be commonly accepted throughout the world, greatly facilitating the exchange of scientific and technical information.

The American Engineering is the more commonly used of the gravitational systems. In this system, by definition, the mass and weight of an object can be assumed to be of equal magnitude *if the acceleration of gravity at that particular location on the earth is 32.2 ft/sec²*. For many analyses such an assumption is a convenience and any inherent error in the assumption (should the acceleration of gravity vary from 32.2 ft/sec²) can be ignored. Because of its widespread use in engineering calculations, this generation of engineering graduates must be equally conversant with the SI and American Engineering systems.

The SI—international system of units

The decimal system of units was first proposed in the sixteenth century as a solution to the great confusion and jumble of units of weights and measures. However, it was not until 1790 that the French National Assembly requested the French Academy of Sciences to work out a system of units suitable for adoption by the entire world. A system, based on the metre as a unit of length and the gram as a unit of mass, was adopted as a practical measure to benefit industry and commerce.

The inherent value of a decimalized system of units has long been recognized in the United States. An act of Congress in 1866 declared that "it shall be lawful throughout the United States of America to employ weight and measures of the metric system."[5]

In 1893, the international metre and kilogram became the fundamental standards of length and mass in the United States, both for metric and customary weights and measures. However, it was not until 1965, when the British Board of Trade announced that the United Kingdom would adopt the metric system, that our government decided to act decisively. After several years of investigation by Congress, legislation was finally enacted in 1974 to implement changeover to the SI system. Although no specific date has been set for mandatory changeover, many of the larger companies have, on their own initiative, adopted plans for changing their operations to the SI system.

SI is a rationalized selection of units from the metric system which, individually, are not new. The great advantage of SI is that there is one and only one unit for each of seven physical quantities—the metre (m) for length, kilogram (kg) for mass, second (s) for time, etc. From these elemental base units, units for all other mechanical quantities are derived. Each derived unit is defined by a simple equation: $F = ma$ (force), $W = Fl$ (work or energy), $P = W/t$ (power), etc. Some of the SI units have only generic names such as metre per second (velocity): others have been assigned special names such as newton (force), joule (work or energy), and watt (power). In each instance, however, there is a unique and well-defined set of symbols and abbreviations.

In the SI system the units of force, energy, and power are the same regardless of whether the process is mechanical, electrical, chemical, or nuclear. Another advantage is the decimal relation between all multiples and submultiples of the base units.

[5] *Metric Practice Guide*, American National Standards Institute, New York, p. 1.

Table 12-2

Prefix	Multiplication factor	SI symbol
tera	10^{12}	T
giga	10^{9}	G
mega	10^{6}	M
kilo	10^{3}	k
hecto	10^{2}	h
deka	10^{1}	da
deci	10^{-1}	d
centi	10^{-2}	c
milli	10^{-3}	m
micro	10^{-6}	u
nano	10^{-9}	n
pico	10^{-12}	p
femto	10^{-15}	f
atto	10^{-18}	a

Symbols for SI units are not capitalized unless the unit is derived from a proper name: thus, N for Sir Isaac Newton, but m for metre. Unabbreviated units are not capitalized. Numerical prefixes (such as milli, kilo, giga, etc.) and their symbols are not capitalized, except for the symbols T, G, and M (mega) (see Table 12-2). Also, periods should not be used after SI unit symbols except at the end of a sentence.

SI prefixes should be used to indicate orders of magnitude, thus eliminating insignificant digits and decimals and providing a convenient substitute for writing powers of 10. A listing of these SI prefixes is given in Table 12-2.

Definitions of the seven fundamental or base units and most of the more important derived units are given below.

Fundamental or base units

Length: 1 metre (m)
Mass: 1 kilogram (kg)
Time: 1 second (s)
Current = 1 ampere (A)
Temperature: 1 kelvin (K)
Amount of substance: 1 mole (mol)
Light source: 1 candela (cd)

Supplementary units

Plane angle: 1 radian (rad)
Solid angle: 1 steradian (sr)

Derived units

Area = 1 m^2
Volume = 1 m^3
Velocity = 1 m/s
Acceleration = 1 m/s^2
Force = 1 newton (N) = 1 $kg\text{-}m/s^2$

Work and energy: 1 joule (J) = 1 kg-m^2/s^2 = N-m
Moment and torque = 1 N-m
Power, 1 watt (W) = 1 kg-m^2/s^3
Pressure, pascal (Pa), 1 N/m^2 = 1 kg/s^2-m (1 bar = 10^{5}N/m)
Thermal conductivity = 1 W/m-°C = 1 kg-m/s^3-°C
Heat transfer coefficient = 1 W/m^2-°C = 1 kg/s^3-°C
Dynamic viscosity = 1 N-s/m^2 = 1 kg/m-s = 1 decapoise
Kinematic viscosity = 1 m^2/s = 1 myriastoke
Density = 1 kg/m^3
Heat coefficient = 1 J/kg°C = 1 m^2/s^2-°C
Enthalpy, heat content, and internal energy = 1 J/kg = 1 m^2/s^2
Electrical charge; 1 coulomb (C) = 1 A-s
Voltage (potential) = 1 volt (V) = 1 W<A = 1 kg-m^2/s^3-A
Resistance = 1 ohm (Ω) = 1 W/A^2 = 1 kg-m^2/s^3-A^2
Capacitance = 1 farad (F) = 1 coulomb/V = 1 A^2-s/W = 1 A^2-s^4/kg-m^2
Inductance = 1 henry (H) = 1 V-s/A = 1 J/A^2 = 1 kg-m^2/s^2-A^2

Capacity or permittivity $\epsilon_0 = \dfrac{10^7}{4\pi c^2}$ = 8.854 × 10^{-12} F/m

Magnetic permeability μ_0 = 4(10)$^{-7}$ = 1.2566(10)$^{-6}$ H/m
Frequency = 1 hertz (Hz)
Conductance = 1 siemens (S) = A/V

The American engineering system of units

Early in the development of engineering analysis a system of units was developed that defined both the units of mass and the units of force. It is perhaps unfortunate that the same word, pounds, was chosen to represent both quantities, since they are physically different. In order to help differentiate the quantities, the pound-mass may be designated as lb$_{mass}$ (or lb$_m$) and the pound-force as lb$_{force}$ (or lb$_f$).

For many engineering applications the numerical values of lb$_m$ and lb$_f$ are very nearly the same. However, in expressions such as $F = Ma$, it is necessary that the difference between lb$_m$ and lb$_f$ be maintained. By definition, a mass of 1 lb$_m$ will be attracted to the earth by a force of 1 lb$_f$ at a place where the acceleration of gravity is 32.2 ft/sec^2. If the acceleration of gravity changes to some other value, the force must change in proportion, since mass is invariant.

Although the pound subscripts, *force* and *mass,* are frequently omitted in engineering and scientific literature, it is nevertheless true that lb$_f$ is not the same as lb$_m$. Their numerical values are equal, however, in the case of sea level, 45°-latitude calculations. However, their values may be widely different, as would be the case in an analysis involving satellite design and space travel.

In Newton's equation, $F = Ma$, dimensional homogeneity must be maintained. If length, force, and time are taken as fundamental dimensions, the dimensions of mass must be derived. This can be accomplished as follows:

$$F = Ma$$

Then

$$M = \frac{F}{a}$$

$$= \frac{(F)}{(L/T^2)} = \frac{FT^2}{L} = FL^{-1}T^2$$

and

$$= \text{lb}_f\text{-ft}^{-1}\text{-sec}^2$$

For convenience, this derived unit of mass (1 lb-sec^2/ft) is called a *slug*. Thus, a force of 1 lb$_f$ will cause a mass of 1 slug to have an acceleration of 1 ft/sec^2.

The relationship between 1 lb$_m$ and 1 slug is given by considering that, whereas 1 lb$_f$ will accelerate 1 lb$_m$ with an acceleration of $g = 32.2$ ft/sec^2, it will accelerate 1 slug with an acceleration of only 1 ft/sec^2. Thus

$$1 \text{ lb}_f = (1 \text{ lb}_m)(32.2 \text{ ft/sec}^2) = (1 \text{ slug})(1 \text{ ft/sec}^2)$$

or
$$1 \text{ slug} = 32.2 \text{ lb}_m$$

It should be noted that with the FPS system a unity conversion factor must be used if a mass unit other than the slug is used. Since the acceleration of gravity varies with both latitude and altitude, the use of a gravitational system is sometimes inconvenient. A 100,000-lb rocket on the earth, for example, would not weigh 100,000 lb$_f$ on the moon, where gravitational forces are smaller. The mass of the rocket, on the other hand, is a fixed quantity and will be a constant amount, regardless of its location in space.

For a freely falling body at sea level and 45° latitude, the acceleration[6] g of the body is 32.174 (approximately 32.2) ft/sec^2. As the mass is attracted to the earth, the only force then acting on it is its own weight.

Then $$F = Ma$$
if $$W = Mg$$
and $$M = \frac{W}{g}$$
where $$a = g \quad \text{and} \quad F = W$$

In this particular system of units, then, the mass of a body in slugs may be calculated by dividing the weight of the body in pounds by the local acceleration of gravity in feet per second squared.

The engineer frequently works in several systems of units in the same calculation. In this case it is only necessary that the force, mass, and acceleration dimensions all be expressed in any valid set of units from any unit system. Numerical equality and unit homogeneity may be determined in any case by applying unity conversion factors to the individual terms of the expression.

Example   Solve for the lb$_m$ which is being accelerated at 3.07 ft/sec^2 by a force of 392 lb$_f$.

Solution $$F = Ma \quad \text{or} \quad M = \frac{F}{a}$$

$$M = \frac{392 \text{ lb}_f}{3.07 \text{ ft/sec}^2} = 127.8 \frac{\text{lb}_f\text{-sec}^2}{\text{ft}}$$

The direct substitution has given mass in the units of slugs instead of lb$_m$ units. This is a perfectly proper set of units for mass, although not in lb$_m$ units as desired. Consequently, the final equation must be altered by applying the unity conversion factor $\left(\dfrac{32.2 \text{ lb}_m}{1 \text{ lb}_f\text{-sec}^2\text{-ft}^{-1}}\right)$. The object, of course, is to cancel units until the desired units appear in the answer. Thus

[6]The value of the acceleration of gravity, g, at any latitude θ on the earth may be approximated from the following relationship: $g = 32.09(1 + 0.0053 \sin^2 \theta)$ ft/sec^2.

$$M = \left(\frac{127.8 \text{ lb}_f\text{-sec}^2}{\text{ft}}\right)\left(\frac{32.2 \text{ lb}_m\text{-ft}}{1 \text{ lb}_f\text{-sec}^2}\right) = \mathbf{(4.11)(10)^3 \text{ lb}_m}$$

Example Solve for the mass in slugs being accelerated at $13.6 \dfrac{\text{m}}{\text{sec}^2}$ by a force of 1782 lb_f.

Solution $F = Ma$

$$M = \frac{F}{a} = \frac{(1782 \text{ lb}_f)}{(13.6 \text{ m/sec}^2)} = \left(\frac{1782 \text{ lb}_f\text{-sec}^2}{13.6 \text{ m}}\right)\left(\frac{1}{3.28}\frac{\text{m}}{\text{ft}}\right)$$

$$= 40 \frac{\text{lb}_f\text{-sec}^2}{\text{ft}} = \mathbf{40 \text{ slugs}}$$

It is recommended that, in writing a mathematical expression to represent some physical phenomena, the engineer should avoid using stereotyped conversion symbols such as g, g_c, k, or J in the equation. If one of these, or any other conversion factor, is needed in an equation to achieve unit balance, it can *then* be added. Since many different unit systems may be used from time to time, it is best to add unity conversion factors *only* as they are needed. Unfortunately, in much engineering literature, the equations used in a particular instance have been written to include one or more unity conversion factors. Considerable care must be exercised, therefore, in using these expressions since they represent a "special case" rather than a "general condition." The engineer should form a habit of always checking the unit balance of all equations.

Remember that

$$1 \text{ slug} = 1 \frac{\text{lb}_f\text{-sec}^2}{\text{ft}} = \mathbf{32.2 \text{ lb}_m}$$

The foregoing discussion has shown that

1. If mass units in slugs are used in the expression $F = Ma$, the force units will come out in the usual units of pounds (lb_f).
2. If mass units in pounds (lb_m) are used in the expression $F = Ma$, force units will come out in an absolute unit called the *poundal* (see Table 12-1).

In engineering calculations the inch is used just as often as the foot to represent the unit of length, and this necessitates the introduction of an additional unit of mass. Consider Newton's law, $F = Ma$, where $F = 1 \text{ lb}_f$ and $a = 1 \text{ in./sec}^2$. Then

$$M = 1 \text{ lb}_f\text{-sec}^2/\text{in.}$$

where lb_f now is lb_{force}.

Example A body weighs $W \text{ lb}_f$ at a place where $g = 386 \text{ in./sec}^2$. Find the mass of the body in units of $\text{lb}_f\text{-sec}^2/\text{in.}$

Solution The relationship between weight and mass is given by

$$W = Mg$$

and if W is given in lb_f and g in in./sec^2, this gives

$$M = \frac{W}{g} = \frac{W}{386} \mathbf{lb}_f \cdot \mathbf{sec}^2/\mathbf{in.}$$

Problems on unit systems

12-49. The kinetic energy of a moving body in space can be expressed as follows:

$$KE = \frac{MV^2}{2}$$

where KE = kinetic energy of the moving body
M = mass of the moving body
V = velocity of the moving body

a. Given: $M = 539 \dfrac{\text{lb}_f\text{-sec}^2}{\text{ft}}$; $V = 2900 \dfrac{\text{ft}}{\text{sec}}$

Find: KE in ft-lb$_f$; in SI units.

b. Given: $M = 42.6 \dfrac{\text{lb}_f\text{-sec}^2}{\text{ft}}$; $KE = 1.20(10)^{11}$ ft-lb$_f$

Find: V in $\dfrac{\text{ft}}{\text{sec}}$; in SI units.

c. Given: $KE = 16{,}900$ in.-lb$_f$; $V = 3960 \dfrac{\text{in.}}{\text{min}}$

Find: M in slugs; in SI units.

d. Given: $M = 143$ g; $KE = 2690$ in.-lb$_f$

Find: V in $\dfrac{\text{mi}}{\text{hr}}$; in SI units.

12-50. The inertia force due to the acceleration of a rocket can be expressed as follows:

$$F = Ma$$

where F = unbalanced force
a = acceleration of the body
M = mass of the body

a. Given: $a = 439 \dfrac{\text{ft}}{\text{sec}^2}$; $M = 89.6 \dfrac{\text{lb}_f\text{-sec}^2}{\text{ft}}$

Find: F in lb$_f$; in SI units.

b. Given: $F = 1500$ lb$_f$; $M = 26.4 \dfrac{\text{lb}_f\text{-sec}^2}{\text{ft}}$

Find: a in $\dfrac{\text{ft}}{\text{sec}^2}$; in SI units.

c. Given: $F = (49.3)(10)^5$ lb$_f$; $a = 32.2 \dfrac{\text{ft}}{\text{sec}^2}$

Find: M in $\dfrac{\text{lb}_f\text{-sec}^2}{\text{ft}}$; in SI units.

d. Given: $M = 9650 \dfrac{\text{lb}_f\text{-sec}^2}{\text{ft}}$; $a = 980 \dfrac{\text{cm}}{\text{sec}^2}$

Find: F in lb$_f$; in SI units.

12-51. The force required to assemble a force-fit joint on a particular piece of machinery may be expressed by the following equation:

$$F = \frac{\pi dlfP}{2000}$$

where d = shaft diameter, in.

l = hub length, in.

f = coefficient of friction

P = radial pressure, psi

F = force of press required, tons

    a. Given: $d = 9.05$ in.; $l = 15.1$ in.; $f = 0.10$; $P = 10,250$ psi

       Find: F in lb_f; in SI units.

    b. Given: $F = 4.21 \times 10^5$ lb_f; $f = 0.162$; $P = 8.32(10^8)$ psf; $l = 1.62$ ft

       Find: d in ft; in SI units.

    c. Given: $d = 25$ cm; $l = 30.2$ cm; $f = 0.08$; $P = 9260$ psi

       Find: F in tons; in SI units.

    d. Given: $F = 206$ tons; $d = 6.23$ in.; $l = 20.4$ in.; $f = 0.153$

       Find: P in lb_f/ft^2; in SI units.

12-52. The dynamic stress in the rim of a certain flywheel has been expressed by the following equation:

$$\sigma = 0.0000284 \rho r^2 n^2$$

where σ = tensile stress, $\dfrac{lb_f}{in.^2}$

       ρ = specific weight of material, $\dfrac{lb_f}{in.^3}$

       r = radius of curvature, in.

       n = number of rpm

    a. Given: $\sigma = 200$ psi; $\rho = 0.282 \dfrac{lb_f}{in.^3}$; $r = 9$ in.

       Find: n in rpm; in SI units.

    b. Given: $\rho = 0.332 \dfrac{lb_f}{in.^3}$; $r = 23.1$ cm; $n = 200$ rpm

       Find: σ in psi; in SI units.

    c. Given: $\rho = 540 \dfrac{lb_f}{ft^3}$; $n = 186$ rpm; $\sigma = (31.2)(10)^3$ psf

       Find: r in ft; in SI units.

    d. Given: $\rho = 326 \dfrac{lb_f}{ft^3}$; $n = 250$ rpm; $r = 0.632$ ft

       Find: σ in psf; in SI units.

12-53. The stress in a certain column may be calculated by the following relationship:

$$\sigma = \frac{F}{A}\left[1 + \left(\frac{l}{k}\right)^2 \frac{R}{\pi^2 n E}\right]$$

where σ = induced stress, psi

       F = applied force, lb_f

       A = cross-sectional area of member, in.^{2}

       l = length of bar, in.

       k = radius of gyration, in.

       R = elastic limit, $lb_f/in.^2$

       E = modulus of elasticity, $lb_f/in.^2$

       n = coefficient for different end conditions

    a. Given: $n = 1$; $E = (3)(10)^7$ psi; $R = (4.2)(10)^4$ psi; $k = 0.29$ in; $l = 20.3$ in.; $A = 17.5$ in.^{2}; $F = 12,000$ lb_f.

       Find: σ in psi; in SI units.

b. Given: $\sigma = 11,500$ psi; $F = 6.3$ tons; $l = 2.11$ ft; $k = 0.41$ in.; $R = 40,000$ psi; $E = (3.16)(10)^7$ psi; $n = 2$
Find: A in ft²; in SI units.

c. Given: $n = \frac{1}{4}$; $E = (2.65)(10)^7$ psi; $R = (3.21)(10)^4$ psi; $k = 0.026$ ft; $A = 102$ cm²; $F = 5.9$ tons; $\sigma = 10,000$ psi
Find: l in ft; in SI units.

d. Given: $\sigma = (1.72)(10)^6$ psf; $F = (1.33)(10)^4$ lb$_f$; $l = 1.67$ ft; $k = 0.331$ in.; $E = (7.87)(10)^7$ psi; $n = 4$; $A = 14.2$ in.²
Find: R in psi; in SI units.

12-54. In the FPS gravitational system, what mass in slugs is necessary to produce 15.6 lb$_f$ at standard conditions?

12-55. In the engineering gravitational system, what mass in lb$_m$ is necessary to produce a 195.3 lb$_f$ at standard conditions?

12-56. Using the FPS gravitational system, calculate the fundamental dimensions of E in Einstein's equation,

$$E = mc^2 \left[\frac{1}{\sqrt{1 - (V^2/c^2)}} - 1 \right]$$

if m is a mass, V is a velocity, and c is the speed of light. What would be the fundamental dimensions of E in the CGS absolute system of units?

12-57. Using the relationship for g on page 248 and the FPS gravitational system of units, determine the weight, at the latitude $0°$, of a stainless steel sphere whose mass is defined as 150 lb$_f$-sec²/ft.

12-58. The mass of solid propellant in a certain container is 5 kg. What is the weight of this material in newtons at a location in Greenland, where the acceleration of gravity is 9.83 m/sec²? What is the weight in newtons?

12-59. Change 100 N of force to lb$_f$.

12-60. If a gold sphere has a mass of 89.3 lb$_m$ on earth, what would be its weight in lb$_f$ on the moon, where the acceleration of gravity if 5.31 fps²? What is the weight in SI units?

12-61. Assuming that the acceleration due to gravitation is 5.31 fps² on the moon, what is the mass in slugs of 100 lb$_m$ located on the moon? In SI units?

12-62. A silver bar weighs 382 lb$_f$ at a point on the earth where the acceleration of gravity is measured to be 32.1 fps². Calculate the mass of the bar in lb$_m$ and slug units.

12-63. The acceleration of gravity can be approximated by the following relationship:

$$g = 980.6 - (3.086)(10)^{-6}A$$

where g is expressed in cm/sec², and A is an altitude in cm. If a rocket weighs 10,370 lb$_f$ at sea level and standard conditions, what will be its weight in dynes at 50,000-ft elevation? In SI units?

12-64. At a certain point on the moon the acceleration due to gravitation is 5.35 fps². A rocket resting on the moon's surface at this point weighs 23,500 lb$_f$. What is its mass in slugs? In lb$_m$? In SI units?

12-65. If a 10-lb weight on the moon (where $g = 5.33$ fps²) is returned to the earth and deposited at a latitude of $90°$ (see page 248), how much would it weigh in the new location?

12-66. A 4.37-slug mass is taken from the earth to the moon and located at a point where $g = 5.33$ fps². What is the magnitude of its mass in the new location?

12-67. Is the equation $F = WV^2/2g$ a homogenous expression if W is a weight, V is a velocity, F is a force, and g is the linear acceleration of gravity? Prove your answer, using the FPS absolute system of units.

12-68. Sir Isaac Newton expressed the belief that all particles in space, regardless of their mass, are each attracted to every other particle in space by a specific force of attraction. For spherical bodies, whose separation is very large compared with the physical dimensions of either particle, the force of attraction may be calculated from the relationship $F = Gm_1m_2/d^2$, where F is the existing gravitational force, d is the distance separating the two masses m_1 and m_2, and G is a gravitational constant whose magnitude depends upon the unit system being used. Using the CGS absolute system of units [$G = 6.67 \times 10^{-8}$ cm^3/g-sec^2], calculate the mass of the earth if it attracts a mass of 1 g with a force of 980 dynes. Assume that the distance from the center of the earth to the gram mass is 6370 km.

12-69. Referring to Problem 12-68, calculate the mass of the sun if the earth (6×10^{24} kg mass) has an orbital diameter of 1.49×10^7 km and the force of attraction between the two celestial bodies is $(1.44)(10)^{25}$ N.

12-70. From Problem 12-68 calculate the acceleration of gravity on the earth in CGS absolute units.

12-71. An interstellar explorer is accelerating uniformly at 58.6 fps^2 in a spherical space ship which has a total mass of 100,000 slugs. What is the force acting on the ship?

12-72. At a certain instant in time a space vehicle is being acted on by a vertically upward thrust of 497,000 lb$_f$. The mass of the space vehicle is 400,000 lb$_m$, and the acceleration of gravity is 32.1 fps^2. Is the vehicle rising or descending? What is its acceleration? (Assume "up" means radially outward from the center of the earth.)

12-73. Some interstellar adventurers land their spacecraft on a certain celestial body. Explain how they could calculate the acceleration of gravity at the point where they landed.

12-74. In a swimming pool manufacturer's design handbook, for a pool whose surface area is triangular, you find the following formula: $V = 3.74Rt\theta$, where V = volume of pool in gallons, R = length of base of triangular-shaped pool in feet, t = altitude of triangular-shaped pool in feet if t is measured perpendicular to R, and θ = average depth of pool in feet. Prove that the equation is valid or invalid.

12-75. You are asked to check the engineering design calculations for a sphere-shaped satellite. At one place in the engineer's calculations you find the expression $A = 0.0872\Delta^2$, where A is the surface area of the satellite measured in square feet, and Δ is the diameter of the satellite measured in inches. Prove that the equation is valid or invalid.

12-76. The U.S. Navy is interested in your torus-shaped lifebelt design and you have been asked to supply some additional calculations. Among these is the request to supply the formula for the volume of the belt in cubic feet if the average diameter of the belt is measured in feet and the diameter of a typical cross-sectional area of the belt is measured in inches. Develop the formula.

12-77. From the window of their spacecraft two astronauts see a satellite with foreign insignia markings. They maneuver for a closer examination. Apparently the satellite has been designed in the shape of an ellipsoid. One of the astronauts quickly estimates its volume in gallons from the relationship $V = 33.8ACE$, where the major radius (A) is measured in meters, the minor radius (E) is measured in feet, and the end-view depth to the center of the ellipsoid (C) is measured in centimeters. Verify the correctness of the mathematical relationship used for the calculations.

12-78. An engineer and his family are visiting in Egypt. The tour guide describes in great detail the preciseness of the mathematical relationships used by the early Egyptians in their construction projects. As an example he points out some peculiar indentations in a large stone block. He explains that these particular markings are the resultant calculations of "early day" Egyptians pertaining to the volume of the pyramids. He says that the mathematical relationship used by these engineers was $\odot = \Box\uparrow$, where $\odot$ was the volume of pyramid in cubic furlongs, $\Box$ was the area of the pyramid base in square leagues, and $\uparrow$ was the height of the pyramid in hectometers. The product of the area and the height equals the volume. The engineer argued that the guide was incorrect in his interpretation. Prove which was correct.

12-79. Develop the mathematical relationship for finding the weight in drams of a truncated cylinder of gold if the diameter of the circular base is measured in centimeters and the height of the piece of precious metal is measured in decimeters.

12-80. If a silver communications satellite has a mass of 126.3 lb_m at Houston, Texas, what would be its weight in newtons on the moon, where the acceleration of gravity is measured to be 162 cm/sec²?

12-81. A volt is defined as the electric potential existing between two points when 1 joule of work is required to carry 1 coulomb of charge from one point to the other. An ampere is defined as a flow of 1 coulomb of charge per second in a conducting medium. From these definitions, derive an expression for power in watts in an electrical circuit.

12-82. An electric light bulb requires 100 W of power while burning. At what rate is heat being produced? What will be the horsepower corresponding to 100 W?

12-83. A 440-V electric motor which is 83 per cent efficient is delivering 4.20 hp to a hoist which is 76 per cent efficient. At what rate can a mass of 1155 kg be lifted?

12-84. How many kilograms of silver will be transferred in an electroplating tank by a passage of 560 amp for 1 hr? (Hint: 96,500 coulombs will deposit a gram-equivalent of an element in a plating solution.)

12-85. A window-mount type of air-conditioning unit is rated at ¾-ton capacity for cooling. If the overall efficiency of the motor and compressor unit is 26 per cent, what electric current will be necessary to operate the unit continuously when connected to a 120-V alternating current power line?

12-86. A large capacitor is rated at 10,000 microfarads. If it is connected to a 6.3-V battery, how many coulombs will be required to charge it?

12-87. A capacitor used in transistor circuits is rated at 5 picofarads. How many coulombs will be required to charge it if it is connected to a 9-V battery?

12-88. The reactance in ohms of a coil of wire is given as $X_L = 2fL$, where f is the frequency of an electric current in cycles per second and L is the coil inductance in henries. Compute the reactance of a small solenoid coil whose inductance is 2.75 millihenries if the coil is connected to a 109-V line whose frequency is 412 hertz.

12-89. A galvanometer has a resistance of 612 ohms and gives a deflection of 18.0 cm for a current of 28 ma. What will be its current sensitivity (amperes per meter), and its voltage sensitivity (volts per meter)?

12-90. A current of 0.63 amp will produce 15.0 cal of heat in a small light bulb in 10.0 sec. What power is expended in the bulb and what is the lamp resistance?

12-91. A coil of resistance wire having a resistance of 3.11 ohms is immersed in a beaker containing 1.150 kg of water. At what rate is heat being produced when a current of 47.5 amp is flowing?

Part Four

Engineering analysis

The analysis of engineering systems
is a very important part
of the engineer's work.

13

Engineering analysis

Analysis has been described as the engineer's toolbox. Just as a carpenter's toolbox may contain several different saws, chisels, and hammers, so the tool of analysis can be divided into several categories such as the following:

Mathematical Tools	Mathematics
	Statistics
	Computer operations
Material Tools	Chemistry
	Materials and metallurgy
Physical Tools	Solid mechanics
	Fluid mechanics
	Electricity
	Thermodynamics
Environmental Tools	Mechanics of man
	Economics
	Social sciences
	Ecological sciences

It is obvious that we cannot describe all these subjects in this text. Many books have been written about each of them. However, the next few chapters will consider some of the tools. Nearly 80 per cent of the engineering curriculum consists of learning how to use them. Here we will present only a taste—*we will let you feel the heft of the hammer.*

Illustration 13-1
The engineer must be especially adept in analyzing design configurations.

The mathematical tools

The language of engineering is *mathematics* and to be an effective engineer one must learn to use its language. Although in theory one can describe everything with words of the English language, English is, in fact, cumbersome and frequently ill adapted to express *precisely* the "physics" of a situation, as can be done with a mathematical equation. Consider, for example, a ball resting on a flat horizontal surface as in Figure 13-1. The force of gravity (W) acts downward and the surface resists upward on the ball (N) so that it does not move. In this case the English language is adequate to describe that the two forces act along the same line and are of equal magnitude. However, consider the same ball resting on an incline against a stop which prevents its moving down the incline, Figure 13-2. Here we must consider the action of three forces: the force of gravity pulling down (W), the normal force from the incline (F_1), and the resisting force from the stop (F_2). Since the ball is not moving, we know that these three forces must somehow balance, and that to do so the magnitudes of the supporting forces will vary depending on the angle of the incline with the horizontal (θ). We can say in general terms that if this angle becomes larger, the force from the plane (F_1) decreases, and the force on the stop (F_2) increases, but just exactly how they are related to one another would be most difficult to describe using words alone. Yet

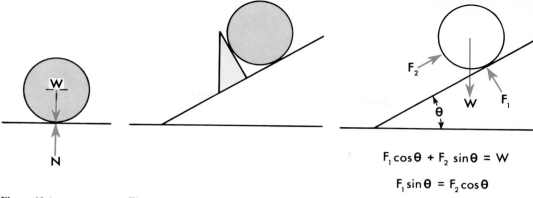

$$F_1 \cos\theta + F_2 \sin\theta = W$$

$$F_1 \sin\theta = F_2 \cos\theta$$

Figure 13-1   Figure 13-2   Figure 13-3

in mathematical terms the relationship between these forces can be expressed simply and easily, Figure 13-3.

The mathematics of engineering is concerned primarily with two subjects: *the relations between forces and body shapes,* including geometry, trigonometry, and vector analysis, and *the description of things that change*—the realm of calculus.

The ball resting on the incline is an example of the first subject. Others will be found later. To get an idea about the second, let us consider a moving object, such as a car on the road. If a person does not have a speedometer, a simple way to tell how fast he is going is to time himself between mileposts. If, for example, he had measured 72 sec between two posts 1 mile apart (Figure 13-4), his *average* speed would have been (1 mi/72 sec)(3600 sec/hr) = 50 mi/hr. Knowing the average speed does not tell him if his speed was steady over the mile or if he had driven at varying speeds (Figure 13-5). If there had been a distance marker every one fifth mile, and he had recorded the time at each, his average speed in each fifth of the mile might have looked like the horizontal bars in the second graph, Figure 13-6. This is closer, but still not a good approximation to the actual curve. The only way that this method would give a good speed reading would be to place markers every few feet and record the time interval at each. Such an action would result in the "fine" averaging shown in Figure 13-7. As the time intervals get smaller and smaller, the actual local velocity of the car is established more accurately by taking the ratio of a very small distance at that location to the very short time used to traverse it. It is called the *local derivative* of position *with respect to time,* or the *local time rate of change* of position.

Note that we now have an expression of *local* (not average) *velocity* in terms of position and time. Mathematicians have developed methods to compute the resultant velocities once the mathematical expression (formula) or graph of the vehicle location is known. This concept can be extended easily to find the *acceleration* which engineers need to compute engine torque and power consumption. This type of mathematical solution is known as *calculus* and is a very powerful concept indeed.

The very small items, whose ratio forms the derivative, are often called *differentials,* and the whole process is called *differentiation.* The inverse process, which for

Figure 13-4

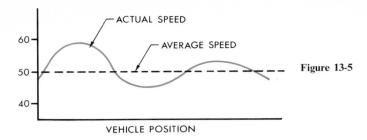

Figure 13-5

VEHICLE POSITION

example allows one to find the vehicle position if he knows its velocity, is called *integration.*

Derivatives need not always be taken "with respect to time." For example, the amount of heat flowing from the hot to the cold side of a piece of metal that is heated on one side depends upon the *derivative* of temperature *with respect to position,* that is, the ratio *dT/dx,* Figure 13-8.

Mathematical analysis is rarely if ever able to depict exactly the physical behavior of *real* materials and devices. In fact, even if it were, such exactitude would be of little value to the engineer, since each piece of material differs from all others (even if these differences are only microscopic); and at best one can talk only about *statistical average behavior* and attempt to estimate how much any one sample may vary from this average.

Statistics relates experiment to analysis. It furnishes the numbers used in the analysis and tells how reliable they are. If, for example, a large number of "pull tests" have shown that the average failure load for the samples tested was 100,000 lb, and the failures were distributed as shown in Figure 13-9, then we can expect ten samples in 100 to fail at as low a load as 75,000 lb and 15 to survive a load of 125,000 lb.

The engineer, who can rarely ask that every production item be tested, must decide what kind of failure rate he can tolerate and proceed with his design accordingly. For unimportant items, easily replaced, he may well accept a 1 or 2 per cent failure rate. However, if failure of the part were to endanger human life, he would want to make sure that its *probability* of failure was very low and that it was tested to full load and beyond before it left the factory.

The *digital computer* has become an almost indispensible tool for today's engineer. Recognizing this, computer manufacturers have built small office units. These are either independent, and thereby limited in speed and in their capacity to store numbers, or they are connected by telephone or cable to a large-sized computer. Some of these large computers work so fast that they can do hundreds of thousands of computations while the designer's answer is being typed out and before he is ready with his next problem. Therefore, several persons can share a single computer—provided that the computer is so constructed that it can keep all the problems separated. This procedure is called *time sharing.*

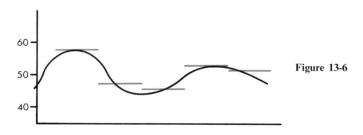

Figure 13-6

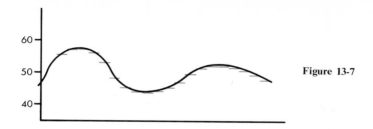

Figure 13-7

But what can a computer do that makes it so valuable to the engineer? Is it some type of ultrafast adding machine or perhaps is it a mysterious device that answers questions that are put to it? It is really neither of these—although it is frequently used for both of these purposes. If one looks beyond the dazzle of flashing lights and whirling tapes, he can identify four primary functions in any modern computer:

1. The ability to store thousands of numbers and to recall them rapidly from storage when needed. The storage is appropriately called *memory*. The memory not only stores the numbers on which the computer operates and the results of its calculations, but also stores the *program* that tells it what kind of calculations to perform. Therefore, the same program can be used again and again with different data, without having to repeat the original instructions to the machine.
2. An *arithmetic* center which works much like a desk calculator, primarily by adding (or subtracting) numbers. It can do this so rapidly—an addition may take as little as a few millionths of a second—that even very lengthy calculations are made in an extremely short time.
3. The capability of comparing two numbers and of making a *decision* based upon the outcome. This is one of the most important attributes of the computer. How this works will be demonstrated in the example below.
4. Communication or *input–output* facilities that allow one to "talk" to the computer, to give it instructions, to feed it numbers with which to compute, and to receive its answers in readable form.

Since the computer works only with numbers, the instructions in the program (such as "add number x to number y") must also be translated into appropriate numbers. In the early days of computers, the programmer was required to accomplish these translations himself. To do so he had to know the internal computer language. Today, computers are able to accomplish their own translation, provided that the program is written in a "standard" language. These standard languages are easy to learn because they are written in almost the same way that one would normally

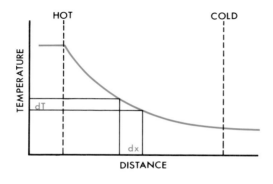

Figure 13-8

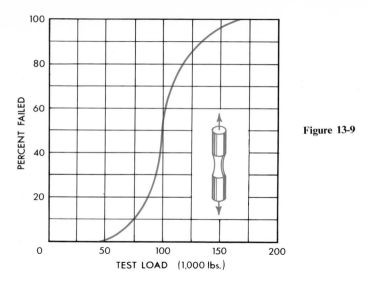

Figure 13-9

express the steps of the program in written English. Among the most frequently used standard languages are FORTRAN, ALGOL, and BASIC.

Let us examine how a typical computer program would work on this problem (reportedly more than 1500 years old).[1]

> According to an Arabic legend, the game of chess was invented by a Brahmian, to show his student, the heir to the throne, that the king is no stronger than his subjects. Invited to name a reward, the Brahmian asked only for grains of rice: one grain on the first square of the chess board, two on the second, four on the third, and so on, always doubling until the sixty-fourth square is reached. Since this request appeared to be modest, it was granted. *Yet all the rice in India could not satisfy this simple request.*

How many grains would there be on each square? How many altogether? Of course one could calculate it easily, *but laboriously.* (Note that the total is one less than twice the amount of the last square.) Here is the computer program (Figure 13-10) that will evelute the desired result. There are nine *statements,* each one numbered for reference. N stands for the number of the square (from 1 to 64), X for the number of grains of rice on that square. Although the statements beginning with LET look like equations, they are more properly called *assignments,* for we ask the computer to calculate the right-hand side and to *assign* the resultant number to the symbol on the left. For example, in statement 4, the computer adds 1 to the present value of N and then assigns this sum to N. The asterisk (*) is used as the symbol for multiplication. Note the decision statement 3. It tells the computer to compare the present value of N with the number 64. If N is equal to 64, then the program switches to statement 7; if not, it goes on to the next statement in line. When it gets to statement 6, the computer is told to go next to statement 2 and then to proceed accordingly.

The result is shown in Figure 13-11. It took a medium-sized computer 7 sec to calculate all these values, and about 2 min to print them. Note that this particular computer does not keep track of more than nine digits. From square 31 on, the

[1] Translated from *Grand Larousse encylopédique,* Librairie Larousse, Paris (1961), vol. 4, p. 321.

```
1 LET X=N=1
2 PRINT N,X
3 IF N=64 THEN 7
4 LET N=N+1
5 LET X=2*X
6 GO TO 2
7 LET X=2*X-1
8 PRINT "TOTAL=";X
9 STOP
```

Figure 13-10

Figure 13-11

1	1
2	2
3	4
4	8
5	16
6	32
7	64
8	128
9	256
10	512
11	1024
12	2048
13	4096
14	8192
15	16384
16	32768
17	65536
18	131072
19	262144
20	524288
21	1048576
22	2097152
23	4194304
24	8388608
25	16777216
26	33554432
27	67108864
28	134217728
29	268435456
30	536870912
31	1.07374E+09
32	2.14748E+09
33	4.29497E+09
34	8.58993E+09
35	1.71799E+10
36	3.43597E+10
37	6.87195E+10
38	1.37439E+11
39	2.74878E+11
40	5.49756E+11
41	1.09951E+12
42	2.19902E+12
43	4.39805E+12
44	8.79609E+12
45	1.75922E+13
46	3.51844E+13
47	7.03687E+13
48	1.40737E+14
49	2.81475E+14
50	5.62950E+14
51	1.12590E+15
52	2.25180E+15
53	4.50360E+15
54	9.00720E+15
55	1.80144E+16
56	3.60288E+16
57	7.20576E+16
58	1.44115E+17
59	2.88230E+17
60	5.76461E+17
61	1.15292E+18
62	2.30584E+18
63	4.61169E+18
64	9.22337E+18

TOTAL = 1.84467E+19

numbers are rounded off and expressed as decimals, with an *exponential* added. For example, square 31 (which should actually be 1,073,741,824) becomes 1.07374×10^9 or, in computer terms, $1.07374E+09$.

The material tools

Whatever an engineer builds, whether it be a suspension bridge, a spaceship, or an electronic circuit, he uses materials of construction. Therefore, he must know something about material properties and the way that they can be shaped into the forms he needs. Different applications demand different types of properties—high strength and rigidity in one case and suppleness and flexibility in another. There is no universal material. Therefore, the engineer uses various metals and polymers, concrete, rubber, wood, ceramics, and special combinations of these.

The most widely used materials for the engineer are the metals. They exhibit strength, toughness, ductility, resistance to both high and low temperatures, and good conductivity of heat and electricity. They can be formed and shaped and joined into a great variety of shapes, and, by suitable treatment, they can be made resistant to most chemical attack. The most important primary metals for the engineer are iron,[2] copper, and aluminum, together with auxiliary metals such as nickel, zinc, tin, cobalt, lead, and manganese. During recent years some of the scarcer metals, such as titanium, tantalum, vanadium, beryllium, and columbium, have gained importance for special applications. Almost without exception metals are not found pure in nature, but as mineral combinations of a metal with such chemical elements as oxygen, sulfur, and carbon in the form of oxides, sulfates, sulfides, and carbonates. The process of separating the metal from the other constituents is called smelting, reducing, or refining. It is usually done either by heating in the presence of other chemicals which have a greater affinity or attractive power for the undesired chemical than the metal has, or by electrolytic methods, such as melting or dissolving the original compound in a liquid and applying an electrical current through two electrodes in such a way that the metal portion of the compound is attracted by one of the electrodes and the less desirable chemical by the other. The relatively pure metal resulting from these refining methods is cast into small bars called "pigs" or into larger forms called "ingots," or left molten for further treatment. With the notable exception of copper,[3] the pure metal is rarely useful as an engineering material. Pig iron, for example, is weak and brittle, and pure aluminum is too soft and ductile for most engineering uses. Later we will discuss how the properties of these pure metals can be improved. But first let us consider the demand and supply and the scarcity of minerals.

Until recently, few people have given any thought to the limitation of our supply of minerals—the raw materials which supply our need for metals. Unfortunately we have been using these metals as if the supply were unlimited. Today industry uses 0.55 tons of steel per year for every person in the United States, as opposed to 0.35 tons of steel per person in the year 1950 (*the average automobile alone requires* 1.5

---

[2] Iron is very important since it is the main constituent of steel.
[3] One of the principal uses for copper is as a conductor of electricity. The purer the copper, the higher its electrical conductivity, i.e., the lower its resistance to the flow of electric current.

tons of steel). The effect of this rate increase is more pronounced when the population increase in the United States (from 150,000,000 to 200,000,000) is considered. Similar increases in use can be demonstrated for most other metals. It is obvious that such a rate of growth cannot be maintained over very many years without using up our mineral resources. In fact, there is already such a shortage of nickel that it has become one of the scarcest and costliest commodities. One answer is to throw less of our used metals away and reuse the maximum amount possible. Already scrap metal is used extensively in the production of new metal, processes are being devised to regain scrap metal from garbage, and aluminum companies are offering a reward for the return of empty beverage cans. Yet all the reuse of scrap is of little help if populations keep increasing and if the amount of metal used per capita continues to rise.

Metals become most useful to the engineer when they are combined with other metals and elements which give them the qualities of strength and toughness for which they are prized. Metallurgy, the science that deals with the improvement of metal properties, is probably the oldest science known to man. Long before he had fires hot enough to work iron, man combined copper and tin to make bronze, which is stronger than either copper or tin alone. (Brass is an alloy of copper and zinc.) When men learned to use coal rather than wood and obtained a flame hot enough to melt iron, it was probably by accident that he discovered the enormous strength that was added to iron by small additions of carbon. One can well imagine coal dust or a small piece of coal entering into the melt and the surprise of the man when his result turned out to be stronger and tougher than any iron that had been produced before. We know today that very little carbon—usually less than 1 per cent—is needed to make steel out of iron. We also know that additional trace amounts of elements such as silicon and manganese can further improve the strength and other qualities of the steel, and that the methods of heating and cooling are very important in determining what compounds are formed and the properties of the final material. We have also learned that there are elements such as phosphorous and sulfur which, when added to steel, are likely to deteriorate its qualities. Every effort must be made to keep these out of the melt.

Alloys such as bronze or brass are mixtures or "solid solutions" of two metals. Among the most useful of the iron alloys are those made with nickel and chromium, which produce the stainless steels, and the high-strength alloys which usually contain cobalt, manganese, nickel, and other elements. Just as the soft metals tin and copper form a stronger alloy, *bronze*, so do small additions of copper to aluminum form a much stronger alloy which the British call *duralumin*.

There are many manufacturing methods by which metals and alloys can be made into useful engineering products. The oldest method (and still widely used) is sand casting. A pattern or model of the desired shape is made, usually of wood, and this pattern is used to make a tightly compacted sand mold. The mold is made in two parts so that the pattern can be removed without destroying the mold. Liquid metal is then poured into the closed mold and allowed to cool and solidify. A sand mold is not reusable so it is broken up to remove the finished product. The main advantages of cast products are (1) they can be formed in intricate shapes and (2) there is almost no limit to the size of a casting. On the other hand, sand castings tend to have rougher surfaces and to be weaker and more brittle than items that have been produced by other methods. It is difficult to hold dimensions accurately and to make sure that the item does not warp during the cooling process. Finally, sand casting does not lend itself well to the making of mass-produced items. However, in the last 50 years the art of casting has been very much improved, especially by the development of permanent metal molds and other special mold materials which improve accuracy and surface finish. Special metals have been developed which, when cast, are no longer brittle and weak but strong and tough, and continuous casting methods have been developed which do lend themselves to certain mass-production practices. Iron, zinc, and aluminum castings especially have found their way into many engineering applications. The engine block in a car, its transmission housing, and carburetor body (as well as many body trim parts) are all made from castings.

We do not know who first discovered that the toughness of steel could be improved appreciably by hammering when hot, but we do know that for thousands of years swords, horseshoes, and ax heads have been produced by hammering or forging. However, the hand forge of old has now been replaced by huge drop forges, presses, and rolling mills. In the rolling mill the hot steel ingot is squeezed again and again through sets of rollers until it is in the desired form of sheet steel, bar stock, or special shapes such as angle iron or I-beams. Although not strictly the same as forging, the rolling process produces the same working of the material as the forge did and produces the same strength and toughness as was achieved in the forge. Many metals are also successfully formed by cold rolling and pressing.

Castings, forgings, and rolled metal shapes can seldom be used without additional machining. Machining means the controlled removal of material, whether it be in cutting, drilling, or the shaving of material off the surface. Metals are cut either by saw or with an intensely hot pencil-shaped flame. Drilling is done on single- or multispindle drill presses or, if the hole is so big that an ordinary drill press cannot handle it, on boring mills. Surface material is removed on shapers, planers, or milling machines if the workpiece is flat or contoured, and on lathes if it is circular. Planers and lathes normally use stationary cutting tools and have the workpiece move past the cutting tool. In milling machines the cutting tool itself rotates and the piece may or may not remain stationary. The surfaces produced by planing, shaping, milling, and lathe work are usually fairly rough unless considerable effort is spent on the cut. To get very fine surfaces and very accurate dimensions, grinding machines are used, but these are not as useful as the other tools in removing substantial amounts of material. Grinding machines are made for flat as well as cylindrical pieces. To achieve a finish even finer than can be obtained by grinding requires mechanical honing, lapping, and polishing, or the use of certain chemical etchants, which can produce a mirror-like surface on metals.

It is the surface of the metal that, like your skin, is in contact with the environment and must be protected against mechanical and chemical attack. Most failures begin at the surface, and there are many special techniques to harden and toughen the surface and to make it more resistant to the environment in which it operates. Among the protective techniques are (1) carburizing, cyaniding, and nitriding, in which the surface is hardened chemically, (2) shot-peening, which strengthens the surface mechanically, and (3) galvanizing or plating with chromium, nickel, or cadmium, which protects the surface from corrosion.

Finally, consideration should be given to how metals are joined together. If two pieces of steel are to be joined permanently, the most common method is to weld them together. In welding, the edges which are to meet are heated until almost molten, and are then brought together with or without the use of a flux or filler (additional metal). A proper weld is as strong as the original material. To accomplish this requires considerable experience and ability because heating and cooling in the vicinity of the weld can cause undesirable changes in the material properties. Brazing and soldering are similar to welding in that a continuous joint is produced. However, the base metal is not heated as high as the melting temperature, but rather to a temperature only high enough to melt a filler of brass or other nonferrous metal in brazing or a lead–tin mixture in soldering. Brazing and soldering are useful where the strength of the joint is not a prime requirement. They can also be used to join dissimilar materials such as aluminum and steel or brass.

The second most common joining method is by means of mechanical fasteners such as rivets, screws, and bolts. Here holes must be drilled through the parts to be joined. Rivets are usually heated before they are inserted in the hole and then hammered so that they fill the hole completely. As the rivets cool they contract and tighten the joint. Bolts and screws are sometimes preferable because they can be removed easily.

Next to the metals, the most useful engineering materials are found among the rubbers and plastics. Where metals are strong, heavy, and relatively inflexible, the rubbers and plastics are relatively weak, light in weight, and pliable. With the single exception of natural or *hevea* rubber, they do not occur in nature and are the product of modern chemistry (even natural rubber must be carefully compounded with other chemicals in order to make a useful material). Metals have a history going back to the beginnings of human history; plastics, however, are primarily an invention of this century.

Rubbers and plastics are long-chain hydrocarbon polymers; that is, they are produced by linking together large numbers of relatively simple molecules called *monomers*. This process is called polymerization. Occasionally a monomer will polymerize with moderate heat (a few hundred degrees Fahrenheit) alone; more often a catalyst[4] is required. The long-chain molecules of a polymer can be visualized as long elastic coiled springs randomly distributed in the polymer. They are interconnected here and there, and it is the degree of or number of interconnections that determines the rigidity and strength of the polymer. In rubbers there are few interconnections; in rigid plastics there are many. In some plastics the interconnections can be broken by moderate heat so that the plastic can be reformed by heating. When cooled the links are reestablished. Such plastics are called "thermoplastic." In other plastics the links are permanent and cannot be broken without destroying the material. Such plastics are called "thermosetting."

The most common way of producing plastic parts is by molding. This is substantially the same as casting except that the low temperatures at which plastics can be formed permit the use of very accurately machined metal molds. These can produce plastic parts of exact dimensions and good surface finish and can be used again and again. We are all aware of the innumerable household products and items of modern life which are produced from molded plastics. Though an original mold may be costly, if its price can be distributed over thousands or even hundreds of thousands of parts, the cost of each part may be remarkably low. This is in part why picnic forks, ball-point pens, and pocket combs cost as little as they do.

In recent years some very strong structures have been built by embedding in the plastics strong reinforcing fibers, such as glass in plastic. Although glass may be brittle, it has very great strength, comparable to some of the strongest metals. Thus a whole group of glass fibers held together in a plastic matrix can produce a material of great strength and flexibility.

The idea of combining two different materials and using the good qualities of each to make a superior product is much older than the glass-reinforced plastics. It

---

[4] A catalyst may be compared to a marriage broker; its presence is essential for the reaction, but it does not end up in the final product.

probably goes back to the use of reinforced concrete, in which the strength of steel and the stiffness of concrete are used to make superior structural members.

Even concrete itself is a mixture of sand, gravel, and cement—each constituent being of relatively little practical importance by itself. Cement, also called Portland cement, is a roasted (calcined) mixture of clay and limestone which combines readily with water to form a rigid, stable solid. Hardened cement is very durable and can resist not only heat, cold, and other variations in climate but also a large number of chemicals. However, it is very brittle and has little physical strength. In concrete, strength is provided by adding the sand and gravel, which are then held together by the cement binder. Concrete, which offers great resistance toward being compressed, offers much less resistance to being pulled apart. Here is where the use of steel reinforcing becomes important. For example, imagine a concrete beam acting as a bridge over a canyon. When a heavy truck travels over this bridge, the beam will be bent downward, with the roadway (or top surface of the beam) being pushed together or compressed, and the bottom part of the beam (the part nearest the canyon) being pulled apart in tension. Since concrete is much stronger in compression than in tension, steel reinforcing bars are cast into the bottom part of the beam, which brings its total strength up to that of the top (compression) side.

Typical concrete consists of 13 parts of cement, 33 parts of sand, and 46 parts of gravel together with 8 parts of water by weight. Since the strength of concrete depends upon how completely the cement and water have reacted, it is important to keep the structure moist for several days after it has been poured if maximum strength is desired.

The cement–water reaction—called *hydration*—develops heat. In most structures this heat is readily dissipated by the surrounding air. However, in large thick sections of concrete, such as dams, special water cooling pipes are often inserted to make sure that this heat can be removed and not cause damage to the structure.

A number of special uses of concrete and cement should also be considered. By adding special chemicals it is possible to make cement harden faster than the usual 12- to 24-hour period. Quick-setting cements are used, for example, in the setting of casings, the metal liners that are put into water or oil wells to prevent collapse of the hole. The cement is pumped into the space between the liner and the soil to prevent earth movement. Consequently, a quick-setting cement is essential. Light-weight concrete can be made either by using a lightweight aggregate such as pumice (the hardened lava from volcanic eruption) or by blowing air into the concrete while it is setting. Of course, the strength of lightweight concrete is much less than that of ordinary concrete. Finally, one can spray a cement mixture under pressure out of a nozzle which resembles a small gun. Hence, the material is called "gunnite." It is used to coat tunnels, chemical vessels, swimming pools, and so on, wherever the hardness and chemical resistance of cement is desired and where necessary strength is provided by the backing onto which the gunnite is sprayed.

For extremely high temperature use (where few of the metals have any strength), refractories or ceramics are used. They resemble cement in that they are generally simple, inorganic chemicals.[5]

[5] Organic compounds contain carbon and hydrogen and can form very large and complicated molecules, such as the rubbers and plastics mentioned above. They are so called because most living organisms, such as plants, animals, and men, are made of compounds containing carbon and hydrogen. Even petroleum and natural gas, the raw material of today's organic chemical industry, come from long-decayed plants and animals. Inorganic compounds are a combination of other chemical elements. These may contain either carbon or hydrogen, but usually not both. Normally inorganic molecules are smaller and simpler than organic molecules.

Among the most common ceramics are the oxides, nitrides, carbides, and borides of metals such as aluminum, beryllium, titanium, tantalum, and many others. Prior to casting, ceramic compounds usually exist in powder form. They can be shaped into useful engineering products by hot forming and sintering. Here the powder is pressed into a suitable mold; at moderately elevated temperature enough interparticle bonds are formed to maintain the desired shape. This is then heated to considerably higher temperatures, at which many of the grains fuze to one another and form a dense, strong body that is highly resistant to heat and abrasion. A common ceramic is porcelain, which is made of sand (silicon oxide) and alumina (aluminum oxide), Ceramics are used in the nozzles of rocket engines, in gas turbine combustors, on the nose cones of missiles, in the heat shield for reentry space vehicles, in furnaces, or wherever extremely high temperature resistance is required. Like concrete, ceramics tend to have much greater strength in compression (being pushed together) than in tension (being pulled apart). To increase their strength in tension, engineers have embedded fine metal fibers such as tungsten or boron in the ceramic, with results comparable to glass fiber-reinforced plastics or steel-reinforced concrete. Graphite is usually included among the ceramics because of its high temperature resistance. Graphite is pure carbon in a form that permits it to be sintered and worked, like other ceramics. Although not of very high strength, it retains its strength to temperatures even higher than many ceramics. Exposure to oxygen must be avoided at high temperatures.

Glass, prized primarily for its transparency and its resistance to most chemicals, is also one of the strongest materials known to man. Unfortunately, it is also very brittle when exposed to air and moisture. Therefore, its strength can be used only when the glass is protected, when, for example, it is embedded in plastic. Glass fiber-reinforced plastics gain their high strength from thousands of fine glass fibers surrounded and held together by epoxy or other plastics. The production process tends to be costly and lends itself primarily to simple shapes such as pressurized fuel tanks, ship hulls, rocket casings, and sporting goods.

Wood, once one of the most widely used engineering materials, is used very little in industry today, partly because of its scarcity, and partly because as a product of nature its properties vary from batch to batch, thus reducing reliability. The two major categories of woods are softwood and hardwood. Softwoods, such as pines and firs, are relatively plentiful, grow fairly rapidly, and are not too costly. However, their strength is low and they tend to warp easily if not prepared properly; consequently, they are used today primarily for building construction and for the production of paper. Softwood is used in making forms for concrete although metal forms are rapidly replacing wood in large-volume construction. Hardwoods, such as oak, ash, walnut, and other hardwoods from the tropics, have become so scarce and costly that they are used almost exclusively for decorative effects or for the production of beautiful and costly furniture. Their major industrial use is the production of patterns for sand castings. We can expect to see the use of wood curtailed more and more as our forests become depleted, and as those that remain are needed more and more for recreation and for the preservation of watersheds.

The physical tools

The engineering curriculum contains more courses in physics and in subjects derived from physics than any of the other pure sciences. *Force, motion, energy,* and *matter*—these are special concerns of the engineer.

Since we can hope to look at only a small selection of the physical tools, let us first consider *mechanics.*

Mechanics is the physical science that describes and predicts the effects of forces acting on material bodies. The condition under study may be one of rest or one of motion. There are three specialized branches into which the general field of mechanics may be divided for more specific studies. These are as follows:

1. Mechanics of rigid bodies
   Statics
   Dynamics
2. Mechanics of deformable bodies
3. Mechanics of fluids
   Compressible flow
   Incompressible flow

Our study of mechanics in this chapter will be concerned with an introduction to static mechanics, and to show how the engineering method is applied to a problem solution.

Fundamental concepts and definitions

Concepts used in our study of static mechanics are *force, space,* and *matter.* These concepts are basic and, as a frame of reference, should be accepted on the basis of our general experience. A *force* is the result of the interaction of two or more bodies and in our study here will be considered to be a localized vector quantity. A force may be evolved as the result of physical contact, or it may be developed at some distance—as is the case with magnetic and gravitational forces. *Space* is a region extending in all directions. It is associated with the location or position of a particle or of particles with respect to one another. *Matter* is a substance that occupies space.

A *particle* may be said to be a negligible amount of matter that occupies a single point in space. A *rigid* body is a body that is constructed entirely of particles that do not change their position in space with respect to each other. No body is truly rigid. However, in many situations the deformation, or change in position of the particles, is very small and therefore would have a negligible effect upon the analysis. Such is the assumption in this chapter.

The principles of statics are most easily learned by using *free-body diagrams.* Here we shall consider in more detail how to draw free-body diagrams and how to use them. Let us start by drawing the free-body diagram of a ship moving in the water, Figure 13-12. It is not necessary that the model of the ship be drawn to scale since the shape of the idealized model is only an imaginary concept.

There are four external forces acting on the model: a forward thrust, which acts at the ship's propeller; a friction drag, which acts so as to retard motion; a buoyant force, which keeps the ship afloat; and the ship's weight, which in simplification may be considered to be acting through the center of gravity of the ship, Figure 13-13.

The symbol ⊗ is used to denote the location of the center of gravity. A coordinate

Illustration 13-2
Structural steel members must be designed in accordance with the fundamentals of the mechanics of rigid and deformable bodies.

system is very useful for purposes of orientation. The diagram shown would make possible an analysis of the relationships between the weight and buoyant force and between the thrust and drag. However, it would not, for example, be useful for determining the loads on the ship's engine mounts. Another model (free-body diagram) of the engine alone would be required for this purpose.

Another example shows the free-body diagram of a four-wheel drive automobile driving up an incline, Figure 13-16. It is customary to show the model in its true position in space. It would appear awkward and be confusing to draw the automobile in a horizontal position when it is actually going uphill.

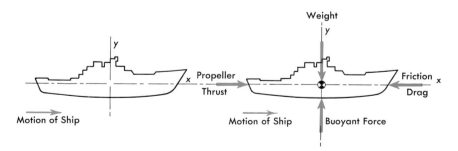

Figure 13-12

Figure 13-13

Figure 13-14

Situation	Free body	Explanation
A box resting on a plane 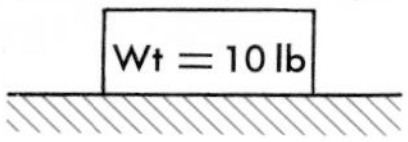	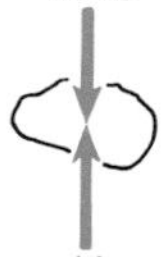 10 lb N	The normal force always acts at an angle of 90° with the surfaces in contact. This force N usually is considered to act through the center of gravity of the body.
A weight hanging from a ring 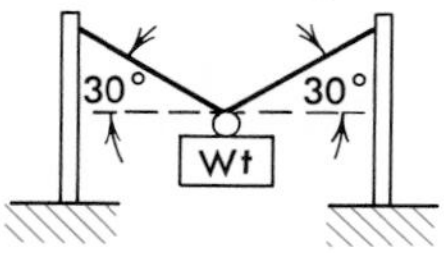	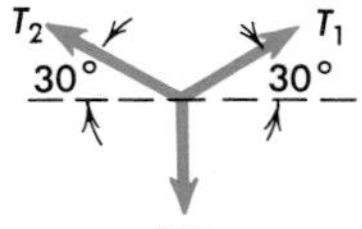 T_2 30° 30° T_1 Wt	Since the ring is of negligible size, it may be considered to be a point. All of the forces would act through this point. The downward force W is balanced by the tensions T_1 and T_2. The numerical sum of these tensions will be greater than the weight. This is true since T_1 is pulling against T_2.
A box on a frictionless surface 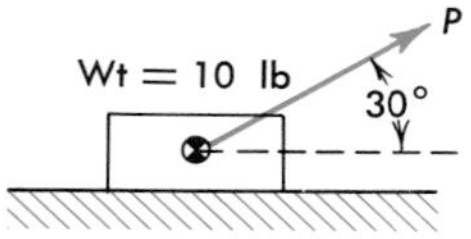	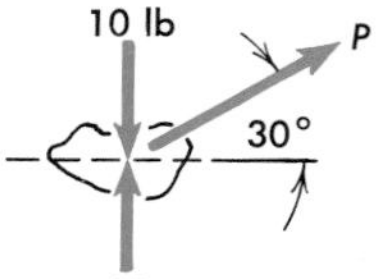 10 lb P 30° N	Some surfaces are considered frictionless although in reality, no surface is frictionless. The force P is an unbalanced force and it will produce an acceleration. The symbol ⊗ denotes the location of the center of gravity of the body.
A small box on a rough surface 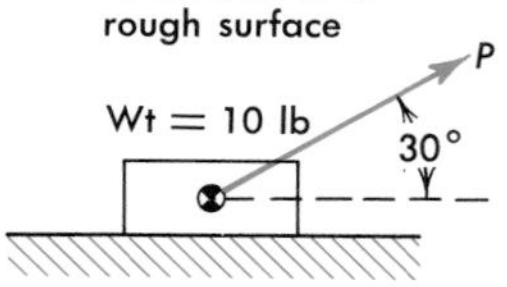	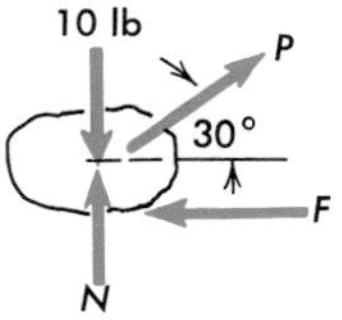 10 lb P 30° F N	The force of friction will always oppose motion or will oppose the tendency to move. For bodies of small size, the *moment effect* of the friction force may be disregarded and the friction and normal forces may be considered to act through the center of gravity of the body.

Figure 13-15

Situation	Free body	Explanation
A beam resting on fixed supports Load ←8 ft→ │ ↓ ← ↓ 2 ft Wt = 50 lb	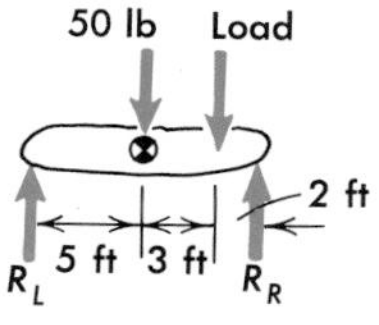 50 lb Load ↓ ↓ 2 ft R_L 5 ft 3 ft R_R	For a uniform beam, the weight acts at the midpoint of the beam regardless of where the supports are located.
A pivoted beam resting on a roller 100 lb 45° ←12 ft→ Wt = 10 lb	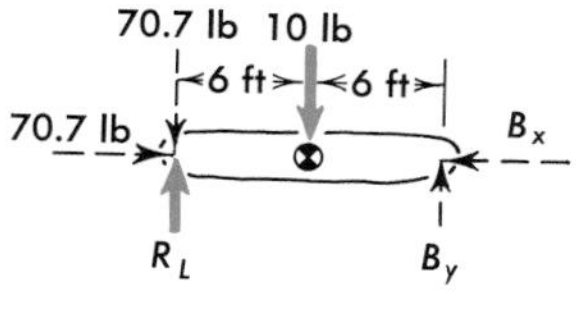 70.7 lb 10 lb ←6 ft→ ←6 ft→ 70.7 lb → B_x R_L B_y	Since a roller cannot produce a horizontal reaction, the horizontal component of any force must be counteracted by the horizontal component of the reaction at the pivoted end.
A ladder resting against a frictionless wall 60°	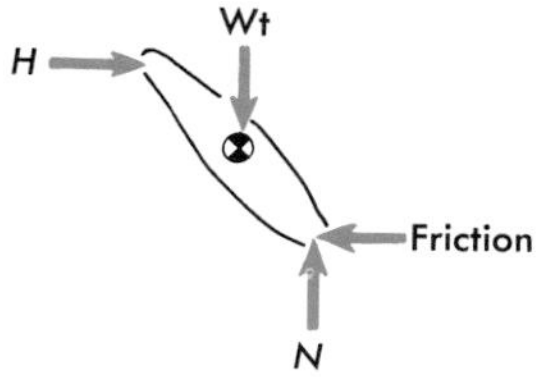 Wt H → ←Friction N	At the upper end of the ladder, the only reaction possible is perpendicular to the wall since the surface is considered to be frictionless.
Sketch	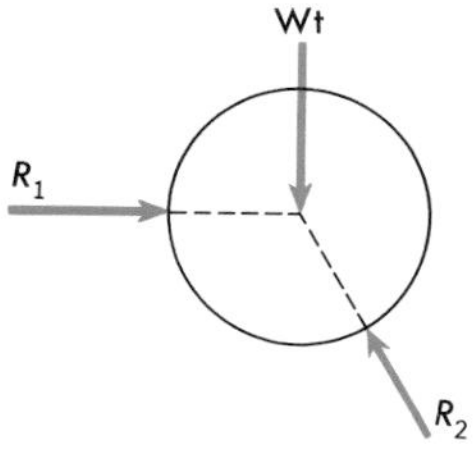 Wt R_1 → R_2 Free body	The reaction between surfaces at rest is perpendicular to the common tangent plane at the point of contact. Thus, if a cylinder rests on a plane, the reaction at the point of contact will pass through the center of the cylinder.
Pulling a barrel over a curb	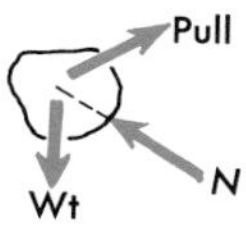 Pull Wt N	All of the forces are acting through the center of the barrel.

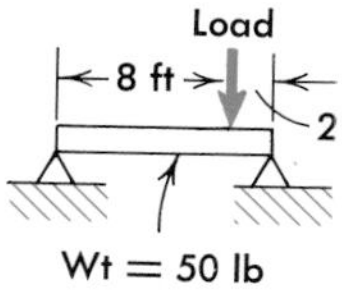

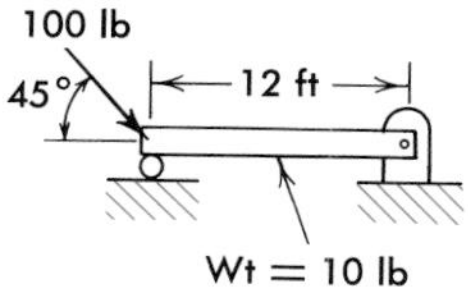

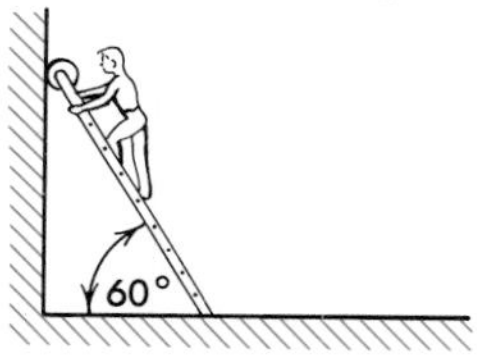

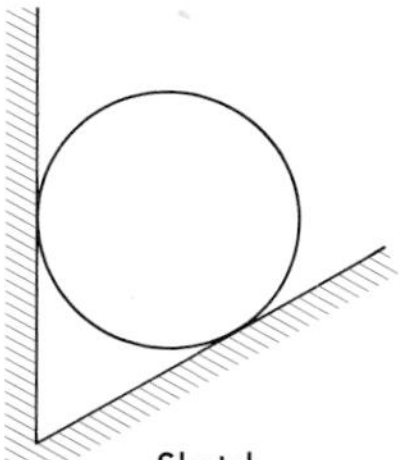

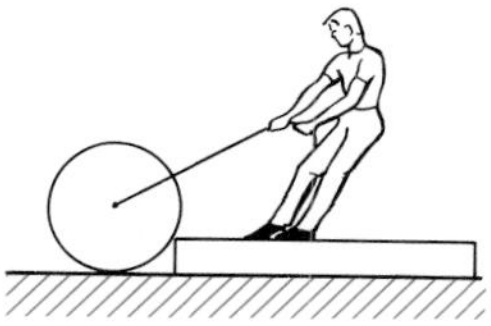

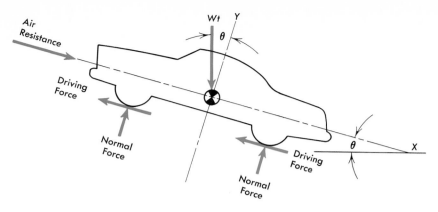

Figure 13-16

General suggestions for drawing free-body diagrams

To aid the student in learning to draw free-body diagrams, the following suggestions are given:

1. Free bodies Be certain that the body is *free* of all surrounding objects. Draw the body so it is *free*. Do not show a supporting surface but rather show only the force vector which replaces that surface. Do not rotate the body from its original position but rather rotate the axes if necessary. Show all forces and label them. Show all needed dimensions and angles.

2. Force components Forces are often best shown in their component forms. When replacing a force by its components, select the most convenient directions for the components. Never show both a force and its components by solid-line vectors; use broken-line vectors for one or the other since the force *and* its components do not occur simultaneously.

3. Weight vectors Show the weight vector as a vertical line with its tail or point at the center of gravity, and place it so that it interferes least with the remainder of the drawing. It should always be drawn vertically.

4. Direction of vectors The free-body diagram should represent the facts as nearly as possible. If a pull on the free body occurs, place the tail of the vector at the actual point of application and let the point of the vector be in the true direction of the pull. Likewise, if a push occurs on the free body, the vector should show the true direction, and the point of the arrow should be placed at the point of application. Force vectors on free-body diagrams are not usually drawn to scale but may be drawn proportionate to their respective magnitudes.

5. Free-body diagram of whole structure This should habitually be the first free body examined in the solution of any problem. Many problems cannot be solved without this first consideration. After the free body of the whole structure or complex has been considered, select such members or subassemblies for further free-body diagrams as may lead to a direct solution.

6. Two-force members When a two-force member is in equilibrium, the forces are equal, opposite, and collinear. If the member is in compression, the vectors should point toward each other; if a member is in tension, they should point away from each other.

7. Three-force members When a member is in equilibrium and has only three forces acting on it, the three forces are always concurrent; that is, they go through the same point if they are not parallel. In analyzing a problem involving a three-force member, one should recall that any set of concurrent forces may be replaced by a resultant force. Hence, if a member in equilibrium has forces acting at three points, it is a three-force member regardless of the fact that the force applied at one or more points may be replaced by two or more components.

8. Concurrent force system For a concurrent force system the size, shape, and dimensions of the body can be neglected, and the body can be considered to be a particle.

Example Draw a free body of point *A*, as shown in Figure 13-17.

Solution See Figure 13-18.
     In describing free-body diagrams we have talked about force *vectors*. Let us look briefly at vectors and their properties. A *vector* quantity has a *direction* as well as a *magnitude* (in contrast to a *scalar* quantity, which has magnitude only). Examples of vector quantities are force, velocity, and acceleration. Examples of scalar quantities are temperature, volume, time, and energy. Vectors are said to be:

1. Coplanar, when all vectors lie in the same plane, Figure 13-19.
2. Collinear, when all vectors act along the same line, Figure 13-20.
3. Concurrent, when all vectors originate or intersect at a single point, Figure 13-21.

Example A force of 150 lb$_f$ is pulling upward from a point at an angle of 30° with the horizontal, Figure 13-22. The length of the arrow was scaled (using an engineer's scale) to 1 in. equals 100 lb$_f$ and is 1½ in. long acting upward at an angle of 30° with

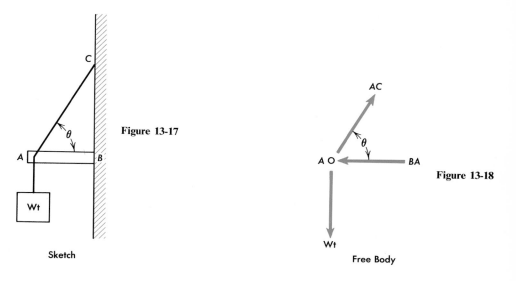

Figure 13-17

Sketch

Figure 13-18

Free Body

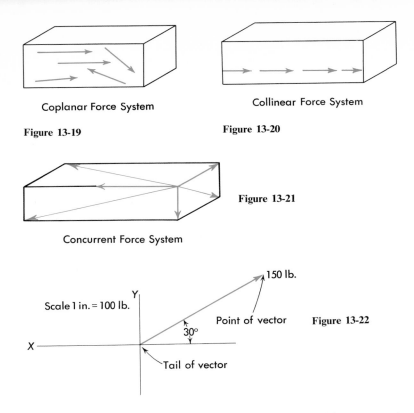

Coplanar Force System

Figure 13-19

Collinear Force System

Figure 13-20

Figure 13-21

Concurrent Force System

Scale 1 in. = 100 lb.

150 lb.

Y

Point of vector   Figure 13-22

30°

X

Tail of vector

the horizontal. In graphic work the arrow point should not extend completely to the end of the vector, since it is very easy to "overrun" the exact length of the measured line in the drawing of the arrowhead.

In rigid-body mechanics the external effect of a force on a rigid body is independent of the point of application of the force along its line of action. Thus it would be considered immaterial whether a tractor pushed or pulled a box from a given position. The total effect on the box would be the same in either case. This may be illustrated as shown in Figure 13-23.

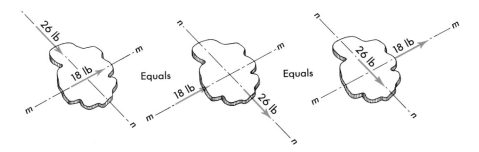

Figure 13-23

Example In each case the body is being acted upon by forces of 26 lb_f and 18 lb_f. The total effect on the body is assumed to be the same for each example, since it is the line of action of a force which is significant, rather than its point of application.

Resolutions of forces

In this initial study of static mechanics we shall deal mainly with concurrent, coplanar force systems. It is sometimes advantageous to combine two such forces into a single equivalent force, which we shall call a *resultant*. The original forces are called *components*.

Example What single force R pulling at point O will have the same effect as components F_1 and F_2 (Figure 13-24)?

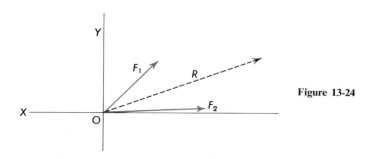

Figure 13-24

There are several methods of combining these two components into a single resultant. Let us examine one of these, the method of rectangular components.

Rectangular components

The method most frequently used by engineers to find the resultant of force systems is the *rectangular component method*. Vector components can be added together or subtracted—always leaving some resultant value. (This resultant value, of course, may be zero.) Also, any vector or resultant value can be replaced by two or more other vectors that are usually called *components*. If the components are two in number and perpendicular to each other, they are called *rectangular components*. Although it is common practice to use space-coordinate axes that are horizontal and vertical, it is by no means necessary to do so. Any orientation of the axes will produce equivalent results.

Figure 13-25 shows a vector quantity F. Figure 13-26 shows F with its rectangular components F_x and F_y. Note that the lengths of the components F_x and F_y can be determined numerically by trigonometry. The components F_x and F_y also can be resolved into the force F by the "polygon of forces" method. Hence, they may replace the force F in any computation.

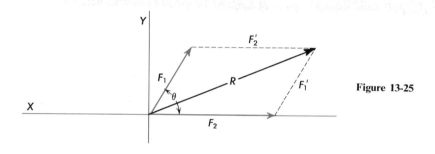

Figure 13-25

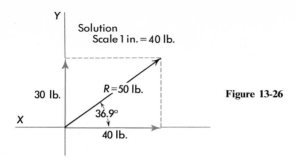

Figure 13-26

Example   Let us examine a concurrent coplanar force system, Figure 13-27, and resolve each force into its rectangular components, Figure 13-28. By trigonometry, F_x can be found, using F and the cosine of the angle θ, or $F_x = F \cos \theta°$. In the same manner $F_y = F \sin \theta°$.

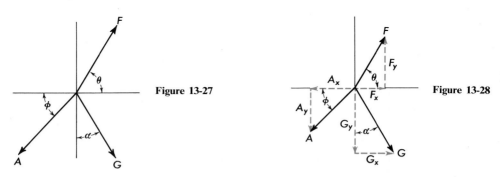

Figure 13-27

Figure 13-28

In order to keep the directions of the vectors better in mind, let us assume that horizontal forces acting to the right are positive and those acting to the left are negative. Also, the forces acting upward may be considered positive and those acting downward negative.

In working such force systems by solving for the rectangular components, a table may be used. When the sums of the horizontal and vertical components have been determined, lay off these values on a new pair of axes to prevent confusion. Solve for the resultant in both magnitude and direction,

Example   Solve for the resultant, R, in Figure 13-29 using the method of rectangular components for the final resolution of the force system.

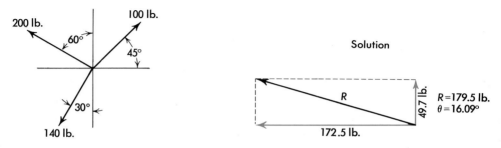

Figure 13-29

Solution

Forces	Horizontal component	Horizontal value		Vertical component	Vertical value
100 lb$_f$	100 cos 45° =	+70.7 lb$_f$		100 sin 45° =	+70.7 lb$_f$
200 lb$_f$	200 sin 60° =	−173.2 lb$_f$		200 cos 60° =	+100 lb$_f$
140 lb$_f$	140 sin 30° =	−70.0 lb$_f$		140 cos 30° =	−121 lb$_f$
Total value	Positive	+70.7 lb$_f$		Positive	+170.7 lb$_f$
Total value	Negative	−243.2 lb$_f$		Negative	−121 lb$_f$
Sum	Horizontal	−172.5 lb$_f$		Vertical	+49.7 lb$_f$

Problems

Solve, using rectangular components.

13-1. Find the resultant, in amount and direction, of the following concurrent coplanar force system: force *A*, 180 newtons acts S 60° W; and force *B*, 158 newtons, acts S 80° W. Check your answer graphically, using a scale of 1 cm equals 50 N.

13-2. Four men are pulling a box. *A* pulls with a force of 115 lb$_f$, N 20°40′E? *B* pulls with a force of 95 lb$_f$ S 64°35′E; *C* pulls with a force of 140 lb$_f$ N 40°20′E; and *D* pulls with a force of 68 lb$_f$ E. In what direction will the box tend to move?

13-3. Determine the amount and direction of the resultant of the concurrent coplanar force system as follows: force *A*, 10 lb$_f$, acting N 55°E; force *B*, 16 lb$_f$, acting due east; force *C*, 12 lb$_f$, acting S 22° W; force *D*, 15 lb$_f$, acting due west; force *E*, 17 lb$_f$, acting N 10° W.

13-4. Find the resultant and the angle the resultant makes with the vertical, using the following data: 10 newtons, N 18° W; 5 newtons N 75° E; 3 newtons, S 64° E; 7 newtons, S 0° W; 10 newtons, S 50° W.

13-5. Five forces act on an object. The forces are as follows: 130 lb$_f$, 0°; 170 lb$_f$, 90°; 70 lb$_f$, 180°; 20 lb$_f$, 270°; 300 lb$_f$, 150°. The angles are measured counterclockwise with reference to the horizontal through the origin. Determine graphically the amount and direction of the resultant by means of the polygon of forces. Check analytically, using horizontal and vertical components. Calculate the angle that *R* makes with the horizontal.

13-6. (*a*) In the sketch in Figure 13-30, using rectangular components, find the resultant of these four forces: *A* = 100 newtons, *B* = 130 newtons, *C* = 195 newtons, *D* = 138 newtons. (*b*) Find a resultant force that would replace forces *A* and *B*. (*c*) Break force *A* into two components, one of which acts N 10° E and has a magnitude of 65 newtons. Give the magnitude and direction of the second component.

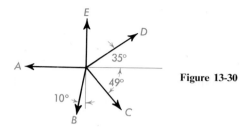

Figure 13-30

13-7. A weight of 1200 newtons is hung by a cable 23 m long. What horizontal pull will be necessary to hold the weight 8 m from a vertical line through the point of support? What will be the tension in the cable?

13-8. A weight of 80 lb$_f$ is suspended by two cords, the tension in *AC* being 70 lb$_f$ and in *BC* being 25 lb$_f$, as shown in Figure 13-31. Find the angles α and θ.

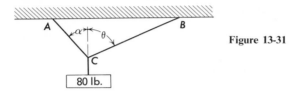

Figure 13-31

Moments

If a force is applied perpendicular to a pivoted beam some distance away from the pivot point, there will be a tendency to cause the beam to turn in either a clockwise or counterclockwise direction (see Figure 13-32). The direction of the tendency will depend on the direction of the applied force. This tendency of a force to cause rotation about a given center is called *moment* (see Figure 13-33).

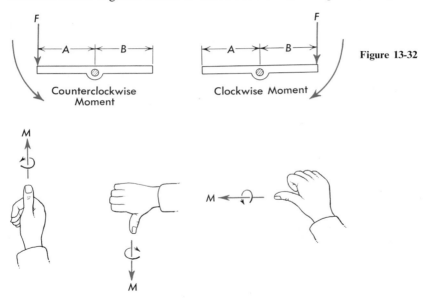

Counterclockwise Moment

Clockwise Moment

Figure 13-32

Figure 13-33

The amount of *moment* will depend upon the magnitude of the applied force as well as upon the length of the moment arm. The moment arm is the perpendicular distance from the point of rotation to the applied force. The magnitude of the moment is calculated by multiplying the force by the moment arm.

The sign convention being used in a given problem analysis should be placed on

Give me a lever long enough, and a fulcrum strong enough, and singlehanded I can move the world.
—Archimedes, ca. 250 B.C.

Illustration 13-3
Leverage is the layman's language for moment.

the calculation sheet adjacent to the problem sketch. In this way no confusion will arise in the mind of the reader concerning the sign convention being used. We shall assume that vectors acting to the right have a positive sign, vectors acting upward have a positive sign, and moments directed counterclockwise have a positive sign. To aid in establishing a system of positive sense, the sketch shown in Figure 13-34 will serve as a basis for problem analysis in this text.

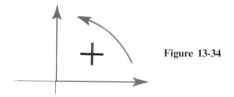

Figure 13-34

Example Solve for the moments in Figure 13-35 that tend to cause turning of the beam about the axle.

$$\text{Counterclockwise moment} = (50 \text{ lb})(2 \text{ ft}) = +100 \text{ lb-ft}$$
$$\text{Clockwise moment} = (100 \text{ lb})(5 \text{ ft}) = -500 \text{ lb-ft}$$

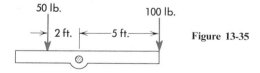

Figure 13-35

Since *moment* is the product of a force and a distance, its units will be the product of force and length units. By convention, moments are usually expressed with the force unit being shown first, as lb_f-ft, lb_f-in., kip-ft (a kip is 1000 lb_f), and so on. This is done because *work* and *energy* also involve the product of distance and force, and the units ft-lb_f, in.-lb_f, and so on are commonly used for this purpose.

The moment of a force about some given center is identical to the sum of the moments of the components of the force about the same center. This principle is commonly called *Varignon's theorem*. In problem analysis it is sometimes more convenient to solve for the sum of the moments of the components of a force rather than the moment of the force itself. However, the problem solutions will be identical.

Example Solve for the total moment of the 1000-lb$_f$ force about point A in Figure 13-36.

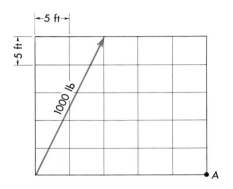

Figure 13-36

Solution A Moment of a force as shown in Figure 13-37.

$$\theta = \text{arc tan } \frac{25}{10} = 68.2°$$

$$\text{Moment arm} = 25 \sin 68.2°$$
$$\text{Total moment} = (1000)(25 \sin 68.2°)$$
$$= \textbf{23,200 lb}_f\textbf{-ft}$$

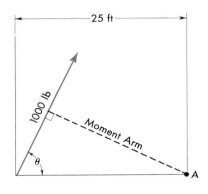

Figure 13-37

Solution B Moments of components of a force as shown in Figure 13-38.

$$\text{Vertical component} = 1000 \sin 68.2°$$
and
$$\text{Moment arm} = 25 \text{ ft}$$
$$\text{Horizontal component} = 1000 \cos 68.2°$$
and
$$\text{Moment arm} = 0$$

(Note that the horizontal component passes through the center A.)

$$\text{Total moment} = (1000 \sin 68.2°)(25) = \textbf{23,200 lb}_f\textbf{-ft}$$

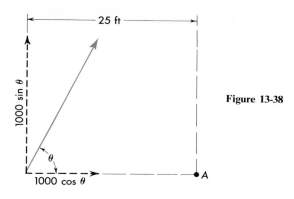

Figure 13-38

Problems

13-9. Solve for the algebraic sum of the moments in pound-feet about A when h is 20 m as shown in Figure 13-39.

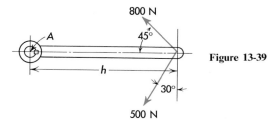

Figure 13-39

13-10. Solve for the algebraic sum of the moments about the center of the axle shown in Figure 13-40.

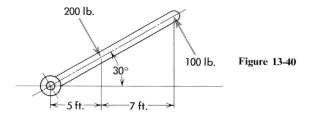

Figure 13-40

13-11. (a) Solve for the clockwise moments about A, B, C, D, and E in Figure 13-41. (b) Solve for the counterclockwise moments about A, B, C, D, and E. (c) Solve for the algebraic sum of the moments about A, B, C, D, and E.

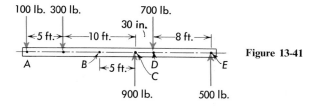

Figure 13-41

13-12. Find the summation of the moments of the forces shown about A in Figure 13-42. Find the moment sum about D.

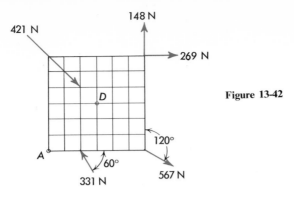

Figure 13-42

13-13. Find the moment of each of the forces shown about O in Figure 13-43.

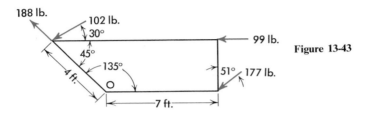

Figure 13-43

13-14. What pull P is required on the handle of a claw hammer to exert a vertical force of 750 newtons on a nail? Dimensions in inches are shown on Figure 13-44.

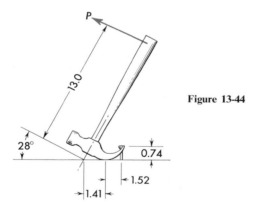

Figure 13-44

Equilibrium

The term *equilibrium* is used to describe the condition of any body when the resultant of all forces acting on the body equals zero. For example, the forces acting upward on a body in equilibrium must be balanced by other forces acting downward on the body. Also, the forces acting horizontally to the right are counteracted by equal forces acting horizontally to the left. Since no unbalance in moment or turning effect can be present when a body is in equilibrium, the sum of the moments of all forces acting on the body must also be zero. The moment center may be located at any convenient place on the body or at any place in space. We may sum up these conditions of equilibrium by the following equations[6]:

$\Sigma F_x = 0$ (the sum of all horizontal forces acting on the body equals zero)
$\Sigma F_y = 0$ (the sum of all vertical forces acting on the body equals zero)
$\Sigma M_o = 0$ (the sum of the moments of all forces acting on the body equals zero)

These equilibrium equations may be used to good advantage in working problems involving beams, trusses, and levers.

Example   A beam of negligible weight is supported at each end by a knife-edge. The beam carries a concentrated load of 500 lb$_f$ and one uniformly distributed load weighing 100 N per metre, as shown in Figure 13-45. Determine the scale readings under the knife-edges.

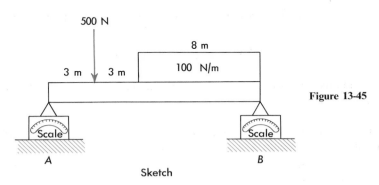

Figure 13-45

Sketch

Solution   The uniformly distributed load is equivalent to a resultant of 8 m × 100 N/m = 800 N acting at the center of gravity of the uniform-load diagram. Therefore, the entire distribution load can be replaced by a concentrated load of 800 N acting at a distance of 10 m from the left end, as shown in Figure 13-46.

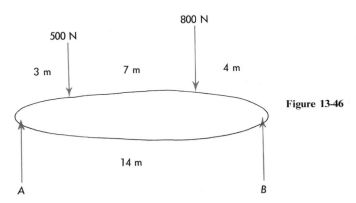

Figure 13-46

1. Draw a free-body diagram of the beam.
2. Since there are no horizontal forces acting on the free body, $\Sigma F_x = 0$ is satisfied.

[6]These equations are applicable for two-dimensional problems—or force systems that lie in the plane of this paper.

3. From $\Sigma F_y = 0$, we know that

$$A + B = 500 \text{ N} + 800 \text{ N}$$
$$= 1300 \text{ N}$$

4. From $\Sigma M_o = 0$, we know that the moments about any point must equal zero. Let us take moments about point A.

$$\Sigma M_A = 0$$
$$(B \text{ N})(14 \text{ m}) - (500 \text{ N})(3 \text{ m}) - (800 \text{ N})(10 \text{ m}) = 0$$
$$B \text{ N} = \frac{1500 \text{ N-m} + 8000 \text{ N-m}}{14 \text{ m}}$$
$$= \frac{9500 \text{ N-m}}{14 \text{ m}}$$
$$B = \textbf{679 N} \uparrow$$

5. From the third step we saw that $A + B = 1300$ N. We can now subtract and obtain

$$A = 1300 \text{ N} - 679 \text{ N} = \textbf{621 N} \uparrow$$

Note The same answer for A could have been obtained by taking moments about B as a moment center.

In this book problems involving trusses, cranes, linkages, bridges, and so on, should be considered to be *pin-connected,* which means that the member is free to rotate about the joint. For simplicity, members also are usually considered to be weightless.

By examining each member of the structure separately, internal forces in the various members may be obtained by the conditions of equilibrium.

Example Solve for the tensions in cables AF and ED and for the reactions at C and R in Figure 13-47.

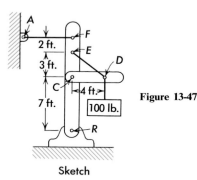

Figure 13-47

Sketch

Equilibrium Equations
$$\Sigma F_x = 0$$
$$\Sigma F_y = 0$$
$$\Sigma M_o = 0$$

Solution
1. Take moments about point R in free body No. 1, Figure 13-48.

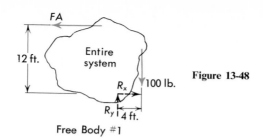

Free Body #1

$$\Sigma M_R = 0$$
$$(12 \text{ ft})(FA) - (100 \text{ lb}_f)(4 \text{ ft}) = 0$$
$$FA = \frac{400 \text{ lb}_f\text{-ft}}{12 \text{ ft}} = 33.3 \text{ lb}_f$$
$$\Sigma F_x = 0$$
$$R_x - FA = 0$$
$$R_x = FA = \textbf{33.3 lb}_f\rightarrow$$

2. Take moments about point C in free body No. 2, Figure 13-49.

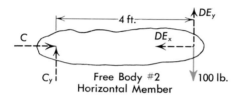

Figure 13-49

Free Body #2
Horizontal Member

$$\Sigma M_c = DE_y(4) - 100(4) = 0$$
$$DE_y = 100 \text{ lb}_f$$

Therefore
$$DE = \frac{100 \text{ lb}_f}{\sin 36.9°} = \textbf{166.8 lb}_f\nwarrow$$

And free body No. 2:

$$\Sigma F_y = 0$$
$$C_y = 100 \text{ lb}_f - 100 \text{ lb}_f$$
$$= \textbf{0}$$

Also free body No. 2:

$$\Sigma F_x = 0$$
$$C_x = DE_x = \frac{100 \text{ lb}_f}{\tan 36.9°}$$
$$= \textbf{133.1 lb}_f\rightarrow$$

3. Consider $\Sigma F_y = 0$, using the third free body (vertical member) as shown in Figure 13-50. Remember that in two-force members, such as cable DE, the reactions at each end will be equal in magnitude but opposite in direction; that is, E_x and E_y are equal to DE_x and DE_y.

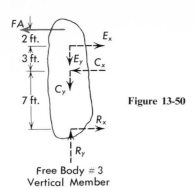

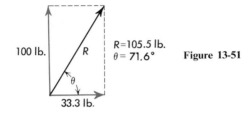

Figure 13-50

$$\Sigma F_y = 0$$
$$R_y - DE_y = 0$$
$$R_y = 100.0 \ \text{lb}_f \uparrow$$

The resultant is indicated as before (Figure 13-51).

100 lb.   R   R=105.5 lb.
$\theta = 71.6°$ Figure 13-51

33.3 lb.

Problems

13-15. A horizontal beam 20 m long weighs 1500 newtons. It is supported at the left end and 4 m from the right end. It has the following concentrated loads: at the left end, 2000 newtons; 8 m from the left end, 3000 newtons; at the right end, 4000 newtons. Calculate the reactions at the supports.

13-16. A horizontal beam 8 ft long and weighing 30 lb_f is supported at the left end and 2 ft from the right end. It has the following loads: at the left end, 18 lb_f; 3 ft from the left end, 22 lb_f; at the right end, 15 lb_f. Compute the reactions at the supports.

13-17. A beam 22 m long weighing 3000 newtons is supporting loads of 7000 newtons 3 m from the left end and 2500 newtons 7 m from the right end. One support is at the left end. How far from the right end should the right support be placed so that the reactions at the two supports will be equal?

13-18. A beam 18 m long is supported at the right end and at a point 5 m from the left end. It is loaded with a concentrated load of 2500 newtons located 2 m from the right end and a concentrated load of 2500 newtons located 9 m from the right end. In addition, it has a uniform load of 200 newtons per linear metre for its entire length. Find the reactions at the supports.

13-19. A 12-ft beam which weighs 10 lb_f per foot is resting horizontally. The left end of the beam is pinned to a vertical wall. The right end of the beam is supported by a cable that is attached to the vertical wall 6 ft above the left end of the beam. There is a 200-lb_f concentrated load acting vertically downward 3 ft from the right end of the beam. Determine the tension in the cable and the amount and direction of the reaction at the left end of the beam.

13-20. A steel I-beam, weighing 75 lb_f per linear foot and 20 ft long, is supported at its left end and at a point 4 ft from its right end. It carries loads of 10 tons and 6 tons at distances of 5 ft and 17 ft, respectively, from the left end. Find the reactions at the supports.

13-21. A horizontal rod 8 m long and weighing 120 newtons has a weight of 150 newtons hung from the right end, and a weight of 40 newtons hung from the left end. Where should a single support be located so the rod will balance?

13-22. A uniform board 22 ft long will balance 4.2 ft from one end when a weight of 61 lb$_f$ is hung from this end. How much does the board weigh?

13-23. An iron beam 12.7 ft long weighing 855 lb$_f$ has a load of 229 lb$_f$ at the right end. A support is located 7.2 ft from the load end. (a) How much force is required at the opposite end to balance it? (b) Disregarding the balancing force, calculate the reactions on the supports if one support is located 7 ft from the left end and the other support is located 4 ft from the right end.

13-24. A horizontal rod 8 ft long and weighing 1.2 lb$_f$ per linear foot has a weight of 15 lb$_f$ hung from the right end, and a weight of 4 lb$_f$ hung from the left end. Where should a single support be located so the rod will balance?

13-25. A 2-ft diameter sphere weighs 56 lb$_f$, is suspended by a cable, and rests against a vertical wall. If the cable AB is 2 ft long, (a) calculate the angle the cable will make with the smooth wall, (b) solve for the tension in the cable and the reaction at C in Figure 13-52.

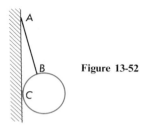

Figure 13-52

13-26. What horizontal pull P will be necessary just to start the wheel weighing 1400 lb$_f$ over the 4-in. block in Figure 13-53.

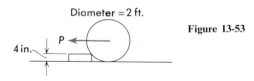

Diameter = 2 ft.

Figure 13-53

4 in.

P

13-27. Find the tension in AB and the angle θ that AB makes with the vertical in Figure 13-54.

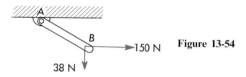

B

→150 N   Figure 13-54

38 N

13-28. If the tension in the cable AB in Figure 13-55 is 1960 N, how much does the sphere B weigh? How much is the reaction of the inclined plane on the sphere?

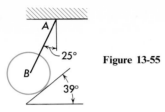

Figure 13-55

13-29. The wheel *B* in Figure 13-56 weighs 1750 N. Solve for the force in member *AB,* the reaction at *C,* and the horizontal and vertical force components at *A.*

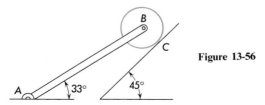

Figure 13-56

13-30. A cylinder weighing 206 lb$_f$ is placed in a smooth trough as shown in Figure 13-57. Find the two supporting forces.

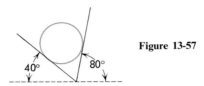

Figure 13-57

13-31. A 7960-N load is supported as shown in Figure 13-58. *AB* equals 8 m, *θ* equals 25°. (*a*) Neglecting the weight of the beam *AB,* solve analytically for the tension in the cable and the reaction at *A.* (*b*) If beam *AB* is uniform and weighs 120 N per metre, solve for the tension in the cable and the reaction at *A.*

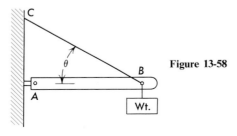

Figure 13-58

13-32. Find the tension in *AB* and the compression in *BC* in Figure 13-59.

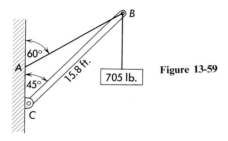

Figure 13-59

13-33. A weight of 13,550 N is supported by two ropes making angles of 30° and 45° on opposite sides of the vertical. What is the tension in each rope?

13-34. Forces are applied on a rigid frame as shown in Figure 13-60. Find the reactions at *A* and *B*.

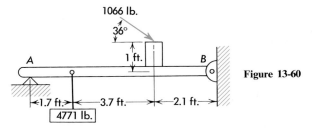

Figure 13-60

13-35. (*a*) What is the tension in *BC* in Figure 13-61? (*b*) What is the amount and direction of the reaction at *A*?

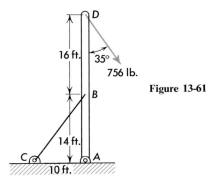

Figure 13-61

13-36. (*a*) Find the tension in *AC* in Figure 13-62. (*b*) Find the amount and direction of the reaction at *B*. *BC* = 10 m, *BD* = 25 m.

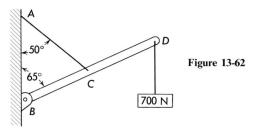

Figure 13-62

13-37. Cylinder No. 1 in Figure 13-63 has a 0.35 m diameter and weighs 840 N. Cylinder No. 2 has a 0.20-m diameter and weighs 270 N. Find the reactions at *A*, *B*, and *C*. All surfaces are smooth.

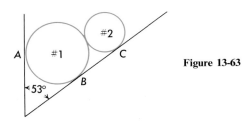

Figure 13-63

13-38. (*a*) Find the force in member *AB* in Figure 13-64 and the reaction at point *E*. (*b*) Find the force in member *CG* and the horizontal and vertical components of the reaction at pin *D*.

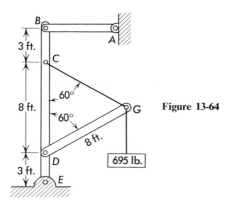

Figure 13-64

13-39. A 5-m ladder leans against the side of a smooth building in such a position that it makes an angle of 60° with the ground (horizontal). A man weighing 790 N stands on the ladder three fourths of the way up the ladder. The bottom of the ladder is prevented from sliding by the ground. Find the horizontal and vertical components of the reaction at the foot of the ladder and the force between the ladder and the wall.

Bodies in uniform motion

Today the theory of motion and moving things is one of our most important studies. In recent years even governments have become vitally interested in all types of motion—from the motion of atoms to that of satellites and celestial bodies. Moving people, animals, fast-moving commuter trains, sleek automobiles, and jet airplanes are all a routine part of our daily life. Even wars are most decidedly wars of motion.

Motion exists when there is a *change of position* of an object with reference to some other object or plane. For example, a passenger in a jet airliner may be sitting still in the opinion of other passengers, but be in rapid motion with reference to a farmer plowing in the field below. The motion of the airliner may be uniform if it has balanced forces acting on it, or accelerated if jet thrust, air resistance, and gravity do not balance each other.

Sometimes we speak of the motion of a body as *speed,* which refers to its rate of motion. The scientific term *velocity,* which refers to *rate of motion in a given direction,* is sometimes used incorrectly as a synonym for speed. Speed is the term used to designate the magnitude of velocity. Thus speed equals distance divided by time.

Examples $\quad \dfrac{m}{s}, \dfrac{ft}{sec}, \dfrac{mi}{hr}, \dfrac{cm}{sec}, \dfrac{yd}{hr}$, etc.

Velocity equals distance divided by time—all expressed in a given direction.

Examples $\dfrac{\text{km}}{\text{hr}}$ west, $\dfrac{\text{mi}}{\text{hr}}$ north, $\dfrac{\text{ft}}{\text{sec}}$ 30° east of north, etc.

Sir Isaac Newton, an English scientist, was the first to generalize the laws of forces and motions. His findings have been set forth in three laws as follows:

Newton's First Law *A body at rest or in motion will continue either at rest or in motion in the same line and at the same speed unless acted on by some external unbalanced force.*

Common experience tells us that a body at rest, such as a billiard ball, will continue in a state of rest unless some force is applied to move it. Similarly, if the ball is struck and begins to roll, it will continue to roll in a straight line until it strikes, for example, another ball at the end of the table. The velocity with which it strikes the second ball will be reduced somewhat from its beginning velocity. The reduction in velocity is caused by the slight but constant friction between the ball and the table. Automobile wrecks result from this tendency of bodies to continue in the same line. The engine and brakes can act against this tendency and slow the car.

Newton's Second Law *When an external unbalanced force does act on a body, the motion of the body will be changed. The body will be accelerated. This change in motion will be in the direction of the unbalanced force and proportional to it.*

Acceleration is the rate of change of velocity. This is a measurement of how much slower (or faster) a body is traveling now than it was 1 sec ago. The rate at which a body slows down is sometimes called negative acceleration or deceleration. For example: an automobile may start from rest and accelerate to a velocity of 48 mi/hr during an 8-sec period. This means that, for every second that the engine acts on the car, there will be an increase in velocity of 6 mi/hr/sec or 8.8 ft/sec/sec. Acceleration is measured as

$$\dfrac{\text{distance}}{(\text{time})^2}; \quad \text{i.e.,} \quad \dfrac{\text{m}}{\text{s}^2}, \ \dfrac{\text{ft}}{(\text{sec})^2}, \ \dfrac{\text{mi}}{(\text{hr})^2}, \ \dfrac{\text{ft}}{(\text{min})^2}, \ \text{etc.}$$

Newton's Third Law *When any force acts on a body, there is created an equal and opposite reaction.*

Again common experience tells us that the mutual actions of two bodies on each other are always equal. If two men are pulling with equal force against each other on a rope, each will sense the same magnitude of resistance. If one presses the button on an electronic computer with his finger, the finger is also pressed by the button with equal force.

The speed–time diagram

The study of motion becomes rather involved if all the situations are represented by mathematical equations. Some better understanding can result if one can picture exactly what is taking place. For this reason extensive use will be made of the *speed–time* diagram as a means of pictorially representing the motions described. In

Illustration 13-4

When an external unbalanced force acts on a body, the motion of the body will be changed. The body will be accelerated. This change of motion will be in the direction of the unbalanced force and proportional to it.

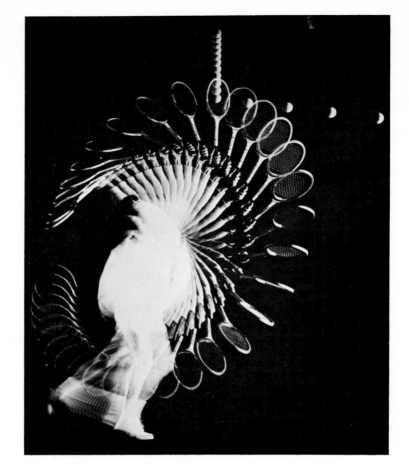

addition, this treatment also reduces the amount of memory work normally associated with the various relations.

In motion problems the total distance traveled is represented by the area which lies under the travel line of a speed–time diagram. For example, if an automobile travels at a uniform velocity of 30 mi/hr for 30 min, it will cover a distance of 15 mi.

$$30\,\frac{\text{mi}}{\text{hr}} \times 30\text{ min} \times \frac{1\text{ hr}}{60\text{ min}} = \textbf{15 min}$$

This may be shown graphically as in Figure 13-65.

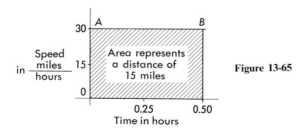

Figure 13-65

If the speed is constant, then the distance traveled may be found by multiplying the ordinate value times the abscissa value. In this case the acceleration is zero as indicated by the straight line *A-B*.

Therefore, in order to work the above problem, the student need only draw the speed–time diagram and then find the area under the line *A-B* by simple arithmetic.

Speed–time diagram principles may be summarized as follows:

1. The ordinate of the line at any instant will give the speed at that instant.
2. Abscissa values give the time consumed during travel.
3. The *area* under the travel line of the speed–time diagram gives the distance traveled during the time interval under consideration.
4. The slope of the line at any point gives the acceleration of the body at that point.

Slope may be defined as the *steepness* of a line and can be calculated by dividing the vertical rise by the corresponding horizontal distance.

Example An automobile accelerates uniformly from a speed V_1 to a speed V_2 in time t (Figure 13-66).

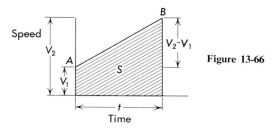

Figure 13-66

A speed–time diagram of the stated problem is drawn. The total distance traveled during time t can be calculated by solving for the area under the line *A-B*. This area is a trapezoid, and by simple arithemetic,

$$\text{Area} = \tfrac{1}{2}h(b_1 + b_2) \quad \text{(see page 503)}$$
or
$$S = \tfrac{1}{2}t(V_1 + V_2) \quad \text{(from speed–time diagram)}$$

The acceleration has been defined as the slope of the travel line. An examination of Figure 13-66 shows that this is also the change in velocity ($V_2 - V_1$) divided by the time (t) that it took to make the change.

Stated algebraically we have

$$a = \frac{V_2 - V_1}{t} \quad \text{(slope of travel line)}$$

Example An automobile starts from rest (Figure 13-67) and accelerates uniformly to 30 mi/hr in 11 sec. What is its acceleration? What distance was covered during the change in velocity?

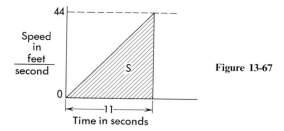

Figure 13-67

$$\left(30\ \frac{mi}{hr}\right)\left(5280\ \frac{ft}{mi}\right)\left(\frac{1\ hr}{3600\ sec}\right) = 44\ ft/sec$$

(a)
$$a = \frac{V_2 - V_1}{t} = \frac{44\ ft/sec - 0}{11\ sec}$$

$$= 4\ \frac{ft}{sec \times sec} \quad or \quad 4\ \frac{ft}{(sec)^2}$$

(b)
$$S = \frac{1}{2}Vt \quad \text{(area of crosshatched triangle)}$$
$$= \frac{1}{2}(44\ ft/sec)(11\ sec) = \mathbf{242\ ft}$$

In some instances the term *average velocity* or *average speed* is used. Average velocity is not necessarily an average of the initial and final velocities. It may be expressed as

$$\text{Average velocity} = \frac{\text{total distance traveled}}{\text{total time during travel}}$$

Example An automobile traveled a total distance of 100 mi at an average speed of 50 mi/hr. During the first 50 mi, the average speed of the automobile was 60 mi/hr. What was the average speed for the last 50 mi?

$$\text{Average speed for trip} = \frac{\text{total distance}}{\text{total time}}$$

$$\text{Total time } (t) = \frac{100\ mi}{50\ mi/hr} = \mathbf{2\ hr}$$

For the first 50 mi:

$$\text{Time} = \frac{50\ mi}{60\ mi/hr} = 0.833\ hr$$

$$\text{Time remaining} = 2\ hr - 0.833\ hr = 1.167\ hr$$

For the last 50 mi:

$$\text{Average speed} = \frac{50\ mi}{1.167\ hr} = \mathbf{42.8\ mi/hr}$$

Many of the situations encountered in linear motion can be solved readily by the use of the speed–time diagram. Some problems involve varied speeds and accelerations during any period under consideration. These changes should be clearly indicated on the speed–time diagram.

Example A train (Figure 13-68) travels 10 mi at a speed of 50 mi/hr and then uniformly increases its speed to 65 mi/hr during a 30-min period. The train continues at this speed for 1 hr before being uniformly slowed to a stop with a deceleration of

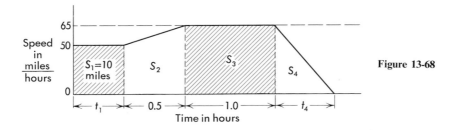

Figure 13-68

650 mi/hr/hr. Find (a) stopping time, (b) distance traveled during acceleration, (c) total time, and (d) total distance traveled.

Solution

(a) $$a = \frac{V_2 - V_1}{t} \quad \text{(slope of travel line as train stops)}$$

$$-650 \text{ mi/(hr)}^2 = \frac{0 - 65 \text{ mi/hr}}{t_4 \text{ hr}}$$

$$t_4 = \textbf{0.10 hr}$$

(b) $$S = \left(\frac{V_1 + V_2}{2}\right) t$$

$$S_2 = \tfrac{1}{2} \times 0.5 \text{ hr } (50 \text{ mi/hr} + 65 \text{ mi/hr})$$
$$= \textbf{28.75 mi}$$

(c) $$t = \frac{S}{V_{av}}$$

$$t_1 = \frac{10 \text{ mi}}{50 \text{ mi/hr}} = 0.20 \text{ hr}$$

$$\text{Total time} = t_1 + t_2 + t_3 + t_4$$
$$= 0.20 + 0.50 + 1.0 + 0.10$$
$$= \textbf{1.8 hr}$$

(d) $$S = (V_{av})(t)$$
$$S_3 = 65 \text{ mi/hr} \times 1 \text{ hr} = 65 \text{ mi}$$

$$S = \left(\frac{V_1 + V_2}{2}\right) t$$

$$S_4 = \tfrac{1}{2}(65 \text{ mi/hr} + 0)\, 0.10 \text{ hr}$$

$$= \frac{6.5 \text{ mi}}{2} = 3.25 \text{ mi}$$

$$\text{Total distance} = S_1 + S_2 + S_3 + S_4$$
$$= 10 + 28.78 + 65 + 3.25$$
$$= \textbf{107.03 mi}$$

Problems

13-40. A ball is thrown vertically upward and in due course of time it falls back to the place of beginning. Starting from the time the ball leaves the hand, sketch a speed–time diagram which shows the motion involved. Add such explanation as you may deem necessary.

13-41. A man in a car travels a certain distance at an average speed of 8.49 m/s. After he arrives, he turns around and returns over the same route at an average speed of 5.81 m/s. What was the man's average speed both going and coming?

13-42. An airplane travels from point A to point B and returns, all at an air speed of 200 mi/hr. If a 50-mi/hr wind blows from point A to B during the entire trip, what was the average ground speed?

13-43. A train moves out from a dead stop at the station and in 12 min has uniformly increased its velocity to 60 mi/hr. It travels at this speed for 15 min and then uniformly decelerates to a stop in 1.5 mi. (a) Draw a speed–time diagram to show the entire movement. (*b*) Find the acceleration in the first 12-min period. (*c*) Find the total distance traveled. (*d*) Find the total time consumed.

13-44. A body moving with a constant acceleration of 16 ft/sec/sec passes an observation post with a velocity of 25 ft/sec. (*a*) What will be its velocity in inches per second after 1 min? (*b*) How far will it have gone in 1 min?

13-45. A car having an initial velocity of 15 mi/hr increases its speed uniformly at the rate of 5.5 ft/sec/sec for a distance of 295 ft. (*a*) What will be its final velocity? (*b*) How long will it require to cover this distance?

13-46. A truck passes station *A* with a speed of 10 mi/hr and increases its speed to 45 mi/hr in 1.8 min. At this time its speed becomes constant and remains so for 8 min. The speed is then decreased to zero in 2 min. (*a*) Draw a speed–time diagram for the truck. (*b*) What total distance does the truck travel? (*c*) What is the acceleration in the first 1.8 min? (*d*) What is the deceleration in the last 2 min?

13-47. An automobile is climbing a 20 per cent grade and has an initial velocity of 12.07 m/s and a final velocity of 26.82 m/s. If the time is 38 sec, find (*a*) acceleration in metres per second per second up the grade, (*b*) distance in metres it moves up the grade.

13-48. An automobile traveling at a speed of 16 mi/hr is given a constant acceleration of 90 ft/sec/min. (*a*) What will be its speed at the end of 10 sec? (*b*) How far will the automobile travel in the 10 sec?

13-49. An automobile which has a speed of 35.76 m/s is decelerated at the rate of 6.10 m/s^2 for 5 sec. What is the speed at the end of the 5 sec and how far in metres did the car travel in this time?

13-50. A 3300-lb automobile is traveling up a steep hill whose grade is 22 per cent at a rate of 31 mi/hr when the power is shut off and the car is allowed to coast. Because of the loose gravel on the hill, the car comes to a stop in a distance of 125 ft. After traveling 75 ft, what will be the velocity of the car?

13-51. A car traveling at 20.12 m/s meets a train which is moving 14.75 m/s and the time required for the car to pass the train is 18 sec. What is the length of the train in feet?

13-52. An elevator in a business block goes down at the rate of 9 mi/hr. If the elevator starts from rest and the maximum permissible acceleration is 22 ft/sec/sec, how many feet are required for the elevator to attain maximum speed? Find the time in seconds required to attain this speed.

13-53. A ball is dropped from the top of a tower 26.2 m high. Its acceleration is 9.81 m/s^2. (*a*) How long does it take it to reach the ground? (*b*) With what velocity does it strike the ground?

13-54. As a train reaches the city limits it reduces its speed uniformly from 60 mi/hr to 18 mi/hr in a distance of ¼ mi. It continues at 18 mi/hr for 6 min and as it leaves the city limits it again increases its speed to 40 mi/hr in 2½ min. Find (*a*) the deceleration while the train is slowing down, (*b*) the acceleration during the last 2½ min, (*c*) the total distance traveled.

13-55. The speed of a ship traveling at the rate of 16 knots is uniformly retarded to 5 knots in a distance of 1 statute mile. If the rate of retardation continues constant, (*a*) what time in minutes will be required to bring the ship to rest? (*b*) How many metres will it have traveled from the point where the speed is 16 knots?

13-56. A train and an automobile are passing in opposite directions. When the automobile passes the front of the train, the automobile has a speed of 33 mi/hr and the train has a speed of 26 mi/hr. When the automobile passes the rear of the train, the speed of the train is 45 mi/hr and that of the automobile is 20 mi/hr. If the train is 4000 ft long find the time in seconds for the two to pass.

13-57. A train running on a straight level track at 26.82 m/s suddenly detaches its caboose, which decelerates uniformly to a stop. After traveling 3218.7 metres, the engineer notices the accident, and he stops the train uniformly in 50 sec. At the instant the train stops, the caboose stops. What distance in metres did the engineer have to back up in order to hook onto the caboose?

13-58. An automobile crashes into a building. The driver contends that he was not exceeding a 30 mi/hr speed limit when he applied the brakes. If the skid marks of the tires extend for 176 ft and the stopping time was 3 sec, would you agree that the driver was telling the truth?

13-59. A rock is dropped into a well. The sound of impact is heard 3 sec after the rock is dropped. Sound travels 335.28 m/s. What was the depth of the well in metres?

13-60. Several small steel balls fall from a tall building at a uniform rate of three every second. After the second ball has fallen for $3\frac{1}{2}$ sec, what distance in inches separates it and the ball following it? What distance separates it and the ball preceding it?

13-61. An automobile is traveling at 15.65 m/s when the driver sees a cow crossing the highway ahead. If 0.9 sec is allowed for reaction time before the brakes take effect, how far away in metres was the cow when the driver first saw her, if the car stopped just as it touched the cow and the time in decelerating was 4 sec?

13-62. An airplane travels from Stephenville to Fort Worth (65 mi) and returns with an overall average ground speed of 196 mi/hr. If there was a 20-mi/hr tailwind on the trip out and a 20-mi/hr headwind on the return trip, what was the air speed of the airplane during the trip? (Assume that the air speed for the trip was constant.)

13-63. An airplane flies from Fort Worth to Amarillo and returns (325 mi each way) at an air speed of 276 mi/hr. If a wind from the northwest (i.e., from Amarillo to Fort Worth) is blowing at 29.6 mi/hr, what is the average ground speed for the return trip? What is the average ground speed for the round trip?

Angular motion

The motion we have just studied was linear motion and was concerned with the movement of a body or particle in straight-line travel. Many machines, however, have parts that do not travel in straight-line motion. For example, flywheels, airplane propellers, turbine rotors, motor armatures, and so on, all travel in curved paths or with angular motion. For purposes of study here all bodies having angular motion will be considered to be rotating about a fixed center. While rotating about this fixed center there may be a *speeding up* or *slowing down* of the body.

Angular distance (usually designated by some Greek letter such as θ, ϕ, β, etc.) may be measured in degrees, radians, or revolutions. A radian is defined as a central angle subtended by an arc whose length is equal to the radius of the circle.

$$1 \text{ revolution} = 360 \text{ degrees}$$
$$1 \text{ revolution} = 2\pi \text{ radians}$$
$$1 \text{ radian} = 57.3 \text{ degrees}$$

Example Point A in Figure 13-69 travels through an angular distance θ while moving to position B. $\theta = 120$ degrees, $\theta = 0.333$ revolutions, $\theta = 2.095$ radians.

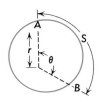

Figure 13-69

Time is measured in the same units as before, that is, seconds, hours, days, and so on. Thus angular velocity, which is an angular distance divided by time, may have such units as radians per second, revolutions per minute, degrees per second, and so on. Angular velocity is usually designated by the Greek letter ω (omega).

Angular acceleration can be found by solving for the slope of the travel line as in linear motion. As in linear motion we must divide the change in angular speed by the time it took to make the change. Problems are worked as before, using the speed–time diagram where applicable. In addition to angular distance being represented by the symbol θ, angular speed, which is θ/t, is usually represented by the symbol ω (omega). Also angular acceleration, represented by α (alpha), is ω/t.

There is a definite relation between angular motion and linear motion. Let us consider a point on the rim of a flywheel. In one revolution the point will travel through an angular distance of 2π radians or a linear distance of $2\pi r$ linear units. All points on a body will travel through the same angular distance during a period of time, but their linear speeds will depend on the radii to the points under consideration. Therefore, linear distance is equal to angular distance in radians multiplied by the radius. Linear speed is found by multiplying the angular speed by the radius.

Length of arc:

$$S = r\theta$$

where θ is measured in radians.

Linear speed:

$$V = r\omega$$

where ω is measured in radians per unit of time.

Linear acceleration:

$$a = r\alpha$$

where α is measured in radians per unit of (time)^2.

Example   Point A in Figure 13-70 is located on the outside of a flywheel 6 ft in diameter. Point B is located on the inside of the rim 1 ft from point A. If the flywheel travels at 300 rev/min for 10 min, find (a) total angular distance traveled by point B in radians, (b) linear speed of point B in feet per minute, (c) linear distance traveled by point A in miles.

Solution   Refer to Figure 13-71.

(a) $\qquad\qquad\qquad\theta = (300 \text{ rev/min})(10 \text{ min}) = 3000 \text{ rev}$
$\qquad\qquad\qquad\quad = (3000 \text{ rev})(2\pi \text{ rad/rev}) = \mathbf{18{,}900 \text{ rad}}$[7]

[7] *Radian* is a name that is given to a ratio, and it is a sterile value. However, in order to be able to show the name in the answer, it has been carried through this analysis as a name rather than as a unit.

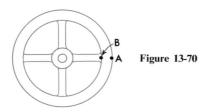

Figure 13-70

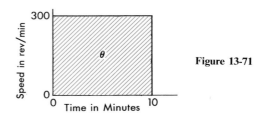

Figure 13-71

(b)
$$V = r\omega$$
$$= (2 \text{ ft/rad})[300 \text{ rev/min}(2\pi \text{ rad/rev})] = \textbf{3780 ft/min}$$
(c)
$$S = r\theta$$
$$= (3 \text{ ft/rad})(18{,}900 \text{ rad})$$
$$= 56{,}700 \text{ ft or } \textbf{10.75 mi}$$

Example The flywheel of a gasoline engine (Figure 13-72) changes its angular velocity from 150 to 300 rev/min during a 5-min period. Solve (*a*) for the total distance traveled by a point on the rim of the flywheel, and (*b*) the angular acceleration of the point during this change.

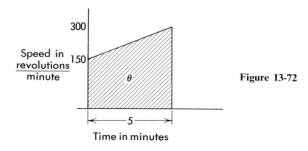

Figure 13-72

Solution

(a)
θ = travel during the change
= area of the speed–time diagram under the travel line
$$= \left(\frac{\omega_1 + \omega_2}{2}\right)t$$
$$= 5 \text{ min}\left(\frac{150 \text{ rev/min} + 300 \text{ rev/min}}{2}\right)$$
$$= (5 \text{ min})(225 \text{ rev/min})$$
$$= \textbf{1125 rev}$$
$$= \textbf{7068.6 rad}$$

(b)
α = angular acceleration
= slope of the travel line
$$= \frac{\omega_2 - \omega_1}{t}$$
$$= \frac{300 \text{ rev/min} - 150 \text{ rev/min}}{5 \text{ min}}$$
$$= \textbf{30 rev/min}^2$$
$$= \textbf{0.0524 rad/s}^2$$

Problems

13-64. While going around a circular curve of 3000-ft radius, a train slows down from 36 to 13 mi/hr in a distance of 850 ft. Find the angular distance covered in degrees.

13-65. Two pulleys 8 and 17 in. in diameter are 10 ft apart on centers and the 17-in. pulley runs 400 rev/min. How many radians per second does the small pulley turn?

13-66. Given $t = 5$ sec; $\omega_1 = 50$ rev/sec; and $\omega_2 = 15$ rad/sec. Draw a speed–time diagram for the motion involved, calibrating both ordinate and abscissa. Is the angular acceleration positive or negative? Why?

13-67. If an automobile engine is accelerated from 1090 to 4600 rev/min in 3.98 sec, what is the angular acceleration of the crankshaft? What is the distance traversed in metres by a point on the circumference of the 0.254-m flywheel?

13-68. Two wheels rolling together without slipping have a velocity ratio of 3 to 1. The driver (which is the smaller of the two) is 9 in. in diameter, and turns at 50 rev/min. (a) What is the speed in feet per second at a point on the surface of the wheels? (b) What is the angular velocity of the larger wheel? (c) What is the diameter of the larger wheel?

13-69. A belt passes around a 15-in. pulley and a 3-ft pulley. If the 15-in. pulley revolves at the rate of 200 rev/min, (a) what is the angular velocity of the 15-in. pulley in radians per second? (b) What is the angular velocity of the 3-ft pulley in radians per second? (c) What is the speed of the belt in feet per second?

13-70. The flywheel of a gas engine is turning at a speed of 300 rev/min. The diameter of the flywheel is 3.61 ft. What is the linear velocity of an oil drop on the edge of the flywheel?

13-71. A belt passes around an 18-in. shaft and a 4-ft pulley. If the shaft revolves at the rate of 300 rev/min, find (a) the angular velocity of the shaft in radians per second, (b) the angular velocity of the pulley in radians per second, (c) the speed of the belt in feet per second.

13-72. A locomotive, having drive wheels 6 ft in diameter, is traveling at the rate of 58 mi/hr. What are the revolutions per minute of the drive wheels?

13-73. An elevator hoisting drum is decelerating at the rate of 15 rev/min/min. If the drum is brought to rest in 9 min, find (a) the total number of revolutions, (b) the initial speed in radians per second.

13-74. A cylindrical drum $2\frac{1}{2}$ ft in diameter is rotated on its axis by pulling a rope wound around it. If the linear acceleration of a point on the rope is 36.9 ft/sec/sec, what will be the angular velocity in (a) revolutions per minute at the end of 6 sec, (b) radians per second? (c) How many turns will it have made during the 6 sec?

13-75. The rotor of a steam turbine is 5 ft $4\frac{3}{4}$ in. in diameter and is turning at the rate of 1850 rev/min when the steam supply is cut off. If it takes the rotor 26 min and 47 sec to come to rest, find (a) the angular deceleration of the shaft in revolutions per minute per second, (b) the angular distance in radians passed through by the shaft before stopping, (c) the average linear velocity of a point on the circumference.

13-76. An engine has a flywheel 49 in. in diameter to which a pulley 16 in. in diameter is attached. The speed of the engine flywheel is 180 rev/min. The engine pulley is connected by a belt to a pump. Assume that the belt does not slip and neglect the thickness of the belt. Find (a) the angular speed of the pulley in revolutions per minute, (b) the angular speed in radians per second. (c) What is the linear speed of the belt in feet per minute? (d) What is the linear speed of a point on the face of the flywheel in feet per minute? (e) If the pump pulley is to turn 105 rev/min, what should be its diameter in feet?

13-77. The flywheel of an engine is 47 in. in diameter and has its speed reduced from 237 to 176 rev/min in 96 rev. Find (a) the average angular velocity in radians per second, (b) the change in angular velocity in radians per second, (c) the average linear velocity in feet per second, (d) the time in seconds required to make the change in speed.

13-78. A wheel turns at an average speed of 50 rev/sec during a total angular distance of 15,900 rad. During the first 300 rev the average speed of the wheel was 75 rev/sec. During the next 300 rev the average speed of the wheel was 53 rev/sec. Since the overall average speed was 50 rev/sec, what was the average speed during the remainder of the distance traveled?

Work, power, and energy

Many words used in physics and engineering have meanings which differ from their common, nontechnical meanings in everyday use. A word such as *work* is an example of this confusion of meanings, as reference to any dictionary will show. In the common use of the word *work,* it may mean anything from merely a thinking process to the hardest sort of physical exertion. It has required the efforts of science for over 200 years to clear up confusion regarding the use of *work* in technical writings, and the handicap of terms loosely used or misused still is a serious factor in concise scientific notations.

Work is defined for our purposes as the product of a force *F* and a distance *S,* both measured along the same line. From this definition we can see that a force executes work on a body when it acts against a resisting force to produce motion of the body. If there is no motion as a result of an applied force, no work is done.

A person who holds a heavy weight soon gets tired and may feel that he has done hard work. Measured in terms of fatigue, he has done work, but fatigue is not a part of our scientific definition of work. If the distance the weight has moved is zero, the work done is zero. While the ideas advanced regarding work may not agree with the everyday usage of the word *work,* the student is encouraged to accept with an open mind the definition given above, which will be the basis of many definitions of other terms.

The units of work will be the product of a unit of force and a unit of length. For example, in English units a common measure of work is the foot-pound. One foot-pound of work is done when a force of 1 pound is exerted in moving an object through a distance of 1 foot in the direction of the force. In the event that force is not in the same direction that distance is measured, work could be calculated by using the component of force in the same direction as distance is measured.

Example A constant force of 50 N acting downward at an angle of 30° with the horizontal moves a box 10 m across a floor (Figure 13-73). How much work is done?

In this example, only a portion of the 50-N force was effective in moving the box from position *A* to position *B.* This effective portion of the 50-N force was evidently $(50)(\cos 30°) = 43.3$ N.

The vertical component $(50)(\sin 30°)$ did not produce any motion but served only to press the box against the floor.

Solution
$$\text{Work} = (F \cos \theta)(S)$$
$$= (50 \text{ N})(\cos 30°)(10 \text{ m})$$
$$= \textbf{433 N-m}$$

Figure 13-73

$F = 50$ N

$30°/\theta$ Position A   Position B

$S = 10$ m

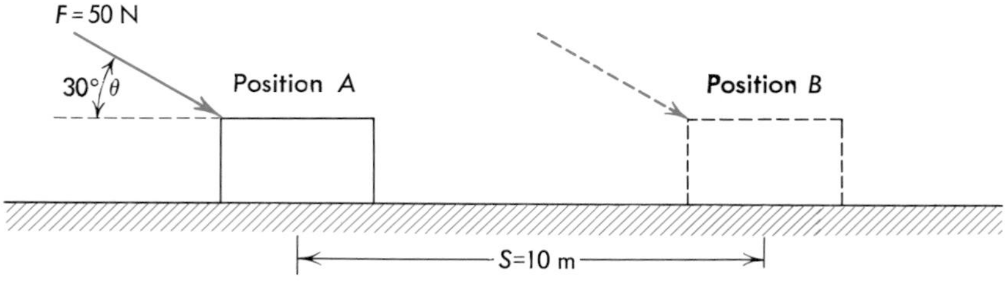

Example A man carries a precision gage weighing 38.5 lb$_f$ up a flight of stairs that has a rise of 8 in. and a tread of 12 in. He climbs at the rate of two steps per second. How much work is done carrying the gage up a stairway of 31 steps?

Analysis Since we are attempting to find only the work done on the gage, we shall ignore the work done in lifting the man's weight. The work done, then, will be the weight lifted times the vertical height. The length of time to move the gage does not enter into the computation for work.

Solution

$$\text{Vertical height} = \frac{(8)(31)}{12} = 20.66 \text{ ft}$$

$$\begin{aligned}
\text{Work} &= (F)(D) \\
&= (38.5)(20.66) \\
&= \textbf{795.6 ft lb}_f
\end{aligned}$$

Example A cable (Figure 13-74) is pulling a wooden crate of electronic computer parts, which weighs 1380 lb$_f$, up a ramp 30.5 ft long that rises to the second floor of

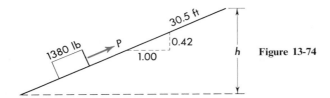

Figure 13-74

a building at the rate of 0.42 ft vertically per foot horizontally. (*a*) If the cable is pulling parallel to the ramp, what work is done on the crate? (*b*) If the friction force between the crate and the ramp is 120 lb$_f$, what work is done in moving the crate up the ramp?

Analysis (*a*) The work done on the crate is its weight times the vertical distance moved. (*b*) The work done by the cable is the product of the pull of the cable and the distance through which this pull or force is exerted.

Solution (*a*) The ramp makes an angle of arctan 0.42, or 22.8°, with the horizontal (see Figure 13-75).

W
67.2°
P

Figure 13-75

F

R

$$\begin{aligned}
h &= (\sin 22.8°)(30.5) \\
&= 11.82 \text{ ft} \\
\text{Work} &= (F)(h) \\
&= (1380)(11.82) \\
&= \textbf{16,310 ft-lb}_f
\end{aligned}$$

(*b*) The pull of the cable is the sum of the friction force and the component of the weight of the crate parallel to the ramp (see Figure 13-75).

$$W_x = (1380)(\sin 22.8°)$$
$$= 534.7 \text{ lb}_f$$
$$\text{Pull} = \text{friction} + W_x$$
$$= 120 + 535$$
$$= 655 \text{ lb}_f$$
$$\text{Work} = (F)(D)$$
$$= (655)(30.5)$$
$$= \mathbf{19{,}980 \text{ ft·lb}_f}$$

Problems

13-79. It requires a constant horizontal force of 300 N to move a table on casters that weighs 4500 N. How much work would be done in moving the table 33 m over a level floor?

13-80. A skip hoist lifts a load of bricks to the third floor of a building under construction. The cable exerts an average pull of 2900 lb for a distance of 25.6 ft. How much work is done in lifting the loaded hoist?

13-81. A man exerts an average force of 133.4 N along the handle of a lawn mower. The handle makes an angle of 41° with the ground. How much work is done in moving the lawn mower 100 m across the lawn?

13-82. An elevated tank in the shape of a cylinder is 13 ft deep and holds 4200 gal. The flat bottom is 30 ft above a ground-level reservoir. How much work is done in filling the tank if the water is pumped in at the bottom of the tank? How much work would be done by the pump if the water is pumped in through a pipe emptying in at the top of the tank?

13-83. A locomotive is pulling a string of 40 box cars, each weighing 45 tons, at a constant speed of 35 mi/hr on a stretch of level track. The frictional resistance of the train is 8 lb/ton weight. How much work is done by the drawbar pull of the engine in moving the train 1 mi?

13-84. A man carries a box weighing 55 lb up a stairway of 17 steps. Each step is 8 in. high and 12 in. wide. How much work does he do in carrying the box up the stairway?

13-85. An automobile engine weighing 2446.4 N is lifted from the floor to a bench 0.787 m high by means of a block and tackle. How much work is done if 10 per cent of the work is used to overcome friction.

13-86. A belt passes over a pulley which is 33 in. in diameter. If the difference in tension on the two sides of the belt is 88 lb and the pulley is turning 530 rev/min, how much work is done per minute by the belt?

13-87. A rope is wrapped around a drum $6\frac{1}{2}$ in. in diameter. A crank handle 14 in. long is connected to the shaft carrying the drum. How much work would be done in lifting a weight of 75 lb a vertical distance of 55 ft? If the length of the crank handle is increased to 20 in., what will be the work done in lifting the weight as before?

13-88. An elevated water storage tank has cylindrical sides and a hemispherical bottom. The tank is 26.8 ft in diameter and the cylindrical part is 36.3 ft high. A pump is located 57.6 ft below the hemispherical bottom. If the tank is filled from the top, what work is necessary to fill it with water? If the hemispherical part of the tank is already filled and water is pumped in from the bottom, how much work is done in filling the cylindrical part of the tank?

13-89. Water is pumped against a constant head of 22 feet at the rate of 710 gallons per minute. How much work is done each minute in pumping the water?

13-90. A tractor is towing a loaded wagon weighing $1.4678(10)^4$ N over level ground and the average tension in the tow cable is 720.58 N. What work is done by the cable in moving the wagon 402.34 metres?

13-91. A man weighing 188 lb seated in a sling is lifted up the side of a building. If he lifts himself by pulling down on the rope passing over the pulley, how much work does he do in lifting himself 27 ft above the ground? How much work would be done by a group of men standing on the ground and pulling the rope to lift him 27 ft?

Power

It is apparent that no interval of time was mentioned in our previous definition of work. In our modern civilization we frequently are as interested in the time of doing work as we are in getting the work done. For this reason the term *power* is introduced, which is the time rate of doing work.

In symbol form:

$$\text{Power} = \frac{\text{work}}{\text{time}}; \qquad P = \frac{W}{T}$$

or it may be expressed as

$$\text{Power} = (\text{force})(\text{velocity}); \qquad P = FV$$

If a pile of bricks is to be moved from the ground to the third floor of a building, the job may be accomplished by moving one brick at a time, ten bricks at a time, or the whole pile of bricks at once. The work done in any case is the same and is the product of the weight of the pile of bricks and the vertical distance through which the pile is moved. However, the time that will be taken will probably vary in each case, as will the capabilities of the lifting mechanism. In order to obtain an indication of the rate at which work can be done, we use the term *power,* which is a measure of how fast a force can move through a given distance.

The units of power in any system can be found by dividing work units by time units. In the FPS gravitational system, power may be expressed as *foot-pounds per second,* or *foot-pounds per minute.* Since the days of James Watt and his steam engine, the horsepower has been a common unit of power and is numerically equal to 550 ft-lb$_f$/sec or 33,000 ft-lb$_f$/min. The SI unit of power is the watt. There are 746 watts in 1 horsepower (hp).

$$\begin{aligned}
1 \text{ hp} &= 550 \text{ ft-lb}_f/\text{sec} \\
&= 33{,}000 \text{ ft-lb}_f/\text{min} \\
&= 746 \text{ watts} \\
&= 0.746 \text{ kw}
\end{aligned}$$

Example A box weighing 1100 lb$_f$ (Figure 13-76) is lifted 15 ft in 3 sec. How much power is necessary?

Analysis
$$\text{Power} = \frac{\text{work}}{\text{time}}$$

Solution

$$\text{Power} = \frac{(1100) \text{ lb}_f(15) \text{ ft}}{3 \text{ sec}}$$

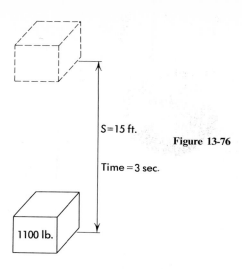

S=15 ft.

Figure 13-76

Time = 3 sec.

1100 lb.

$$P = 5500 \, \frac{\text{ft-lb}_f}{\text{sec}}$$

$$\text{Horsepower} = \frac{\text{work in ft-lb}_f}{(\text{time in sec})(550)}$$

$$\text{hp} = \frac{(1100)(15)}{(3)(550)} = \textbf{10 hp}$$

$$\text{Power} = 746 \, \frac{\text{watts}}{\text{hp}} \, (10 \, \text{hp}) = \textbf{7460 watts}$$

Electric power usually is expressed in watts or kilowatts. A kilowatt is 1000 watts. When electric rates are prepared by utility companies, they customarily base their rates on the kilowatt-hour. Since the kilowatt-hour is the product of power and time, charges for electric services actually are charges for work or energy. When you pay your electric utility bill, you actually are paying for work performed electrically rather than for electric power.

The kilowatt-hour is simply power consumed at the rate of 1 kw for 1 hr.

Example How much will it cost to operate a 150-watt electric light for 2.5 hr when the utility company charges are 6.5 cents per kilowatt-hour.

Analysis Work (or energy) in kwh = (power in kilowatts)(time in hours).

Solution
$$\text{Energy} = \left(\frac{150}{1000}\right)(2.5)$$

$$E = 0.375 \, \text{kwh}$$
$$\text{Cost of electric work (or energy)} = (\text{kwh})(\text{cost per kwh})$$
$$= (0.375)(6.5)$$
$$= \textbf{2.43 cents}$$

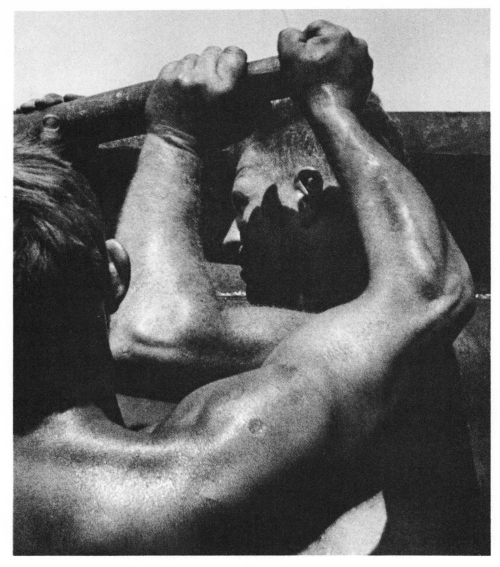

Illustration 13-5
Prior to this century muscle power reigned supreme throughout the world. In some areas there has been little change, but in others the engineer has brought about its demise as a result of his designs.

Efficiency

The efficiency of any machine is expressed as the ratio of work output to the work input, or as a ratio of power output to power input. While efficiency has no units, it is usually expressed as a percentage.

$$\text{Efficiency of a machine} = \frac{\text{work output}}{\text{work input}} = \frac{\text{power output}}{\text{power input}}$$

$$\text{Per cent efficiency} = \frac{\text{work output}}{\text{work input}}(100 \text{ per cent})$$

$$= \frac{\text{power output}}{\text{power input}}(100 \text{ per cent})$$

Example What is the per cent efficiency of a 12-hp electric motor that requires 9.95 kw of electric power when running at full load?

Analysis The units of power input and power output must both be the same in order to calculate efficiency. We shall convert 12 hp to kilowatts and compute the ratio of power output to power input.

Solution
$$\begin{aligned}\text{Power output in kw} &= (\text{power output in hp})(0.746)\\ &= (12)(0.746)\\ &= 8.95 \text{ kw output}\end{aligned}$$

$$\begin{aligned}\text{Per cent efficiency} &= \frac{\text{power output}}{\text{power input}}(100 \text{ per cent})\\ &= \frac{8.95}{9.95}(100 \text{ per cent})\\ &= \mathbf{90.0} \text{ per cent efficiency}\end{aligned}$$

The result would be the same if the power input in kilowatts had been converted to horsepower, and efficiency had been obtained as a ratio of horsepower output to horsepower input.

The power rating of motors and engines is the maximum output power that they are expected to deliver constantly, unless specifically stated otherwise. The input power will always be greater than the output power. For instance, a 100-hp electric motor can develop 100 hp at its pulley, but more than 100 hp will have to be supplied by the electric power line connected to the motor.

In some situations account must be taken of the efficiency of several machines as we trace the flow of power through them. As an example, let us consider a case in which an electric motor is connected to a pump which is pumping water. The motor obtains its power from electric lines which run through a switchboard. The data are given in Figure 13-77, in which blocks are used to represent parts of the system.

Figure 13-77

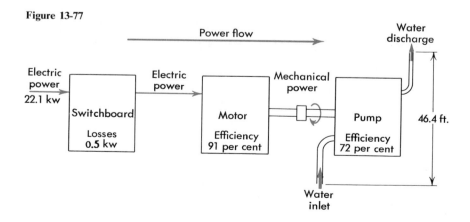

Example Power supplied to the system, as indicated by electric meters, is 22.1 kw. Find the amount of water delivered by the pump in cubic feet per second.

Analysis Compute the output power of each part of the system in order, beginning with the switchboard.

Solution

Power supplied to the motor = 22.1 − 0.5
                            = 21.6 kw   (This is the power input to the motor.)

Power output of the motor = (21.6)(0.91)
                          = 19.66 kw

If we assume no losses in the coupling between the motor and the pump, the power output of the motor is the same as the power input to the pump.

Power output of the pump = (19.66)(0.72)
                         = 14.15 kw

Converting 14.15 kw to foot-pounds per second:

$$\text{Power} = \frac{(14.15)(550)}{0.746}$$

$$= \textbf{10,434 ft lb}_f/\textbf{sec}$$

The amount of water delivered now may be found if we remember that

$$\text{Power} = \frac{\text{work}}{\text{time}}$$

Then

$$\text{Weight of water per unit time} = \frac{10,434 \text{ ft-lb}_f/\text{sec}}{46.4 \text{ ft}}$$

$$= 225 \text{ lb}_f \text{ per sec}$$

Converting 225 lb_f/sec to cubic feet per second:

$$\text{Volume of water per second} = \frac{225 \text{ lb}_f/\text{sec}}{62.4 \text{ lb}_f/\text{ft}^3}$$

$$= \textbf{3.60 ft}^3/\textbf{sec}$$

One additional item of information should be called to the student's attention. We notice that the efficiency of the motor is 91 per cent and the efficiency of the pump is 72 per cent. Considering the overall efficiency of both machines, the input to the motor is 21.6 kw, and the output of the pump is 14.16 kw. The overall efficiency of both machines can be found as follows:

$$\text{Overall efficiency} = \frac{\text{output}}{\text{input}}(100 \text{ per cent})$$

$$= \frac{14.15}{21.6}(100 \text{ per cent})$$

$$= 65.5 \text{ per cent}$$

The overall efficiency of both machines could also be determined by finding the product of the individual efficiencies:

$$\text{Overall efficiency} = (0.91)(0.72)(100 \text{ per cent})$$
$$= \textbf{65.5 per cent}$$

Problems

13-92. An automobile requires 47 hp to maintain a speed of 55 mi/hr. What force in newtons is being exerted on it by the engine?

13-93. In a recent experiment a student weighing 168 lb ran from the first floor to the third floor of a building, a vertical distance of 26 ft, in 9 sec. How much horsepower did he develop?

13-94. If a horse can actually develop 1 hp while pulling a loaded wagon at 3.5 mi/hr, what force does he exert on the wagon?

13-95. A car weighing 2900 lb is moving at constant speed up a hill having a slope of 17 per cent. Neglecting friction, how fast will the car be moving when it is developing 25 hp?

13-96. An airplane engine which develops 2000 hp is driving the plane at a speed of 250 mi/hr. What thrust is developed by the propeller?

13-97. An elevator and its load weigh 5300 lb. What will be the maximum upward velocity of the elevator when the driving motor is developing 15 hp?

13-98. A diesel engine runs a pump which pumps 18,000 gal of water per hour into a tank 65 ft above the supply. How many horsepower are required at the pump?

13-99. A bulldozer exerts a force of 7200 lb on its blade while moving 6.5 mi/hr. What horsepower is necessary?

13-100. A car weighing 3900 lb is being towed by another car at a rate of 35 mi/hr. The average force exerted by the tow cable is 200 lb. What horsepower is necessary to tow the car?

13-101. A tank holding 3500 gal of water is to be emptied by a small centrifugal pump. The water is 6 ft deep and is to be pumped to a height of 13.5 ft above the bottom of the tank. The pump is 65 per cent efficient and is driven by a motor which develops $\frac{1}{4}$ hp. How long will it take to empty the tank?

13-102. In a certain industrial plant it was necessary to pump 120,000 gal of water per day an average height of 12 ft. The pump used was 68 per cent efficient and was direct connected to an electric motor having an efficiency of 85 per cent. While running, the motor develops 1 hp. (*a*) How many hours per day would the pump need to run? (*b*) What would be the kilowatt input to the motor? (*c*) Electrical energy costs 3 cents per kilowatt-hour. What will it cost to operate the pump 30 days per month?

13-103. A belt conveyor is used to carry crushed coal into a hopper. The belt carries 13 tons/hr up a 13 per cent slope 45 ft long. The friction losses in the belt and rollers amount to 22 per cent of the power supplied. How many horsepower are needed to operate the belt?

13-104. A $\frac{1}{2}$-hp electric motor drives a pump that lifts 1200 gal of water each hour to a height of 46 ft. What is the efficiency of the pump? While running, the motor requires 440 watts of power. What is the efficiency of the motor? What is the overall efficiency of the motor-pump combination?

13-105. A pump having an efficiency of 55 per cent is used to pump gasoline. If the pump delivers 50 gal of gasoline per minute through an average height of 38 ft, what horsepower is needed to run the pump?

13-106. A pulley 10 in. in diameter is on the shaft of a motor which is running 1730 rev/min. If the motor is developing 25 hp, what is the difference in tension on the sides of the belt that passes over the pulley?

13-107. The piston of a steam engine is 12 in. in diameter and moves through a distance of

20 in. each stroke. The average pressure on the piston is 75 lb/in.² and the piston makes 250 power strokes per minute. How much horsepower is developed?

Energy

Another expression much used in mechanics is the term *energy*. The energy of a body is its ability to do work. Energy is measured in terms of work and has the same dimensions and units. Of all the various ways in which energy is produced, such as chemical, electrical, light, heat, sound, and mechanical, the forms in which we shall be concerned are *potential* and *kinetic*.

The potential energy (*PE*) of a body is its ability or capacity to do work because of its position or location.

$$Potential\ energy\ (PE) = (weight)(vertical\ height)$$

Example   A 100-lb$_f$ box (Figure 13-78) is on a platform 10 ft above the ground. What is its potential energy with respect to the ground?

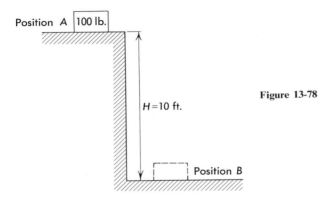

Figure 13-78

To analyze the problem, let us assume the box is initially in position *B*. The work necessary to raise the box to position *A* is (10 ft)(100 lb$_f$) or 1000 ft-lb$_f$. Since energy and work are convertible, the work of lifting the box evidently has gone to increasing its *PE*. The *PE* can then be found as the product of weight and the vertical distance above some reference plane. In this problem:

$$PE = (W)(H)$$
$$= (100\ lb_f)(10\ ft)$$
$$= 1000\ ft\text{-}lb_f$$

The kinetic energy of a body is its ability to do work because of its motion. The dimensions of kinetic energy must be the dimensions of work. The usual expression for determining *KE* is

$$KE = \frac{(mass)(velocity)^2}{2}$$

The derivation of this expression is as follows: If a force acts on a body that is free to move, the body will accelerate. From Newton's laws this force will produce an acceleration which is proportional to the mass of the body.[8]

$$F = (\text{mass})(\text{acceleration}) = Ma$$

If the force is constant and acts through a distance S while the body is accelerating, the work done is $(F)(S)$. Substituting the above value of F in the expression for work:

$$\text{Work} = (Ma)S$$

From the expression of accelerated motion,[9] the velocity acquired by a body starting from rest is $V^2 = 2aS$, or $S = V^2/2a$. Substituting this value of S in the expression for work, we get

$$\text{Work} = (Ma)\left(\frac{V^2}{2a}\right) = \frac{MV^2}{2}$$

Since this is the work to give the body a velocity V, the work must have gone into increasing its KE, so

$$KE = \frac{MV^2}{2}$$

The dimensional equation using FPS gravitational units is

$$KE = \left[\frac{\text{lb}_f\text{-sec}^2}{\text{ft}}\right]\left[\frac{\text{ft}}{\text{sec}}\right]^2 = \text{ft-lb}_f$$

The units of KE are identical with the units of work. It should be remembered that in the FPS gravitational system of units the mass of a body in slugs can be calculated by dividing the weight of the body in pounds by the local acceleration of gravity in feet per second per second. In the SI system the unit of energy is the joule, which is one newton-metre.

Example   A 10-lb$_f$ box (Figure 13-79) is moving with a velocity of 12 ft/sec. What is its kinetic energy?

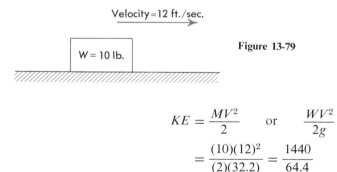

Figure 13-79

$$KE = \frac{MV^2}{2} \quad \text{or} \quad \frac{WV^2}{2g}$$
$$= \frac{(10)(12)^2}{(2)(32.2)} = \frac{1440}{64.4}$$
$$= 22.36 \text{ ft-lb}_f$$
$$= \mathbf{30.37 \text{ J}}$$

---

[8] For a discussion of *unit homogeneity* refer to Chapter 12.
[9] Refer to page 520.

These relations of the equivalence of work and energy can be summed up in what is known as the *Law of Conservation of Energy*. This principle states that energy can neither be created nor destroyed but is only transformed from one kind to another (neglecting mass–energy transformations). As an example, let us take a problem which was previously solved.

The 100-lb$_f$ box in Figure 13-78 when in position A has a *PE* of 1000 ft-lb$_f$. Its *KE* is zero because it is not moving. However, if we push the box to the edge of the platform so that it falls, we can see that, just as the box reaches position B, the height of the box above the ground is zero and its *PE* is zero. Let us calculate its *KE* as the box reaches position B. From the expression for motion of a freely falling object starting from rest, the velocity of the box after falling 10 ft will be

$$V^2 = 2\,gS \qquad\qquad V = \sqrt{664\,\frac{\text{ft}^2}{\text{sec}^2}}$$

$$V^2 = (2)(32.2)(10) \qquad\qquad = 25.4 \text{ ft/sec}$$

Then, solving for the KE of the box as it reaches position B,

$$KE = \frac{MV^2}{2} \quad \text{or} \quad \frac{WV^2}{2g}$$

$$= \frac{(100)(644)}{(2)(32.2)}$$

$$= \textbf{1000 ft lb}_f$$

which is the same as the potential energy in position A.

Example   A 1000-N pile-driving hammer falls 6 m onto a pile and drives the pile 6 cm. What is the average force exerted?

Analysis   Using the equivalence of energy and work, the energy of the moving hammer was transformed into work by moving the pile 0.06 m.

$$KE \text{ (of hammer)} = \text{work of driving pile}$$

$$KE = \frac{WV^2}{2g}$$

Let $S_1 = 6$ m and $S_2 = 0.06$ m.

Solution   Since the hammer is assumed to fall freely,

$$V_2^2 - V_1^2 = 2aS$$

$$V^2 = 2gS$$

$$= (2)\left(9.807\,\frac{\text{m}}{\text{s}^2}\right)(6 \text{ m})$$

$$= 117.68\,\frac{\text{m}^2}{\text{s}^2}$$

$$KE = \frac{(1000)(1030)}{(2)(32.2)} = \frac{(1000)(117.68)}{(2)(9.807)}$$

$$= 6000 \text{ N-m} = \text{work of driving the pile}$$

Let FS_2 represent the work of driving the pile. Then

$$6000 \text{ N-m} = (F) \text{ N } (0.06) \text{ m}$$

or
$$F = \frac{(6000)}{(0.06)} = 6(10)^5 \text{ N average force}$$

It may be seen also in the above example that the *PE* of the hammer at the beginning of the 6 m drop is equal to the *KE* at the end of the travel of the hammer.

Another example of an energy–work conversion is in the use of springs. Using a coil spring as an example, if we compress a coil spring in our hands, we exert a force in order to shorten the spring. This means we have exerted force through a distance and have done work. Also we have stored energy in the spring due to its change in shape. This energy is in the form of potential energy. We know from experience that, as the spring is compressed more and more, an increasing amount of force is required. The work done, then, evidently must be the average of the initial and final forces multiplied by the distance the average force has acted.

In diagram *A* in Figure 13-80, there is no force on the 6-in. spring. As we slowly add weight to the spring, it will shorten. In diagram *B* the weight has been increased to 12 lb$_f$, and the spring has been compressed until it is only 4 in. long. The applied force, which initially was zero, has been increased to 12 lb$_f$, which is an average force of 6 lb$_f$.

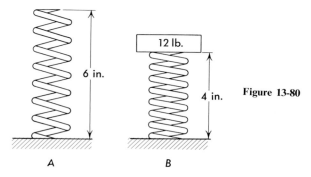

Figure 13-80

We may take the average force, 6 lb$_f$, times the 2-in. movement of the spring as the work done, rather than take the small change of length due to each increase of force from zero pounds to 12 lb$_f$, and then add all the small increments of work. It can be shown by advanced mathematics that the increment method may be used, but for our purpose we shall use the average force multiplied by the distance the average force will act. The expression for work will then be

$$\text{Work} = (\text{average force})(\text{distance}) = \text{energy in the spring}$$

Using the data in Figure 13-80,

$$\text{Work} = (F_{av})(S)$$
$$= \frac{F_1 + F_2}{2}(S)$$
$$= \frac{0 + 12 \text{ lb}_f}{2}(2) \text{ in.}$$
$$= \textbf{12 in.-lb}_f$$

In diagrams A and B it is shown that a force of 12 lb_f changes the length of the spring 2 in. A common way of rating springs is by giving the force necessary to change their length a unit distance, such as an inch. In the example used, it will take 6 lb_f to compress the spring 1 in., so we speak of the spring as being a 6-lb_f spring. This value of 6 lb_f/in. is called the *force constant* or the *spring rate* of the spring and is substantially independent of the applied force if the elastic limit of the material is not exceeded. If we let K represent the force constant of the spring, then

$$K = \frac{F}{S}$$

where F is the force applied and S is the distance the end of the spring moves. Rearranging the above expression,

$$F = KS$$

In the expression for work, when the initial force is zero

$$\text{Work} = \frac{F}{2}(S)$$

Substituting KS for F,

$$\text{Work} = \frac{KS}{2}(S)$$

$$= \frac{KS^2}{2}$$

This shows that we can find the work done on a given spring if we know its force constant and the distance through which a given force has moved it.

Example   A spring has a scale (force constant) of 600 lb_f/ft. How much work is done by a force that stretches it 3 in.? What force was acting to stretch the spring 3 in.?

Analysis   We must convert our different units of distance into the same units. A distance of 3 in. is $\frac{3}{12}$ or $\frac{1}{4}$ ft.

Solution   The work done is

$$\text{Work} = \frac{KS^2}{2}$$

$$= \frac{600 \ (lb_f/ft)}{2}(0.25 \ ft)^2$$

$$= 300 \ lb/ft \ (0.0625 \ ft^2)$$

$$= \textbf{18.75 ft·lb}_f$$

The force to stretch the spring 3 in., or 0.25 ft, is found as follows:

$$F = KS$$
$$= (600 \ lb_f/ft)(0.25 \ ft)$$
$$= \textbf{150 lb}_f$$
In SI units: $$F = \textbf{667.2 N}$$

Problems

13-108. A car weighing 17 kN is moving 9 m/s. What is its kinetic energy? If the speed is doubled, by how much will the kinetic energy be increased?

13-109. How much potential energy is lost when a cake of ice weighing 1300 N slides down an incline 30 m long that makes an angle of 25° with the horizontal?

13-110. A box weighing 50 lb starts from rest and slides down an inclined plane with an acceleration of 7 ft/sec/sec. What will be its kinetic energy at the end of the first, second, third, and tenth seconds?

13-111. A train weighing 1100 tons is moving fast enough to possess $(1.5)(10^8)$ ft-lb$_f$ of kinetic energy. What is its speed in miles per hour?

13-112. A car weighing 12 kN is moving with a speed of 13 m/s. What average force is needed to stop it in 19 m?

13-113. A hammer weighing 1 lb and moving 30 ft/sec strikes a nail and drives it $\frac{3}{4}$ in. into a block of wood. What was the average force exerted on the nail?

13-114. A 22-caliber rifle fires a bullet weighing $\frac{1}{15}$ oz with a muzzle velocity of 1020 ft/sec. The barrel is 26 in. long. Assuming the force on the bullet is constant while it moves down the barrel, what force was exerted on the bullet? What is the kinetic energy of the bullet as it leaves the muzzle?

13-115. A ball weighing 11.12 N is dropped from the top of a building 38.1 m above the ground. After the ball is dropped, how long will it take for the kinetic energy and potential energy to be equal?

13-116. It requires a force of 12 N to stretch a spring 3 cm. How much work is done in stretching the spring 9 cm?

13-117. A coil spring has a scale of 70 lb/in. A weight on it has shortened it 2.5 in., and when more weight is added it is shortened by an additional 0.75 in. What work was done by the added weight?

13-118. The floor of a car is 13.6 in. from level ground when no one is in the car. When several people whose combined weight is 573 lb get in the car, the floor is 11.9 in. from the ground. Assuming that the load was equally distributed to the front and rear wheels, what would be the force constant of the front spring system?

13-119. A weight of 130 N stretches a spring 1.6 cm. What energy is stored in the spring? What is the scale of the spring?

13-120. One end of a screen door spring is fastened to the door 17 in. from the hinge side of the door, and the other end is fastened on the door jamb 2.5 in. from the screen door. It requires a force of 11 oz on the door 32 in. from the hinge side to start the door to open. What is the initial tension in the spring? If the force constant of the spring is 12 lb/in., what force is necessary to open the door through an angle of 65°?

13-121. An iron ball weighing 7.5 lb is dropped on a spring from a height of 10 ft. The spring has a force constant of 70 lb/in. How far is the end of the spring deflected?

The general law of work and energy

A generalization of work and energy relationships may be stated in what is known as the *law of work and energy* and is general enough to apply to almost any problem involving accelerated motion. The law of work and energy may be stated as

$$\begin{bmatrix} \text{Initial} \\ KE \text{ of a} \\ \text{body} \end{bmatrix} + \begin{bmatrix} \text{Work done by} \\ \text{forces tending} \\ \text{to increase the} \\ \text{velocity of the} \\ \text{body} \end{bmatrix} - \begin{bmatrix} \text{Work done by} \\ \text{forces tending} \\ \text{to decrease the} \\ \text{velocity of the} \\ \text{body} \end{bmatrix} = \begin{bmatrix} \text{Final } KE \text{ of} \\ \text{the body} \end{bmatrix}$$

Example   A box weighing 200 lb$_f$ is on a plane which makes an angle of 20° with the

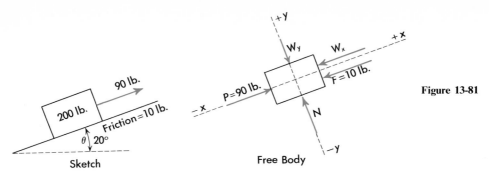

Figure 13-81

Sketch: 90 lb., 200 lb., Friction = 10 lb., θ 20°

Free Body: +Y, W_y, W_x, +X, P = 90 lb., F = 10 lb., -X, N, -Y

horizontal, Figure 13-81. A force of 90 lb$_f$ is applied parallel to the plane and moves the box up the plane. Friction between the box and the plane is 10 lb$_f$. If the box starts from rest, what will be its velocity at the end of 5 sec?

Let us first solve for the weight components W_x and W_y.

$$W_x = W \sin \theta \qquad\qquad W_y = W \cos \theta$$
$$= (200) \sin 20° \qquad\quad = (200) \cos 20°$$
$$= (200)(0.342) \qquad\quad = (200)(0.94)$$
$$= 68.4 \text{ lb}_f \qquad\qquad = 188 \text{ lb}_f$$

The component W_y is perpendicular to the plane and therefore cannot produce motion along the plane. The work done by the component W_y will be zero. The other forces may produce motion parallel to the plane and must be included in the work and energy law expression.

Taking one term at a time:

Initial kinetic energy $= 0$ (since the body is starting from rest).

Work done by forces tending to increase the velocity. This will be the work done by the 90-lb$_f$ force, as it is the only one that tends to make the body increase its speed up the plane. This work is

$$\text{Work} = (90)(S)$$

Since the numerical value of S is not known, it will have to be included as a letter symbol and solved for later.

Work done by forces tending to decrease the velocity. The component W_x and friction both tend to slow the box as it moves up the plane. The work due to these forces is

$$\text{Work} = (68.4)(S) + (10)(S)$$
$$= (78.4)(S)$$

Final kinetic energy of the box is

$$KE = \frac{MV^2}{2} \qquad \text{or} \qquad \frac{WV^2}{2g}$$
$$= \frac{(200)(V^2)}{(2)(32.2)}$$

Note that the kinetic energy of the box is found by using all the weight of the box and not just a component of the weight.

Combining all the terms into a single expression, we have

$$0 + (90)(S) - (78.4)(S) = \frac{(200)(V^2)}{(2)(32.2)}$$

$$11.6S = 3.11V^2$$

In the initial statement of the problem, the velocity at the end of 5 sec was required. From the expressions of motion of a body:

$$S = (\text{average velocity})(\text{time}) = \frac{V_1 + V_2}{2}(t)$$

Since the initial velocity V_1 is zero,

$$S = \left(\frac{V_2}{2}\right)(t)$$

and since $t = 5$ sec,

$$S = \left(\frac{V}{2}\right)(5)$$

Substituting for S in the expression $11.6S = 3.11V^2$,

$$(11.6)\left[\left(\frac{V}{2}\right)(5)\right] = 3.11V^2$$

$$29V = 3.11V^2$$

$$29 = 3.11V$$

$$V = \textbf{9.33 ft/sec}$$

Example A cart and its contents (Figure 13-82) weigh 4260 lb$_f$. It is sitting on a ramp that makes an angle of 11° with the horizontal. The coefficient of friction between the cart and the ramp is 0.2. What horizontal force will be needed to give the loaded cart an acceleration of 3.70 ft/sec/sec up the ramp?

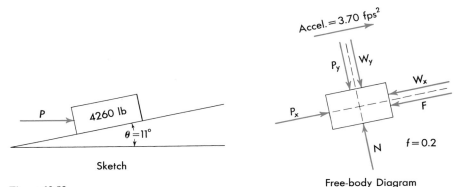

Sketch

Free-body Diagram

Figure 13-82

Analysis Determine the components of forces parallel and perpendicular to the ramp and solve, using the work and energy law.

Solution   Solve for the weight components W_x and W_y.

$$W_x = W \sin \theta \qquad\qquad W_y = W \cos \theta$$
$$= (4260)(\sin 11°) \qquad = (4260)(\cos 11°)$$
$$= (4260)(0.1908) \qquad = (4260)(0.982)$$
$$= 812 \text{ lb} \qquad\qquad = 4190 \text{ lb}$$

From $\Sigma F_y = 0$, $N = W_y + P_y$,

and friction force
$$F = 0.2N$$
$$= 0.2(4260 + P_y)$$
$$= 852 + 0.2P_y$$

Since velocities are not given, we should solve for a velocity at some assumed time to provide a value of velocity from which to solve for work and energy relations. Assume that the cart starts from rest and travels for 1 sec.

$$\text{Velocity} = at$$
$$= (3.7)(1)$$
$$= \textbf{3.7 ft/sec}$$

Similarly solve for the distance traveled in 1 sec:

$$S = \tfrac{1}{2}at^2$$
$$= \frac{(3.7)(1^2)}{2}$$
$$= 1.85 \text{ ft}$$

Substitute in each part of the work and energy equation.

The *initial kinetic energy* is zero, since the cart starts from rest.
Work done by forces tending to increase the velocity:

$$\text{Work} = (P_x)(1.85)$$

Since the value of P_x is not known, it will be solved for later.
Work done by forces tending to decrease the velocity:

$$\text{Work} = (W_x)(S) + (F)(S)$$
$$= (812)(1.85) + (1.85)(852 + 0.2P_y)$$
$$= 1502 + 1578 + 0.37P_y$$

Final kinetic energy of the cart is

$$KE = \frac{MV^2}{2} \quad \text{or} \quad \frac{WV^2}{2g}$$
$$= \frac{(4260)(3.7)^2}{(2)(32.2)}$$
$$= 907 \text{ ft-lb}_f$$

Combining all the terms into a single expression, we have

$$0 + (P_x)(1.85) - 1502 - 1578 - 0.37(P_y) = 907$$

To solve for the force P:

$$P_x = P(\cos \theta) \quad \text{and} \quad P_y = P(\sin \theta)$$
$$= P(0.982) \qquad\qquad = P(0.1908)$$

Substituting:

$$P(1.85)(0.982) - P(0.37)(0.1908) = 3987$$
$$P = \mathbf{2285\ lb}_f$$

Problems

13-122. An electric motor is delivering 18 hp to a water pump. How many gallons of water per minute will be pumped to a height of 32 ft if the efficiency of the motor is 70 per cent and the efficiency of the pump is 55 per cent?

13-123. A bullet weighing 0.289 N and traveling with a velocity of 335 m/s strikes a large tree. Assuming the bullet meets a constant resistance to motion of 18 kN, how far will the bullet go into the tree?

13-124. What horsepower motor is necessary to raise a 1200-lb elevator at a constant velocity of 12 ft/sec? (Assume no loss of power in the hoisting cables.) If the motor is 85 per cent efficient, what is the kilowatt input?

13-125. A 15 kN elevator is raised 12 m vertically at a constant velocity of 0.50 m/s. (a) How much work is done? (b) If the elevator hoist is 90 per cent efficient, what power motor is required to operate the hoist?

13-126. A block weighing 200 lb is sitting on an incline that makes an angle of 25° with the horizontal. What force parallel to the incline is necessary to give the block an acceleration of 4 ft/sec² up the incline? Friction amounts to 12 lb.

13-127. How much energy does a 2-lb hammer have if it is moving 52 ft/sec? How far will it drive a nail into a piece of wood if the nail meets a constant resistance of 3000 lb?

13-128. Water flows into a mine which is 100 m deep at the rate of 0.472 m³/s. What power should be supplied to a pump that is 60 per cent efficient if it is to keep the mine pumped out?

13-129. An electric motor is driving a pump which is delivering 750 gal of water per minute to a height of 83 ft. The motor has an efficiency of 81 per cent and the pump has an efficiency of 73 per cent. What power in kilowatts is supplied to the motor?

13-130. A small boat is powered by a 12-hp outboard motor. At full throttle the speed is 15 mi/hr. Find the resistance to motion of the boat.

13-131. A certain city has a water consumption of 2,500,000 gal per 24-hour day. The average pressure on the discharge side of the pump is 125 lb/in.². If the efficiency of the pump and engine together is 60 per cent, what is the horsepower supplied to the motor driving the pump? If electric current costs 3 cents per kilowatt-hour, how much does the electricity for running the motor cost per month of 30 days?

13-132. A freight train consisting of 60 cars, each weighing 50 tons, starts up a 1.5 per cent grade with an initial speed of 15 mi/hr. The drawbar pull is 90 tons and the train resistance including rolling resistance and air resistance is 15 lb per ton of weight. At the top of the grade the speed is 30 mi/hr. (a) How long is the grade? (b) How much is the work of the drawbar pull? (c) How much work is done against gravity?

13-133. A cylindrical water tank 15 ft high and 10 ft in diameter is filled in $2\frac{1}{2}$ hr by a pump located 30 ft below the bottom of the tank. What horsepower motor is required to operate the pump if the pump is 72 per cent efficient? The water is pumped into the tank through a 4-in. pipe opening into the tank at the top. Neglect friction in the pipe and other friction and head losses.

13-134. A 5-hp electric motor having an efficiency of 80 per cent is directly coupled to a centrifugal pump having an efficiency of 70 per cent. (a) If the pump is delivering 600 gal of water per minute against a head of 18 ft, what horsepower is being supplied by the motor? (b) If the amount of water delivered by the pump is changed so that the motor takes 3.2 kW, what horsepower is the motor putting out at its shaft?

13-135. Water is supplied to a Pelton waterwheel from a lake whose surface is 810 ft above the wheel. Water from the lake flows through a conduit and discharges through a nozzle, and 10 per cent of the energy of the flowing water is lost in the conduit and nozzle. When the flow of water is 7.75 ft^3/sec, the efficiency of the waterwheel is 80.0 per cent. The waterwheel drives an electric generator on the same shaft and the generator efficiency is 90.0 per cent. How much electric power in kilowatts is delivered by the generator under the above conditions?

13-136. A certain city has a water consumption of 5,600,000 gal per day of 24 hours, and a pressure gage on the delivery side of the water pump reads 135 lb/in.^{2} pressure. If the efficiency of the pump is 80 per cent, and if the motor is 90 per cent efficient, how many kilowatts of electrical power are supplied to the motor?

13-137. It is desired to install a hydroelectric station on a certain stream. The cross-sectional area of the stream is 800 ft^2. There is a fall of 48 ft obtainable and the velocity of the stream is 5 mi/hr. What would be the horsepower output assuming an overall efficiency of 75 per cent?

13-138. A wooden box weighing 2000 N starts from rest and slides down a wooden inclined plane with an acceleration of 1.60 m/s? What will be its kinetic energy when it reaches a speed of 6.5 m/s? 1.60 m/s?

13-139. A 1.75-ton car coasting at 15 mi/hr comes to the foot of a 2 per cent slope. If it meets a resistance of 12 lb per ton on the slope caused by friction and windage, how far up the slope will it go before it stops?

13-140. A 3400-lb automobile is traveling 63 mi/hr up a 3 per cent grade. The brakes are suddenly applied and the car is brought to a standstill. If the average air resistance is 54 lb and the rolling resistance is 20 lb per ton, what must the braking force be to stop the car in 300 ft?

13-141. A 5-hp motor is operated at full load for 6.3 hr per day, 25 days out of each month. How many kilowatt-hours will be consumed in a month if the motor is 75 per cent efficient?

13-142. An electric iron requires 550 watts of electric power. How much will it cost to operate the iron for an hour if energy costs 6.0 cents per kilowatt-hour?

13-143. What acceleration will be given an elevator weighing 20 kN if the pull on the supporting cables is 23.5 kN?

13-144. A 3000-lb car is moving with a velocity of 88 ft/sec over level ground. The car is brought to a stop in a distance of 300 ft. Find the braking force. Consider the frictional force to be 50 lb.

14

The modeling of engineering systems— mechanical, electrical, fluid, and thermal

The basic work of the practicing engineer is the design and analysis of physical systems. This requires conceiving of a total system as an arrangement of component parts and then representing these components as idealized elements which possess important characteristics of the *real* components and whose behavior can be described. It further involves the expression of the behavior of the different parts of the system and, finally, of the system as a whole in terms of mathematical equations.[1] These two steps are called modeling and they are among the most important tasks performed by today's engineer.

In the modeling process the engineer's first task is to devise a conceptual model that schematically represents the real system being studied. The next step is to express this conceptual model mathematically. This expression of physical components by mathematical equations is referred to as developing a "mathematical" model. Once an engineer has chosen a mathematical model of a real system, he can begin to predict how the system would behave in various conditions and under different constraints without actually building sample systems and testing them experimentally, which can be a very slow and expensive process. In other words, he can use his mathematical model to analyze and revise his system design.

> When we mean to build, we first survey the plot, then draw the model; And
> when we see the figure of the house, then must we rate the cost of the erection.
> —King Henry IV, Part II, Act I, Sc. 3, Line 41

[1] A more extensive treatment of the material presented in this chapter can be found in reference 1.

The construction of a mathematical model of a real system is usually based on the relations between the *inputs* and *outputs* of certain idealized basic elements which make up the chosen conceptual model. As an example, in mechanical systems, a spring is a commonly used element. The engineer expresses mathematically the relation between a force applied to the spring (input) and the resulting displacement or stretching of the spring (output). Other elements commonly used are electrical resistances, mechanical dampers, fluid capacitors, etc. Finally, the engineer must combine all the basic elements so that mathematical equations can be obtained which relate the system inputs to the system outputs.

The engineer must understand the concept of system *inputs* and *outputs*. Figure 14-1b is a schematic representation of system inputs and outputs. If an aircraft is the system that we are attempting to model, the *inputs* to the system would be such factors as wind gusts, throttle position, and the position of the aircraft control surfaces, such as the elevator, aileron, and rudder. *Outputs* from the system that the engineer would like to be able to *predict* would include such quantities as the aircraft's altitude, speed, and acceleration; that is, the *inputs* act on the system to produce the *outputs*.

Figure 14-2 provides an example of the modeling process for each of the four basic types of systems: electrical, mechanical, fluid, and thermal. In each case the actual physical system is first conceptualized as an idealized physical model.[2] Then this idealized model is represented by a mathematical model. In other words, the action performed by the system, or the relation between its input and output, is finally described in terms of mathematical equations.

If we were modeling the vertical dynamics of an automobile as it travels down a road, an input to the actual system would be the roughness of the road. An output might be the acceleration felt by the passenger in the car. The conceptual model of

[2] The symbols and schematic that are used to represent the various elements will be explained in a later section.

Figure 14-1

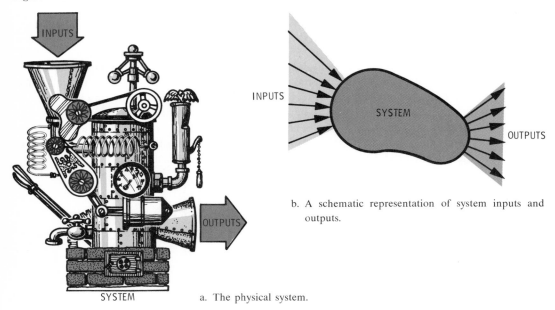

b. A schematic representation of system inputs and outputs.

a. The physical system.

Figure 14-2 Examples of the modeling of physical systems.

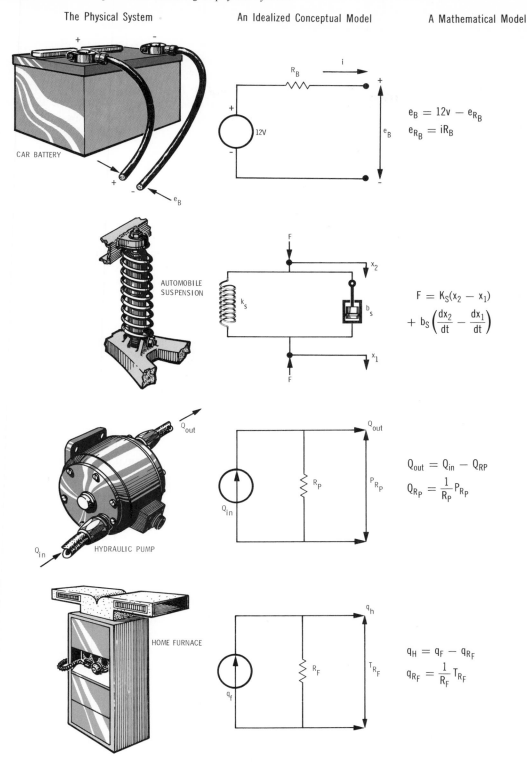

The Physical System   An Idealized Conceptual Model   A Mathematical Model

CAR BATTERY

$$e_B = 12v - e_{R_B}$$
$$e_{R_B} = iR_B$$

AUTOMOBILE SUSPENSION

$$F = K_S(x_2 - x_1)$$
$$+ b_S\left(\frac{dx_2}{dt} - \frac{dx_1}{dt}\right)$$

HYDRAULIC PUMP

$$Q_{out} = Q_{in} - Q_{RP}$$
$$Q_{R_P} = \frac{1}{R_P}P_{R_P}$$

HOME FURNACE

$$q_H = q_F - q_{R_F}$$
$$q_{R_F} = \frac{1}{R_F}T_{R_F}$$

our suspension system would be the series of springs and shock absorbers which comprise a car's suspension system. Finally, our mathematical model would consist of a group of mathematical equations which represent the behavior of these springs and shock absorbers. By means of our mathematical model we are now able to test the suspension system that we have modeled by mathematically determining how it would respond to a variety of irregularities in the road surface. Hence we can determine whether our original system will behave properly. If its performance varies from our expectations, we can easily make changes in our conceptual model, which will hopefully improve its behavior. Finally, we can actually build a new physical system which resembles our revised model.

Problems

14-1. What is the primary input and the primary output of an automobile power steering system?
14-2. The purpose of one type of aircraft autopilot is to return the aircraft to a desired altitude if it is forced up or down by wind gusts. The autopilot works by sensing altitude changes and moving the elevator surface automatically to return the aircraft to the desired altitude. What is the input and output of the autopilot?
14-3. Your home heating system consists of a thermostat that is sensitive to the difference between the room's temperature and some preselected desired temperature and a furnace that produces heat energy. Define the input and output of this system.

In this chapter the systems that we will learn to model are "dynamic" systems. This implies that their inputs and outputs are changing with time. Generally the equations by which we describe the behavior of physical systems will contain *variables* and their *time derivatives*. These are called *differential equations*. For simple systems these differential equations can sometimes be solved with pencil and paper, but often their solution requires the use of computers.

A concept that is inherent in the study of system dynamics is that of "rate of change." Let us consider a system which involves variables and their time derivatives. An automobile traveling over a bumpy road has a changing vertical position, velocity, and acceleration. In mechanical systems we define the *velocity* of a point as the time rate of change of the point's *position*. Similarly, *acceleration* is defined as the time rate of change of the point's velocity. In graphical form, Figure 14-3 illustrates how a point's position (s), velocity (v), and acceleration (a) will vary with *time*. The average velocity during a particular time interval ($\Delta t = t_2 - t_1$) is defined as the change in position ($\Delta s = s_2 - s_1$) divided by the time interval; that is,

$$v_{av} = \frac{s_2 - s_1}{t_2 - t_1} = \frac{\Delta s}{\Delta t} \qquad (14\text{-}1)^3$$

The *instantaneous* velocity of the point is the average velocity of the point during an *infinitesimal* time interval, that is, as Δt approaches zero. We will define this velocity as

$$v = \lim_{\Delta t \to 0} \frac{\Delta s}{\Delta t} = \frac{ds}{dt} \qquad (14\text{-}2)^4$$

The notation ds/dt means the *time derivative* of s. By a similar argument the average

---

[3] The notation Δ means "change in."
[4] The notation d means an "infinitesimal change in."

Figure 14-3 Concept of time rate of change.

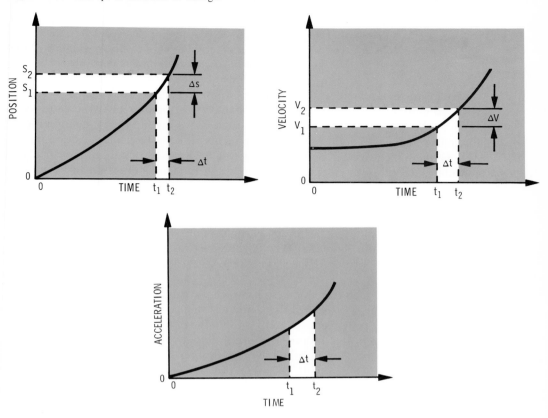

acceleration is defined as the change in velocity divided by the change in time,

$$a_{av} = \frac{v_2 - v_1}{t_2 - t_1} = \frac{\Delta v}{\Delta t} \tag{14-3}$$

and the instantaneous acceleration is the *time derivative* of the velocity,

$$a = \frac{dv}{dt} \tag{14-4}$$

Consider a driver traveling from Paris to LeMans. At a given instant his speedometer reads 72 km/hr, his watch reads 3:30, and his odometer reads 64,360 km. Ten minutes later he notes that his odometer reads 64,375 km, his speedometer reads 88 km/hr, and his watch reads 3:40. Using equations 14-1 and 14-3, we can compute his average velocity and acceleration. For example, his average velocity is

$$V_{av} = \frac{(64{,}375 - 64{,}360)\ \text{km}}{10\ \text{min}} \times \frac{60\ \text{min}}{\text{hr}} = \textbf{90 km/hr}$$

and his average acceleration is

$$a_{av} = \frac{(88 - 72)\ \text{km/hr}}{10\ \text{min}} \times \frac{60\ \text{min}}{\text{hr}} = \textbf{96 km/hr}^2$$

Problems

14-4. In Figure 14-3, $s_1 = 300$ km, $s_2 = 350$ km, $v_1 = 70$ km/hr, $v_2 = 90$ km/hr, $t_1 = 3$ hr, and $t_2 = 3$ hr 15 min. Calculate the average velocity and average acceleration.

14-5. A rock falling from rest off a mountain is observed to have a velocity of 2.45 m/s after 0.25 sec. Find the average acceleration of the rock during this time interval.

By means of mathematical expressions known as differential equations we can write equations which will describe the behavior of models of dynamic physical systems. We are now familiar with the basic elements used in the modeling process: the conceptualization of the physical components of the system, and the development of a mathematical model by which they can be described. We are now ready to consider a more extensive example of how the modeling process can be used in solving a real engineering problem.

By 1990, it has been predicted that high-speed ground transportation vehicles may exist that travel as fast as 500 kilometres per hour (Illustration 14-1). At these speeds even small irregularities in the guideway can produce uncomfortable accelerations in the passenger cabin. Therefore, the vehicle's suspension system must be carefully designed to (1) support the vehicle's weight, and (2) isolate the passenger cabin from excessive vibrations.

At the start of such a major design project, it is not feasible to construct scale models of a number of possible designs and test them experimentally. However, it is possible to *model* a system conceptually and mathematically, as illustrated in Figure 14-2. The first step in this process is to formulate a conceptual model of an actual system. This can be accomplished by assembling various masses, springs, and dampers which, together, will resemble the behavior of the actual system. A simplified conceptual model of a design that could meet the two stipulated design requirements is shown in Figure 14-4.

The second step is to formulate a mathematical model that will represent the characteristics of each component in the conceptual model. In this case the input to the mathematical model is the ground irregularity, the output is the passenger compartment acceleration, and the mathematical model itself is the particular set of mathematical expressions (differential equations) that will include the suspension and mass components. Once this mathematical model has been formulated the designer

Illustration 14-1

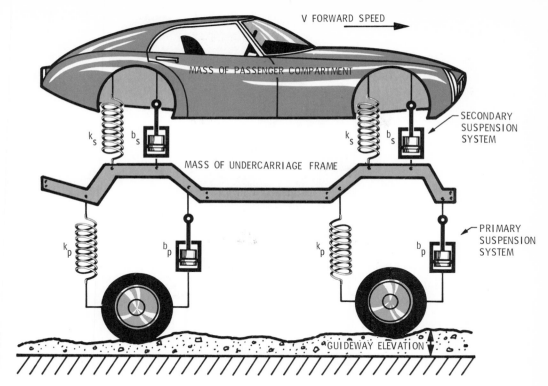

Figure 14-4 Conceptual model of a vertical suspension design.

can vary the characteristics of each system component until he is satisfied with the *performance* of the system as a whole.

In some cases our engineer may decide not to use springs and shock absorbers at all. For high-speed ground systems it has been proposed that a noncontacting suspension such as an "air cushion" suspension be used. Figure 14-5 is a schematic sketch of the fundamental components of such a system. In order to evaluate the feasibility of such a *fluid* suspension system, we will need to formulate a new conceptual model. Notice that in this new system the conceptual model will involve fluid component relationships such as the *volume* flow through the flow control valve and the *pressure drop* across it. It will be important to determine the relationship between the system inputs, such as the road irregularities, and the system outputs, such as the vertical acceleration of the passenger compartment.

Clearly the fluid suspension in this design is performing the same *function* as would a mechanical spring and dashpot[5] suspension. We will find that this similarity applies not only to the function but also to the mathematical modeling. In fact, each of the four basic types of physical systems (mechanical, electrical, fluid, and thermal) can be used to perform similar functions and operate in analogous ways. More than at any time in the past, the modern engineer must have a broad understanding of all physical systems and not limit his knowledge to one or two specialized areas. The successful practicing engineer—the group leader, the section chief, the chief engineer—must understand, for example, how energy is transferred between mechanical, electrical, thermal, and fluid systems and must be able to design interfaces between

---

[5] A dashpot is a mechanical damper, like the shock absorber on a conventional automobile.

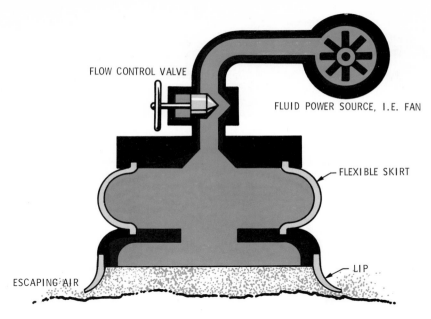

FLOW CONTROL VALVE

FLUID POWER SOURCE, I.E. FAN

FLEXIBLE SKIRT

LIP

ESCAPING AIR

Figure 14-5 Schematic sketch of air cushion suspension.

these different media. Also, an understanding of the basic and underlying analogies that exist will enable the specialist, educated primarily in one field, to apply his knowledge to other fields.

When an engineer wants to transfer energy from one type of system to another, he has available to him many familiar devices. An electric motor transforms electric energy into mechanical energy; an electric generator transforms mechanical energy into electrical energy; a hydraulic actuator transforms fluid energy into mechanical energy; a hydraulic pump converts mechanical energy into fluid energy; and a steam engine converts heat energy into mechanical energy. All these devices are called *transducers* because they transfer energy from one media to another, and they are used in many engineering applications.

To perform his job, the engineer must be familiar with all four basic types of physical systems: mechanical, electrical, fluid, and thermal. However, because there are basic conceptual similarities between all four, he does not need to learn a separate set of rules for dealing with each. By understanding their underlying similarities, specialized knowledge of one type of system can be applied to any of the others.

Some of the analogies between real physical systems are familiar to us all and are quite simple to understand. Each of the four basic types of systems can be conceived of as possessing "through" and "across" characteristics. We can think of current flowing *through* a wire, liquid flowing *through* a tube, force being transmitted *through* a spring, and heat flow (thermal energy) being transmitted *through* a conductor. Electric current, fluid flow, mechanical force, and thermal heat flow are both physical and conceptual analogies. We will call them "through" variables because they are measured in terms of a quantity transmitted *through* a respective element.

Each of these four basic types of systems also possesses "across" characteristics. Electric voltage, fluid pressure, the velocity of mechanical elements, and temperature are analogous in that they all are measured between or *across* two terminals, one at the "entrance to" and one at the "exit from" their respective systems. The "across"

Figure 14-6

variables—voltage, pressure, velocity, and temperature—have a different value at each terminal. It is this difference that we measure. When we speak of a 12-volt battery, we mean a battery in which we can measure a 12-volt difference between or *across* the two battery terminals.

When we talk about the value of an "across" variable, we need to define a reference value. Generally this reference value is arbitrary but constant; that is, voltage is measured with respect to "ground," velocity is measured with respect to a fixed (generally zero) velocity, pressure is measured with respect to a zero pressure (absolute) or atmospheric pressure (gauge), and temperature is measured with respect to the zero point on a fixed scale (0° Celsius, 0° Fahrenheit, 0° Kelvin, or 0° Rankine).

To illustrate the difference between "through" and "across" variables, let us look at a hypothetical model of a simple physical system. Figure 14-6 could represent either a mechanical, a fluid, an electrical, or a thermal system. In Figure 14-6 the "through" variable f (current, fluid flow, force, or heat flow) is shown flowing through the element from terminal 2 to terminal 1. The "across" variable v (voltage, pressure, velocity, or temperature) is a measure of the difference between the values of v_2 and v_1 at the two terminals.

If Figure 14-6 represented an electrical system, f would represent electrical current, and $v = v_2 - v_1$ would represent voltage. If Figure 14-6 represented a mechanical system, f would represent force, and $v = v_2 - v_1$ would represent velocity. If Figure 14-6 represented a fluid system, f would represent fluid flow (volume flow rate), and $v = v_2 - v_1$ would represent the pressure difference. If Figure 14-6 represented a thermal system, f would represent heat flow and $v = v_2 - v_1$ would represent the temperature difference between the terminals.

This basic concept of the difference between "through" and "across" variables and their analogous roles in all four types of systems will prove to be a very useful tool for us in understanding physical systems, for it will allow us to apply our knowledge of one type of system to any of the others.

Elements of dynamic systems

Now let us develop the basic building blocks of model dynamic systems. In general these building blocks will be idealizations or approximations of physical realities. However, this lack of precision will be compensated for by increased model simplicity. Therefore, we will model our systems using *ideal elements,* that is, elements that are represented as simplified mathematical relationships. More complex models should be utilized in those cases where the simpler models do not adequately describe the phenomena being investigated.

Power and energy

In general, the ideal elements that we will consider will store, dissipate, or transform *energy.* Energy can be defined as the ability to do *work*, and *power* is defined as the *time rate* of doing work. Equivalently power can be defined as the rate at which energy *flows* into or out of an ideal element. If we define the energy contained in an ideal element as E, then the power is $\mathbf{P} = dE/dt$, where positive ($+$) power represents energy flowing *into* the element.

Choice of dynamic variables

Earlier in this chapter analogies were made between mechanical, electrical, fluid, and thermal systems by defining *through* and *across* variables. If an engineer has a good understanding of any one of the different media, then by understanding the analogies between systems he can transfer his knowledge to other media. To make this easier, we will describe each ideal element of the various media in terms of its *through* and *across* variables.

Traditionally *through* and *across* variables have been called *power variables*[6] because, in all except thermal systems, the product of an element's through and across variables is the power delivered to the element. If f represents a generalized through variable and v represents a generalized across variable, the power is

$$\mathbf{P} = fv$$

In a mechanical system, power is the product of force and velocity; in an electrical system, power is the product of current and voltage; in a fluid system, power is the product of volume flow rate and pressure. In thermal systems power is simply the heat flow rate. From Chapter 12 the units of power are FLT^{-1}. It is clear from this expression that the fundamental dimensions of power will not vary from system to system. The SI unit of power is the watt, which is a joule/second or a newton-meter/second.

Mechanical systems

In this section we will analyze mechanical systems in which the motion of the system remains parallel to fixed axes. The ideal elements that will be most useful in analyzing mechanical translational systems are the ideal mass, spring, and damper.

Ideal translational mass Newton's Second Law of Motion expresses the relationship that exists between a force (f_m) that is applied to a mass and the time rate of change of the mass's *linear momentum*,[7] dp_m/dt, that is,

$$f_m = \frac{dp_m}{dt}$$

where p_m is the linear momentum of the mass. An *ideal* mass is defined as one in which the linear momentum is linearly proportional to its velocity,

$$p_m = mv_m \tag{14-5}$$

where v_m is the inertial *velocity* of the mass, that is, with respect to a "fixed" reference.

---

[6]There are other variables that can be used to describe dynamic systems; however, the analogies between systems are more easily understood by the use of power variables.

[7]The rate of change of momentum of a body is proportional to the resultant force acting on the body and is in the direction of that force.

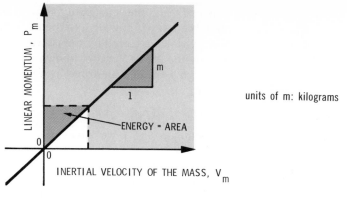

units of m: kilograms

Figure 14-7 Momentum–velocity relationship for an ideal mass.

Substituting this ideal relationship into Newton's second law yields the familiar result,

$$f_m = m\frac{dv_m}{dt} = ma_m \qquad (14\text{-}6)$$

where dv_m/dt is the inertial acceleration of the mass. Figure 14-7 shows this linear relationship for an ideal mass.[8] Note that for ideal elements the energy can be found from either the area above or below the curve, as indicated. For non-ideal elements these areas are not equal and considerable care must be taken to choose the correct one. However, a consideration of non-linear elements is beyond the scope of this chapter.

Figure 14-8 is a conceptual model of a translational mass. On conceptual models a dashed line is used to represent a connection between one terminal of the element and a reference terminal for those situations where a material linkage does not exist. This would be true, for example, for the acceleration of a mass in relation to an inertial reference, for the temperature of a thermal capacitor relative to absolute zero, or for gravitational force acting through a distance.

The power or rate at which energy is flowing into the mass (Figure 14-8) is given by

$$\mathbf{P}_m = f_m v_m \qquad (14\text{-}7)$$

[8] The theory of relativity has shown that in general there is a nonlinear relationship between linear momentum and velocity. However, in nearly all engineering applications the assumption of an ideal mass is valid.

Figure 14-8 Conceptual model of a translational mass.

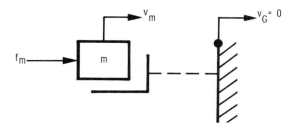

The force, f_m, is transmitted *through* the mass, and the velocity, v_m, is measured *across* the mass (with respect to a fixed frame). If energy is flowing into the mass there must be some way of *storing* this energy. Energy which is stored by a mass due to its motion is called its kinetic energy and is given by the triangular area shown in Figure 14-7. Since the area of a triangle is $\frac{1}{2}$(base)(height), the energy stored by a mass that has a linear momentum, p_m, and a velocity, v_m, is

$$\text{Energy} = \text{area} = \frac{1}{2}(p_m)(v_m)$$

Using the expression for an ideal mass ($p_m = mv_m$), we have

$$E_m = \frac{1}{2}mv_m^2 \tag{14-8}$$

Because the energy stored by the mass is a function of its across variable,[9] velocity, we can categorize it as an *across variable energy storage element.*

Example A mass of 2 kilograms is subjected to a force of 5 newtons. Calculate the time rate of change of the mass' velocity.

Solution

$$a = \frac{dv_m}{dt} = \frac{1}{m}f_m = \frac{5 \text{ newtons}}{2 \text{ kilograms}} = \mathbf{2.5 \text{ metres/sec}^2}$$

Summarizing the properties of an ideal translational mass:

Elemental equation $\qquad\qquad \dfrac{dv_m}{dt} = \dfrac{1}{m}f_m$

Energy stored $\qquad\qquad\qquad E_m = \frac{1}{2}mv_m^2$

Conceptual model

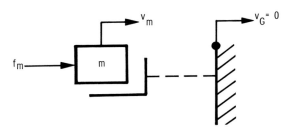

Problem

14-6. A rocket has a mass of 400,000 kg and is traveling in outer space. The rocket's thrustors have a thrust capability of 40,000 newtons. If they are turned on, calculate the acceleration, dv/dt, of the rocket.

Ideal translational spring Springs are a common part of our daily experience. For example, scales, car suspensions, and elastic materials of all kinds exhibit the property of compressing or elongating under load. Figure 14-9 is a conceptual model of a coiled spring. An ideal translational spring is defined as one that (1) has no mass, and

[9] We could have just as easily expressed the energy in terms of p_m rather than $v(E_m = \frac{1}{2}p_m^2/m)$. However, p_m is neither an across nor a through variable (it is an integrated through variable).

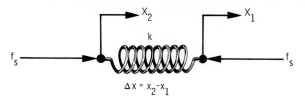

Figure 14-9 Conceptual model of a translational spring.

(2) has a linear force–displacement elemental representation as shown in Figure 14-10.

In Figure 14-10, Δx represents the deflection of the spring measured from its relaxed state. The force–displacement characteristic is also not usually linear as shown in Figure 14-10 but will curve upward (hardening spring) or downward (softening spring). However, if the deflections (Δx) are small, such a linear approximation is often acceptable. The spring stiffness (k) is a function of the design of the spring and the material used in its manufacture. The flexibility of the spring is the reciprocal of its stiffness ($1/k$) and is denoted by the symbol A. Moreover, real springs have some mass; however, any real spring can be modeled as a combination of an ideal mass and an ideal spring.

The defining equation for the ideal spring is

$$f_s = k\,\Delta x = \frac{\Delta x}{A} \tag{14-9}$$

where $\Delta x = x_2 - x_1$. If we define v_s as the time rate of change of Δx, then

$$\frac{df_s}{dt} = \frac{k\,d(\Delta x)}{dt} = kv_s = \frac{v_s}{A} \tag{14-10}$$

Equation 14-10 is called the elemental equation for the spring, which is the equation that relates its across variable (v_s) to its through variable (f_s). The power or rate of change of energy of the spring is $\mathbf{P} = f_s v_s$. The energy stored in the spring can be expressed either in terms of the force, f_s, or the deflection, Δx. However, since we have chosen to use power variables to describe our ideal elements, we will use f_s. The expression for the stored energy in terms of the *through* variable f_s is given by the

Figure 14-10 Force–displacement relationship for an ideal translational spring.

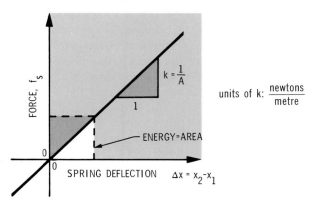

triangular area shown in Figure 14-10. Expressing this area in terms of the force, f_s, we have

$$E_s = \frac{1}{2}\frac{f_s^2}{k} = \frac{1}{2}Af_s^2 \qquad (14\text{-}11)$$

Thus we can categorize an ideal spring as a *through variable energy storage element*.

Again we note that the stored energy in a spring could be expressed in terms of its displacement; however, displacement is neither an across nor a through variable (it is an integrated[10] across variable).

Example A spring whose spring constant is $k = 100$ newtons/metre is supporting a weight of 200 newtons. Find the energy stored in the spring.

Solution

$$E_s = \frac{f_s^2}{2k} = \frac{(200 \text{ N})^2}{2(100)\text{N/m}} = \textbf{200 newton-metres} \quad \text{(joules)}$$

Summarizing the properties of an ideal translational spring:

Elemental equation $\qquad\qquad \dfrac{df_s}{dt} = kv_s = \dfrac{v_s}{A}$

Energy stored $\qquad\qquad\qquad E_s = \dfrac{f_s^2}{2k} = \dfrac{1}{2}Af_s^2$

Conceptual model[11]

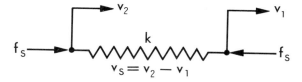

$$v_s = v_2 - v_1$$

Problem

14-7. An automobile suspension spring is supporting 4448 newtons and is compressed 0.025 m. Find the spring constant, k, of the spring and the energy stored in the spring.

Ideal translational damper The ideal mass and ideal spring comprise two energy storage elements in our modeling "arsenal." It is now time to add an ideal element that is capable of dissipating energy, the *ideal damper*. The common shock absorber used in automobiles is a good example of an element that dissipates rather than stores energy. If energy is added to a spring by compressing it, the spring can return energy to the system by returning to its original position. However, if a shock absorber is compressed, it cannot return energy to the system. An ideal force–velocity relationship for an ideal damper is shown in Figure 14-11 where v_b is the relative velocity across the damper. A conceptual model of this system is shown in Figure 14-12. We assume in modeling the ideal damper that (1) it has no mass or elastic properties, and

[10] If $v_s = d\,\Delta x/dt$, then Δx is the integral of v_s.
[11] The conceptual model is shown using through and across variables; thus v_2 and v_1 have replaced x_2 and x_1 of Figure 14-9.

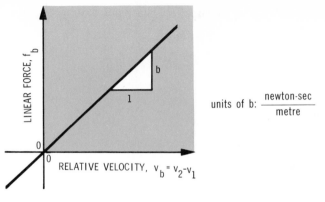

Figure 14-11 Force–velocity relationship for an ideal translational damper.

(2) it has a linear force–velocity relationship. If this idealized model is not adequate, it can be improved by the addition of an ideal mass and spring to model the physical damper.[12] The second assumption is valid if the damping mechanism is viscous, and if the velocity difference across the damper is relatively small. The elemental equation for the ideal damper is

$$f_b = b(v_2 - v_1) = bv_b \tag{14-12}$$

The energy flow rate or power into the damper is

$$\mathbf{P} = f_b v_b = bv_b^2 \geq 0 \tag{14-13}$$

Note that this expression is always positive (for nonzero velocity), which means that energy is *always being dissipated.* The energy is dissipated as thermal energy, causing a temperature rise in the damper.

Example The damper shown in Figure 14-12 has its right end fixed ($v_1 = 0$) and its left end has a velocity of 10 m/s. The damping coefficient, b, has a value of 20 newton-seconds/meter. Find the force, f_b.

Solution

$$f_b = b(v_2 - v_1) = \left(20 \frac{\text{N-s}}{\text{m}}\right)(10 - 0)\frac{\text{m}}{\text{s}}$$

$$= \textbf{200 newtons}$$

[12] A realistic mechanical damper will be modeled later as a combination of ideal elements.

Figure 14-12 Conceptual model of an ideal translational damper.

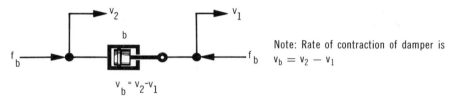

Note: Rate of contraction of damper is $v_b = v_2 - v_1$

Summarizing the properties of an ideal translational damper:

Elemental equation $\qquad\qquad\qquad f_b = bv_b$

Power dissipated $\qquad\qquad\qquad \mathbf{P} = bv_b^2$

Conceptual model

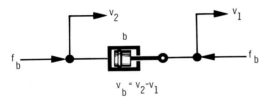

Problem

14-8. An automobile shock absorber is subjected to a force of 1000 N and has a relative velocity across it of 0.5 m/s. Find the damping coefficient, b, and the power dissipated in watts.

Ideal sources To complete our "arsenal" of modeling tools for mechanical translational systems, we need to define the concept of an ideal source. Since we are using power variables we will need both an ideal *through* variable source (force source) and ideal *across* variable source (velocity source). An ideal source is defined as a device that is capable of supplying a prescribed variable level (either constant or time varying) that is independent of other system variables. Theoretically this is not possible; however, in practice, it can be a useful and very nearly correct assumption. For example, the force on a mass near the surface of the earth due to gravity is very nearly constant, independent of its position. The voltage across a car battery is very nearly constant for a small level of current drawn from it. The pressure at the bottom of a large reservoir of water is approximately constant for small changes of depth, independent of the volume flow rate out of the reservoir. Figure 14-13 illustrates a typical relationship that could be approximated as an ideal source for small changes in a system parameter. If we expect the system variable to have values outside of the

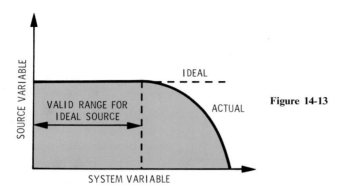

Figure 14-13

valid range for an ideal source, we can usually model the device as a combination of an ideal source and an ideal element.

Thus, in the examples cited, gravity can be modeled as a *force source*, a car battery as a voltage source, and a reservoir as a pressure source. Other examples of ideal sources are the following:

Ice (temperature source)
Atmosphere (temperature source)
Furnace (heat flow, or more precisely *thermal energy,* source)
Motor (velocity source)
Pump (volume flow rate source)

The schematic representation that is used for a mechanical ideal source is shown in Figure 14-14. The sign convention used for the velocity source is indicated by the + and − signs on the circle; thus the velocity at the top of Figure 14-14a is greater than at the bottom. The sign convention for the force source shown in Figure 14-14b is to point the arrow in the direction of the applied force. These two sign conventions will be illustrated by examples.

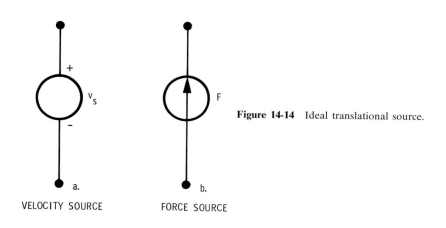

Figure 14-14 Ideal translational source.

VELOCITY SOURCE   FORCE SOURCE

Example The spring–mass system shown in Figure 14-15a exists in a gravity field. The notation x represents the displacement of the spring from its undeflected position. Figure 14-15b shows the system using ideal elements and sources. Note that the direction of the force source to model gravity is upward since it tends to *increase* the velocity of the mass with respect to the reference velocity ($v_g = 0$).

Example: Typical Shock Absorber Figure 14-16 is a sketch of a typical automobile shock absorber. In order to model this *dynamic system,* we will need most of the ideal mechanical elements that have been introduced. Both the outside casing and the plunger can be modeled as ideal masses. The spring connecting the plunger to the casing can be modeled as an ideal spring, and the hole in the plunger to allow fluid passage can be modeled as an ideal viscous damper.

Depending on how the shock absorber is to be used, we can have a variety of different dynamic models. First, suppose that the case mass is attached to a fixed reference and that the plunger mass is subjected to an applied force, F, that is directed

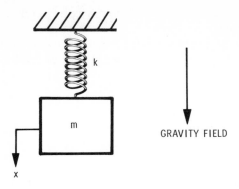

Figure 14-15a Pictorial representation of a simple spring–mass system.

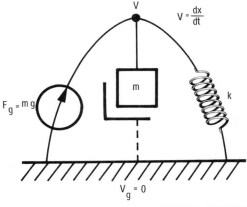

DYNAMIC MODEL USING CONCEPTUAL ELEMENTS

Figure 14-15b Conceptual model of a simple spring-mass system using ideal elements.

Figure 14-16 Typical shock absorber.

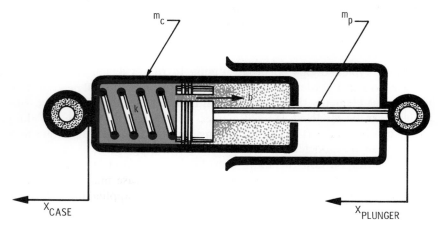

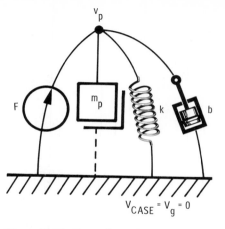

Figure 14-17 Dynamic model of shock absorber with case fixed.

in the x_p direction. Figure 14-17 is a schematic representation of this situation using ideal elements. Note that the force is modeled as an ideal source with the arrow pointing upward since the force tends to increase the velocity, v_p. It is also important to note that the ideal mass, spring, and damper are connected in parallel since they all share the same across variable, v_p.

Next, suppose that we wish to *analyze* how the shock absorber will perform when it is installed in an automotive suspension system. To do this we will assume that the plunger is rigidly connected to a quarter of the automobile's mass since there are four suspension groups in an automobile. Next, we will assume that the case is rigidly attached to an axle which is supported by a rubber tire, as shown in Figure 14-18.

We will model the tire as an ideal spring and the road's vertical irregularities as a known function of time; thus

$$x_{\text{road}} = f(t)$$

or

$$v_{\text{road}} = \frac{dx_{\text{road}}}{dt} = \frac{df(t)}{dt}$$

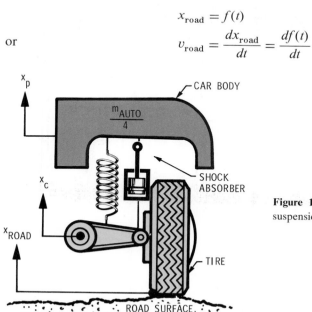

Figure 14-18 Schematic of an automobile suspension.

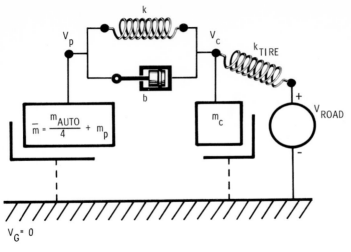

Figure 14-19 Dynamic model of an automobile suspension.

A dynamic model of the automobile suspension system using ideal elements[13] is shown in Figure 14-19.

It is important to note that the effect of gravity has not been included in this model; it could be included by the addition of two force sources on the two masses as was done in Figure 14-15b. Also, note the sign of the velocity source; we are defining v_{road} to be positive with respect to our reference velocity, $v_G = 0$.

After we present the remaining fundamental building blocks for all the different media, we will develop a rational procedure to formulate the mathematical *equations* that predict the behavior of a system, such as that shown in Figure 14-19.

Problems

14-9. A constant force is applied to a mass as shown in Figure 14-20. The *dynamic model* for this system is composed of the *conceptual* models of a force source and a mass. Draw the dynamic model of this system.

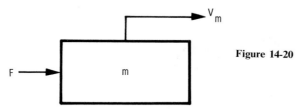

Figure 14-20

14-10. A constant force is applied to a damper as shown in Figure 14-21. Sketch the dynamic model.

Figure 14-21

---

[13] Since the plunger mass and the automobile mass are rigidly attached, we can combine them as $\bar{m} = \dfrac{m_{auto}}{4} + m_p$.

14-11. A constant force is applied to the mass–damper system shown in Figure 14-22. Sketch the dynamic model.

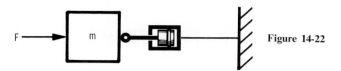

F → m

Figure 14-22

14-12. A constant force is applied to the mass–spring–damper system shown in Figure 14-23. Sketch the dynamic model.

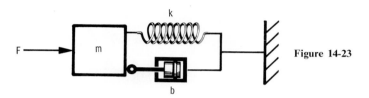

k

F → m

b

Figure 14-23

14-13. Figure 14-24 shows a simple vibration isolator. It is desired to keep the machine mass, m_m, from vibrating too severely when the floor is subject to a vibratory velocity input, $v_f(t)$. The attached spring–mass system is designed so that the combination of the mass, m_a, and the spring, k_a, will *absorb* the energy while the machine mass, m_m, remains relatively motionless. Sketch the dynamic model of the system.

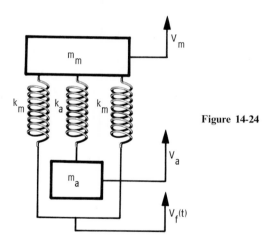

V_m

m_m

k_m k_a k_m

Figure 14-24

V_a

m_a

$V_f(t)$

14-14. Assume in Problem 14-13 that the base of the machine is fixed ($v_f = 0$) and that the machine mass, m_m, is subject to a force $F(t)$ which is in the direction of v_m. Sketch the dynamic model.

14-15. Vibration isolators (also called vibration absorbers) are sometimes installed on the top floor of large skyscrapers. The purpose of these devices is to absorb the energy of gusting winds and thus to relieve the load on the building's structural elements. Figure 14-25 shows a pictorial representation of one of these systems. Assume that the wind imparts a velocity $v_b(t)$ to the top floor. Sketch the dynamic model of the vibration absorber assuming $v_b(t)$ to be a velocity source.

14-16. Another important consideration in the design of large skyscrapers is their ability to withstand the effect of earth tremors or earthquakes. The building's walls have a

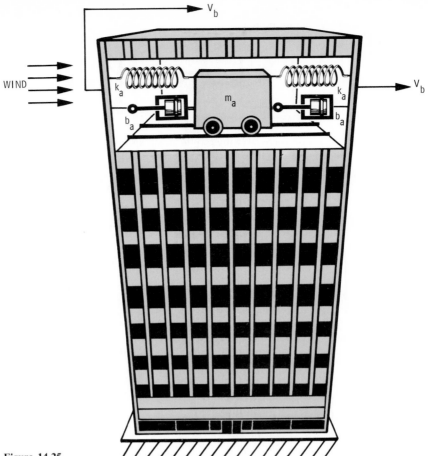

Figure 14-25

certain stiffness and damping. A very simple pictorial model of the building's resistance to a side to side velocity caused by an earth tremor is shown in Figure 14-26. Sketch the dynamic model.

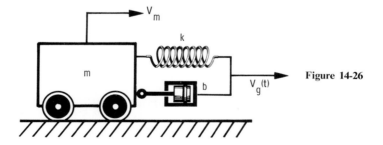

Figure 14-26

14-17. A simplified model of a vehicle suspension system is shown in Figure 14-27. The passenger compartment has a mass, m_p; the secondary suspension system is composed of a spring, k_s, and a damper, b_s; the undercarriage mass is m_u; and the tire has a spring constant, k_t. The passenger compartment is subjected to a force, $F(t)$, and the ground excitation is modeled by a velocity source, $v_g(t)$. Sketch the dynamic model (neglect the influence of gravity).

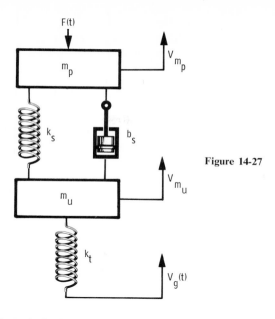

Figure 14-27

14-18. Include the effect of gravity in Problem 14-17 by modeling the gravity force on each mass by a force source.

Electrical systems

In this section we will analyze electrical systems and model them by constructing simple ideal elements, much in the same way that we considered mechanical systems. In fact, the analogies between mechanical and electrical systems should become clear when we see the similarities between the functions of *energy storage and dissipation*.

The electrical power delivered to an element can be computed by multiplying the current passing *through* the element by the voltage drop *across* it, as shown in Figure 14-28; that is, $\mathbf{P} = i(e_2 - e_1)$. The *voltage* (potential difference) between points 1 and 2 is defined as the work which would be done in moving a positive charge from point 2 to point 1. The *current* is defined as the *rate of flow* of positive charge across a given area during a unit time.

We will use power variables to describe the ideal electrical elements, with the

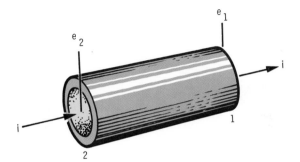

Figure 14-28 Power expression for an element.

through variable being the current and the across variable being the voltage. Current is measured in amperes (A) and the voltage in volts (V). The power is expressed in watts (W). If the element in Figure 14-28 had a current $i = 1$ amp, and a voltage drop $e = 1$ volt, then the power *into* the element is $\mathbf{P} = (1 \text{ amp})(1 \text{ volt}) = 1$ watt.

Ideal capacitance

Charged particles can be stored on a conducting material if two pieces of the conducting material are separated by a medium that does not allow charge to flow through it. An element that has the ability to store charge is a capacitor. Figure 14-29 shows two sheets of conducting material separated by a different medium that act as a capacitor. This ability to store charge is called *capacitance*. The physical arrangement shown in Figure 14-29 is not the only useful geometric arrangement; other possibilities exist, such as concentric spheres.

An *ideal* capacitance is one in which the charge that is stored on the capacitance is linearly proportional to the voltage between the two plates. Figure 14-30 illustrates this linear relationship, which can be expressed as

$$q_c = C(e_2 - e_1) = Ce_c \tag{14-14}$$

where e_c is the voltage across the capacitance. The constant C is called the capacitance and is in general a function of the geometry and material properties of the capacitor. We need to express this equation in terms of power variables. Therefore, by differentiating equation 14-14, we obtain

$$\frac{dq_c}{dt} = C\frac{de_c}{dt} \tag{14-15}$$

and since current is the rate of charge flow (dq_c/dt), this elemental equation becomes

$$i_c = C\frac{de_c}{dt} \tag{14-16}$$

The electric energy stored in the capacitor is the triangular area shown in Figure 14-30 and can be expressed as

$$E_c = \frac{Ce_c^2}{2} \tag{14-17}$$

Therefore, the energy stored is a function of the *across* variable, voltage. Thus it is an across variable storage element, just as the ideal mass stores energy by virtue of its across variable, velocity.

Figure 14-29 Simple capacitor.

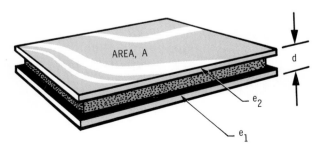

AREA, A

d

e_2

e_1

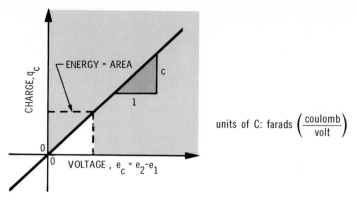

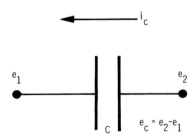

units of C: farads $\left(\dfrac{\text{coulomb}}{\text{volt}}\right)$

Figure 14-30 Charge–voltage relationship for an ideal capacitor.

Example In the capacitor shown in Figure 14-29 the voltage across the capacitor $(e_c = e_2 - e_1)$ is changing with a time rate of change of 500 volts/sec. The value of the capacitance is $C = 0.50$ microfarad $(0.50 \times 10^{-6}$ farad). Find the current passing through the capacitance.

Solution

$$i_c = C\frac{de_c}{dt} = (0.5 \times 10^{-6}\ \text{farad})(500\ \text{volts/sec})$$

$$= \mathbf{2.5 \times 10^{-4}\ ampere}$$

Summarizing the properties of an ideal capacitor:

Elemental equation $\dfrac{de_c}{dt} = \dfrac{1}{C}i_c$

Energy stored $E_c = \dfrac{Ce_c^2}{2}$

Conceptual model

The analogy between the ideal mass and the ideal capacitance can be seen more clearly by comparing the respective elemental and energy storage equations. The analogous quantities are the mechanical velocity and the electrical voltage (across variables), the mechanical force and the electrical current (through variables), and the mechanical mass and electrical capacitance.

Problems

14-19. At a particular instant a capacitor has a capacitance $C = 1 \times 10^{-7}$ farad, and a current flowing through it of $i = 1 \times 10^{-4}$ amp. Find the rate of change of voltage across it.

14-20. At a particular instant a capacitor has a current flowing through it of $i = 1 \times 10^{-5}$ amp and a rate of change of voltage across it of $de_c/dt = 100$ volts/s. Find the capacitance of the capacitor.

14-21. The capacitor of Problem 14-19 has a voltage across it of $e_c = 0.1$ volt. Find the energy in joules stored in the capacitor and the power in watts being supplied to it.

14-22. The capacitor of Problem 14-20 has a stored energy of 1×10^{-6} joule. Find the voltage across it.

Ideal inductance

The next ideal element that we need is one that stores energy by virtue of its through variable, the current, which is the electrical equivalent of the mechanical spring. When current flows through a conductor a magnetic field is established. The energy that is stored in this field is called electromagnetic energy, and the magnitude of this stored energy depends on the magnitude of the current flowing through the element. This property of storing energy by virtue of current flow is called *inductance*.

An ideal *inductance* is defined as an element in which the voltage across the element is linearly proportional to the *rate of change of current* through it, or

$$e_2 - e_1 = e_L = L\frac{di_L}{dt} \tag{14-18}$$

The constant L is called the inductance of the coil. Figure 14-31 is a conceptual model of an ideal inductor; Figure 14-32 shows the ideal inductor's linear characteristic.

Figure 14-31 Conceptual model of an ideal inductor.

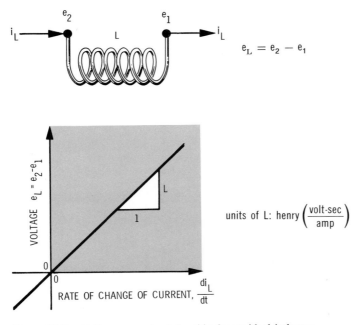

$$e_L = e_2 - e_1$$

units of L: henry $\left(\dfrac{\text{volt-sec}}{\text{amp}}\right)$

Figure 14-32 Voltage–current relationship for an ideal inductor.

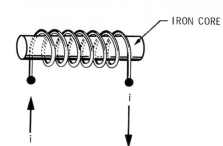

IRON CORE

Figure 14-33 Solenoid coil.

An example of a practical inductor is the solenoid coil shown in Figure 14-33. The inductance of the coil is a function of the number of turns of wire around the rod, the type of rod material, and the length of the rod.

The energy stored in an inductor is

$$E_L = \tfrac{1}{2}Li_L^2 \tag{14-19}$$

Example The current through the inductor shown in Figure 14-32 is changing at a rate of 100 amp/sec. The inductance of the coil is $L = 2 \times 10^{-3}$ henry. Find the voltage drop across the inductor.

Solution

$$e_L = L\frac{di_L}{dt} = (2 \times 10^{-3} \text{ henry})(100 \text{ amp/sec}) = \mathbf{0.2 \ volt}$$

Summarizing the properties of the ideal inductor:

Elemental equation $\qquad\qquad \dfrac{di_L}{dt} = \dfrac{1}{L}e_L$

Energy stored $\qquad\qquad\quad E_L = \tfrac{1}{2}Li_L^2$

Conceptual model

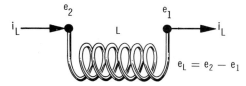

$$e_L = e_2 - e_1$$

By comparing the summaries of the electrical inductor with those of the mechanical spring we see that the analogous variables are again the force and current, the velocity and voltage, and the inductance L and the spring flexibility A (or reciprocal of the spring constant, $1/k$).

Problems

14-23. A coil has an inductance of $L = 1 \times 10^{-3}$ henry and a voltage across it, $e_L = 0.1$ volt. Find the rate of change of current through it.

14-24. At a particular instant a coil has a voltage across it of 1 volt and a rate of change of current through it of 200 amp/s. Find the inductance of the coil.

14-25. The coil of Problem 14-23 has a current flowing through it of $i_L = 60$ amp. Find the energy stored in the coil and the power flowing into it.

14-26. The coil of Problem 14-24 has a stored energy of 10^{-6} kilowatt hour (or 36 joules). Find the current flowing through it.

Ideal resistance

We have formulated the energy storage elements, one for across variable energy storage (capacitors), and one for through variable energy storage (inductors). We now need an element that dissipates energy. The electrical analogy to the mechanical damper is the electrical resistor. When charged particles flow through a conductor, there is a *resistance* to this flow, called simply the resistance of the conductor. Figure 14-34 is a conceptual model of current flowing through a section of a conductor.

Figure 14-34   Conceptual model of an ideal resistance.

An ideal resistor is defined as one in which the voltage drop across it is linearly proportional to the current through it, or

$$e_R = e_2 - e_1 = Ri_R \tag{14-20}$$

This is known as Ohm's Law; Figure 14-35 shows this linear relationship.

The power or rate of energy flow into the resistor is

$$\mathbf{P} = e_R i_R = \frac{e_R^2}{R} = i_R^2 R \geq 0$$

Note that just as in the case of the mechanical damper, the power always flows into the damper; that is, it is always dissipating.

Example   The resistance shown in Figure 14-34 has a value of $R = 200$ ohms. If the voltage drop across the resistance is 50 volts, find the current through it.

Figure 14-35   Voltage–current relationship for an ideal resistor.

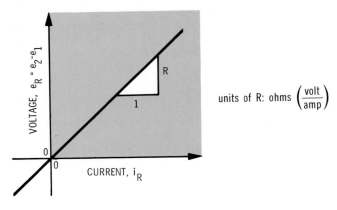

units of R: ohms $\left(\dfrac{\text{volt}}{\text{amp}}\right)$

Solution

$$i_R = \frac{e_r}{R} = \frac{50 \text{ volts}}{200 \text{ ohms}} = \mathbf{0.25 \text{ amp}}$$

In summarizing the properties of the ideal resistor:

Elemental equation $e_R = R i_R$

Power dissipated $\mathbf{P} = i_R^2 R$

Conceptual model

$$e_R = e_2 - e_1$$

Problems

14-27. A wire of resistance $R = 100$ ohms has a current of 10 amps passing through it. Find the voltage across it and the power dissipated by it.

14-28. In order to experimentally determine the resistance of a given length of wire, a constant voltage of 6 volts was maintained across it, and the current through it was measured to be 0.01 amp. What is the resistance of the wire?

Ideal sources

Just as we needed to include ideal elements to model mechanical system *inputs,* we also need ideal electrical input models. Again, since we are using power variables, the two types of sources that we need are (1) a *through* variable source (current source) and (2) an *across* variable source (voltage source). An ideal source is one that is capable of maintaining either a prescribed through or across variable level, independent of other system variables. Figure 14-36 shows the schematic of the ideal electrical sources.

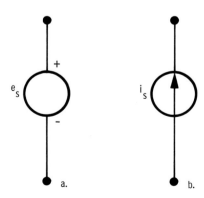

Figure 14-36   Electrical sources.

IDEAL VOLTAGE SOURCE   IDEAL CURRENT SOURCE

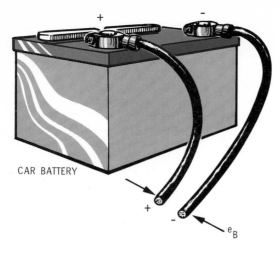

Figure 14-37 Standard 12-volt car battery.

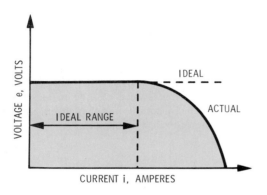

Figure 14-38 Voltage–current relationship for a standard automobile battery.

Figure 14-37 shows a standard car battery. If we model this as an ideal voltage source that supplies a constant 12 volts to any system, it must have a voltage–current relation such as that shown for the ideal range in Figure 14-38. In practice, however, as the system draws more and more current from the battery, the voltage drops off as shown. We will see later how the actual car battery performance shown in Figure 14-38 can be modeled as a combination of an ideal voltage source and an ideal resistor.

Example Figure 14-39 shows a simple electrical circuit which is called a "low-pass filter." The voltage, e_{in}, is a prescribed function of time; therefore, we can model it as an ideal voltage source. Figure 14-40 shows a slightly rearranged form of Figure 14-39.

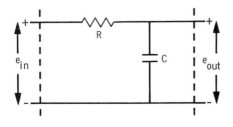

Figure 14-39 Low-pass filter.

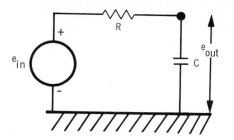

Figure 14-40 Dynamic model of a low-pass filter.

Problems

14-29. Sketch the dynamic model of a circuit composed of a voltage source that is in series with an inductor and a resistor.

14-30. Draw the dynamic model of the pictorial system shown in Figure 14-41. (*Hint:* The light bulb can be modeled as a resistance.)

Figure 14-41

14-31. Sketch the dynamic model of the circuit shown in Figure 14-42. The symbols R, C, and L represent resistance, capacitance, and inductance.

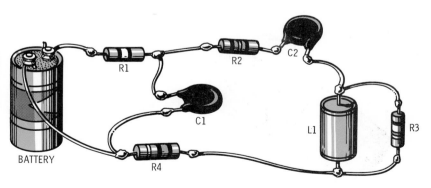

Figure 14-42

14-32. An automobile battery is tested and its voltage–current relationship is found to be like that in Figure 14-38. An ideal voltage source would behave like that shown in the ideal range. A good approximation for such a realistic curve as shown in Figure 14-38 is a voltage source in series with a resistance. Sketch the dynamic model of such a realistic battery.

14-33. Suppose the realistic automobile battery described in Problem 14-32 is connected to a resistor in series with a solenoid coil (inductor). Sketch the dynamic model of this system.

14-34. An analog computer may be composed of electrical elements that are connected in such a way that they simulate the system of interest. Suppose it is desired to simulate the mechanical system described in Problem 14-12. Sketch a dynamic model composed of *electrical* elements that would simulate the mechanical system.

14-35. Sketch the dynamic model of the electrical analog of Problem 14-13.

14-36. Sketch the dynamic model of the electrical analog of Problem 14-15.

14-37. Sketch the dynamic model of the electrical analog of Problem 14-16.

14-38. Sketch the dynamic model of the electrical analog of Problem 14-17.

Fluid systems

In this section we will consider simple fluid elements that are often used in models of real systems. In particular we will consider fluids that are constrained to move in pipes or tubes or are stored in containers under pressure. The operation of machine tool equipment, movement of aircraft control surfaces, actuation of automotive power brakes, and so on, are applications of using moving fluids to do work. The study of working fluids in motion is one aspect of the field of hydraulics. In general, fluid elements tend to be more nonlinear than mechanical or electrical system elements, but valid preliminary design decisions can often be made on the basis of idealized linear behavior. In this section we will only consider the ideal behavior of fluid elements. More advanced nonlinear treatment can be found in such books as reference 3.

Figure 14-43 shows a section of tubing of area, A, with a quantity of fluid passing through it. At any instant in time the volume of fluid contained in the small element of length dx is $d\text{Vol} = A\, dx$. We will assume that the pressure acting over the area A, is a constant, where pressure is defined as the force (F) on the area of fluid divided by the area (A); that is,

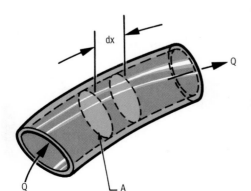

Figure 14-43 Section of tubing.

$$P = \frac{F}{A} \quad \text{or} \quad A = \frac{F}{P}$$

By substitution we have

$$\frac{F}{P}(dx) = d\text{Vol} \quad \text{and} \quad F\,dx = P\,d\text{Vol}$$

The volume flow rate (Q) is the quantity of fluid passing through the element (dx) in the time dt, or $Q = d\text{Vol}/dt$. Now since $F\,dx = P\,d\text{Vol}$, we can divide both sides of the equation by dt and obtain

$$F\frac{dx}{dt} = P\frac{d\text{Vol}}{dt}$$

or

$$Fv = PQ \tag{14-21}$$

The expression Fv (force times velocity) represents the mechanical power or time rate of change of energy. The expression for fluid power is PQ or the product of pressure (N/m^2) and volume flow rate (m^3/s).

Following our convention of using power variables to describe system behavior we will consider pressure to be the *across variable* since pressure is measured between two points. We will consider volume flow rate as the *through variable* since it is the quantity of fluid passing *through* an element in a small period of time, dt.

Ideal elements

Our next task is to define a set of simple fluid elements so that we can model practical systems with a *conceptual model* that is composed of simple elements. Just as we did in the mechanical and electrical cases, we will need to define the following:

1. An element that stores energy by virtue of its across variable, pressure.
2. An element that stores energy by virtue of its through variable, volume flow rate.
3. An element that dissipates energy.
4. An ideal across variable source (pressure source).
5. An ideal through variable source (volume flow rate source).

Ideal fluid capacitance

An ideal fluid capacitance is an element which stores energy by virtue of the pressure difference across it. For example, Figure 14-44 shows a fluid tank open to the atmospheric pressure with area, A, height of fluid, H, and volume flow rate into the tank, Q. If the fluid is modeled as incompressible,[14] the volume flow rate entering is equal to $A(dH/dt)$. The pressure difference across the tank, $P_2 - P_1$, is equal to the weight of the fluid in the tank divided by the area of the tank,[15] or

$$P_2 - P_1 = \frac{\rho g H A}{A} = \rho g H \tag{14-22}$$

---

[14] Most liquids can be modeled as incompressible, while gases must often be modeled as compressible.
[15] ρ is the mass density of the fluid, kg/m^3; g is acceleration due to gravity, $9.8\ \text{m/s}^2$.

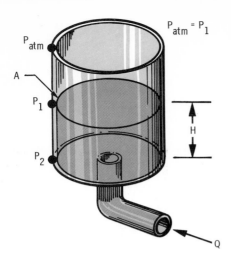

P_{atm}

$P_{atm} = P_1$

A

P_1

H

P_2

Q

Figure 14-44 Ideal fluid capacitance.

If we adopt the nomenclature that $P_c = P_2 - P_1$ (where $P_1 = $ constant reference pressure), we can rewrite equation 14-22 as

$$H = \frac{P_c}{\rho g} \tag{14-23}$$

If we substitute this expression into the expression for volume flow rate, we have

$$Q = A\frac{d}{dt}\left(\frac{P_c}{\rho g}\right) = \frac{A}{\rho g}\frac{dP_c}{dt} \tag{14-24}$$

Next we can define the quantity $A/\rho g$ (which has units of m⁵/N) as a capacitance, C, and can substitute and rewrite equation 14-24 as

$$Q = C\frac{dP_c}{dt} \tag{14-25}$$

Note the similarity of this expression with that derived for the electrical capacitance in equation 14-16. In fact, if we replaced the electrical through variable, i_c, with the fluid through variable, Q, and the electrical across variable, V_c, with the fluid across variable, P_c, we would have identical expressions.

In general, an ideal linear fluid capacitance is defined as an element that has a volume–pressure characteristic as shown in Figure 14-45. The energy stored in the tank is equal to the area under the V vs P_c curve. For the linear range, this is

$$E = \frac{1}{2}VP_c = \frac{1}{2}CP_c^2 \tag{14-26}$$

By substituting the previously defined values for C and P_c into equation 14-26, the energy is seen to be equal to the weight of fluid, $\rho g h A$, in the tank times the height to the fluid mass center, $h/2$.

An open tank is not the only fluid element that can store energy by virtue of its pressure; others are shown in Figure 14-46. All these elements can be modeled as an ideal fluid capacitance.

Example The tank shown in Figure 14-44 has a cross-sectional area of 1 m². The density of water is 1000 kg/m³, and the acceleration due to gravity is $g = 9.8$ m/s². Find the capacitance of the tank.

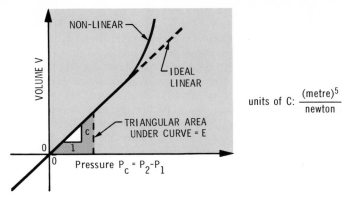

units of C: $\dfrac{(\text{metre})^5}{\text{newton}}$

Figure 14-45 Volume–pressure relationship for an ideal fluid capacitance.

Solution

$$C = \frac{A}{\rho g} = \frac{1\ \text{m}^2}{(1000\ \text{kg/m}^3)(9.8\ \text{m/s}^2)} = \mathbf{1.02 \times 10^{-4}\ m^5/N}$$

Summarizing the properties of an ideal fluid capacitance:

Elemental equation $\dfrac{dP_c}{dt} = \dfrac{1}{C} Q_c$

Energy stored $E_c = \frac{1}{2} C P_c^2$

Conceptual model

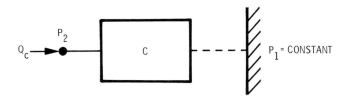

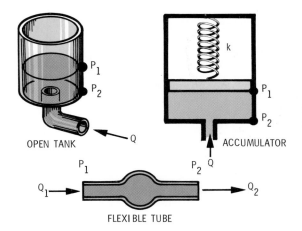

Figure 14-46 Examples of ideal fluid capacitors.

Problems

14-39. The tank shown in Figure 14-44 has a cross-sectional area of 5 m². If water is the fluid medium, calculate the fluid capacitance of the tank.

14-40. The tank of Problem 14-39 has a net volume flow rate into it of $Q_c = 100$ m³/sec. Calculate the rate of change of pressure across it. If at this particular instant the pressure across it is 9.8×10^4 N/m², find the stored energy in the tank.

14-41. A fluid capacitor has a capacitance of 1×10^{-3} m⁵/N. At a particular instant it has a pressure drop across it of 9.6×10^4 N/m². Find the energy stored in the capacitor.

14-42. A flexible tube has the property of fluid capacitance due to its elasticity. In order to determine the value of the capacitance of a certain tube, an experiment was devised. It was observed that at a particular instant the rate of change of pressure across the section of tubing was 100 N/m²-s, while the net flow through it was 6×10^{-3} m³/s. Calculate the capacitance of the tube.

Ideal fluid inertance

A fluid element which stores energy by virtue of its through variable, volume flow rate, is called fluid inertance (I). The term inertance is used because inertia forces are associated with the acceleration of the fluid in a pipe. Figure 14-47 shows a fluid in a pipe of length, ℓ, and area, A. The net force on the mass of fluid in the length, ℓ, is

$$F = A(P_2 - P_1) = AP_I$$

By applying Newton's second law and noting the fact that the velocity of the fluid is equal to the flow rate, Q, divided by the area, A, we have

$$P_I = \frac{\rho \ell}{A} \frac{dQ}{dt} \tag{14-27}$$

Again, if we define $I = \rho \ell / A$, equation 14-27 becomes

$$P_I = I \frac{dQ}{dt} \tag{14-28}$$

where I is called the fluid inertance and has units of N-s²/m⁵. Note the similarity between equations 14-28 and 14-18. The analogous quantities are voltage (pressure), current (volume flow rate), and inductance (inertance). The kinetic energy of the fluid in the pipe is equal to one half the fluid mass times the velocity squared, where the fluid velocity is Q_I / A.

$$E_I = \tfrac{1}{2} \text{ mass (velocity)}^2$$
$$= \tfrac{1}{2}(\rho A \ell)\left(\frac{Q_I}{A}\right)^2 = \tfrac{1}{2}\left(\rho\frac{\ell}{A}\right)Q_I^2 = \tfrac{1}{2}IQ_I^2 \tag{14-29}$$

The term Q_I is the fluid volume flow rate through the fluid element. The units of fluid inertance are N-s²/m⁵.

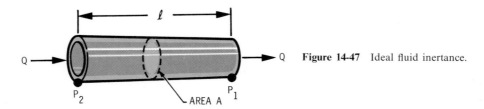

Q →   → Q   **Figure 14-47**   Ideal fluid inertance.

P_2 AREA A P_1

Example The pipe shown in Figure 14-33 has a length of 10 meters and an area of 0.01 m². The density of water is 1000 kg/m³. Find the inertance of the pipe if water is flowing through it.

Solution

$$I = \frac{\rho \ell}{A} = \frac{(1000 \text{ kg/m}^3)(10 \text{ m})}{0.01 \text{ m}^2} = \mathbf{1 \times 10^6 \text{ N-s}^2/\text{m}^5}$$

Summarizing the properties of an ideal fluid inertance:

Elemental equation

$$\frac{dQ_I}{dt} = \frac{1}{I} P_I$$

Energy stored

$$E_I = \tfrac{1}{2} I Q_I^2$$

Conceptual model

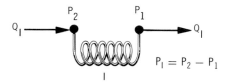

$$P_I = P_2 - P_1$$

Problems

14-43. The pipe shown in Figure 14-43 has a length of 15 m and an area of 0.02 m². If water is the fluid flowing through it, find the inertance of the pipe.

14-44. The inertance of a given section of tubing is known to be 1×10^5 N-s²/m⁵. If the pressure drop across it is 1×10^5 N/m², find the rate of change of volume flow rate through it.

14-45. Experimental measurements have determined that the pressure drop across a section of tubing is 2×10^6 N/m² while the rate of change of volume flow rate through it is 9×10^{-2} m³/s². Find the inertance of the section.

Ideal fluid resistance

There are many different mechanisms by which energy can be dissipated by a fluid in motion. Some of these include friction losses in a long tube (length of tube is much greater than the tube diameter), turbulent flow, flow through a porous medium, and flow through a sharp-edged orifice. We may recall from the previous sections on mechanical and electrical systems that the energy dissipator was defined as an element in which the through and across power variables are related by a constant ($f = bv, e = Ri$). Similarly in a fluid system an ideal linear resistance will be defined as an element in which the fluid flow rate is linearly proportional to the pressure; that is,

$$P_R = RQ_R \geq 0 \tag{14-30}$$

This relationship is illustrated in Figure 14-48.

Most fluid resistors are highly nonlinear in nature and the use of an ideal linear resistor is only valid over a small range of either pressure or flow rate. An example of

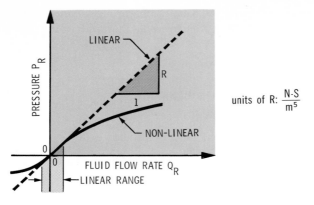

Figure 14-48   Pressure–flow rate relationship for a fluid resistance.

an ideal fluid resistance is the flow through a long tube. It can be shown that the flow resistance of a long tube for an incompressible fluid is (reference 1):

$$R = \frac{128\mu\ell}{\pi d^4} \text{ N-s/m}^5 \qquad (14\text{-}31)$$

where μ = absolute viscosity of the fluid, N-s/m²
      ℓ = length of tube, m
      d = diameter of tube, m

Example   A long thin tube has a length of 30 m and a diameter of 1 cm. The absolute viscosity of the fluid flowing through the tube is $\mu = 0.36$ N-s/m². Find the fluid resistance of the pipe.

Solution

$$R = \frac{128\mu\ell}{\pi d^4} = \frac{128(0.36 \text{ N-s/m}^2)(30 \text{ m})}{3.14(1 \times 10^{-2} \text{ m})^4} = \mathbf{4.4 \times 10^{10} \text{ N-s/m}^5}$$

Summarizing the properties of an ideal fluid resistance:

Elemental equation $p_R = RQ_R$
Power dissipated $\mathbf{P} = P_R Q_R = RQ_R^2 \geq 0$
Conceptual model

$$Q_R \longrightarrow \underset{P_2}{\bullet}\!\!-\!\!\!\wedge\!\!\wedge\!\!\wedge\!\!-\!\!\underset{P_1}{\bullet} \longrightarrow Q_R$$

$$P_R = P_2 - P_1$$

Problems

14-46. A tube has a length of 100 m and a diameter of 0.1 m. The viscosity of the fluid flowing through the tube is $\mu = 0.30$ N-s/m². Find the resistance of the pipe.

14-47. For the pipe of Problem 14-46 it is known that the volume flow rate flowing through it is constant and equal to 4×10^{-1} m³/s. Find the pressure drop across the 100-m section.

14-48. The arteries of the cardiovascular system present a resistance to fluid flow due to the viscosity of the blood. Suppose it has been experimentally determined that during a certain time interval the volume flow rate and pressure drop across a given length of artery are approximately constant and equal to $\bar{Q}$ and $\bar{p}$, respectively. Find the resistance and the power being dissipated.

Ideal sources

In this section we need to define an ideal through variable source (volume flow rate source) and an ideal across variable source (pressure source). An example of a volume flow rate source is a constant displacement pump; examples of pressure sources are the pressure at the bottom of a large reservoir and a constant pressure pump. Figure 14-49 shows the schematic representation of these ideal sources.

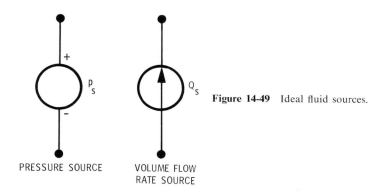

PRESSURE SOURCE VOLUME FLOW RATE SOURCE

Figure 14-49 Ideal fluid sources.

Example: water supply system Figure 14-50 shows a sketch of a city water supply system that is composed of a mountain reservoir, a long section of pipe, a storage tank and a constant displacement pump. Figure 14-51 shows a dynamic model of the water system. In drawing this circuit diagram we have made several important modeling assumptions. First, the reservoir has been modeled as a constant pressure source. This assumption is reasonable if the flow through the system does not appreciably affect the height of the water in the reservoir. Second, the long pipe has been modeled to

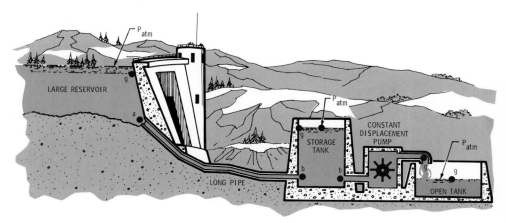

Figure 14-50 Water supply system.

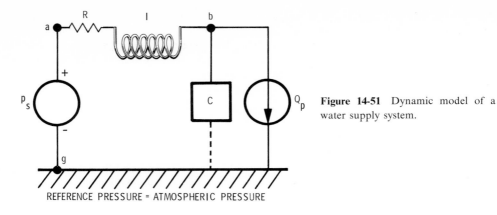

Figure 14-51 Dynamic model of a water supply system.

REFERENCE PRESSURE = ATMOSPHERIC PRESSURE

include both a fluid resistance and a fluid inertance that are connected in series since they have a common through variable (volume flow rate). The fluid capacitance of the storage tank and the constant displacement pump are connected in parallel since they have a common cross variable (pressure difference).

Problems

14-49. Sketch the dynamic model of a fluid system composed of a pressure source that is in series with a pipeline that has both resistance and inertance.

14-50. Draw the dynamic model of the fluid system shown in Figure 14-52.

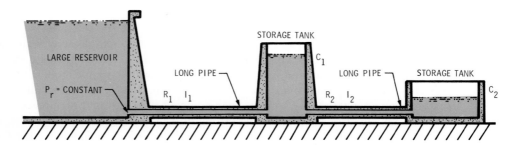

Figure 14-52 Fluid system.

14-51. Draw the dynamic model of the heart system shown in Figure 14-53.

Note The arterial system should be modeled as a single tube that has the properties of resistance, inertance, and capacitance. The heart can be modeled as a volume flow rate source.

14-52. Draw the dynamic model of the drainage control system shown in Figure 14-54.

14-53. It is possible to build a fluidic "analog" computer using fluid elements rather than electrical elements. Such a computer would be considerably slower in response but would be less sensitive to magnetic and electrical fields encountered, for example, by missiles and aircraft. Sketch both a *pictorial* and a dynamic fluid analog model of Problem 14-12.

14-54. Sketch the pictorial and dynamic fluid analog of Problem 14-13.

14-55. Sketch the pictorial and dynamic fluid analog of Problem 14-15.

14-56. Sketch the pictorial and dynamic fluid analog of Problem 14-16.

14-57. Sketch the pictorial and dynamic fluid analog of Problem 14-17.

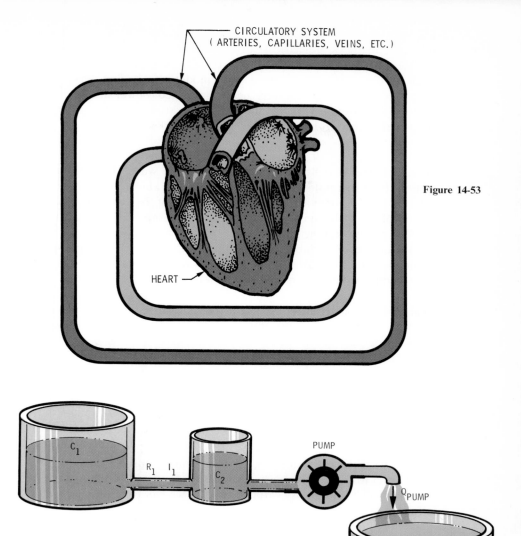

Figure 14-53

HEART

C_1

R_1 I_1

C_2

PUMP

Q_{PUMP}

Figure 14-54

Thermal systems

The final types of systems that will be considered in this chapter are those associated with thermal energy. The physical quantities associated with thermal energy are heat[16] and temperature. However, only simple thermal systems will be considered here, and the more complex interrelations of fluid flow, thermal energy, and mechanical motions will be reserved for study in subsequent courses. In some ways thermal systems are much like the other systems that we have considered. For example, heat flow rate (actually energy flow rate) is related to a temperature

[16] *Heat* is energy transferred from one system to another solely by reason of a temperature difference between the systems.

difference in much the same way that fluid volume flow rate is related to a pressure difference or that an electrical current flow is related to the voltage.

There are some unusual characteristics of thermal systems that make them the "odd-balls" of our systems approach to modeling. The first reason is that, in all the other types of systems that have been considered, power has been expressed as the product of a through variable and an across variable. In thermal systems, however, the power *is the through variable;* thus

$$\mathbf{P} = q \tag{14-32}$$

where q is the heat flow rate and has units of watts. The second reason for the peculiarity of thermal systems is that there is no known means of storing energy by virtue of the through variable, that is, by virtue of the heat flow through it. Thus our arsenal of ideal elements will not include a thermal "inductor," that is, a through variable energy storage element.

Our choice of variables will be the heat flow rate (through variable) and the temperature (across variable). *The units of temperature are in* °C (Celsius scale) or equivalently in K (Kelvin absolute scale).

Ideal thermal capacitance An ideal thermal capacitance is defined as an element which stores energy by virtue of its across variable, temperature. The heat (flow rate), q, is the time rate of change of the thermal energy; that is, $q = dE_T/dt$, where E_T is the thermal energy in joules. Figure 14-55 shows the relationship for an ideal thermal capacitor where T_c represents the temperature drop across the capacitor.

Thus, for an ideal linear thermal capacitance,

$$E_T = C_T T_C \tag{14-33}$$

where C_T is the thermal capacitance joule/K and T_C is the temperature in K of the element with respect to a fixed, constant temperature.

In order to express this *elemental equation* for the ideal thermal capacitor in terms of our through and across variables (q, T), we can differentiate equation 14-33, as follows:

$$\frac{dE_T}{dt} = q_c = C_T \frac{dT_c}{dt} \tag{14-34}$$

Figure 14-55 Thermal energy–temperature difference relationship for an ideal thermal capacitance.

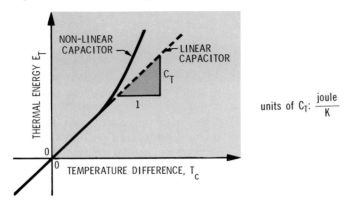

units of C_T: $\dfrac{\text{joule}}{\text{K}}$

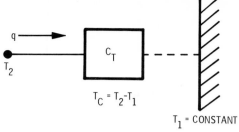

Figure 14-56 Conceptual model of a thermal capacitor.

$T_C = T_2 - T_1$

$T_1 = \text{CONSTANT}$

where q_c is the heat flow through the capacitor and T_c is the temperature of the capacitor with respect to a constant reference temperature. The thermal capacitance, C_T, is the product of the mass of the element and its specific heat,[17] C_p; that is, $C_T = mC_p$. Figure 14-56 is a conceptual model of a thermal capacitor. Our homes are full of thermal capacitors, such as drapes, couches, windows, and metallic objects.

Example The thermal capacitance of a metal block is known to be 1000 joules/K. If the instantaneous heat flow rate through the block is 300 watts, find the rate of change of temperature across it.

Solution

$$\frac{dT_c}{dt} = \frac{1}{C_T} q_c = \frac{300 \text{ joules/s}}{1000 \text{ joules/K}} = \textbf{0.3 K/s}$$

Summarizing the properties of an ideal thermal capacitance:

Elemental equation $\qquad\qquad q_c = C_T \dfrac{dT_c}{dt}$

Energy stored $\qquad\qquad E = C_T T_c$

Conceptual model

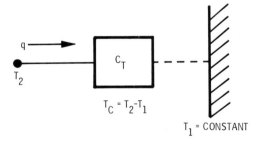

$T_C = T_2 - T_1$

$T_1 = \text{CONSTANT}$

Problems

14-58. A metal block has a constant heat flow rate through it of 500 watts. The thermal capacitance is known to be 500 joules/°K. Find the rate of change of temperature across it.

14-59. A solid object is subjected to a heat flow rate of 60 watts, and a rate of change of

---

[17] The specific heat of a given material is a function of many factors. Reference 4 provides more detailed information.

temperature of 0.05 K/s is measured across it. What is the thermal capacitance of the object?

Ideal thermal resistance The property of a material to oppose the flow of heat (thermal energy) through it is called its thermal resistance. In general there is a temperature drop in the direction of the heat flow through it. Materials that offer little resistance to heat flow are called heat conductors (metals) and materials which offer large resistance are called insulators (fiberglass, wood).

An ideal thermal resistance is defined as one in which the heat flow, q, is linearly proportional to the temperature difference across it. Figure 14-57 shows this linear relationship. Thus the elemental equation for an ideal resistance is

$$q_r = \frac{T_r}{R_t} \tag{14-35}$$

where R_t is the thermal resistance and has units of K/watt.

There are many ways thermal energy can be transferred from one element to another. These basic mechanisms of heat transfer are conduction, convection, and radiation.

Heat conduction occurs through a material on a microscopic level from atom to atom. Figure 14-58 shows an element of length ℓ and area A through which heat conduction is occurring. The relationship between heat flow and temperature difference (reference 1) can be expressed as

$$q = \frac{\sigma A}{\ell}(T_2 - T_1)$$

Thus the thermal resistance is $R_t = \ell/\sigma A$, where σ is the thermal conductivity of the material, which can be found in many textbooks (reference 4).

Heat convection is the transport of energy by movement of a mass after heat has been transferred to the mass, for example, the movement of hot air or water. A very useful concept to help analyze systems in which convection is occurring is that of a heat transfer coefficient, C_h, where

$$C_h = \frac{q}{(T_2 - T_1)A}$$

and has units of watt/m²-K. The heat transfer coefficient for a given arrangement of

Figure 14-57 Heat–temperature differential relationship for an ideal thermal resistor.

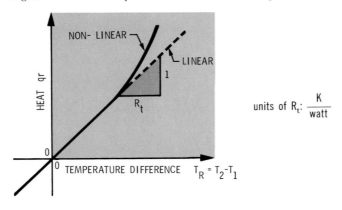

units of R_t: $\dfrac{K}{watt}$

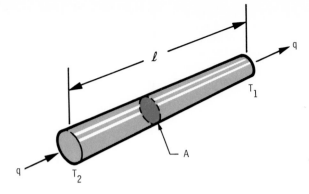

Figure 14-58 Heat conduction through an elemental length.

elements and materials is often determined experimentally. From our definition of thermal resistance we see that $R_t = 1/C_h A$.

Heat radiation is a very nonlinear phenomena; in fact, all thermal energy transfer mechanisms[18] are very nonlinear and complex. However, for *small differences in temperature* they often can be modeled as linear ideal thermal resistors. Often in practice an approximate value is obtained *experimentally* by measuring the heat flow and temperature difference; thus

$$R_t \simeq \frac{\text{temperature difference}}{\text{heat flow rate}}$$

Example   Air at 327 K flows over a plate of area 0.23 m² that is maintained at 333 K. The heat transfer coefficient has been measured experimentally and has been found to be $C_h = 372.5$ watt/K(m²). Find the heat flow rate from the plate.

Solution

$$R_t = \frac{1}{C_h A} = \frac{1}{(372.5)(0.23)} = 0.01167 \; \frac{\text{K}}{\text{watt}}$$

$$q = \frac{T_2 - T_1}{R_t} = \textbf{514 watts}$$

Summarizing the properties of an ideal thermal resistance:

Elemental equation $q_r = \dfrac{T_r}{R_t}$

Conceptual model

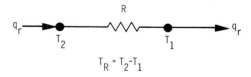

$$T_R = T_2 - T_1$$

Problems

14-60. Air at 400 K flows over a plate of area 0.5 m² that is maintained at 425 K. The heat

[18] For a more complete treatment of thermal systems, see references 1 and 4.

transfer coefficient is known to be $C_h = 400$ watt/K(m²). Find the heat flow rate from the plate.

14-61. It is desired to measure experimentally the thermal resistance of a solid object. The material is subjected to a constant heat flow rate of 100 watts, and a constant temperature difference of 0.1 K is measured across it. What is its thermal resistance?

14-62. A house is maintained at a constant temperature of $21°C = 294$ K while the outside temperature is $0°C = 273$ K. It is known that the furnace is providing a constant heat flow rate of 100 watts. Find the effective thermal resistance of the combined house walls and windows.

Ideal sources

The ideal sources that we will need to complete our arsenal of thermal elements are the ideal heat flow rate source (through source) and the ideal temperature source (across source). A furnace could be modeled as a heat flow rate source, while the atmosphere could be modeled as a temperature source. It must be remembered that ideal sources do not have to be constant but they do have to be independent of the system variables. For example, if we model the atmosphere as a temperature source and a house as the system, we are assuming that the temperature changes of the house will not influence the atmospheric temperature. Schematic representations for the ideal heat flow source and temperature source are shown in Figure 14-59.

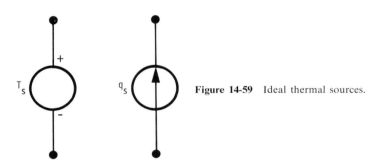

Figure 14-59 Ideal thermal sources.

Example: home heating system   The house shown in Figure 14-60 has been modeled by a thermal capacitance (walls, furniture, draperies, etc.), a thermal resistance (windows, doors), an ideal heat flow source (furnace), and an ideal temperature source (atmosphere). Of course, the walls, furniture, and the like, have some thermal

Figure 14-60 Home heating system.

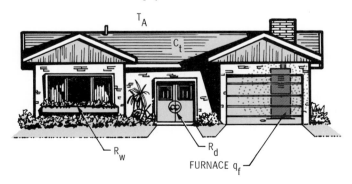

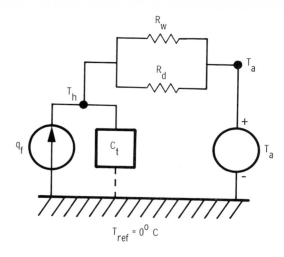

Figure 14-61 Dynamic model of a home heating system.

$$T_{ref} = 0°\ C$$

resistance and the windows and doors have some thermal capacitance. However, it is reasonable to analyze the problem by lumping all the capacitance into certain elements and all the resistance into others.

Figure 14-61 shows the dynamic model of the home heating system using ideal thermal elements. Notice that the furnace and house capacitance are connected in parallel since they have the same across variable (T_h, temperature of the inside of the house), and similarly that the two resistances are connected in parallel since they also have the same across variable ($T_h - T_A$).

Problems

14-63. Sketch the dynamic model of a heat flow rate source that is in series with a thermal resistance.

14-64. Sketch the dynamic model of a temperature source, a resistance, and a capacitance that are in parallel.

14-65. Draw the dynamic model of the thermal system shown in Figure 14-62. The metal has a capacitance, C_t, and is initially at a very high temperature. The surrounding insulating material has a thermal resistance of R, and the atmosphere remains at a constant temperature T_A.

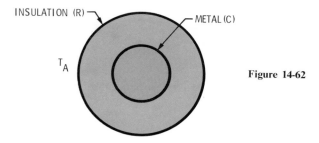

Figure 14-62

14-66. Draw the dynamic model of the thermal system shown in Figure 14-63. The metal ingot has a thermal capacitance, c; the bath solution has a thermal resistance, R_1; the liner has a thermal resistance, R_2; the atmospheric temperature remains constant at T_A.

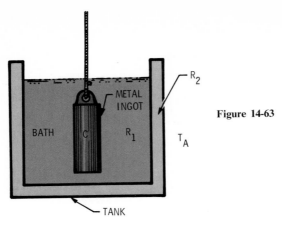

Figure 14-63

14-67. The figure shown in Figure 14-64 is a mechanical system. Sketch the dynamic model of the thermal analog of this system.

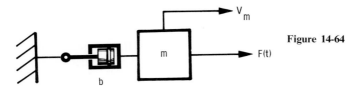

Figure 14-64

14-68. Sketch the dynamic model of the thermal analog of the electrical system shown in Figure 14-65.

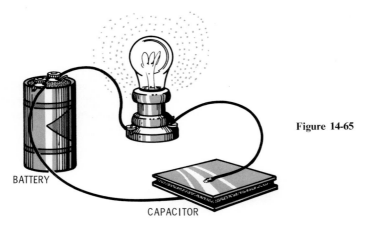

Figure 14-65

14-69. The figure shown in Figure 14-66 is a mechanical spring–mass system. Can a thermal analog of this system be drawn? Why or why not?

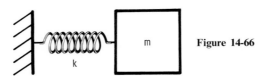

Figure 14-66

System analogies

We have just completed an elementary study of mechanical, electrical, fluid, and thermal systems. In all systems we found that the concept of through and across variables was useful in describing the behavior of ideal elements. In order to emphasize these similarities, we can define a generalized through variable (f) and a generalized across variable (V). In all cases except for thermal systems the power flow can be represented as

$$\mathbf{P} = fV \tag{14-36}$$

It is also useful to define a generalized capacitance (C), a generalized inductance (I), and a generalized resistance (R).

In all the systems we have elements that store energy by virtue of the across variable. This stored energy can be represented by (except thermal)

$$E_A = \tfrac{1}{2}CV^2 \tag{14-37}$$

We also have elements that store energy by virtue of their through variable. This stored energy can be represented by

$$E_T = \tfrac{1}{2}If^2 \tag{14-38}$$

In addition we have elements that dissipate energy at all times. The rate of change of energy (power) dissipated by these elements can be represented by

$$\mathbf{P} = Rf^2 = \frac{1}{R}V^2 \geq 0 \tag{14-39}$$

Finally we have ideal sources that maintain specified values of either the through variable or across variable independent of other system variables.

The various dynamic diagrams that were developed for all the examples can be greatly simplified by using unified schematic representations called linear graphs.[19] For example, the conceptual model for the mass of a mechanical system is shown in Figure 14-67a. The corresponding linear graph is shown in Figure 14-67b. The direction of the arrow in Figure 14-67b shows that the velocity of v_m is positive with respect to the reference velocity, v_G.

Figure 14-68 illustrates how the through, across, and dissipative elements, as well as the ideal sources, can be represented by simplified schematic diagrams consisting of directed line segments.[20] There is an associated through and across variable with each element of Figure 14-68. The through variable flows from point 2 to point 1, and the across variable is measured across the two *terminals*. For example, in Figure 14-68a, $v_C = v_2 - v_1$. These concepts will be illustrated later by a number of examples. The simplified schematic diagrams of Figure 14-68 are called *linear graphs* because they are composed of a simple line with an arrow on it.

Table 14-1 summarizes all the ideal elements that we have covered so far. This table contains an abundant amount of information and points out the analogies that

---

[19] Adopted from reference 1.
[20] In general the linear graph elements are drawn slightly curved so that a number of elements can be easily connected together.

Table 14-1 *Summary of Ideal Elements**

	Mechanical systems	Electrical systems	Fluid systems	Thermal systems	Generalized system†
Across variable energy storage elements					
Conceptual element	Ideal mass	Ideal electrical capacitance	Ideal fluid capacitance	Ideal thermal capacitance	Ideal capacitance
Conceptual model	$V_G = 0$		$P_1 = \text{CONSTANT}$	$T_1 = \text{CONSTANT}$	
Linear graph	$V_G = 0$; $P_1 = \text{CONSTANT}$		$P_1 = \text{CONSTANT}$	$T_1 = \text{CONSTANT}$	
Elemental equation	$\dfrac{dV_m}{dt} = \dfrac{1}{m} f_m$	$\dfrac{de_C}{dt} = \dfrac{1}{C} i_C$	$\dfrac{dP_C}{dt} = \dfrac{1}{C} Q_C$	$\dfrac{dT_C}{dt} = \dfrac{1}{C_t} q_C$	$\dfrac{dV_C}{dt} = \dfrac{1}{C} f_C$
Energy stored	$E_m = \dfrac{1}{2} m V_m^2$	$E_C = \dfrac{1}{2} C e_C^2$	$E_C = \dfrac{1}{2} C P_C^2$	$E_C = C_t T_C$	$E_C = \dfrac{1}{2} C V_C^2$
Through variable energy storage elements					
Conceptual element	Ideal spring	Ideal inductance	Ideal inertance	—	Ideal inductance
Conceptual model				—	
Linear graph				—	

	Ideal damper	Ideal electrical resistance	Ideal fluid resistance	Ideal thermal resistance	Ideal resistance
Elemental equation	$\dfrac{df_s}{dt} = \dfrac{1}{A}V_s = kV_s$	$\dfrac{di_L}{dt} = \dfrac{1}{L}e_L$	$\dfrac{dQ_i}{dt} = \dfrac{1}{I}P_i$	—	$\dfrac{df_i}{dt} = \dfrac{1}{I}V_i$
Energy stored	$E_S = \dfrac{1}{2}Af_s^2 = \dfrac{f_s^2}{2K}$	$E_L = \dfrac{1}{2}Li_L^2$	$E_i = \dfrac{1}{2}IQ_i^2$	—	$E_i = \dfrac{1}{2}If_i^2$
Conceptual element	Ideal damper	Ideal electrical resistance	Ideal fluid resistance	Ideal thermal resistance	Ideal resistance
Conceptual model	(diagram: f_b, V_2, V_1, f_b)	(diagram: i_R, e_2, R, e_1, i_R)	(diagram: Q_R, P_2, R, P_1, Q_R)	(diagram: q_r, T_2, R_t, T_1, q_r)	(diagram: f_R, V_2, R, V_1, f_R)
Linear graph	(b, f_b, V_1, V_2)	(R, i_R, e_1, e_2)	(R, Q_R, P_1, P_2)	(R_t, q_R, T_1, T_2)	(R, f_R, V_1, V_2)
Elemental equation	$f_b = bv_b$	$e_R = Ri_R$	$P_R = RQ_r$	$T_R = R_t q_R$	$V_R = Rf_R$
Power dissipated	$\mathbf{P} = bv_b^2$	$\mathbf{P} = Ri_R^2$	$\mathbf{P} = RQ_R^2$	$\mathbf{P} = q_R$	$\mathbf{P} = Rf_R^2$
Across variable sources — Linear graph	($+$, $-$) V_S	($+$, $-$) e_S	($+$, $-$) P_S	($+$, $-$) T_S	($+$, $-$) V_S
Through variable sources — Linear graph	F_S	i_S	Q_S	q_S	f_S

Energy dissipators

*Adapted from reference 1.
†Excluding thermal systems.

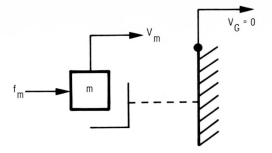

Figure 14-67 Alternative element schematic.

(a) CONCEPTUAL DIAGRAM OF A TRANSLATIONAL MASS

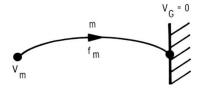

(b) SIMPLIFIED LINEAR GRAPH MODEL OF A TRANSLATIONAL MASS

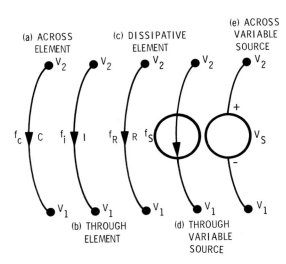

Figure 14-68 Generalized schematic diagrams (linear graphs).

exist between the various types of systems, as well as places all the elemental equations in one place for easy reference. The table also shows that the linear graph representation of the ideal elements is simpler to use than the original conceptual model.

In order to illustrate how these linear graphs can be used to replace the conceptual models, we will formulate several example problems in terms of the linear graphs. Figure 14-69b is a dynamic model using linear graph elements. The arrow on the mass symbol is usually shown pointing down since the velocity of the mass, v, is taken to be positive with respect to $v_g = 0$. The direction of the arrows on the spring is arbitrary and merely defines whether we are calling a tension or compression force as positive. Figure 14-70 is a linear graph version of the automobile suspension (Figure 14-19). Note that the arrows on the masses point toward the reference velocity,

Mechanical example: (Figure 14-15b)

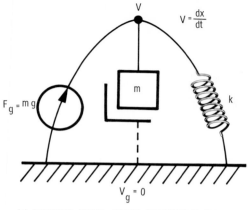

$$V = \frac{dx}{dt}$$

(a) DYNAMIC MODEL USING CONCEPTUAL ELEMENTS

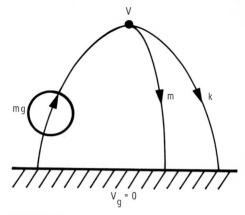

(b) DYNAMIC MODEL USING LINEAR GRAPH ELEMENTS

Figure 14-69

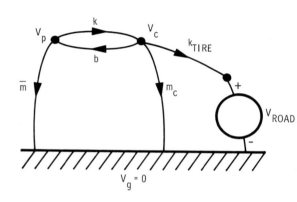

Figure 14-70 Linear graph model of an automobile suspension.

Electrical example: (Figure 14-40)

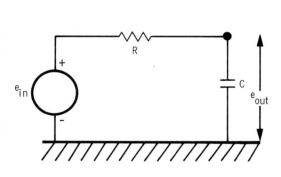

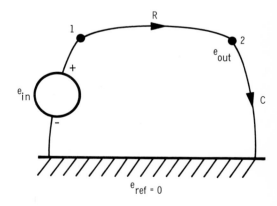

Figure 14-71a Dynamic model of a low-pass filter.

Figure 14-71b Linear graph of a low-pass filter.

Fluid example: (Figure 14-51)

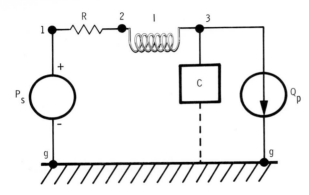

Figure 14-72a Dynamic model of a water supply system.

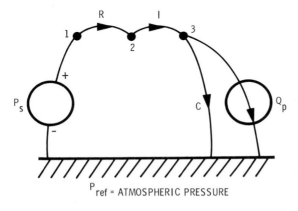

Figure 14-72b Linear graph of a water supply system.

P_{ref} = ATMOSPHERIC PRESSURE

Thermal example: (Figure 14-61)

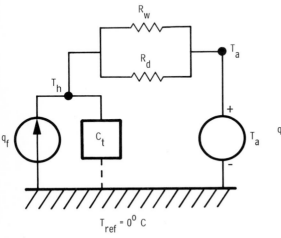

Figure 14-73a Dynamic model of a home heating system.

$T_{ref} = 0^0$ C

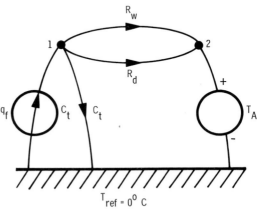

Figure 14-73b Linear graph of a home heating system.

$T_{ref} = 0^0$ C

$v_G = 0$. The directions of the arrows on the two springs and the damper are arbitrary as long as the correct elemental equation is used. For example, the direction of the arrow on the tire spring requires that

$$\frac{df_{\text{tire}}}{dt} = (k_{\text{tire}})(v_c - v_{\text{road}})$$

Figures 14-70 to 14-73 show how the use of linear graphs can simplify the formulation of conceptual models. The real power of the linear graph formulation, however, is that it also simplifies *equation formulation*. Up to this point we have identified a number of ideal elements and have shown how they can be connected together to model physical systems. If we are to use these models to *predict how the system will behave,* we must be able to *formulate the system equations.*

Problems

14-70. Sketch the linear graph of Problem 14-12.
14-71. Sketch the linear graph of Problem 14-13.
14-72. Sketch the linear graph of Problem 14-15.
14-73. Sketch the linear graph of Problem 14-16.
14-74. Sketch the linear graph of Problem 14-17.
14-75. Sketch the linear graph of Problem 14-30.
14-76. Sketch the linear graph of Problem 14-31.
14-77. Sketch the linear graph of Problem 14-50.
14-78. Sketch the linear graph of Problem 14-51.
14-79. Sketch the linear graph of Problem 14-52.
14-80. Sketch the linear graph of Problem 14-65.
14-81. Sketch the linear graph of Problem 14-66.

Equation formulation

Figures 14-69 to 14-73 represent *linear graph* models of practical physical systems and are composed of *ideal elements* that are interconnected. From the previous development and Table 14-1 we know the *elemental equations* for these ideal elements. For instance, in Figure 14-70 the elemental equation for the ideal mass tells us how the force acting through the case mass is related to the velocity across it; that is, $f_{m_c} = m_c \frac{dv_{m_c}}{dt}$. However, a more interesting problem concerning the typical shock absorber (shown in Figure 14-16) is to determine what the vertical velocity (v_p) of the passenger compartment will be when the tire is subjected to a vertical input, v_{road}. Figure 14-70 shows these two variables, *input* (v_{road}) and *output* (v_p). Likewise in Figure 14-71, we might like to be able to determine what the voltage across the capacitor, e_{out}, is when an arbitrary input voltage, e_{in}, is applied.

In order to determine these relationships we need to develop two additional concepts, the concepts of *compatibility* and *continuity*.

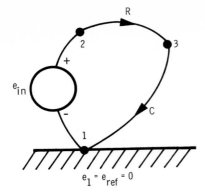

Figure 14-74 Linear graph of a low-pass filter.

$$e_1 = e_{ref} = 0$$

Compatibility

The concept of compatibility is rather simple but very necessary and useful. Simply stated, the principle of compatibility states that the across variable at a specified point equals itself. For example, revising Figure 14-71 as Figure 14-74[21], the voltage of point 2 with respect to point 1 is merely the difference in the voltages; that is,

$$e_{21} = e_2 - e_1 \tag{14-40}$$

where the subscript notation e_{21} means 2 with respect to 1. Similarly,

$$e_{23} = e_2 - e_3 \tag{14-41}$$

and

$$e_{31} = e_3 - e_1 \tag{14-42}$$

We know from our element definitions that[22]

$$e_{21} = e_{in}$$
$$e_{23} = e_R$$
$$e_{31} = e_C$$

If we substitute the expression for e_1 in equation 14-40 into equation 14-42 and the expression for e_2 in equation 14-41 into equation 14-42, we find that

$$-e_{in} + e_R + e_C = 0 \tag{14-43}$$

This expression, although very useful, simply says that $e_1 = e_1$. We could obtain the same expression by starting at any point on the *loop*[23] of Figure 14-70 and sum the across variables until we return to where we originally started. The net change in the across variable will be zero. For example, if we start at *node*[24] *1* and proceed around the loop in a clockwise fashion, summing voltages until we return to node 1, we have

$$-e_{in} + e_R + e_C = 0 \tag{14-44}$$

---

[21] Note here that a single "node" has been formed from the base of the linear graph since there is no element between them. Also the nodes have been renumbered.

[22] Remember that the direction of the arrow implies an across variable *decrease* in that direction. Also the notation e_R means the voltage across the resistor.

[23] A loop is defined as any closed path of a linear graph.

[24] A node is defined as any point where two or more elements are connected.

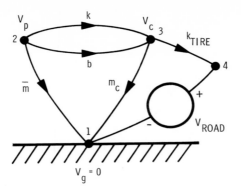

Figure 14-75 Linear graph of an automobile suspension.

Notice that, when we pass through an element in the direction of the arrow, there is a *positive* across variable drop, and, similarly, if we pass through an element with an arrow in the opposite direction, there is a negative across variable drop. Summarizing, we see that the *compatibility* relations are equations which relate the *across* variables around any loop of a linear graph. In electrical systems the compatibility relation is called Kirchhoff's voltage law.

In order to obtain some practice with compatibility, the four examples discussed previously will be treated.

Example: automobile suspension   Figure 14-75 is a repeat of Figure 14-70 with the four nodes numbered. In this problem we have four nodes and a possibility of six simple[25] loops. Figure 14-76 illustrates these simple loops. We can write compatibility equations for all the loops shown in Figure 14-76. Doing this, we obtain

loop (a)	$-v_p + v_k + v_c = 0$
loop (b)	$-v_p + v_b + v_c = 0$
loop (c)	$v_k - v_b = 0$
loop (d)	$-v_p + v_k + v_{k_{\text{tire}}} + v_{\text{road}} = 0$
loop (e)	$-v_p + v_b + v_{k_{\text{tire}}} + v_{\text{road}} = 0$
loop (f)	$-v_c + v_{k_{\text{tire}}} + v_{\text{road}} = 0$

$$(14\text{-}45)$$

By looking at equations 14-45 we see that only three of the six equations are *independent*. For example, by subtracting the second equation from the first we obtain the third, and by subtracting the fifth from the fourth we also obtain the third.

Example: water supply system   Figure 14-72b is the linear graph of the water supply system. There are three simple loops for this system, and writing a compatibility equation for each one yields

$$-P_S + P_R + P_I + P_C = 0$$
$$-P_S + P_R + P_I + P_P = 0$$
$$-P_C + P_P = 0$$

$$(14\text{-}46)$$

The term P_P represents the pressure drop across the pump. Note that, although the flow through the pump is specified, the pressure across it is not. Thus unless we are

[25] A simple loop is a closed path that does not pass through the same node more than once.

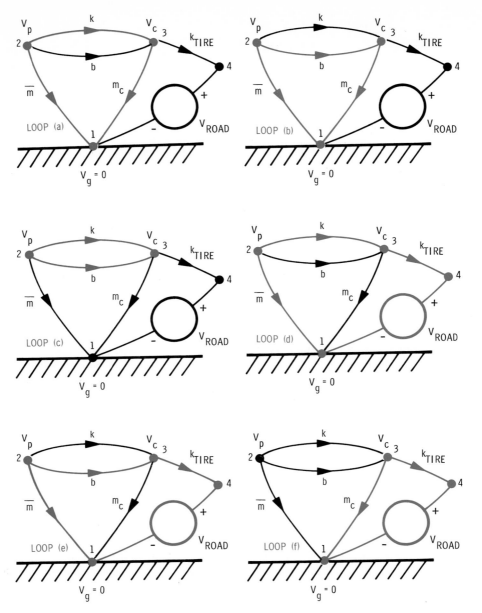

Figure 14-76 Simple loops.

interested in this quantity the last two of equations 14-46 provide no new information and in general only the first equation need be written. Generalizing this we can say that *it is unnecessary to write the compatibility equation for loops that contain a through variable source.* The first equation relates the constant pressure of the reservoir (P_S), the pressure drop across the resistance of the pipe (P_R), the pressure drop across the inertance of the pipe (P_I), and the pressure drop across the storage tank (P_C). Note that the pipe has been modeled as a series combination of a resistance and a fluid inertance; thus the pressure drop across the pipe is the sum of the two pressure drops ($P_R + P_I$).

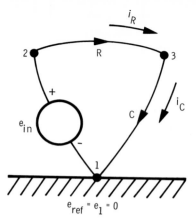

Figure 14-77 Linear graph of an electrical low-pass filter.

Example: home heating system Figure 14-73b is the linear graph of a home heating system. The necessary compatibility relations[26] are

$$-T_{C_t} + T_{R_d} + T_A = 0 \qquad (14\text{-}47)$$
$$T_{R_w} - T_{R_d} = 0 \qquad (14\text{-}48)$$

The first equation states that the temperature of the house ($T_{C_t} = T_h$) is equal to the outside temperature (T_A) plus the temperature drop across the door (T_{R_d}). The second equation simply states that the temperature drop across the window (T_{R_w}) is equal to the temperature drop across the door (T_{R_d}).

Continuity

We have developed the concept of *compatibility,* which relates the *across variables* of the system. We will now develop the concept of *continuity,* which relates the *through variables* of the system. Figure 14-77 shows the linear graph of a low pass filter. Notice at *node 3* of the graph that the resistor current, i_R, is passing *into* the node, while the capacitor current, i_C, is passing *out* of the node. Since we know that electric charge (and consequently the current) cannot be created or destroyed, we find that $i_R = i_C$.[27] The concept of continuity is a statement of *conservation;* it simply says that at every *node* in a linear graph the sum of the *through* variables *into* the node must equal the sum of the *through* variables *out* of the node. For the electrical system, continuity is known as Kirchhoff's current law.

We can apply the continuity principle at every node in the linear graph; however, just as we did not obtain any useful information by applying compatibility around a loop that contained a through variable source, in like manner we do not obtain any useful information by applying the principle of continuity to a node that has an across variable source connected to it. Thus in Figure 14-77, if we applied continuity to node 2, we would find that the current through the voltage source was equal to the current through the resistor. Although this is certainly true, it is the voltage across the source that is known, not the current through it.

[26] The compatibility equations for the loops containing a through variable source have been left out.
[27] The sign convention of the linear graph is that the through variable flows in the direction of the arrow.

Example: automobile suspension We will apply the principle of continuity to nodes 2 and 3 of Figure 14-75. Note that since node 4 has an across variable source (v_{road}) connected to it there is no need to apply continuity to it.

node 2	$-f_{\overline{m}} - f_b - f_k = 0$	(14-49)
node 3	$f_k + f_b - f_{m_c} - f_{k_{\text{tire}}} = 0$	(14-50)

The signs are all negative in equation 14-49 since all the forces at node 2 in Figure 14-75 are leaving the node.

Example: water supply system Applying continuity to nodes 2 and 3 of Figure 14-72b yields

node 2	$Q_R - Q_I = 0$	(14-51)
node 3	$Q_I - Q_C - Q_P = 0$	(14-52)

Equation 14-52 expresses the fact that the net flow into the storage tank (Q_C) is equal to the flow from the pipe (Q_I) minus the flow through the constant displacement pump (Q_P).

Example: home heating system Applying continuity to the only node of Figure 14-73b *that does not contain an across variable* source connection (node 1) yields

node 1	$q_f - q_{R_w} - q_{R_d} - q_{C_t} = 0$	(14-53)

This equation states that the net heat flow rate into the house (q_{C_t}) is equal to the heat flow rate of the furnace (q_F) minus the heat flow rate through the window (q_{R_w}) minus the heat flow rate through the door (q_{R_d}).

Equation formulation

We have now developed all the concepts needed to be able to formulate the equations that describe a system's *dynamic* behavior. We will now develop a unified way to express them. The three required steps are as follows:

1. Write all the *elemental equations.*
2. Write the necessary *compatibility equations.*
3. Write the necessary *continuity equations.*

Example: automobile suspension (Figure 14-75) We have already developed all the pieces to this problem.

1. *Elemental equations*

$$\frac{dv_c}{dt} = \frac{1}{m_c} f_{m_c} \tag{14-54}$$

$$\frac{dv_p}{dt} = \frac{1}{m} f_{\overline{m}} \tag{14-55}$$

$$\frac{df_k}{dt} = kv_k \tag{14-56}$$

$$f_b = bv_b \tag{14-57}$$

$$\frac{df_{k_{\text{tire}}}}{dt} = k_{\text{tire}}v_{k_{\text{tire}}} \tag{14-58}$$

2. *Compatibility equations*

loop (a)	$-v_p + v_k + v_c = 0$	(14-59)
loop (c)	$v_k - v_b = 0$	(14-60)
loop (d)	$-v_p + v_k + v_{k_{\text{tire}}} + v_{\text{road}} = 0$	(14-61)
loop (f)	$-v_c + v_{k_{\text{tire}}} + v_{\text{road}} = 0$	(14-62)

3. *Continuity equations*

node 2	$-f_{\bar{m}} - f_b - f_k = 0$	(14-63)
node 3	$f_k + f_b - f_{m_c} - f_{k_{\text{tire}}} = 0$	(14-64)

Equations 14-54 through 14-64 represent a complete set of *differential equations,* that is, a set of equations in terms of the system variables and their *derivatives.* These equations can be expressed more compactly by manipulating the equations so that they contain only the variables whose derivative appears in the elemental equations and the input sources. From equations 14-54 through 14-58 we see that the variables whose derivatives appear are v_c, v_p, f_k, and $f_{k_{\text{tire}}}$. Thus we can use all the other equations to eliminate v_b, v_k, $v_{k_{\text{tire}}}$, $f_{\bar{m}}$, f_{m_c}, and f_b. By making the appropriate *algebraic* substitutions, we have

$$\frac{dv_c}{dt} = \frac{1}{m_c}[f_k - f_{k_{\text{tire}}} + b(v_p - v_c)] \tag{14-65}$$

$$\frac{dv_p}{dt} = \frac{1}{\bar{m}}[b(v_c - v_p) - f_k] \tag{14-66}$$

$$\frac{df_k}{dt} = k(v_p - v_c) \tag{14-67}$$

$$\frac{df_{k_{\text{tire}}}}{dt} = k_{\text{tire}}(v_c - v_{\text{road}}) \tag{14-68}$$

Equations 14-65 through 14-68 are the final mathematical equations that can be used to predict how the automobile suspension will *behave dynamically* when subjected to an arbitrary road input, v_{road}. The *solution* of these *equations* to a specified forcing function (v_{road}) is beyond the scope of this chapter. However, differential equations of this form can sometimes be solved *analytically,* and they can always be solved using a digital or analog computer.

Example: electrical low pass filter (Figure 14-71b)

1. *Elemental equations*

$$\frac{de_c}{dt} = \frac{1}{C}i_C \tag{14-69}$$

$$e_r = Ri_R \tag{14-70}$$

2. *Compatibility*

$$-e_{\text{in}} + e_R + e_C = 0 \tag{14-71}$$

3. *Continuity*

$$\boxed{\text{node 2}} \qquad i_R - i_c = 0 \qquad \text{(14-72)}$$

As in the case of the shock-absorber equations, we can simplify equations 14-69 and 14-70 by retaining only the differentiated variable (e_C) and the source variable (e_{in}). Applying algebra to reduce the number of equations yields

$$\frac{de_c}{dt} = \frac{1}{RC}(e_{in} - e_C) \qquad \text{(14-73)}$$

Equation 14-73 is the fundamental equation that describes the low-pass filter behavior. Thus, once $e_{in}(t)$[28] is specified, we can solve for $e_C(t)$.

Example: water supply system (Figure 14-72)

1. *Elemental equations*

$$\frac{dQ_I}{dt} = \frac{1}{I}P_I \qquad \text{(14-74)}$$

$$\frac{dP_C}{dt} = \frac{1}{C}Q_C \qquad \text{(14-75)}$$

$$P_R = RQ_R \qquad \text{(14-76)}$$

2. *Compatibility*

$$-P_S + P_R + P_I + P_C = 0 \qquad \text{(14-77)}$$

3. *Continuity*

$$\boxed{\text{node 2}} \qquad Q_R - Q_I = 0 \qquad \text{(14-78)}$$
$$\boxed{\text{node 3}} \qquad Q_I - Q_C - Q_P = 0 \qquad \text{(14-79)}$$

Expressing these equations in terms of the differentiated variables (Q_I, P_C) and the source variables (P_S, Q_P) yields

$$\frac{dQ_I}{dt} = \frac{1}{I}(P_S - RQ_I - P_C) \qquad \text{(14-80)}$$

$$\frac{dP_C}{dt} = \frac{1}{C}(Q_I - Q_P) \qquad \text{(14-81)}$$

Equations 14-80 and 14-81 are the fundamental equations that define how the water supply system will respond to a specified reservoir pressure (P_S) and a specified pump flow (Q_P).

Example: home heating system (Figure 14-73b)

1. *Elemental equations*

$$\frac{dT_{C_t}}{dt} = \frac{1}{C_t}q_{C_t} \qquad \text{(14-82)}$$

[28] The notation $e_{in}(t)$ means that e_{in} is a function of time.

$$T_{R_w} = R_w q_{R_w} \qquad (14\text{-}83)$$

$$T_{R_d} = R_d q_{R_d} \qquad (14\text{-}84)$$

2. *Compatibility*

$$-T_{C_t} + T_{R_d} + T_A = 0 \qquad (14\text{-}85)$$

$$T_{R_w} - T_{R_d} = 0 \qquad (14\text{-}86)$$

3. *Continuity*

$$\boxed{\text{node } 1} \qquad q_f - q_{R_w} - q_{R_d} - q_{C_t} = 0 \qquad (14\text{-}87)$$

Again, we will express these equations in terms of the differentiated variable (T_C) and the source variables (q_f, T_A). The appropriate algebraic manipulations yield

$$\frac{dT_{C_t}}{dt} = \frac{1}{C_t}\left(\frac{1}{R_w} + \frac{1}{R_d}\right)(T_A - T_{C_t}) + \frac{q_f}{C_t} \qquad (14\text{-}88)$$

Equation 14-88 is the fundamental equation for the home heating system. It describes how the temperature of the house ($T_h = T_{C_t}$) will change when subject to a varying atmospheric temperature (T_A) and an adjustable furnace heat flow rate (q_f).

Example: qualitative model for the human heart Figure 14-78 is a schematic representation of the heart and associated circulatory system. Suppose it is desired to investigate in a qualitative way the effect of smoking, hardening of the arteries, and

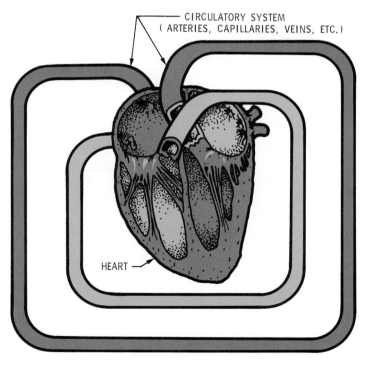

Figure 14-78 Heart–circulatory system.

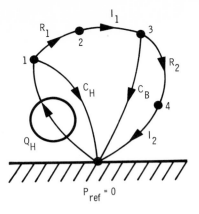

Figure 14-79 Linear graph representation of an idealized model of a heart–artery system.

cholesterol on the blood pressure. We will need to develop an engineering model that is as simple as possible, and yet it will need to incorporate as many of the important characteristics of the circulatory system as possible. We know that primarily the heart acts as a pump; therefore, we will need to include a *volume flow rate source, Q_H.* The heart also is elastic and can store energy as a *fluid capacitor, C_H.* The artery and capillary system can be modeled as small pipes that offer *resistance* to flow and also *inertance* (store energy due to the flow through it). We also know that the blood vessels are elastic and therefore our model should include some *capacitance.*

The modeling of the artery and capillary system requires careful evaluation. A long thin elastic tube appears to possess all three ideal element properties. It can store energy by virtue of a pressure difference; in this case it is the elastic nature of the vessel walls that expand under pressure (see Figure 14-46). Clearly the capacitance of the blood vessel is proportional to its elasticity since the capacitance decreases as the artery stiffens. The blood vessel also offers a resistance to flow, which for long thin tubes is given by equation 14-31; that is, $R = G_1/d^4$, where G_1 is a constant[29] and d is the diameter of the tube. The vessel also can store energy by virtue of the volume flow rate through it, that is, by its inertance. The inertance of a long tube is given by equation 14-28:

$$I = \frac{\rho l}{A} = \frac{\rho l}{\pi d^2/4} = \frac{G_2}{d^2}$$

Now that we have assembled a set of ideal fluid elements that characterize the heart system, we need to connect them together properly. Figure 14-79 shows one possible way that we might piece together the ideal elements. Note that the blood vessel system has been modeled as five lumped elements: a series connection of a resistance and inertance (R_1, I_1), and a capacitance (C_B) connected to the reference pressure, followed by another series connection of a resistance and an inertance (R_2, I_2). Looking at node 1, the blood flow entering the arterial system is equal to the heart flow (Q_H) minus the flow stored by the expanding heart walls; that is, $Q_{R_1} = Q_H - Q_{C_H}$. Looking at node 3, the blood flow returning to the heart (Q_{R_2}) is equal to the blood flow entering the arterial system ($Q_{R_1} = Q_{I_1}$) minus the flow stored by the expanding arterial walls; that is, $Q_{R_2} = Q_{I_1} - Q_{C_B}$. This model, although very simplistic, can yield some interesting insight into how the heart behaves.

[29] For a particular situation the value $128\mu l/\pi$ can be represented by a constant, for example G_1.

To obtain the complete set of equations for the heart system the elemental, compatibility, and continuity equations are written as follows:

Elemental Equations	Compatibility	Continuity
$\dfrac{dP_{C_H}}{dt} = \dfrac{1}{C_H}Q_{C_H}$	$-P_{C_H} + P_{R_1} + P_{I_1} + P_{C_B} = 0$	$Q_H = Q_{C_H} + Q_{R_1}$
$\dfrac{dQ_{I_1}}{dt} = \dfrac{1}{I_1}P_{I_1}$		$Q_{R_1} = Q_{I_1}$
$\dfrac{dP_{C_B}}{dt} = \dfrac{1}{C_B}(Q_{C_B})$	$-P_{C_B} + P_{R_2} + P_{I_2} = 0$	$Q_{I_1} = Q_{C_B} = Q_{R_2}$
$\dfrac{dQ_{I_2}}{dt} = \dfrac{1}{I_2}P_{I_2}$		$Q_{R_2} = Q_{I_2}$
$P_{R_1} = R_1 Q_{R_1}$		
$P_{R_2} = R_2 Q_{R_2}$		

Again by expressing these equations in terms of the variables whose derivative appears in the elemental equations, we obtain

$$\frac{dP_{C_H}}{dt} = \frac{1}{C_H}(Q_H - Q_{I_1}) \tag{14-89}$$

$$\frac{dQ_{I_1}}{dt} = \frac{1}{I_1}(P_{C_H} - P_{C_B} - R_1 Q_{I_1}) \tag{14-90}$$

$$\frac{dP_{C_B}}{dt} = \frac{1}{C_B}(Q_{I_1} - Q_{I_2}) \tag{14-91}$$

$$\frac{dQ_{I_2}}{dt} = \frac{1}{I_2}(P_{C_B} - R_2 Q_{I_2}) \tag{14-92}$$

Applying compatibility around the heart yields $P_{C_H} = P_H$; thus equations 14-89 to 14-92 determine the system output (P_H) as a function of Q_H and the blood vessel diameter, d.

We will make the assumptions that smoking increases the heart flow rate, Q_H, that hardening of the arteries decreases the blood vessel capacitance, C_A, and that cholesterol decreases the blood vessel diameter, d.

A careful analysis of equations 14-89 to 14-92 would indicate to us that increasing Q_H, decreasing C_H, and decreasing d would *increase* the heart pressure. A definitive solution of equations 14-89 to 14-92 is beyond the scope of this chapter. However, it is important that we recognize that a complex system, like the heart, can be modeled by the principles described. It should be emphasized that the heart model developed above is a *very simple* and idealized model, and it should be used only to gain a qualitative insight into how the heart model is affected by certain stipulated parameters.

An example of an analytical solution

In general the differential equations that are formulated for a given system need to be solved by the use of a computer. However, if the model is simple enough the solution often can be found by analytical means. For example, the fundamental

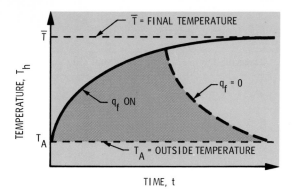

Figure 14-80 Temperature–time history.

equation for the home heating system (equation 14-88) is a *first-order* differential equation. If we assume that the atmospheric temperature and the furnace heat flow rate are constant, then the complete solution is given by

$$T_h = R_{eq}q_f(1 - e^{-t/R_{eq}C_t}) + T_A$$

where $R_{eq} = R_w R_d/(R_w + R_d)$, and is defined as the equivalent resistance of the combined parallel window and door resistances. This solution assumes that the initial temperature of the house is the outside atmospheric temperature (T_A) before the furnace is turned on (q_f). A typical *time history* of the house temperature is shown in Figure 14-80.

The final temperature $(\bar{T})$ that the house will eventually reach can be found by setting $dT_H/dt = 0$ in equation 14-88, or equivalently by setting $t = \infty$ in the temperature solution. This yields

$$\bar{T} = T_A + R_{eq}q_f$$

Therefore, the final temperature is a function of the atmospheric temperature, the equivalent resistance, and the furnace heat flow rate. The speed with which this final temperature is achieved is dependent on the equivalent resistance and the thermal capacitance, C_t.

If at any time the furnace is turned off $(q_f = 0)$, the temperature will begin to decrease, as shown in Figure 14-80.

Suppose it is desired to determine what the effect of decreasing the equivalent resistance (e.g., opening the window slightly) would be on the house temperature. All

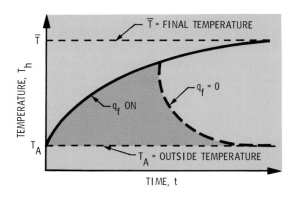

Figure 14-81 Temperature–time history for reduced R_{rq}.

we have to do is to decrease the term R_{eq} in the complete time solution. Figure 14-81 shows a sketch of the resulting temperature time history.

Note that the *final temperature* achieved is lower than before and that the *rate* of increase and decrease of temperature is greater than before. Thus, if for a given furnace we want a high maximum temperature capability and also want the house to dissipate its heat slowly, it is important to have as high an R_{eq} as possible.

Problems

14-82. The fundamental equation for the electrical low pass filter is given by equation 14-73. Suppose $e_{in} = $ constant $= 6$ volts. What will be the *steady-state* value of e_c?

14-83. At $t = 0$ seconds, the voltage across the capacitor in Figure 14-71b is equal to zero volts. Assuming that $e_{in} = 6$ volts, sketch the *time response of the system*. (*Hint:* notice the similarity between this problem and the thermal example.)

14-84. Assuming the same conditions as in Problem 14-83, sketch how the time response of $e_c(t)$ would change if (*a*) RC is increased; (*b*) RC is decreased.

14-85. Using equation 14-88, predict what effect on the final temperature of the house the following parameter changes would have (assuming T_A and q_f are constant): (*a*) an increase in R_d; (*b*) decrease in C_t.

14-86. Using equation 14-88, predict what effect on the *rate of change of house temperature* (dT_h/dt) the following parameter changes would have: (*a*) increase in R_w; (*b*) increase in C_t.

14-87. The fundamental equations for the water supply system are 14-80 and 14-81. Find the steady-state values of Q_I and P_C assuming the values of C, Q_P, P_S, R, and I are known. (*Hint:* Set rate terms equal to zero.)

14-88. What will be the steady-state value of the storage tank pressure (P_C) if the pump flow (Q_P) is equal to zero?

14-89. What would be the effect of increasing the storage tank capacitance (c) on the *rate of change of pressure* (dP_C/dt)?

14-90. What would be the effect on the steady-state tank pressure (P_C) if the pipe diameter were increased?

Summary and conclusions

Several important concepts have been developed in this chapter, as follows:

1. The modeling of physical systems using simple ideal elements.
2. There are analogies between mechanical, electrical, fluid, and thermal systems that can be used to simplify the modeling task.
3. Four basic ideal elements are the following:
    (a) *Generalized capacitance:* elements that store energy by virtue of their *across* variable.
    (b) *Generalized inductance:* elements that store energy by virtue of their *through* variable.
    (c) *General resistance:* elements that dissipate energy.
    (d) *Ideal sources:* elements that maintain through or across variables at values independent of any system variables.

4. By applying the elemental equations, compatibility equations, and continuity equations, a set of differential equations that describe the system's behavior can be formulated.

Problems

In Problems 14-91 through 14-102, (1) sketch the linear graph, (2) write the elemental equations, (3) write the compatibility equations, (4) write the continuity equations, and (5) formulate the *system equations* in terms of only the variables that appear as derivatives in the elemental equations and the input sources.

14-91. Use the system shown in Problem 14-12.
14-92. Use the system shown in Problem 14-13.
14-93. Use the system shown in Problem 14-15.
14-94. Use the system shown in Problem 14-16.
14-95. Use the system shown in Problem 14-17.
14-96. Use the system shown in Problem 14-30.
14-97. Use the system shown in Problem 14-31.
14-98. Use the system shown in Problem 14-50.
14-99. Use the system shown in Problem 14-51.
14-100. Use the system shown in Problem 14-52.
14-101. Use the system shown in Problem 14-65.
14-102. Use the system shown in Problem 14-66.

In the following four problems, (1) formulate a simple dynamic model, (2) sketch the linear graph, (3) find the fundamental equations, and (4) analyze the effect of the basic elements in your model.

14-103. A new vehicle is to be tested for its crashworthiness. Important elements to include in your model are the front bumper (stiffness k_b and damping b_b), the driver (mass m_d), the restraining harness (stiffness k_H), and the vehicle mass (m_v).

14-104. A drainage system has been proposed for a city. The primary components of the system are as follows. A large storage tank (C_1) is located at one part of the city. After a large rainfall the water (Q_1) is channeled into this storage tank. A long underground pipe (I, R) connects this storage tank to a second storage tank (C_2) at the outskirts of the city. A pumping station (Q_{out}) is connected to this second storage tank and pumps the water out of the tank into an open area.

14-105. It is desired to build an electrical circuit that can simulate the behavior of the new bumper described in Problem 14-103. Illustrate how this could be done.

14-106. A home heating system contains the following elements: a furnace (q_f), a radiator (thermal resistance R_r), the room (C_t, R) and the external temperature (T_A).

References

1. J. L. Shearer, A. T. Murphy, and H. H. Richardson. *Introduction to System Dynamics.* Reading, Mass.: Addison-Wesley Publishing Company, Inc., 1967.
2. *M.E. 2.02 Course Notes,* Department of Mechanical Engineering, Massachusetts Institute of Technology, Cambridge, Mass.

3. J. F. Blackburn, G. Reethof, and J. L. Shearer. *Fluid Power Control*. Cambridge, Mass.: The MIT Press, 1960.

4. J. H. Keenan, *Thermodynamics*. New York: John Wiley & Sons, Inc., 1941.

5. R. H. Cannon, *Dynamics of Physical Systems*. New York: McGraw-Hill Book Company, 1967.

Part Five

Introduction to engineering design

Engineering design is exciting.

15

The engineer—
a creative person

As man counts time, the first act of recorded history was one of creation. When God created man, he endowed him with some of this ability to bring new things into being. Today the ability to think creatively is one of the most important assets that all men possess. The accelerated pace of today's technology emphasizes the need for conscious and directed imagination and creative behavior in the engineer's daily routine. However, this idea is not new.

For centuries primitive man fulfilled his natural needs by using the bounty nature placed about him. Since his choice was limited by terrain, climate, and accessibility, he was forced to choose from his environment those things which he could readily adapt to his needs. His only guide—trial and error—was a stern teacher. He ate whatever stimulated his sense of smell and taste, and he clothed himself and his family in whatever crude materials he could fashion to achieve warmth, comfort, and modesty. His mistakes often bore serious consequences, and he eventually learned that his survival depended upon his ability to think and to act in accordance with a plan. He learned the importance of imaginative reasoning in the improvement of his lot.

In recent years archaeologists have discovered evidence of early civilizations that made hunting weapons and agricultural tools, mastered the use of fire, and improvised fishing equipment from materials at hand—all at an advanced level of complexity. These remains are silent reminders of man's ingenuity. Only his cunning and imagination protected him from his natural enemies. The situation is much the same even today, centuries later. Many believe that in this respect man may not have improved his lot substantially over the centuries. The well-being of our civilization still depends upon how successfully we can mobilize our creative manpower. As a profession, engineering must rise to meet this challenge.

In scientific work the term *creativity* is often used interchangeably with *innovation*.

Illustration 15-1
"In the beginning God created the heavens and the earth." (Genesis i:1.)

However, the two are not synonymous although they do have some similarities. Both creativity and innovation refer to certain processes within an individual or system. Innovation is the discovery of a new, novel, or unusual idea or product by the application of logic, experience, or artistry. This would include the recombination of things or ideas already known. Creativity is the origination of a concept in response to a human need—a solution that is both satisfying and innovative. It is reserved for those individuals who originate, make, or cause something to come into existence for the first time or those who originate new principles. Innovation, on the other hand, may or may not respond to a human need, and it may or may not be valuable. In effect, creativity is innovation to meet a need.

Creativity is a human endeavor. It presupposes an understanding of human experience and human values, and it is without doubt one of the highest forms of mental activity. In addition to requiring innovation, creative behavior requires a peculiar insight that is set into action by a vivid but purposeful imagination—seemingly the result of a divine inspiration that some often call a "spark of genius." Indeed, the moment of inspiration is somewhat analogous to an electrical capacitor that has "soaked up" an electrical charge and then discharges it in a single instant. To sustain creative thought over a period of time requires a large reservoir of innovations from which to feed. Creative thought may be expressed in such diverse things as a suspension bridge, a musical composition, a poem, a painting, or a new type of machine or process. Problem solving, as such, does not necessarily require creative thought, because many kinds of problems can be solved by careful, discriminating logic.

A given engineer may or may not be a creative thinker, although all engineers should have mastered the basic techniques of problem solving (see Chapter 13). For

---

Imagination is more important than knowledge.
—Albert Einstein

More today than yesterday and more tomorrow than today, the survival of people and their institutions depends upon innovation.
—Jack Morton, *Innovation,* 1969

The age is running mad after innovation. All the business of the world is to be done in a new way. Men are to be hanged in a new way.
—Samuel Johnson, 1777

Illustration 15-2
Creative behavior requires a peculiar insight that is set into action by a vivid but purposeful imagination; for example, for a public transportation conveyance, what type of seating would be most appropriate?

problem solving, the engineer must be intelligent, well informed, and discerning, so that he can apply the principles of deductive reasoning to the various innovational alternatives when he encounters them.

Every new or original thought may not be a creative thought. A psychotic's hallucinations might well be unique even though they have no intrinsic value. Such thoughts are neither innovational nor creative. Although all creative thinkers must be innovators, it does not necessarily follow that all innovators have to be creative thinkers. Innovation occurs daily, on every hand and in every walk of life. True creative behavior is much rarer and usually requires the fulfillment of some deliberate contemplation.

All persons of normal intelligence possess some ability to think creatively and to engage in imaginative and innovative effort. Unfortunately the vast majority of people are only partially aware of the range of their creative potential. This potential seldom is attained, even if recognized. This is true partially because one's social environment, home life, and education experiences either stimulate or depress the urge to be creative. Even at a very early age, children are often urged to conform to group standards. Any deviations may bring immediate rebuke from the adult in charge. As an example, in the first grade little Johnny may be assigned to color inside the boundaries of his outlined and predrawn horse. He must color his horse

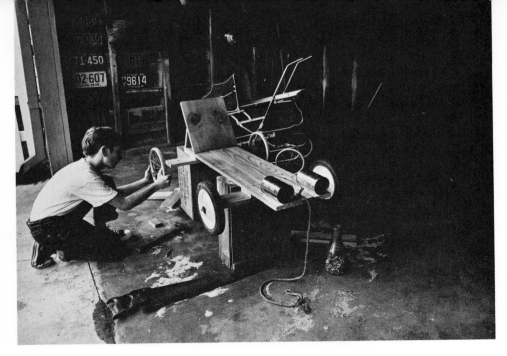

Illustration 15-3
All persons of normal intelligence possess some ability to think creatively and to engage in imaginative and innovative effort.

brown—because the teacher likes brown horses. Black horses, white horses, green horses, or other choices which might occur to Johnny are ruled out. The outline has been predrawn because, in this way, all the children's work will appear to be reasonably good to the parents on PTA night. No child's work will have the obvious appearance of poor quality or extreme excellence—a very important item to please the majority of parents. Also the teacher will not be embarrassed by horses with horns or wings. And so it goes as Johnny grows up to assume his place in adult life. As a teenager he is considered "different" or an "oddball" unless he always joins in with the majority. As a citizen, he is criticized as "anti-American" unless he affiliates with the political party that is in power at the moment. His neighbors "wonder" about him if he refuses to join a neighborhood drive to "achieve the Community Chest goal." Hie coworkers believe he is a "threat to our way of life" if he prefers independent action to letting some union speak in his behalf. And on and on we might continue. . . . It is no wonder that many well-informed persons today are creatively sterile, whereas others in former years (like Franklin and Edison) accomplished seemingly impossible results in spite of a poor formal education.

> Creativity is man's most challenging frontier!
>
> A child is highly creative until he starts to school
> —Stanley Czurles, Director of Art Education, New York State College for
> Teachers

> An inventor is simply a fellow who doesn't take his education too seriously.
> —Charles F. Kettering
>
> Everybody is ignorant, only in different subjects.
> —Will Rogers

Years ago most American youths were accustomed to using innovative and imaginative design to solve their daily problems. Home life was largely one of rural experience. If tools or materials were not available, they quickly improvised some other scheme to accomplish the desired task. Most people literally "lived by their wits." Often it was not convenient, or even possible, to "go to town" to buy a clamp or some other standard device. Innovation was, in many cases, "the only way out." A visit to a typical midwestern farm or western ranch today, or to Peace Corps workers overseas, will show that these innovative and creative processes are still at work. However, today most American youths are city or suburb dwellers who do not have many opportunities to solve real physical problems with novel ideas.

A person is not born with either a creative or noncreative mind, although some are fortunate enough to have exceptionally alert minds that literally feed on new experiences. Intellect is essential, but it is not a golden key to success in creative thinking. Intellectual capacity certainly sets the upper limits of one's innovative and creative ability; but nevertheless, motivation and environmental opportunities determine whether or not a person reaches this limit. Surprisingly, students with high I.Q.'s are not necessarily inclined to be creative. Recent studies reveal the fact that over 70 per cent of the most creative students do not rank in the upper 20 per cent of their class on traditional I.Q. measures.[1]

Everyone has some innovative or creative ability. For the average person, due to inactivity or conformity, this ability has probably been retarded since childhood. If we bind our hand or foot (as was practiced in some parts of the Orient) and do not use it, it soon becomes paralyzed and ineffective. But unlike the hand or foot, which cannot recover full usefulness after long inactivity, the dormant instinct to think creatively may be revived through exercise and stimulated into activity after years of near suspended animation. Thus everyone can benefit from studying the creative and innovative processes and the psychological factors related to them.

Imaginative thinking can be stimulated, and the basic principles of innovative thinking can be mastered. Parnes and Meadow[2] have shown that deliberate education in innovative thinking can significantly increase innovative and creative potential. In reporting this research, Osborn[3] notes that, for an experimental sample of 330 students, the subjects who enrolled in courses in creative problem solving produced 94 per cent more good ideas than subjects who did not get such training. Even if these results are somewhat optimistic, we certainly cannot deny that even a 50 per cent improvement in our own individual creative abilities would be worth achieving. Many organizations—including du Pont, General Electric, Aluminum Company of America, Westinghouse, Aerojet General Corporation, General Motors, and the

[1] E. Paul Torrance, "Explorations in Creative Thinking in the Early School Years," in Calvin W. Taylor and Frank Barron (eds.), *Scientific Creativity,* Wiley, New York (1963), p. 182.
[2] Sidney J. Parnes and Arnold Meadow, "Development of Individual Creative Talent," ibid.
[3] Alex F. Osborn, *Applied Imagination,* Scribner's, New York (1963), p. xii.

Illustration 15-4
Where does the firefly get its light?

Armed Forces—believe the fundamental principles of creative thinking and problem solving can be taught, and give their personnel such training. Therefore, all young engineering students should profit from studying the principles used to spark innovative and creative effort.

> Creativity is the art of taking a fresh look at old knowledge.
>
> All men are born with a very definite potential for creative activity.
> —John E. Arnold

Development of creative effort

Associated with innovative and creative thought are imagination, curiosity, and intuitive insight. As suggested above, the desire to use these faculties begins at an early age. Thwarting or suppressing this individuality of thought may change a child's personality. It is unfortunate that many of our mental resources are wasted in this way. Creative talent should be sought out, developed, and utilized wherever possible. But doing so is far from easy; although psychologists have described some general attributes and traits of the creative personality, it may be difficult to measure an individual's potential to perform creatively.

Although everyone has some capacity to be innovative and creative, "creative ability" is usually a scarce commodity. It need not be, however, because we can enumerate and measure the influence of the mental attitudes and thought processes that are most conducive to producing innovative and creative effort. Using some of these fundamental processes will certainly return valuable dividends. But first, one must have a proper mental attitude.

An attitude for innovative and creative thought

Unfortunately, there is no *one* set of ideal conditions that will always give the most effective imaginative and creative thought. The best conditions vary with personality and circumstances. However, it is important to approach all problems with an open mind—one as free from restrictions and preconceived limiting conditions as possible. Sentiments such as fear, greed, and hatred must be put aside. Try to approach problem situations with a clear mind that has been stimulated *but not restrained* by past experiences. In general, your thought processes are influenced by *how* and *what* you have already learned, but tradition may hinder rather than help, especially if you have made incorrect or irrelevant assumptions. This is particularly true where certain attitudes, convictions, or feelings have stimulated your emotions excessively. In such instances reasoning tends to be influenced so it will harmonize with these convictions. The engineer must learn to be receptive to new ideas, even though they may depart from conventional practices. He must always seek authenticity and truth, rather than trying to verify preconceived ideas or existing procedures.

The engineer must be *motivated* to use imaginative and innovative thought. Basically most creative persons—whether they are artists, musicians, poets, scientists, or engineers—are motivated to work at a particular task partly because of the exhilaration, thrill, special satisfaction, pride, and pleasure they get from completing a creative task. It is perhaps natural that man should emulate his Creator in this

> Reason can answer questions, but imagination has to ask them.

> Important ideas are those that lie within the allowable scope of nature's laws.

respect. For, in each case, after creating the heavens and the earth, after adorning the earth with plant and animal life, and again after creating man and woman, God gave expression of His pleasure.

And God saw everything that He had made, and behold, it was very good.[4]

But besides the sense of satisfaction that comes from the creative process itself, other factors also stimulate and motivate the engineer toward creative design efforts. These may be classified into two groups.

1. *Basic motives:* food and preservation, faith, love, aspiration for fame or freedom.
2. *Secondary motives:* competition, pride, loyalty.

Motivation is the power source that drives all engineers forward in their role as problem solvers, innovators, and creators. Some factors and circumstances will reinforce and stimulate natural motive power. Others will weaken and depress motive power. Engineers should be acquainted with both positive and negative motivating factors.

Conditions that stimulate creative thinking

There are a number of conditions and circumstances that stimulate creative thinking. Some of these are general *conditions of circumstance* that are related to individual *personality* and *philosophy* and are apart from any particular or specific action. Other conditions are related to the individual's *state of mind.* In addition, the engineer must have particular personal qualities and attitudes to achieve maximum motivational stimulation.

The engineer must understand both nature and his environment. He must learn to evaluate carefully the results and consequences of his work. Many times it will be easy for him to draw an incorrect, though seemingly obvious, conclusion. The story is told of a young biologist who was investigating the sensitivity of a frog's sensory system. He devised an experimental apparatus with blinking lights and screeching sirens and positioned his frog for testing. He reasoned that the frog would become frightened by the noise and lights and attempt to escape. Beginning with the right rear leg, he carefully severed each leg in turn, and noted how far the frog could jump. When the frog did not move after its fourth leg was severed, he noted the following in his laboratory report:

> All frogs are very sensitive to light and sound. However, at the moment the left foreleg is removed, they become deaf and blind.

[4] Genesis i:31.

Illustration 15-5
The mind can die from inactivity.

We laugh at the young man's foolish statement, but daily we react in a similar manner as time after time we draw incorrect conclusions.

The engineer will devote a lifetime to changing and modifying his environment. His daily work will affect social and economic life. His actions and designs will be reflected in the lives of all people everywhere as their habits and customs change. He must recognize and assume a special responsibility in this regard because *all* of his innovative designs will probably not be used for the *betterment* of his fellow men and for uplifting their culture. He must recognize that the products of his imagination may, in fact, be used in ways that are detrimental to society. Generally his designs are, within themselves, morally neutral. However, it is the use that people make of his designs that become recognized as forces for "good" or "evil." In addition, he must realize that his failures and his successes, all the fruits of his labor, are always on public display. President Herbert Hoover, himself an engineer, stated these conditions well:

Engineering training deals with the exact sciences. That sort of exactness makes for truth and conscience. It might be good for the world if more men had that sort of mental start in life, even if they did not pursue the profession. But he who would enter these precincts as a life work must have a test taken of his imaginative faculties, for engineering without imagination sinks to a trade.

It is a great profession. There is the fascination of watching a figment of the imagination emerge through the aid of science to a plan on paper. Then it moves to realization in stone or metal or energy. Then it brings jobs and homes to men. Then it elevates the standards of living and adds to the comforts of life. That is the engineer's high privilege.

The great liability of the engineer compared to men of other professions is that his works are out in the open where all can see them. His acts, step by step, are in hard substance. He cannot bury his mistakes in the grave like the doctors. He cannot argue them into thin air or blame the judge like the lawyers. He cannot, like the architect, cover his failures with trees and vines. He cannot, like the politicians, screen his shortcomings by blaming his opponents and hope that the people will forget. The Engineer simply cannot deny that he did it. If his works do not work, he is damned. That is the phantasmagoria that haunts his nights and dogs his days. He comes from the job at the end of a day resolved to calculate it again. He wakes in the morning. All day he shivers at the thought of the bugs which will inevitably appear to jolt its smooth consummation.

On the other hand, unlike the doctor, his is not a life among the weak. Unlike the soldier, destruction is not his purpose. Unlike the lawyer, quarrels are not his daily bread. To the engineer falls the job of clothing the bare bones of science with life, comfort and hope. . . .

The engineer performs many public functions from which he gets only

---

A man must have a certain amount of intelligent ignorance to get anywhere.
—Charles F. Kettering

Illustration 15-6

Engineers are motivated to work at a particular task partly because of the exhilaration, thrill, special satisfaction, pride, and pleasure they get from completing a creative task, and partly because of. . . .

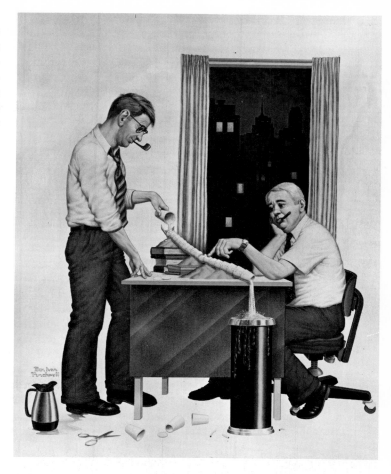

philosophical satisfactions. Most people do not know it, but he is an economic and social force. Every time he discovers a new application of science, thereby creating a new industry, providing new jobs, adding to the standards of living, he also disturbs everything that is. New laws and regulations have to be made and new sorts of wickedness curbed . . . . But the engineer himself looks back at the unending stream of goodness which flows from his successes with satisfactions that few professions may know.[5]

No one, regardless of his profession, is likely to be motivated to creative effort unless he has a strong and undiminishing love for his work. With this love, each day's

[5] Herbert Hoover, *Memoirs of Herbert Hoover, Vol. 1, Years of Adventure,* Macmillan, New York (1951), p. 132.

> Our doubts are traitors and make us lose the good we oft might win by fearing to attempt.
> —William Shakespeare

task becomes more than a means of providing a better standard of living. Each successfully accomplished design provides a special satisfaction that comes only to those who have a strong ambition to succeed. The habit of work will become a part of the individual's personality until even his subconscious mind becomes saturated with the problem. These general conditions and circumstances provide very important climates for creative and imaginative thought.

The proper attitudes or states of mind can also contribute much to creative and imaginative thought. To be most effective, a person should certainly have a healthy body and a clear, intelligent mind, although a high I.Q. or a strong physique by no means guarantees innovative or creative ideas. Psychological freedom, in which the mind is unrestricted by past or present evaluations and judgements, is also very important. In fact, where the "fear of being wrong" has been removed, innovative and creative thought usually increases significantly, for groups of engineers working together as well as for individuals.

Significant and imaginative thought processes are usually rare when the conscious mind becomes fatigued or when there is intense emotion (joy, sorrow, or fear). The relationship between *effective creative behavior* and *physiological stress* might be illustrated (Figure 15-1). Each personality would have its own individual pattern or mathematical expression relating these variables.

The primary curve (A) indicates that, when the mind is without tension and the emotions are at rest, there is considerable possibility that imaginative and creative ideas can emerge. As supporting evidence for this conclusion, many creative people testify that their most novel ideas have appeared when they were engaged in such mild mental activities as bathing, listening to a musical concert, walking on the golf course, or riding the subway. Although this is not the usual case, such a situation might be represented by curve D. The secondary curves, B and C, indicate typical alternative paths that two different individuals might show. Thus there is an ideal condition or emotional zone for each individual; for him it is most conducive to creative thought.

Other mental attributes that contribute positively toward creative thought are (a) an inquiring and questioning mind, (b) abilities to concentrate and communicate, (c) ability to accept conflict and tension without becoming frustrated, and (d) willingness to consider a new idea *even though it may seem to be in conflict* with previous experience.

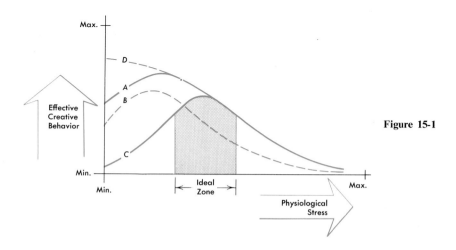

Figure 15-1

In addition, there are personal qualities which are frequently associated with creative individuals. Developing them will enhance the likelihood that the individual will express himself creatively. They are as follows:

1. Intellectual curiosity.
2. Acute powers of observation.
3. Sensitivity to recognize that a problem exists.
4. Directed imagination.
5. Initiative.
6. Originality.
7. Memory.
8. Ability to analyze and synthesize.
9. Intellectual integrity.
10. Ability to think in analogies and images.
11. Intuition.
12. Being articulate in verbal response and alert in mental processes.
13. Patience, determination, and persistence.
14. Understanding of the creative process.

Conditions that depress creative thinking

Just as certain conditions stimulate creative thinking, certain conditions also depress creative thinking and creative behavior. Thus, although the engineer may have high creative potential and intellectual ability to analyze, synthesize, and evaluate, he still may not be creative and innovative. These "road blocks to creative behavior" may be classified into three categories: (*a*) barriers resulting from experience and perception, (*b*) emotional barriers, and (*c*) social and cultural barriers. Each of these will be considered briefly here.

Barriers to creative behavior resulting from experience and perception

A recent experiment vividly illustrates the limitations that can be imposed by habit. This experiment involved a problem-solving situation where two groups were asked to extract a Ping-Pong ball from the bottom of a long, small-diameter pipe standing vertically. When the members of the first group entered the experimental room, they saw assorted objects, including a screwdriver, pliers, string, thumbtacks, and a bucket of dirty water. None of the tools seemed useful, but after some time, about half of the group realized that the Ping-Pong ball could be recovered by pouring water into the pipe until the ball floated to the top.

> Observation, not old age, brings wisdom.
>
> Behold the turtle, he makes progress only when his neck is out.
> —James B. Conant, President, Harvard University

A second group attacked the same problem. The small articles were displayed again. In this case, however, the container of water was missing. In its place was a dinner table which had been set with china and silverware. On the table were a large pitcher of milk, and a bucket of ice cubes. No one was able to solve the problem because the subjects could not relate the liquid (milk) or "solid water" (ice) used for dining to the totally different mechanical problem.

This experiment illustrates the danger of blind reliance on *restricted experience* in problem solving. In some instances, one may assume artificial restrictions that limit and bind his thought processes. As an example, consider the puzzle of the trees and the cows (Figure 15-2).

```
x   x   x   x

x   O   O   x        Figure 15-2

O   x   O   x

O   O   x   x
```

Problem   Six cows (shown as circles) are standing in a grove of trees (shown as crosses). Draw three straight *connected* lines *to join* all the trees without touching any cows.

Nothing in the problem statement implies that the lines represent fences or that the three lines must be restricted to the boundaries of the imaginary rectangular plot containing the cows and trees. However, most people automatically restrict themselves within this field and, under these artificial conditions, the problem becomes impossible to solve (Figure 15-3). This puzzle also illustrates the point that, in many instances, workable solutions to a problem are suggested by someone with minimal technical background related *directly* to the problem, but who has a broad fundamental understanding of the principles governing the situation.

Strange as it may seem, it is nevertheless true that the more original and novel an idea is the more vulnerable it is to criticism. Often the people most apt to prejudge a situation and allow past experiences to strangle a new idea are the ones whose analytical abilities have carried them to prior success. Certainly such skills *are essential* for minimum accomplishment in engineering. However, it sometimes seems easier to rely upon a previously successful mathematical model than to consider the problem anew. It may well be that the original conditions have changed. We are all familiar with this tendency to "overconfidence" which sometimes overcomes those who excel in their field. It is particularly evident in athletics, where a less-able person may be the eventual victor because he never recognizes that he is supposed to suffer defeat. The moral: Never prejudge; consider each situation on its own merits.

We are told that a man dying of thirst has little difficulty in seeing a mirage of a lake off in the distance. The image he *is* seeing seems to be affected in large measure by what his mind tells him that he *needs* to see. He is expectant and thirsty; therefore, it is easy for him to see the lake of water. In a sense, then, believing is seeing. The mind's recall of events and experiences also tends to influence one's observation, discernment, and judgment. For example, look at the two arrangements of straight lines (Figures 15-4 and 15-5).

In Figure 15-4 do the two vertical lines appear equal in length? Probably not. Are the two lines in the Figure 15-5 parallel to each other? "No," you say. Most people will think that these two simple questions are strange indeed. The answers appear obvious in both cases. However, if we tell you that the first is actually a picture of two telephone poles standing in an abandoned field, and that the second is a picture of a railroad track receding into the distance, you might quickly change your answers to "yes" or "maybe." In addition, the brief verbal descriptions that were added have given each picture a quality of depth that you did not recognize originally, demonstrating that we must sharpen our powers of observation, be alert to alternative explanations, and avoid the pitfalls of prejudgment and presumption that so often stifle our thought processes.

Another example might show how prior experience can artificially limit our thinking. Once two medical doctors were riding down the street together when they observed a "head-on" collision ahead. Upon arriving at the scene, they ran to the wrecked automobiles to render aid. After looking into one vehicle, the first physician moaned, "My wife and child!" Hearing this exclamation, the second physician pulled out a gun and killed the first physician. What is your analysis of the motive for this murder?

Writers of novels are skillful in maneuvering fiction plots so that the reader makes an invalid assumption. Perhaps you did this in the above example. Did you assume that both physicians were men? If both physicians were men, the story seems confused and no plausible explanation appears possible. However, the familiar triangular plot of secret love and consequential murder quickly unfolds when you realize that the second physician is a young woman.

Other barriers resulting from experience and perception are the following:

1. Limited scope of basic knowledge.
2. Failure to recognize all the conditions relating to the problem—failure to get all the facts.
3. Preconception and reliance upon the history of other events.

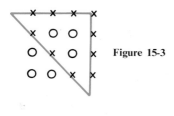

Figure 15-3

Figure 15-4

Figure 15-5

4. Failure to investigate both the obvious and the trivial.
5. Artificial restriction of the problem.
6. Failure to recognize the *real* problem.
7. Inclusion of extraneous environmental factors.
8. Failure to distinguish between cause and effect.
9. Inability to manipulate the abstract.

Emotional barriers

The graph of effective creative behavior versus physiological stress (Figure 16-1) illustrates that everyone's creative behavior diminishes to insignificance under high emotional stress. When under emotional strain, one is likely to narrow his field of observation, to make "snap judgments" that are not well thought out, and thus to disregard alternative and more valuable solutions. Overmotivated people are also likely to choose unrealistic and overambitious objectives. Emotional constraints are perhaps more damaging than other types because they can have such lasting influence upon one's personality. The emotional constraint most difficult to cope with is fear—fear of failure, of criticism, of ridicule, of embarrassment, or of loss of employ-

Illustration 15-7

Illustration 15-8

ment. The fear of social disapproval can stifle initiative and reduce the flow of imaginative idea. Controlled psychological experiments have shown that groups produce up to 70 per cent more innovative and novel ideas when group members do not *judge and evaluate each other's ideas* until later, thus largely removing the fears of ridicule and criticism. Brainstorming, a technique developed in the advertising business to produce more imaginative ideas, is one type of group effort that receives its stimulus by deferring judgment. Ways to implement this technique will be discussed later.

An inferiority complex, resistance to change from the status quo, and a lack of reward stimulus can also be barriers to creative and innovative thought.

Social and cultural barriers

The history of civilization is essentially the record of man's creative behavior or lack thereof. Ancient cultures rose to great heights in Egypt (2700–1800 B.C.), Greece (600–300 B.C.), and Rome (400 B.C.–A.D. 400), but these civilizations eventually fell because of laxity of purpose, moral decay, and the people's overall lack of initiative. These conditions frequently arise when complacency, comfort, and luxury become primary objectives. When one must live or die by his wits, so to speak, his mind is stimulated to function more clearly than it would in a sheltered society. Younger generations who *inherit* the advantages of prosperity generally neither know nor

> There is no experiment to which a man will not resort to avoid the real labor of thinking.
> —Sir Joshua Reynolds

> Daring ideas are like chessmen moved forward; they may be beaten, but they may start a winning game.
> —Goethe
>
> The probability is that tomorrow will not be an extrapolation of today.
> —Ernest C. Arbuckle

appreciate the discipline of work. All these conditions reduce the motivation for creative thought.

Today, America faces a challenge much like those other cultures have faced in ages past. The physical frontiers that inspired our pioneer forefathers are fast disappearing. Fortunately, however, there are new frontiers, such as proper use and reuse of natural resources, ecological balance, outer space, ocean exploration, improved human relations and communications, disease eradication, new food and power sources, and waste and pollution elimination, which challenge our best and our maximum effort. These twentieth-century goals are, in many ways, more challenging than the frontier-day obstacles of a few hundred years ago. Even more significant, perhaps, is the fact that these new frontiers cannot be conquered successfully by applying known procedures or processes routinely. Individual initiative and motivation must continue to be the keys that unlock new ideas and stimulate creative thought. However, as with other cultures, intellectual decay must inevitably result if we desire security too strongly, choose undirected leisure instead of work, or deviate from our fundamental ideals.

Man is a social being and, as such, he needs the companionship of other men. His emotions, habits, and thoughts are strongly affected by the cultural influences that surround him. At an early age, he learns that his associates disapprove of some of his actions and reward others with accolades and commendation. Such rewards may motivate him to make supreme efforts to gain recognition, but condemnation may make him afraid of deviating from his comrades' "group opinion" and thus stifle his creative and imaginative thought.

Overconformity to a group seems especially unfortunate since a group or a committee as a discrete entity cannot, as such, produce creative thoughts. Creative thoughts come *only* from the minds of individuals. However, a committee or group can very definitely possess a unique personality that has strengths, weaknesses, and abilities—just as is the case for an individual. This fact does not discredit the accomplishments of teams, where the team members stimulate each other to produce novel and imaginative ideas. Not uncommonly, one team member's inspired idea will set off a chain reaction of ideas from other team members, whose subconscious memories have been awakened and stimulated into action. Team action is particularly effective in producing a large *volume* of ideas or getting a moderate course of action based on a *consensus*. Remember that hundreds of thousands of statues have been erected around the world . . . all to honor individuals. But so far, not one has been raised to honor a committee. It has been said, perhaps too harshly, that a

Illustration 15-9
"Nothing can withstand the power of the mind. Barriers, enormous masses of matter, the remotest recesses are conquered; all things succumb; the very heaven itself is laid open" (*Marcus Manilus, ca. 40* B.C.).

committee never accomplishes anything unless it has three members, one of whom is always absent and one of whom is always ill. Winston Churchill is said to have once remarked that a committee was the organized result of a group of the incompetent who have been appointed by the uninformed to accomplish the unnecessary. Although his statement brings a smile to the lips of anyone who has served on very many committees, we must recognize that Churchill's own life showed there is no substitute for bold, imaginative, individual thought.

Cultural restraints may be intangible, but they are very real. For example, someone assigned to reduce hunger in India might logically begin by looking at the availability of edible and nourishing foodstuffs in India. He would soon discover that India has a higher ratio of cows to people than any other country. Many of these cows could be slaughtered to provide enough bouillon, or clear meat broth, to sustain millions of people. Yet Indian culture, reinforced by the country's predominant religion, considers the cow a sacred animal that must not be harmed—certainly not killed and made into steaks and bouillon cubes.

In modern society most people are reluctant to accept change. Generally they are either indifferent or negative to proposed ideas. This is why creative people like Leonardo da Vinci, Copernicus, Galileo, and Mozart never lived to see mankind accept the products of their imaginations. Modern civilizations have frequently been no more charitable to those who dared challenge contemporary mores. For example, John Kay was assaulted by weavers who feared his flying shuttle would destroy their means of livelihood; farmers scoffed at Charles Newbold's iron plow and insisted it would contaminate the soil; and the medical profession censured Dr. Horace Wells for using "gas" when extracting teeth.[6]

In more recent times, when the motel was first proposed as a new concept in innkeeping, the idea was greeted with scorn by leading hotel executives. However, the test of time has shown the immense value of the idea. Because of this built-in resistance to change, many new developments must necessarily originate outside the specialized area of endeavor.

Cultural blocks to creative behavior are not always as obvious as in these situations. For example, few of us would doubt the validity of a statement if we read it in a school textbook or in the daily newspaper. Under other circumstances, however, the same people might greet the same statement with considerable debate, for example, if it appears to be the casual observation of a friend or associate with equal or lower social or intellectual standing. Yesterday everyone admired the young person who showed initiative in thinking for himself. Unfortunately, today we may not. Too often teachers, parents, and friends value the young person's ability to adapt himself to associates' dictates and his willingness to think and act in accordance with crowd sentiments above everything else. These social constraints tend to stifle and suppress our desire and ability to think independently and imaginatively and to behave creatively.

All these constraints therefore are detrimental to creative thought processes.

[6] Alex F. Osborn, *Applied Imagination,* Scribner's, New York (1963), p. 54.

We do not have to teach people to be creative; we just have to quit interfering with their being creative.
—Ross L. Mooney

BEWARE! Don't become victimized by habit.

The stimulation of ideas

The engineer, as a professional person, must have keen analytical skill and the ability to synthesize. Without it he would be as handicapped as a boat without a rudder. His education must, necessarily, concentrate on this important part of the engineer's development. Both the engineer and his client must have confidence that his design calculations are both pertinent and accurate. However, an engineer who cannot produce a continuous flow of imaginative ideas is analogous to a boat without an engine. On the one hand, he may wander aimlessly, stumbling over his errors. But on the other hand, he may never get started at all. Therefore, engineering education must consider procedures for stimulating ideas. Certain of these procedures will work satisfactorily in one situation, yet at other times, different methods may be needed.

It is highly desirable for the engineer to maintain the proper mental attitude toward the problem under study. High emotional stress, preconceived ideas based upon habit, or overemphasis on some assumed evaluation of an idea's ultimate value—all these are particularly damaging in the initial stages of idea development. Freedom of thought is essential. If it is restricted, intentionally or not, it makes little difference, the results are the same—reduced imaginative effort. Ancient history reveals that, before 3000 B.C. societies allowed the individual considerable freedom; artisans worked to enhance their own well-being rather than to expand the dominion of some ruler, king, or god-king. Inventions, like the ax, the wheel, the plow, sailboats, writing, irrigation, the arch, pottery, spinning, and metallurgy, are all examples of new ideas that appeared in this *free* environment. By about 1000 B.C. artisans found themselves working primarily to enhance the power of the ruler or king. Under these conditions, they produced considerably fewer new ideas.

Customs may also block imaginative thought. For example, archaeologists now believe that the pyramid-shaped structures built to protect Egyptian tombs were originally oblong buildings with sun-dried brick walls. Since rain deteriorated these walls, the Egyptians learned to slope the walls inward at the top, thus improving drainage and increasing durability. This custom persisted many years later, long after stone had replaced the primitive clay bricks. Even though the outside walls no longer needed to be sloped for protection against erosion, *custom* dictated that they should slope inward at the top.

Other examples have emphasized that the people who have imaginative ideas are the ones who "see with their minds" as well as their eyes. Many times we will think of an idea that seems particularly exciting and innovative. When such a thought occurs, the substance of the idea should be recorded immediately so that it will not be lost. The engineer always should have a small notebook or card that can be carried easily in a shirt or coat pocket and a pencil or pen. He must be continually sensitive to impressions and to their significance. It is said that Galileo was walking about a

---

If you want to kill an idea, assign it to a committee for study.

. . . every idea is the product of a single brain.
—Bishop Richard Cumberland

Society is never prepared to receive any invention. Every new thing is resisted, and it takes years for the inventor to get people to listen to him and years more before it can be introduced.
—Thomas Alva Edison

Illustration 15-10
See with your mind as well as with your eyes.

cathedral one day when he noticed a large lamp swinging from side to side. From this observation he conceived his idea of the pendulum. There are similar possibilities for imaginative thought today, perhaps even more than in the ancient past.

Chapter 17 describes in some detail the several methods that are used in industry to deliberately stimulate ideas.

General principles

The engineer who masters the fundamental principles of mathematics and science is able to understand the laws of nature. If this were the total requirement, the task of the engineer would be simplified. However, he never operates in a free environment where he is limited only by the laws of nature. The engineer always must endeavor to bridge between the "desires of man" and the "realities of nature." He must work both with nature and with people. Because of these practical considerations he is limited by artificial or man-made restrictions such as time, money, or personal preference. These restrictions necessitate compromises on the part of the engineer. Such is the nature of the real world, and the engineer must live and make his livelihood in it.

The engineer may be able to produce a novel solution that is seemingly desirable and economically justified, but it does not follow that his fellowman will always accept or implement it. People of all civilizations have resisted change; today's world is no exception. Although he will not suffer being thrown into jail, being whipped, shot, hanged, or burned at the stake as he might have been years ago, the engineer with a radical idea may find that he is ignored, demoted, transferred to another part of the company, or even fired. Such is life in the real world.

It is important for the young engineer to recognize the importance of being able to sell his idea. Some suggestions to keep in mind are the following:

1. People resist change. The status quo is comfortable and familiar. Any alterations or modifications to existing patterns must be "sold" to those who have the authority to approve decisions of change.

2. Never belittle a current practice or procedure in order to enhance the position of your own idea. Remember that your superior may have been responsible for implementing the technique that is now in use. Give him an opportunity to help you refine any improvement. If the idea is successful, there will be honor for all.

3. Present your design in a professional manner. Do not use sloppy sketches and poorly prepared commentary. Rather, take pride in your work. *Remember that its worth may be judged solely upon its clarity and appearance.*

4. Be prepared for all types of criticism. Try to think up as many reasons as you can why your idea *should not* be adopted. Prepare an answer for each objection.

5. Do not boast. It is better to minimize the overall effect of your idea and let others sell its virtue as a major contribution.

6. Do not become discouraged if you fail to sell your idea immediately. Time frequently acts as ointment to injured pride.

Since God created man in His own image, it is only natural for man to express himself in creative ways. The history of civilization is a history of man's creative efforts through the centuries. Man alone possesses the capacity for creative thought, and everyone has some capability for creative thinking. Remember that the real

Illustration 15-11
You oaf! You misread the scale again. I wanted a toy for my son. Now what could we ever do with a wooden horse this big?

> The mind is not a vessel to be filled but a fire to be kindled.
> —Plutarch
>
> It takes courage to be creative. Just as soon as you have a new idea, you are a minority of one.
> —E. Paul Torrance
>
> Disciplined thinking focuses inspiration rather than constricts it.
>
> It is better to wear out than to rust out.
> —Bishop Richard Cumberland

world is not always predictable, and that the art of compromise is in many cases the difference between success and failure. Remember also that creative behavior is a function of the individual personality rather than of organization, luck, or happenstance. For this reason, it is important to understand the characteristics of the creative person and to develop the attributes basic to imaginative and creative thought.

Exercises in creative thinking

15-1. How can engineering help solve some of the major world problems?

15-2. Discuss some of the inventions that have contributed to the success of man's first lunar exploration.

15-3. Write a paragraph entitled "Fiction Today, Engineering Tomorrow."

15-4. Propose a method and describe the general features of a value system whereby we could replace the use of money.

15-5. List five problems that might now confront the city officials of your home town. Propose at least three solutions for each of these problems.

15-6. Cut out five humorous cartoons from magazines. Recaption each cartoon such that the story told is completely changed. Attach a typed copy of your own caption underneath the original caption for each cartoon.

15-7. Propose a title and theme for five new television programs.

15-8. The following series of five words are related such that each word has a meaningful association with the word adjacent to it. Supply the missing words.

Example	girl	*blond*	*hair*	*oil*	rich
a. astronaut		_____	_____	_____	engineer
b. pollution		_____	_____	_____	automobile
c. college		_____	_____	_____	textbook
d. football		_____	_____	_____	radio
e. food		_____	_____	_____	energy

15-9. Suggest several "highly desirable" alterations that would encourage personal travel by rail.

15-10. What are five ways in which you might accumulate a crowd of 100 people at the corner of Main Street and Central at 6 A.M. on Saturday?

15-11. "As inevitable as night after day"—using the word "inevitable," contrive six similar figures of speech. "As inevitable as . . . ."

15-12. Name five waste products, and suggest ways in which these products may be reclaimed for useful purposes.

15-13. Recall the last time that you lost your temper. Describe those things accomplished and those things lost by this display of emotion. Develop a strategy to regain that which was lost.

15-14. You have just been named president of the college or university that you now attend. List your first ten official actions.

15-15. Describe the best original idea that you have ever had. Why has it (not) been adopted?

15-16. Discuss an idea that has been accepted within the past ten years but which originally was ridiculed.

15-17. Describe some design that you believe defies improvement.

15-18. Describe how one of the following might be used to start a fire: (*a*) scout knife, (*b*) baseball, (*c*) pocket watch, (*d*) turnip, (*e*) light bulb.

15-19. At night you can hear a mouse gnawing wood inside your bedroom wall. Noise does not seem to encourage him to leave. Describe how you will get rid of him.

15-20. Write a jingle using each of these words: cow, scholar, lass, nimble.

15-21. You are interviewing young engineering graduates to work on a project under your direction. What three questions would you ask each one in order to evaluate his creative ability?

15-22. Describe the most annoying habit of your girlfriend (boyfriend). Suggest three ways in which you might tactfully get this person to alter that habit for the better.

15-23. Suggest five designs that are direct results of ideas that have been stimulated by each of the five senses.

15-24. "A man's mother is his misfortune; his wife is his own fault." *The London Spectator.* Write three similar epigrams on boy–girl relations.

15-25. Put a blob of ink on a piece of paper and quickly press another piece of paper against it. Allow it to dry, and then write a paragraph describing "what you see in the resulting smear."

16

The phases of design

Much of the history of man has been influenced by developments in engineering, science, and technology. When progress in these fields was impeded, the culture of the era tended to stagnate and decline; the converse was also true. Although many definitions have been given of "engineering," it is generally agreed that *the basic purpose of the engineering profession is to develop technical devices, services, and systems for the use and benefit of man*. The engineer's design is, in a sense, a bridge across the unknown between the resources available and the needs of mankind (Figure 16-1).

Regardless of his field of specialization or the complexity of the problem, the method by which the engineer does his work is known as the *engineering design process*. This process is a creative and iterative approach to problem solving. It is creative because it brings into being new ideas and combinations of ideas that did not exist before. It is iterative because it brings into play the cyclic process of problem solving, applied over and over again as the scope of a problem becomes more completely defined and better understood.

> . . . the process of design, the process of inventing physical things which display new physical order, organization, form, in response to function.
> —Christopher Alexander, *Notes on the Synthesis of Form*
>
> A scientist can discover a new star but he cannot make one. He would have to ask an engineer to do it for him.
> —Gordon L. Glegg

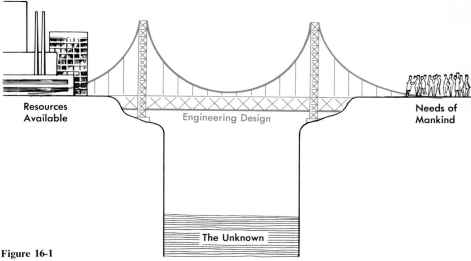

Resources
Available

Engineering Design

Needs of
Mankind

The Unknown

Figure 16-1

Thus a design engineer must be a creative person—an idea man—and he must be able to try one idea after another without becoming discouraged. In general he learns more from his failures than from his successes, and his final designs usually will be compromises and departures from the "ideal" that he would like to achieve.

A final engineering design usually is the product of the inspired and organized efforts of more than one person. The personalities of good designers vary, but certain characteristics are strikingly similar. Among these will be the following:

1. Technical competence.
2. Understanding of nature.
3. Empathy for the requirements of his fellowmen.
4. Active curiosity.
5. Ability to observe with discernment.
6. Initiative.
7. Motivation to design for the pleasure of accomplishment.
8. Confidence.
9. Integrity.
10. Willingness to take a calculated risk and to assume responsibility.
11. Capacity to synthesize.
12. Persistence and sense of purpose.

Certain design precepts and methods can be learned by study, but the ability to design cannot be gained solely by reading or studying. The engineer also must grapple with real problems and apply his knowledge and abilities to finding solutions. Just as an athlete needs rigorous practice, so an engineer needs practice on design problems as he attempts to gain proficiency in his art. Such experience must necessarily be gained over a period of years, but now is a good time to begin acquiring some of the requisite fundamentals.

Phases of engineering design

Most engineering designs go through three distinct phases:

1. The feasibility study.
2. The preliminary design.
3. The detail design.

In general, a design project will proceed through the various phases in the sequence indicated (Figure 16-2). The amount of time spent on any phase is a function of the complexity of the problem and the restrictions placed upon the engineer—time, money, or performance characteristics.

The feasibility study

The feasibility study is concerned with the following:

1. Definition of the elements of the problem.
2. Identification of the factors that limit the scope of the design.
3. Evaluation of the difficulties that can be anticipated as probable in the design process.
4. Consideration of the consequences of the design.

The objectives of the feasibility study are to discover possible solutions and to determine which of these appear to have promise and which are not feasible, and why.

Let us see how this might work in a situation where you, as the chief engineer of an aircraft company, have been asked to diversify the company's product line by designing a small, low-power passenger vehicle for town driving with substantially less pollution than present cars.

What are the elements of the problem; what factors limit its scope? Where and by whom will such vehicles be used? Are they to carry people individually and randomly to and from work, school, shopping areas, or places of amusement like the present car, or are they to be links in a more comprehensive transportation system? Are they to be privately or publicly owned? If privately owned, perhaps the emphasis should be on low cost, simple upkeep, and ease of parking. If publicly owned—a car which a licensed driver can pick up at one parking place and leave at another—then ease of

Figure 16-2

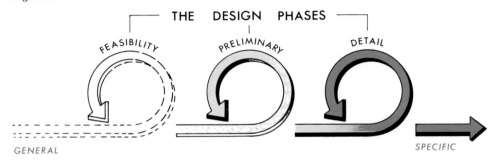

THE DESIGN PHASES

FEASIBILITY   PRELIMINARY   DETAIL

GENERAL   SPECIFIC

handling, reliable operation, and long life might be the major considerations. You will want to know how fast it is to go, how far between fuelings, and how many people it is to carry.

Assume it is decided to design a vehicle for public ownership. What difficulties can be anticipated? Probably the major ones will have to do with people and what they might do. How do you make such a vehicle safe and nearly foolproof? People must be prevented from driving it too far and from abandoning it anywhere except at designated parking places. Provision must be made to redistribute the vehicles if for some reason—a ball game, a sale, a happening—too many people converge on one area. Maintenance and repair will present many problems.

Assuming that such a vehicle can be built and sold to cities, what would be the consequences? Some of the desirable ones are obvious: less traffic, lower air pollution, fewer parking problems, and more efficient vehicle utilization. But what about the uncertainty of finding a car when and where you want it, particularly on a wet, cold night or during the rush hour? The new rules and regulations that would have to be devised? The risk of nonacceptance by the public? These would be some of the early considerations during such a feasibility study.

The ideas and possibilities which are generated in early discussions should be checked for the following:

1. Acceptability in meeting the specifications.
2. Compatibility with known principles of science and engineering.
3. Compatibility with the environment.
4. Compatibility of the properties of the design with other parts of the system.
5. Comparison of the design with other known solutions to the problem.

Each alternative is examined to determine whether or not it can be physically achieved, whether its potential usefulness is commensurate with the costs of making it available, and whether the return on the investment warrants its implementation. The feasibility study is in effect a "pilot" effort whose primary purpose is to seek information pertinent to all possible solutions to the problem. After the information has been collected and evaluated, and after the undesirable design possibilities have been discarded, the engineer still may have several alternatives to consider—all of which may be acceptable.

During the generation of ideas, the engineer has intentionally avoided making any final selection so as to leave his mind open to all possibilities and to give free rein to his thoughts. Now he must reduce this number of ideas to a few—those most likely to be successful, those that will compete for the final solution. How many ideas he keeps will depend on the complexity of ideas and the amount of time and manpower that he can afford to spend during the preliminary design phase. In most design situations the number of ideas remaining at the end of the feasibility study will vary from two to six.

At this point no objective evaluations are available; the discarding of ideas must

> The successful producer of an article sells it for more than it cost him to make, and that's profit. But the customer buys it only because it is worth *more* to him than he pays for it, and that's his profit. No one can long make a profit *producing* anything unless the customer makes a profit *using* it.
> —Samuel B. Pettengill

depend to a large extent upon experience and judgment. There are few substitutes for experience, but there are ways in which judgment can be improved. For example, decision processes based on the theory of probability can be employed effectively. Analog and digital simulations are particularly useful to the engineer in this early comparison of alternatives.

In some instances, it will be more convenient for the engineer to compare the expected performance of the component parts of one design with the counterpart performances of another design. When this is done, he must be very careful to consider if the component parts create the optimum effect in the overall design. Frequently it is true that a simple combination of seemingly ideal parts will not produce an optimum condition. It is not too difficult to list the advantages and disadvantages of each alternative, but the proper evaluation of such lists may require the wisdom of Solomon.

The consideration of *value* is very important in the early selection process. From whose point of view should a particular alternative be appraised? Performance characteristics that may be advantageous in one situation may be equally disadvantageous in another. As an example, automatic redistribution of cars would increase the efficiency of the public car system and save driver cost. However, such an automatic system would almost surely not be possible on public streets, and the cost of extra rights-of-way may make it prohibitive. How does one select the location and proximity of parking places? How far should people be asked to walk, and how many parking places can be serviced effectively? How does one select the maximum speed of the cars and reconcile the conflicting demands of safety and service? Where danger to human life is a possibility, the measurement of value becomes exceedingly difficult. There is great reluctance to place a "cost" or value on the life of a human being. If the engineer assumes an infinite cost penalty, the design may be impossible, but to ignore this factor would effectively assign a cost factor of zero to a life. The engineer must face his responsibilities with honesty and realism.[1]

Engineers engaged in a feasibility study must be able to project the future effectiveness of the alternative designs. In many cases the preliminary design stage of a product will precede its manufacture by several years. Conditions change with time, and these changes must be anticipated by the engineer. Many companies have become eminently successful because of the accuracy of their projections, whereas others have been forced into bankruptcy.

The preliminary design phase

With alternatives narrowed to a few, the engineer must select the design he wishes to develop in detail. The choice is easy if only one of the proposed designs fulfills all requirements. More often, several of the concepts appear to meet the specifications equally well. The choice then must be made on such factors as economics, novelty, reliability, and the number and severity of unsolved problems.

Since it is difficult to make such comparisons in one's head without introducing personal bias, it is useful to prepare an evaluation table. All the important design criteria are listed, and each is assigned an importance factor. There always will be both positive and negative criteria. Then each design is rated as to how well it meets

---

[1] By assigning financial damages to families whose breadwinner has lost his life in an industrial accident, the courts have effectively placed a monetary value on human life. Damages as high as $250,000 have been awarded.

each criterion. This rating should be done by somebody who is not aware of the value assigned to each importance factor, so that he is not unduly influenced.

Let us apply this procedure to our city transportation problem, and particularly to the selection of the propulsion system. Let us assume that the ordinary automobile engine has already been discarded because it is unable to meet air pollution requirements, and that the choice has narrowed to one of three types of engines: the gas turbine, the electric motor, and the steam engine. We will then enter these as Designs (1), (2), and (3) in a table and assign values to the various positive and negative design criteria (Figure 16-3). For example, the gas turbine and electric motor rate low on "novelty" for they are well developed, but an automobile steam engine could rate high if it used modern thermodynamic principles. On "practicability" the electric motor rates higher than the others, for it requires the least service and provides the easiest and safest way to power a small vehicle. This table is completed to the best ability of the engineer for each of the criteria.

Then the engineer "blanks out" the ratings and assigns "importance" factors to each of the criteria (Figure 16-4). For example, he may rate practicability much higher than novelty.

Finally the ratings and importance factors are multiplied and added, yielding a final rating for the three systems, Figure 16-5, which, in this case, favors the electric

Figure 16-3 Evaluation of propulsion systems. Importance (I) varies from 1 (small importance) to 5 (extreme importance). Rating (R) values are 3 (high), 2 (medium), 1 (low), and 0 (none).

Design criteria	Importance I	Design (1) gas turbine R	R × I	Design (2) electric R	R × I	Design (3) steam R	R × I
Positive							
a. Novelty		0		1		3	
b. Practicability		1		3		2	
c. Reliability		2		3		1	
d. Life expectancy		2		2		2	
e. Probability of meeting specifications		2		3		2	
*f. Adaptability to company expertise (research, sales, etc.)		1		1		1	
*g. Suitability to human use							
*h. Other							
Total positive score		—	—	—	—	—	—
Negative							
a. Number and severity of unresolved problems		1		2		3	
b. Production cost		3		1		2	
c. Maintenance cost		1		1		2	
d. Time to perfect		1		1		3	
*Environmental effects		1		0		1	
*Other							
Total negative score		—		—		—	
Net score		—		—		—	

*Such factors may not always be pertinent.

Design criteria	Importance I
Positive	
a. Novelty	2
b. Practicability	5
c. Reliability	5
d. Life expectancy	3
e. Probability of meeting specifications	4
*f. Adaptability to company expertise (research, sales, etc.)	3
*g. Suitability to human use	N.A.
*h. Other	
Total positive score	
Negative	
a. Number and severity of unresolved problems	3
b. Production cost	4
c. Maintenance cost	4
d. Time to perfect	4
*Environmental effects	4
*Other	
Total negative score	
Net score	

*Such factors may not always be pertinent.

Figure 16-4 Evaluation of propulsion systems. Importance (I) varies from 1 (small importance) to 5 (extreme importance). Rating (R) values are 3 (high), 2 (medium), 1 (low), and 0 (none).

motor drive. Although others may come up with different ratings, the method minimizes personal bias.

After selecting the best alternative to pursue, the engineer should make every effort to refine the chosen concept into its most elementary form. Simplicity in design has long been recognized as a hallmark of quality. Simple solutions are the most difficult to achieve, but the engineer should work to this end. He should also learn that such timeless ideas as the lever, the wedge, the inclined plane, the screw, the pulley, and the wheel are still basic ingredients of good design.

In terms of the electric drive vehicle, this means that initially he will strive for a single motor, directly driving the rear wheels, and a battery that can be recharged in each parking area. He may later find that a smaller motor at each wheel is preferable, that a geared-down, high-speed motor is more efficient than a direct-drive motor, or that an on-board electric generator is preferable to a rechargeable battery. He will start with the simplest ideas.

Once the design concept has been selected, the engineer must consider all the component parts—their sizes, relationships, and materials. In selecting materials, he must consider their strengths, dimensions, and the loads to which they will be exposed. In this sense, he is analogous to the painter who has just chosen his subject and now must select his colors, shapes, and brush strokes and put them together in a pleasing and harmonious arrangement. The engineer, having selected a design

Design criteria	Importance I	Design (1) gas turbine R	R × I	Design (2) electric R	R × I	Design (3) steam R	R × I
Positive							
a. Novelty	2	0	0	1	2	3	6
b. Practicability	5	1	5	3	15	2	10
c. Reliability	5	2	10	3	15	1	5
d. Life expectancy	3	2	6	2	6	2	6
e. Probability of meeting specifications	4	2	8	3	12	2	8
*f. Adaptability to company expertise (research, sales, etc.)	3	1	3	1	3	1	3
*g. Suitability to human use	N.A.						
*h. Other							
Total positive score		—	32	—	53	—	38
Negative							
a. Number and severity of unresolved problems	3	1	3	2	6	3	9
b. Production cost	4	3	12	1	4	2	8
c. Maintenance cost	4	1	4	1	4	2	8
d. Time to perfect	4	1	4	1	4	3	12
*Environmental effects	4	1	4	0	0	3	12
*Other						1	4
Total negative score		—	27	—	18	—	41
Net score		—	5	—	35	—	−3

* Such factors may not always be pertinent.

Figure 16-5 Evaluation of propulsion systems. Importance (I) varies from 1 (small importance) to 5 (extreme importance). Rating (R) values are 3 (high), 2 (medium), 1 (low), and 0 (none).

concept that fulfills the desired functions, must organize his components to produce a device that is not only pleasing to the eye but is economical to build and operate.

The engineer must make sure that his design does not interfere with or disturb the environment, that it agrees with man and nature. We are especially reminded of these responsibilities when we encounter foul air, polluted streams, and eroded watersheds. Environmental effects are increasingly important criteria in the design of engineering structures, as evidenced by the voluble concern about such projects as the trans-Alaska pipeline, the supersonic jet transport, and facilities for the disposal or reclamation of industrial and human waste. As the earth's natural resources are depleted, the engineer will be under increasing pressure to provide technical assurances that no harm is done to the environment.

The designer must consider such factors as heat, noise, light, vibration, acceleration, air supply, and humidity, and their effects upon the physical and mental well-being of the user. For example, while it would be desirable to accelerate to top speed as quickly as possible, there are human comfort limits on acceleration that should not be exceeded. Controls must respond rapidly, have the right "feel," and not tire the driver. The suspension system must be "soft" for a comfortable ride, but stiff enough for good performance on curves. Automatic heating and air conditioning will probably be required in most parts of the country.

By now the picture of the vehicle has become clearer, and the chief engineer can delegate the preliminary design of components to various engineers or designers in his organization. Someone will be working on the drive train, another on the wheels and suspension, a third on the battery. Then there are the speed control systems, the interior layout, and perhaps three or four other components, such as access protection, recharging, and systems for redistributing the cars that must be developed.

The detail design phase

Detailed design begins after determination of the overall functions and dimensions of the major members, the forces and allowable deflections of load-carrying members, the speed and power requirements of rotating parts, the pressures and flow rates of moving fluids, the aesthetic proportions, and the needs of the operation—in short, after the principal requirements are determined. The models that were devised during the preliminary selection process should be refined and studied under a considerably wider range of parameters than was possible originally. The designer is interested not only in normal operation, but also in what happens during start-up and shutdown, during malfunctions, and in emergencies. He will study the range of the loads which act on his design and how these loads are transmitted through its parts as stresses and strains. He will look at the effects of temperature, wind, and weather, of vibrations and chemical attack. In short, he will determine the range of operating conditions for each component of the design and for the entire device.

He must have an understanding of the mechanisms of engineering: the levers, linkages, and screw threads that transfer and transform linear and rotating motion; the shafts, gears, belts, and chain drives that transmit power; the electrical power generating systems and their electronic control circuits.

With today's wide range of available materials, shapes, and manufacturing techniques, with the growing array of prefabricated devices and parts, the choices for the design engineer are vast indeed. How should he start? What guidelines are available if he wants to produce the best possible design? It is usually wise to begin investigating that part or component which is thought to be most critical in the overall design—perhaps the one that must withstand the greatest variation of loads or other environmental influences, the one that is likely to be most expensive to make, or the most critical in operation. He may find that operating conditions limit his choices to a few possibilities.

At this stage the designer will encounter many conflicting requirements. One consideration tells him that he needs more power, another that the motor must be smaller and lighter. Springs should be stiff to minimize road clearance; they should be soft to give a comfortable ride. Windows should be large for good visibility, but small for safety and high body strength. The way to resolve this type of conflict is called optimization. It is accomplished by assigning values to all requirements and selecting that design which maximizes (optimizes) the total value.

Materials and stock subassemblies are commercially available in a specific range of sizes. Sheet steel is commonly available in certain thicknesses (gages), electric motors in certain horsepower ratings, and pipe in a limited range of diameters and wall thicknesses. Generally, the engineer should specify commonly available items;

A civilization is both developed and limited by the materials at its disposal . . . .
—Sir George Paget Thomson

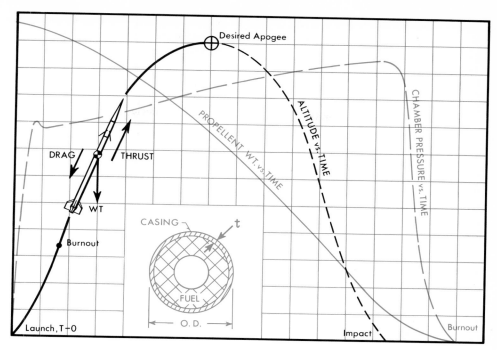

Figure 16-6

only rarely will the design justify the cost of a "special mill run" with off-standard dimensions or specifications. When available sizes are substantially different from the desired optimum size, the engineer may have to revise his optimization procedure.

To illustrate, let us look at the design of a meteorological rocket. At an earlier point in the design process the fuel for this rocket will have been chosen. Let us assume that it is a solid fuel, a material that looks and feels like rubber, burns without air, and when ignited produces high-temperature, high-pressure gases which are expelled through the nozzle to propel the rocket. The rocket consists principally of the payload (the meteorological instruments that are to be carried aloft), the nose cone which houses the instruments, the fuel, the fuel casing, and the nozzle. If we can estimate the weight of the rocket and how high it is to ascend, then we can calculate the requirements.

The most critical design part is the fuel casing, that is, the cylindrical shell which must contain the rocket fuel while it burns. It must be strong enough to withstand the pressure and temperature of the burning fuel, and strong enough to transmit the thrust from the nozzle to the nose cone without buckling and without vibrating. The shell must also be light. If the casing weighs more than had been estimated originally, then more fuel will be needed to propel the rocket. More fuel will produce higher pressures and higher temperatures inside the casing. This, in turn, will require a stronger casing and even more weight. This additional weight requires still more fuel, and the spiral continues. (See Figure 16-6.)

Let us assume we decided to use a high-strength, high-temperature-resistant steel for our casing. Our calculations indicate its wall thickness to be not less than 0.28 in. Our steel catalog tells us this steel is generally available in sheet form only in thicknesses of $\frac{1}{4}$ and $\frac{3}{16}$ in. If we use the thicker sheet, the casing weight will increase by 2.7 per cent; then we must recalculate the amount of fuel required, the pressures and stresses in the casing, and consequent changes in the dimensions of the rocket. Will the $\frac{1}{4}$-in. material withstand the resultant higher stresses? Can we improve its strength by heat treating? If we choose the thinner material, must we provide the casing with extra stiffeners (rings which will reduce the stresses in the casing shell)? In either case, the original design must be altered until the stresses, weights, pressures, and dimensions are satisfactory.

Similar design procedures will be followed in designing the nose cone, the nozzle, and the launching gear for the rocket.

It is important to understand that this example is typical of the design process. Design is not a simple straightforward process but a procedure of *trial and error and compromise* until a well-matched combination of components has been found. The more the engineer knows about materials and about ways of reducing or redistributing stresses (in short, the more alternatives he has) the better the structural design is likely to be.

Consider, as another example, that the engineer has been asked to design the gear shift lever for a racing automobile. The gear box has already been designed, so he knows how far the shifting fork (the end that actually moves the gears in the gear box) must travel in all directions. He also knows how much force will be required at the fork under normal and abnormal driving conditions. He will need to refer to anthropometric[2] data to learn how much force the healthy driver can provide forward, backward, and sideways, and what his reach can be without distracting his eye from the road. With all this information he can choose the location of the ball joint, the fulcrum of the gear shift lever, and the length of each arm of the lever. He may decide to use a straight stick or he may find that a bent lever is more convenient for the driver. Before he finalizes this decision he may build a mock-up and make experiments to determine the most convenient location. Next he must select the material and the cross-sectional shape and area of the lever. Since it is likely to be loaded evenly in all directions, he may find that a circular or a cruciform cross section is most suitable. He must decide between a lever of constant thickness and a lighter, tapered stick (with the greater strength where it is needed—near the joint) which is more costly to manufacture.

Next he will consider the design of the ball joint, which transmits the motion smoothly to the gear box and provides vibration isolation so that the hand of the driver does not shake. It is difficult to find just the right amount of isolation which will retain for the driver the "feel" that is so essential during a race. The engineer needs a complete understanding of lubricated ball joints and proficiency in testing a series of possible designs.

The final component in this design is the handle itself, which should be attractive to look at and comfortable to grip. Here again anthropometric data can tell him much, yet he will be well advised to make several mock-ups and to have them tested for "feel" by experienced drivers.

During the design process, the engineer will have made a series of sketches (somewhat like those in this book) to illustrate to himself the relative position of the parts that he is designing. Now he or his draftsman will use these sketches to make a finished drawing. This will consist of a separate detail drawing for each individually machined item, showing all dimensions, the material from which it is to be made, the type of work to be performed, and the finish to be provided. There also will be subassembly and assembly drawings showing how these parts are to be put together.

---

[2]*Anthropometry* is the study of human body measurements, especially on a comparative basis.

The detail design phase will include the completion of an operating physical model or prototype (a model having the correct layout and physical appearance but constructed by custom techniques), which may have been started in an earlier design phase. The first prototype usually will be incomplete and modifications and alterations will be necessary. This is to be expected. Problems previously unanticipated may be identified, undesirable characteristics may be eliminated, and performance under design conditions may be observed for the first time. This part of the design process is always a time of excitement for everyone, especially the engineer.

The final phase of design involves the checking of every detail, every component, and every subsystem. All must be compatible. Much testing may be necessary to prove theoretical calculations or to discover unsuspected consequences. Assumptions made in the earlier design phases should be reexamined and viewed with suspicion. Are they still valid? Would other assumptions now be more realistic? If so, what changes would be called for in the design?

As one moves through the design phases—from feasibility study to detail design—the tasks to be accomplished become less and less abstract and consequently more closely defined as to their expected functions. (See Figure 16-7.) In the earlier phases, the engineer worked with the design of systems, subsystems, and components. In the detail design phase he also will work with the design of the parts and elementary pieces that will be assembled to form the components.

In the previous phase of engineering design, a large majority of the people involved were engineers. In the detail phase this is not necessarily the case. Many people—metallurgists, chemists, tool designers, detailers, draftsmen, technicians, checkers, estimators, manufacturing and shop personnel—will work together under the direction of engineers. These technically trained support people probably will outnumber the engineers. The engineer who works in this phase of design must be a good manager in addition to his technical responsibilities, and his successes may be measured largely by his ability to bring forth the best efforts of many people.

The engineer should strive to produce a design which is the "obvious" answer to everyone who sees it, *once it is complete*. Such designs, simple and pleasing in

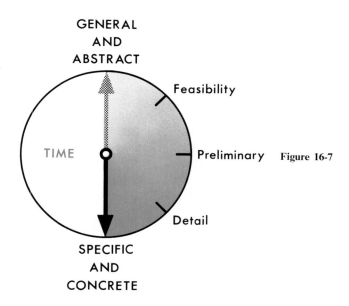

Figure 16-7

appearance, are in a sense as beautiful as any painting, piece of sculpture, or poem, and they are frequently considerably more useful to his well-being.

The planning of engineering projects

In every walk of life, we notice and appreciate evidence of well-planned activities. You may have noticed that good planning involves more than "the assignment of tasks to be performed" although this frequently is the only aspect of planning that is given any attention. Planning in the broad sense must include the enumeration of all the activities and events associated with a project and a recognition and evaluation of their interrelationships and interdependencies. The assignment of tasks to be performed and other aspects of scheduling should follow.

Since "time is money," planning is a very important part of the implementation of any engineering design. Good planning is often the difference between success and failure, and the young engineering student would do well, therefore, to learn some of the fundamental aspects of planning as applied to the implementation of engineering projects.

In 1957 the U.S. Navy was attempting to complete the Polaris Missile System in record time. The estimated time for completion seemed unreasonably long. Through the efforts of an operations research team, a new method of planning and coordinating the many complex parts of the project was finally developed. The overall saving in time for the project amounted to more than 18 months. Since that time a large percentage of engineering projects, particularly those which are complex and time consuming, have used this same planning technique to excellent advantage. It is called PERT (Program Evaluation and Review Technique).

PERT enables the engineer in charge to view the total project as well as to recognize the interrelationships of the component parts of the design. Its utility is not limited to the beginning of the project but rather it continues to provide an accurate measure of progress throughout the work period. Pertinent features of PERT are combined in the following discussion.

How does PERT work?

Basically PERT consists of events (or jobs) and activities arranged into a *time-oriented network* to show the interrelationships and interdependencies that exist. One of the primary objectives of such a network is to identify where bottlenecks may occur that would slow down the process. Once such bottlenecks have been identified, then extra resources such as time and effort can be applied at the appropriate places to make certain that the entire process will not be slowed. The network is also used to portray the events as they occur in the process of accomplishing missions or objectives,

> Though this be madness, yet there is method in it.
> —Shakespeare

together with the activities that necessarily occur to interconnect the events. These relationships will be discussed more fully below.

The network A PERT network is one type of pictorial representation of a project. This network establishes the "precedent relationships" that exist within a project. That is, it identifies those activities which must be completed before other activities are started. It also specifies the time that it takes to complete these activities. This is accomplished by using *events* (points in time) to separate the project *activities*. In other words, project events are connected by activities to form a project network. Progress from one event to another is made by completing the activity which connects them. Let us examine each component of the network in more detail.

Events An event is the *start* or *completion* of a mental or physical task. It does not involve the actual performance of the task. Thus, events are *points in time* which require that action be taken or that decisions be made. Various symbols are used in industry to designate events, such as circles, squares, ellipses, or rectangles. In this book circles, called *nodes,* will be used, Figure 16-8.

Figure 16-8

Events are joined together to form a project network. It is important that the events be arranged within the network in logical or time sequence from left to right. If this is done, the completion of each event will occupy a discrete and identifiable point in time. An event cannot consume time and it cannot be considered to be completed until all activities leading to it have been completed. After all events have been identified and arranged within the network, they are assigned identification numbers. Since events and activities may be altered during the course of the project, the logical order of the events will not necessarily follow in exact numerical sequence, 1, 2, 3, 4, 5, and so on. The event numbers, therefore, serve only for identification purposes. The final or terminal node in the network is usually called the *sink,* while the beginning or initial node is called the *source.* Networks may have varying numbers of sources and sinks.

Activities An activity is the actual performance of a task and, as such, it consumes an increment of time. Activities separate events. An activity cannot begin until all preceding activities have been completed. An arrow is used to represent the time span of an activity, with time flowing from the tail to the point of the arrow, Figure 16-9. In

Activity ⟶ Figure 16-9

a PERT network an activity may indicate the use of time, manpower, materials, facilities, space, or other resources. A *phantom* activity also may represent waiting time or "interdependencies." A phantom activity, represented by a dashed arrow, Figure 16-10, may be inserted into the network for clarity of the logic, although it

— — Phantom — ➤ Figure 16-10
    Activity

represents no real physical activity. Waiting time would also be noted in this manner. Remember that:

Events "happen or occur."
Activities are "started or completed."

The case of Mr. Jones getting ready for work each morning can be examined as an example.

Events	Activities
1. The alarm rings.	
	A. Jones stirs restlessly.
2. Jones awakens.	
	B. Jones nudges his wife.
	C. Jones lies in bed wishing that he didn't have to go to work.
3. Wife awakens.	
	D. Wife lies in bed wishing that it were Saturday.
4. Jones's wife gets up and begins breakfast.	
Meanwhile	*E.* Wife cooks breakfast.
5. Jones begins morning toilet.	
	F. Jones shaves, bathes, and dresses.
6. The Joneses begin to eat breakfast.	
	G. The Joneses eat part of their breakfast.
7. Jones realizes his bus is about to pass the bus stop.	
	H. Jones jumps up, grabs his briefcase, and runs for the bus.
	I. Wife goes back to bed.
8. Jones boards bus.	
9. Wife falls asleep.	

His PERT network can now be drawn as shown in Figure 16-11. This is a very

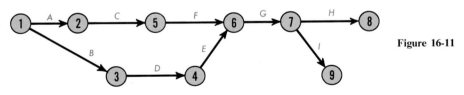

Figure 16-11

elementary example, but it does point up the constituent parts of a PERT network. Note that Jones and his wife must wait until he is dressed (*F*) and the breakfast is cooked (*E*) before they can eat.

In a PERT network each activity should be assigned a specified time for expected accomplishment. The time units chosen should be consistent throughout the network, but the size of the time unit (years, work-weeks, days, hours, etc.) should be selected by the engineer in charge of the project. The time value chosen for each activity should represent the mean (see page 208) of the various times that the activity would take if it were repeated many times.

By using the network of events and activities and by taking into account the times consumed by the various activities, a *critical path* can be established for the project. It is this path that controls the successful completion of the project, and it is important that the engineer be able to isolate it for study. Let us consider the PERT network in Figure 16-12, where the activity times are represented by arabic numbers and are

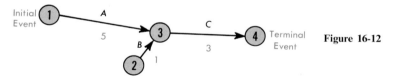

Figure 16-12

indicated in days. Activities represent the expenditure of time and effort. For example, activity *A* (from event 1 to event 3) requires 5 days and is likely devoted to planning the project, while activity *B* requires 1 day and may represent the procurement of basic supplies. Event 1 is the beginning of the project and event 4 is the end of the project. The first step in locating the *critical path* is to determine the "earliest" event times (T_E), the "latest" event times (T_L), and the "slack" time $(T_L - T_E)$.

Earliest event times (T_E)

The earliest expected time of an event refers to the time, T_E, when an event can be expected to be completed. T_E for an event is calculated by summing all the activity duration times from the beginning event to the event in question *if the most time-consuming route is chosen*. To avoid confusion, the T_E times of events are usually placed near the network as arabic numbers within rectangular blocks. For reference purposes the beginning of the project is usually considered to be "time zero." In Figure 16-13, T_E for event 3 would be $\boxed{0} + 5 = \boxed{5}$ and T_E for event 4 would be

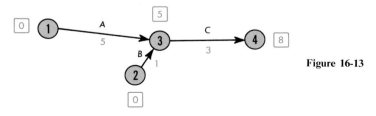

Figure 16-13

$\boxed{0} + 5 + 3 = \boxed{8}$. However, there are two possible routes to event 4 ($A + C$, or $B + C$). The *maximum* duration of these event times should be selected as the T_E for event 4. Summing the times, we find

By path $A + C$: $\boxed{0} + 5 + 3 = \boxed{8}$ ← Select as T_E for event 4
By path $B + C$: $\boxed{0} + 1 + 3 = \boxed{4}$

Method is like packing things in a box; a good packer will get in half as much again as a bad one.
—Cecil

Latest event times (T_L)

The latest expected time, T_L, of an event refers to the longest time which can be allowed for an event, assuming that the entire project is kept on schedule. T_L for an event is determined by beginning at the terminal event and working backward through the various event circuits, subtracting the value T_E at each event *assuming the least time-consuming route is chosen*. The resulting values of T_L are recorded as arabic numbers in small ellipses located near the T_E times. Thus, in Figure 16-14, T_L for event 3 would be ⑧ − 3 = ⑤; for event 2, ⑧ − 3 − 1 = ④; and for event 1, ⑧ − 3 − 5 = ⓪.

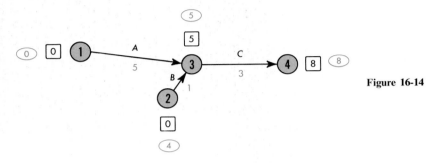

Figure 16-14

Remember that T_L is determined to be the *minimum* of the differences between the succeeding event T_L and the intervening activity times. Also, in calculating T_L values one must always proceed backward through the network—from the point of the arrows to the tail of the arrows.

Slack times

The *slack* time for each event is the difference between the latest event time and the earliest possible time ($T_L - T_E$). Intuitively, one may verify that it is the "extra time that an event can slip" and not affect the scheduled completion time of the project. For example, in Figure 16-14 the slack time for event 2 is ④ − ⓪ = 4. For this reason activity B may be started as much as 4 days late and still not cause any overall delay in the minimum project time of 8 days.

The critical path

The *critical path* through a PERT network is a path that is drawn from the initial event of the network to the terminal event by connecting the events of zero slack. The *critical path* is usually emphasized with a very thick line. Color is sometimes used. In the example problem above the *critical path* would be shown connecting events 1–3–4, Figure 16-15. Slack times for each event are indicated as small arabic numbers that are located in triangles adjacent to the events.

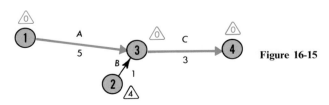

Figure 16-15

Illustration 16-1
A proper evaluation of PERT will help the engineer to schedule all subcontracts in proper sequence and especially to not allow one work assignment to be pushed ahead of others prematurely or to lag behind unnecessarily.

Remember that the *critical path* is the path that controls the successful completion of the project. It is also the path that requires the most time to get from the initial event to the terminal event. Any event on the critical path that is delayed will cause the final event to be delayed by the same amount. Conversely, putting an extra effort on noncritical activities will not speed up the project.

Although calculations in this chapter have been done manually, it is conventional practice to program complex networks for solution by digital computer. In this way thousands of activities and events may be considered, and one or more critical paths can be located for further study. Finally, the PERT network should be updated periodically as the work on the project progresses.

The following example will show how a typical PERT diagram is analyzed. It should be noted here, however, that in real-life situations the most difficult task is to identify the precedent relationships that exist and to draw a realistic network of the events and activities. After this is accomplished, following through with a solution technique becomes a relatively routine task.

Example In the PERT network diagram of Figure 16-16, assume that all activity

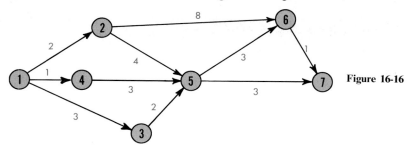

Figure 16-16

times are given in months and that they exist as indicated on the proper activity branch. Find the earliest times, T_E, the latest times, T_L, and the slack times for each event. Identify the critical path through the network.

Solution   See Figure 16-17.

Figure 16-17

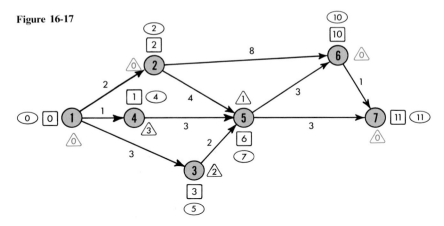

It is usually advisable to construct a summary table of the calculations.

Event ○	Path	T_E ☐	Path	T_L ○	Slack, $T_L - T_E$ △	On critical path
1	—	0	7–6–2–1	0	0	✓
2	1–2	2	7–6–2	2	0	✓
3	1–3	3	7–6–5–3	5	2	
4	1–4	1	7–6–5–4	4	3	
5	1–2–5	6	7–6–5	7	1	
6	1–2–6	10	7–6	10	0	✓
7	1–2–6–7	11	—	11	0	✓

The critical path then is 1–2–6–7, Figure 16-18. This means that as the project is now organized it will take 11 months to complete.

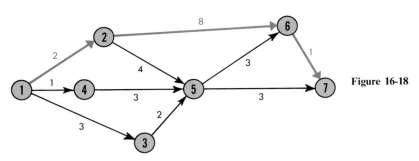

Figure 16-18

Problems

16-1.   Consider the network in Figure 16-19. Find T_E, T_L, slack times, and the critical path through the network.

Figure 16-19

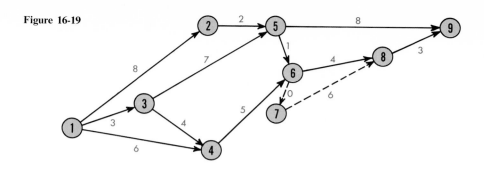

16-2. In Figure 16-20, what effect on project length would the following changes have:

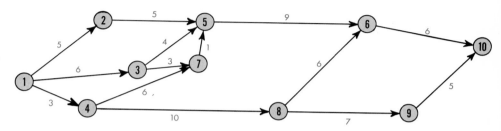

Figure 16-20

        a. Decrease activity 1–2 to 6 days.
        b. Decrease activity 5–9 to 1 day.
        c. Decrease activity 3–4 to 2 days.

16-3. Explain why "phantom activities" are necessary, and give an example of one.

16-4. Given the following tabular information, determine the PERT network and its critical path.

Activity	Precedent relationships	Time
A	None	5
B	None	3
C	A	1
D	B	4
E	B	3
F	E	7

16-5. For some general process with which you are familiar, construct a PERT network. Be sure to label all events and activities.

16-6. Find the critical path in Figure 16-20 and explain its significance here.

16-7. *a.* Does a decrease in an activity time on the critical path always decrease the project time correspondingly? Why or why not? (*Hint:* see Problem 16-6.)
        b. Does an increase in an activity time on the critical path always increase the project time correspondingly? Why or why not? (*Hint:* See Problem 16-4.)

16-8. Given the PERT network in Figure 16-21, when is the earliest possible project completion time?

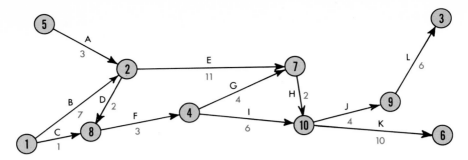

Figure 16-21

16-9. If you have extra resources to allocate to one activity in Figure 16-21, where would you put those resources and why? How might this affect the expected duration time of the project?

16-10. One thing that a PERT analysis does not consider is the allocation of limited resources (see Problem 16-9). How can this inability affect the usefulness of a PERT analysis?

Design philosophy

Design philosophy is an important factor with many industries, companies, and consulting firms. The aircraft industry, for example, generally would support a design philosophy that includes (a) lightweight components, (b) safety, (c) limited service life, (d) a wide range of loading conditions and temperature extremes, and (e) concern about vibration and fatigue. The automobile industry, on the other hand, would be more likely to support a design philosophy that stresses (a) consumer price consciousness, (b) long life with minimum service and maintenance, (c) customer appeal, (d) safety for the occupants, and (e) design for mass production.

Some companies are concerned that their products have a "family-like" image and that a responsiveness to customer appeal be designed into all their products. In some instances the image is safety, in some efficiency, in some quality. Public relations should be an important factor with all companies, and the engineer should not be insensitive to the effects that his design will have upon the total company image. The appearance of the product is particularly important in consumer-oriented industries. In such cases the engineer must take this into account in all phases of his design.

Any engineering design is but *one* answer to an identified problem. For this reason few designs have withstood the test of time without undergoing substantial revisions. One need but look at the continuous parade of modifications, alterations, changes, and complete redesigns that have taken place within the automobile industry to see how the product of a single industry has been changed thousands and thousands of times. Each change, it was believed at the time, was an improvement over the existing model, even if in appearance only. In some instances this assumption proved to be false, and other modifications were quickly made.

In some situations the pressure for a quick solution has led to the adoption of designs of minimum acceptability. Generally the handicaps and pressures under which the engineer works are of little interest to the customer, who tends to judge the quality of a device or machine by its performance. This emphasis on the product places an additional responsibility upon the shoulders of the engineer to release only those designs which he believes are good designs, those to which he will have no hesitancy in affixing his signature. As a professional person he must be equally aware of his responsibilities to his fellowman and to his employer or client. He must perceive *when he knows,* he must realize *when he does not know,* and he must assume the final responsibility in either case.

An example of the development of an engineering design

There are many well-known engineering designs on the American scene that have, over a period of many years, almost become a "way of life." Although the authors of this text are reluctant to pick any one of these as being superior to another, they are eager that the students who study this text gain an appreciation for the concepts that "good ideas plus good engineering design practice equals success" and "all good engineering designs can be improved." For this reason the story of the developing design of the safety razor is given here.[3]

The idea for the safety razor that the American public knows today was the brainchild of a traveling salesman, King C. Gillette, who on a summer morning in 1895 became irritated and exasperated at his inability to shave with a dull straight razor. In an instant the idea of a replaceable flat blade secured in a holder for maximum safety was born. In Gillette's own words,

> I saw it all in a moment, and in that same moment many unvoiced questions were asked and answered more with the rapidity of a dream than by the slow process of reasoning. A razor is only a sharp edge and all back of that edge is but a support for that edge. Why do they spend so much material and time in fashioning a backing which has nothing to do with the shaving? Why do they forge a great piece of steel and spend so much labor in hollow grinding it when they could get the same result by putting an edge on a piece of steel that was only thick enough to hold an edge? At that time and in that moment it seemed as though I could see the way the blade could be held in a holder. Then came the idea of sharpening the two opposite edges on the thin piece of steel that was uniform in thickness throughout, thus doubling its service; and following in sequence came the clamping plates for the blade with a handle equally disposed between the two edges of the blade. All this came more in pictures than in thought as though the razor were already a finished thing and held before my eyes. I stood there before that mirror in a trance of joy at what I saw.

Previous to this time, men of wealth and influence of all nationalities frequented barbershops in which the customer was lathered from the community mug and

---

[3] Much of the material presented here was made available by and is used with the permission of the Gillette Safety Razor Company, Boston, Massachusetts.

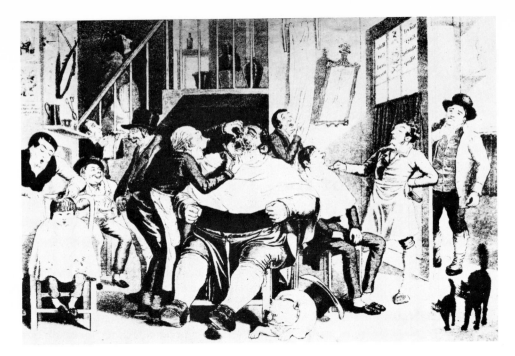

Illustration 16-2
*The morning shave in 1768 was not without its difficulties, as shown in this
English woodcut.*

shaved with an unsterilized razor, Illustration 16-2. Such a barbershop shave was a
luxury that poor people could not afford, but many men of modest means did
purchase their own straight razors. Ladies of respect would not think of using a razor
to remove unsightly hair, although it is reported that such practice was not uncommon for burlesque queens.

Several years previous to the inspirational moment of 1895, King Gillette was
talking with a successful inventor friend who advised him:

> King, you are always thinking and inventing something; why don't you try to
> think of something like the Crown Cork, when used once, it is thrown away,
> and the customer keeps coming back for more—and with every additional
> customer you get, you are building a permanent foundation of profit.

Although Gillette often thought of this advice, he never was able to capitalize on
it until that moment—holding in his hand a dull razor which was beyond the point of
successful stropping and in need of honing—that the idea in his subconscious
emerged to reinforce his need for a new and novel solution.

Gillette knew very little about razors and practically nothing about steel, and he
could not foresee the trials and frustrations that were to come his way before the
"safety razor" was a success. On the same day that he received the inspiration to
devise a razor which could use interchangeable and disposable blades, and which
were safe to use, Gillette went to a local hardware store and purchased several pieces
of brass, some steel ribbon used for clock springs, and some hand tools. Using some
rough pencil sketches and the recently purchased hand tools, he fashioned a crude
model of his new design, Illustration 16-3. Gillette's invention did not consist pri-

Illustration 16-3

King Gillette's moment of triumph came when he discovered that his idea could be made to work.

marily in a particular form of blade or design of a blade holder, but in the conception of a blade so cheap as to be discarded when dull. To obtain such a blade he abandoned the forged type and fashioned one of thin steel, so that it might be cut from a strip, avoiding the expense of forging or hollow grinding. Prior to this invention, razor makers produced an expensive blade that was expected to give service as long as possible, even a lifetime, and to be honed and stropped indefinitely. The new idea was a complete reversal of this practice and was a really unique invention.

In his new razor Gillette carried his theory to great completeness. The blade was to be made of relatively thin steel and thereby achieve economy through the saving of both material and labor. It was to have two edges, one on each side, thus giving double shaving service. The adjustment of the blade edge in relation to the guard was to be obtained by flexing the blade so as to bring the edge nearer to or farther from the guard teeth, in order to obtain a finer or coarser cut, Figure 16-22.

However, all was not bright for this new idea. No one but Gillette had any faith in a razor the blades of which were to be used once and then wasted. Such a proposal did not seem to be within the bounds of reason, and even Gillette's friends looked upon it as a joke. Actually he had thought originally that the blades might be made very cheaply from a thin ribbon of steel, but he was, of course, aware that new machines and processes would need to be invented and developed before such "ribbon blades" could be manufactured cheaply. This did not seem to be a likely

Figure 16-22

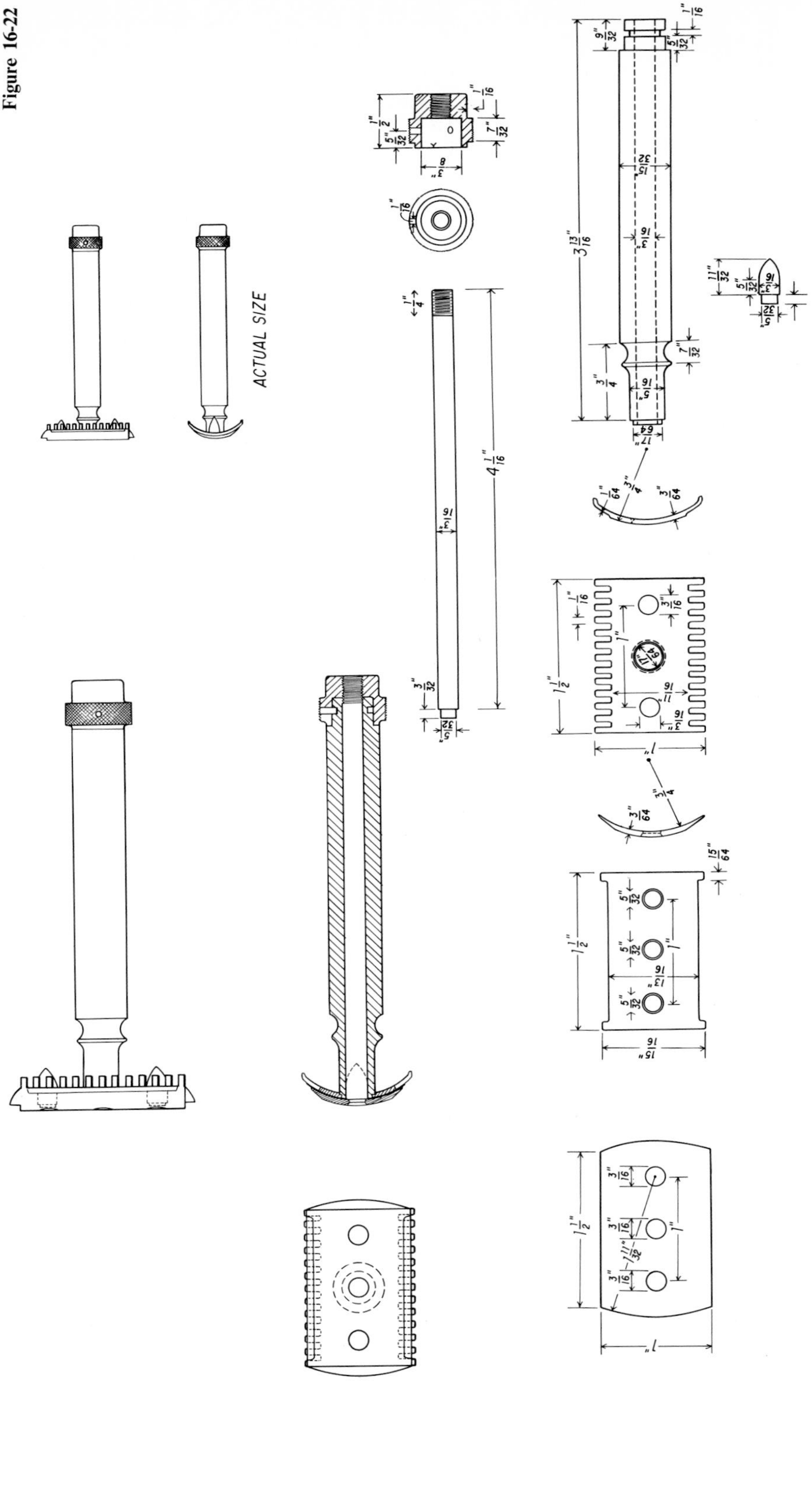

ACTUAL SIZE

prospect. For more than five years Gillette clung tenaciously to his razor theories. He made a number of models with minor variations and sought through others to get blades made with shaving qualities. He got very little encouragement either from his helpers and advisers or from the results of his experiments. People who knew most about cutlery and razors in particular were most discouraging. Years later Gillette said, "They told me I was throwing my money away; that a razor was only possible when made from cast steel forged and fashioned under the hammer to give it density so it would take an edge. But I didn't know enough to quit. If I had been technically trained, I would have quit or probably would never have begun." In spite of this discouragement, Gillette did not falter in his faith and persistence.

Faced with an inability to cope with the technical difficulties surrounding his idea, Gillette began to search for others to help him. He associated himself with several men, one of whom—W. E. Nickerson—was a mechanical engineering graduate of the Massachusetts Institute of Technology. The design capability of Nickerson soon became apparent. A notation from the Gillette Safety Razor Company silver jubilee history relates the following:

> . . . after a very urgent plea, he [Nickerson] agreed to turn the problem over in his mind and give a decision within a month. On giving the problem serious thought, he began to see the proper procedure and felt that he could develop the razor into a commercial proposition. Things began to take definite shape in his mind, he could visualize the hardening process and sharpening machines, and definite ideas were developed as to the type of handle necessary to properly hold the blade.
>
> Hardening apparatus and sharpening machines could not be properly designed until the form and size of the blade were known, so the first step was to decide just what the blade and the handle were to be like. Mr. Gillette's models were amply developed to disclose the fundamental ideas, but there was left a wide range of choice in the matter of carrying out these ideas; and furthermore, the commercial success of the razor was sure to depend very much upon the judgment used in selecting just the right form and thickness of blade and the best construction in the handle.
>
> Mr. Nickerson's fundamental thought in relation to the remodeled razor was that the handle must have sufficient stability to make possible very great accuracy of adjustment between the edge of the blade and the protecting guard. Here is a point upon which he laid great stress, and which we are constantly endeavoring to drive home today: "No matter how perfect the blade is, you cannot get the best result unless the handle is perfect also." The Gillette handle is made to micrometric dimensions and is an extremely accurate instrument. If damaged or thrown out of alignment, poor shaves are likely to result. This idea of great stability led Mr. Nickerson to design a handle to be "machined" out of solid metal, in contra-distinction to one stamped from relatively thin sheet metal. To this fact much of the Gillette commercial success is due. In fact, it is doubtful if great success could have been achieved without it.
>
> The shape and thickness of the blade were determined as follows: Sheet steel thinner than six one-thousandths of an inch appeared to lack sufficient firmness to make a good blade, and a thickness greater than that seemed too difficult to flex readily; so six thousandths was chosen. In the matter of width, one inch was thought to be unnecessarily wide and three-quarters of

an inch was found to be too narrow, especially when flexing was considered. Thus seven-eights of an inch was adopted. As to contour, a circle one and three-quarters inches in diameter if symmetrically crossed by two parallel lines seven-eights of an inch apart gives chords corresponding to the cutting edges, one and one-half inches long, which was thought to be the right length for the edges. The rounded ends to the blade form thus produced strengthened the blade along the center where holes were to be and gave the blade its well-known and pleasing shape. After twenty-five years of use nothing has transpired to cause regret that some other shape was not selected. These early decisions were of the utmost importance and almost seemed inspired.

On September 9,1901, Mr. Nickerson sent a report of his findings and recommendations to Mr. Heilborn of which the following is an exact copy:

Boston, Mass., Sept. 9, 1901.

Jacob Heilborn, Esq.,
Boston, Mass.

Dear Sir:

I have had your proposition, in regard to the manufacture of the Gillette Safety Razor, under consideration for rather more than a month and desire to report as follows:

It is my confident opinion that not only can a successful razor be made on the principles of the Gillette patent, but that if the blades are made by proper methods a result in advance of anything known can be reached. On the other hand, to put out these razors with blades of other than the finest quality of temper and edge would be disastrous to their reputation and to their successful introduction.

With an almost unlimited market, and with such inducements as are offered by this razor, in the way of cheapness of manufacture and of convenience and effectiveness in use, I can see no reason why it cannot easily compete for popular favor with anything in its line ever put before the public.

I wish to reiterate that in my opinion the success of the razor depends very largely, if not almost wholly, on the production at a low price of a substantially perfect blade. This blade must possess an edge that shall, at least, be equal of any rival on the market, and should combine extreme keenness with a hardness and toughness sufficient to stand using a number of times without much deterioration.

For the past month I have been giving much thought to the subject of manufacturing these blades, and I now feel justified in offering to undertake the construction of machines and apparatus to that end. I am confident that I have grasped the situation and can guarantee, as far as such a thing can be guaranteed, a sucessful outcome. Your knowledge of my long experience with inventions and machine building will, perhaps, cause you to attach considerable weight to my opinion in this matter. You are of course aware that special machines will have to be designed and built for putting on the blades that delicate edge which is necessary for easy shaving. The problem is entirely different from that involved in the tempering and grinding of ordinary razors and other keen tools, not only on account of the thinness of the blades, but also on account of the cheapness with which it must be done. I believe that with the machines which I have in mind, an edge can be put upon these blades which will be unapproachable by ordinary hand sharpened razors. The machinery and methods for making the blades will naturally be of a novel character and admit of sound patents, which would become the property of the Company and would be of great advantage in disposing of foreign rights. It is not unlikely that the machines for honing these blades may be adapted for any of the present form of razors and do away with hand honing. I will also add that I have in mind a convenient and simple method of adjusting the position of the blade for different beards.

In reply to your questions as to the probable expense of fitting up to manufacture the razor on a scale suitable for a beginning on a commercial bases, I will make the following approximation:

Drawings for machines for tempering, grinding, honing and stropping	$ 100
Patterns for ditto	250
Materials for machines (one each)	300
Cost of building (one each)	700
Special dies and tools	150
Tools for making holders { Small turret lathe, Power punch, Small plain milling machine, Sensitive drill, Bench lathe, Bench tools, etc. }	1500
Foreign patents: England, Germany, Belgium, France, Canada, Spain, Italy, Austria—about	800
Labor services, etc.	1200
	$5000

I have made what seems to me to be fairly liberal but by no means extravagant figures. It may cost considerably less or possibly a little more, but I think the sum given will not come out very far from the truth.

I should recommend that the machines for making the blades be built in some shop already established, and when they are completed, a suitable room be engaged and they and the holder tools set up in it. It is not easy to say just how long it would take to be ready for manufacturing, but if there are no serious delays it is possible that four months might cover it.

In conclusion let me add that so thoroughly am I satisfied that I can perfect machinery described on original lines which will be patentable, that I am ready to accept for my compensation stock in a Company which I understand you propose forming.

Very truly yours,
(Signed) Wm. E. Nickerson

Nickerson did design a machine for sharpening the blades and an apparatus for hardening the blades in packs. Thus through the application of fundamental engineering principles a successful new industry was born.

Success was not immediate because two years later, in 1903, when Gillette put his first razor on the market only 51 razors and 168 blades were sold. Barbers, who believed that their business would be ruined if this new fad caught on, were particularly scathing in their reproof. However, the new razor caught on, sales soared, and by 1905 manufacturing operations had to be moved to larger quarters. By 1917 razor sales had risen to over a million a year, and blade sales averaged 150 million a year. As a result of World War I, self-shaving became widespread and returning servicemen carried the habit home with them. While World War I taught thousands of men the self-shaving habit, World War II introduced millions of men to daily shaving practice.

In the 62 years since Gillette razors first went on sale, the company has produced over one half billion razors and over one hundred billion blades. Throughout this period of time, however, many modifications and redesigns have been made, Figures 16-23 and 16-24.

The latest of these designs, the Trac II, Figure 16-26, followed the Techmatic Razor with razor band, Figure 16-25, which was itself a complete departure from the blade-changing routine which has been so successfully sold to the American public. Interestingly enough, the idea of shaving with a "ribbon of steel" (the Techmatic) is a simple adaptation of the original material purchased by King Gillette on that summer day in 1895. It took, however, 70 years for engineering design to make possible mass-produced "ribbon blades." Other improvements will continue to follow in the years ahead.

Many other American industries have equally exciting engineering histories. In many respects the engineering students of today live in the most challenging period of history ever, and a *good idea,* together with the application of sound engineering design principles, will still produce *success.*

Exercises in design

16-11. Estimate the number of drugstores in the United States. Give reasons for your estimate.

16-12. Estimate the number of liters of water of the Mississippi River that pass New Orleans every day. Show your analysis.

Figure 16-23

Year	New Design	Improvements
1932	Gillette Blue Blade	Better shaving edge
1934	One-Piece Razor	Convenience, more exact edge exposure
1938	Thin Gillette Blade	Reduced cost by one half
1947	Blade Dispenser	Blade edges protected, simplified blade changing
1957	Adjustable Safety Razor	Variable cut, ease of adjustment
1960	Super Blue Blade	Longer life, first coated edge, less pull
1963	Stainless Steel Blade	Comfort, coated edge, durability
1963	Lady Gillette Razor	Designed expressly for women
1965	Super Stainless Steel Blade	Better steel, longer life, new coating
1965	Techmatic Razor with Razor Band	Cartridge load, convenience, no blades
1968	Injector-Type Single-Edge Blade	Provides alternative shaving method
1969	Platinum-Plus Double-Edge Blade	Stronger, harder, corrosion-resistant edges
1970	Platinum-Plus Injector Blade	Improved blade for alternative shaving method
1971	TRAC II Razor	Tandem blade system

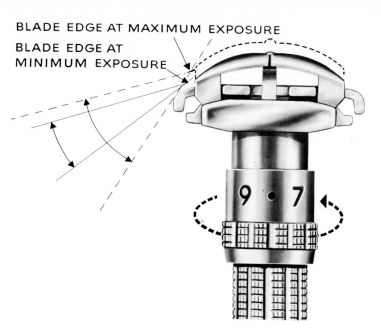

BLADE EDGE AT MAXIMUM EXPOSURE

BLADE EDGE AT
MINIMUM EXPOSURE

Figure 16-24 The adjustable safety razor.

16-13. In 100 words or less describe how a household water softener works.

16-14. By the use of simple sketches and a brief accompanying explanation, describe the mechanical operation of a household toilet.

16-15. By the use of a diagrammatic sketch show how plumbing in a home might be installed so that hot water is always instantly available when the hot water tap is opened.

16-16. Analyze and discuss the economic problems involved in replacing ground-level railroad tracks with a suspended monorail system for a congested urban area.

16-17. Discuss the feasibility of railroads offering a service whereby your automobile would be carried on a railroad car on the same train on which you are traveling so that you might have your car available for use upon arrival at your destination.

16-18. Discuss the desirability of assigning an identifying number to each person as soon as they are born. The number could, for example, be tatooed at some place on the body to serve as a social security number, military number, credit card number, and so forth.

16-19. Using local gas utility rates, electric utility rates, coal costs, fuel oil costs, and wood costs, what would be the comparative cost of heating a five-room house in your home community for a winter season?

16-20. Discuss the advantages and disadvantages of having a channel of television show nothing but market quotations, except for brief commercials, during the time the New York stock market and the Chicago commodity market are open.

16-21. You are called to Alaska to consider the problem of public buildings that are sinking in permafrost due to warm weather. What might you do to solve this problem?

16-22. You are located on an ice cap. Ice and snow are everywhere but no water. Fuel and equipment are available. How can you prepare a well from which water can be pumped?

16-23. Assemble the following items: an ink bottle, a marble, a yardstick, an engineer's scale of triangular cross section, five wooden matches, a pocket knife, a candle, a pencil, and a key. Now, using as few of the objects as possible, balance the yardstick across

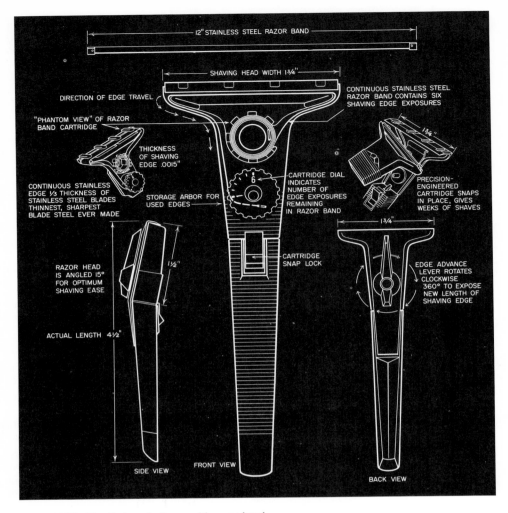

Figure 16-25 The Techmatic Razor with razor band.

the top "knife-edge" of the engineer's scale in such manner that soon after being released, and without being touched again, it unbalances itself.

16-24. Explain the operation of the rewind mechanism for the hand cord of a home gasoline lawnmower.

16-25. Devise a new method of feeding passengers on airplanes.

16-26. List the consequences of everyone being able to read everyone else's mind.

16-27. At current market values determine the number of years that would be necessary to regain the loss of money (lost salary plus college expense) if one stayed in college one additional year to obtain a master's degree in engineering. What would be the number of years necessary to regain the loss by staying three years beyond the bachelor's degree to obtain a doctorate in engineering?

16-28. Estimate the number of policemen in (*a*) New York City, and (*b*) the United States.

16-29. Estimate the number of churches of all faiths in the United States.

16-30. Explain how the following work:
   a. An automobile differential.
   b. A toggle switch.
   c. An automatic cutoff on gasoline pumps.
   d. A sewing machine needle when sewing cloth.

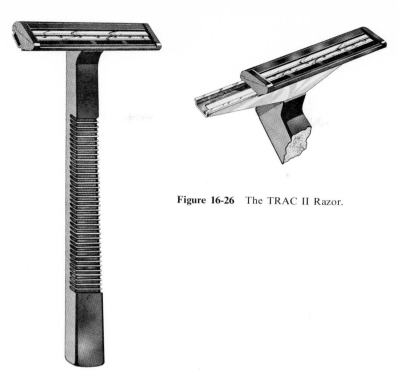

Figure 16-26 The TRAC II Razor.

   e. A refrigeration cycle which does not depend upon electricity.

16-31. With six equal-length sticks construct four equilateral triangles.

16-32. Estimate the number of aspirin tablets now available in the United States.

16-33. A cube whose surface area is 6 mi^2 is filled with water. How long will it take to empty this tank using a 1000 gal/min pump?

16-34. From memory sketch (*a*) a bicycle, (*b*) a reel-type lawnmower, (*c*) a coffee pot, (*d*) a salt-water fishing reel, and (*e*) a rifle.

16-35. Make something useful from the following items: a piece of corrugated cardboard 12 in. × 24 in., 6 ft of string, 3 pieces of chalk, 10 rubber bands, a small piece of gummed tape, 3 tongue depressors, 5 paper clips, and 7 toothpicks.

16-36. Propose some way to eliminate the need for bifocal glasses.

16-37. Design a device that can measure to a high degree of accuracy the wall thickness of a long tube whose ends are not accessible.

16-38. Design a man's compact travel kit that can be carried in the inside coat pocket.

16-39. Design a home-type sugar dispenser for a locality where the average rainfall is 100 in./yr.

16-40. Design a new type of men's apparel to be worn around the neck in lieu of a necktie.

16-41. Design a new type of clothespin.

16-42. Design a new fastener for shirts or blouses.

16-43. Design a personal monogram.

16-44. Design a device to aid federal or civil officers in the prevention or suppression of crime.

16-45. Design a highway system and appropriate vehicles for a country where gasoline is not obtainable and where motive power must be supplied external to the vehicle.

16-46. Design an electrical system for a home that does not receive its energy from a power company or a storage battery.

16-47. Design a device for weighing quantities of food for astronauts who are enroute to the moon.

16-48. Design a machine or process to remove Irish potato peelings.

16-49. Design a "black-eyed pea" sheller.

16-50. Design a corn shucker.

16-51. Design a trap to snare mosquitoes alive.

16-52. Design the "ideal" bathroom, including new toilet fixtures.

16-53. Design a toothpaste dispenser.

16-54. Design a woozle.

16-55. Design a device that would enable paralyzed people to read in bed.

16-56. Design a jig-like device that an amateur "do-it-yourself" home workman could use to lay up an acceptably straight brick wall.

16-57. Design a device to retail for less than $20.00 to warn "tailgaters" that they are too close to your automobile.

16-58. Devise a system of warning lights connected to your automobile that will warn drivers in cars following you of the changes in the speed of your car.

16-59. You live in a remote community near the Canadian border, and you have a shallow well near your home from which you can get a copious supply of water. Although the water is unfit for drinking or irrigation, its temperature is a constant 64°F. Design a system to use this water to help heat your home.

16-60. Design and build a prototype model of a small spot welder suitable for use by hobby craftsmen. Prepare working sketches and make an economic study of the advisability of producing these units in volume production.

16-61. Design some device that will awaken a deaf person.

16-62. Design a coin-operated hair-cutting machine.

16-63. Design a two-passenger battery-powered Urbanmobile for use around the neighborhood, for local shopping center visits, to commute to the railway station, and so on. The rechargeable battery should last for 60 mi on each charge. Provide a complete report on the design, including a market survey and economic study.

16-64. Design some means of visually determining the rate of gasoline consumption (mi/gal) at any time while the vehicle is in operation.

16-65. Design a device to continuously monitor and/or regulate automobile tire pressures.

16-66. Design a novel method of catching and executing mice that will not infringe the patent of any other known system now on the market.

16-67. Design a new toy for children ages 6 to 10.

16-68. Design a device to replace the conventional oarlocks used on all rowboats.

16-69. Devise an improved method of garbage disposal for a "new" city that is to be constructed in its entirety next year.

16-70. Design and build a simple device to measure the specific heat of liquids. Use components costing less than $3.00.

16-71. Design for teenagers an educational hobby kit that might foster an interest in engineering.

16-72. Design a portable traffic signal that can be quickly put into operation for emergency use.

16-73. Design an egg breaker for kitchen use.

16-74. Design an automatic dog-food dispenser.

16-75. Design a device to automatically mix body soap in shower water as needed.

16-76. Design an improved keyholder.

16-77. Design a self-measuring and self-mixing epoxy glue container.

16-78. Design an improved means of cleaning automobile windshields.

16-79. Design a noise suppressor for a motorcycle.

16-80. Design a collapsible bicycle.

16-81. Design a tire-chain changer.

16-82. Design a set of improved highway markers.

16-83. Design an automatic oil-level indicator for automobiles.

16-84. Design an underwater means of communication for skin divers.

16-85. Design a means of locating lost golf balls.

16-86. Design a musician's page turner.

16-87. Design an improved violin tuning device.

16-88. Design an attachment to allow a motorcycle to be used on water.

16-89. Design a bedroll heater for use in camping.

16-90. Design a portable device for student use in keypunching computer cards.

16-91. Design an improved writing instrument.

16-92. Design a means of disposing of solid household waste.

16-93. Design a type of building block that can be erected without mortar.

16-94. Design a means for self-cleaning of sinks and toilet bowls.

16-95. Design some means to replace door knobs or door latches.

16-96. Design a simple animal-powered irrigation pump for use in developing nations.

16-97. Design a therapeutic exerciser for use in strengthening weak or undeveloped muscles.

16-98. Design a Morse-code translator that will allow a deaf person to read code received from radio receivers.

16-99. Design an empty-seat locator for use in theaters.

16-100. Design a writing device for use by armless people.

16-101. Design and build an indicator to tell when a steak is cooked as desired.

16-102. Design a device that would effectively eliminate wall outlets and cords for electrical household appliances.

16-103. Design the mechanism by which the rotary motion of a 1-in. diameter shaft can be transferred around a 90° corner and imparted to a $\frac{1}{2}$-in. diameter shaft.

16-104. Design a mechanism by which the vibratory translation of a steel rod can be transferred around a 90° corner and imparted to another steel rod.

16-105. Design a device or system to prevent snow accumulation on the roof of a mountain cabin. Electricity is available, and the owner is absent during the winter.

16-106. Using the parts out of an old spring-wound clock, design and fabricate some useful device.

16-107. Out of popsicle sticks build a pinned-joint structure that will support a load of 50 lb.

16-108. Design a new device to replace the standard wall light switch.

16-109. Design and build a record changer that will flip records as well as change them.

16-110. Design a wheelchair that can lift itself from street level to a level 1 ft higher.

16-111. Design a can opener that can be used to make a continuous cut in the top of a tin can whose top is of irregular shape.

16-112. Design and build for camping purposes a solar still that can produce 1 gallon of pure water per day.

16-113. Design and construct a working model of a powered and self-controlled surface vehicle which will negotiate a "figure 8" course on a smooth, horizontal surface of 2 square meters. Quantities and types of materials that may be used in the construction are:

a. Balsa wood or cardboard ≤ 5 mm thick, not more than one standard sheet of 750 cm². Calculate cost at $1.00/cm³.

b. Cotton thread (no nylon), not more than 30 cm in length and not larger than 20 gauge. Calculate cost at $3.00/cm.

c. One small tube of balsa wood cement. Calculate cost at $10.00 for use if tube is opened.

d. Four standard size rubber bands for use in controls and power source. Calculate cost at $50.00 each.

e. Four standard paper clips. Calculate cost at $25.00 each.

The overall value of your model will be calculated in accordance with the following formula:

$$ V = \left[\frac{f^3 (10)^3}{w(10)^2 + C} \right] 10^3 $$

where V = design value
   f = fraction of "figure 8"
     successfully negotiated

w = weight of vehicle, newtons
C = cost of materials used, $

16-114. Design, build, and demonstrate a device that will measure and indicate 15 seconds of time as accurately as possible. The device must not use commercially available timing devices.

16-115. Few new musical instruments have been invented within the last 100 years. With the availability of modern materials and processes, many novel and innovative designs are now within the realm of possibility. To be marketable over an extended period of time such an instrument should utilize the conventional diatonic scale of eight tones to the octave. It could, therefore, be utilized by symphonies, in ensembles, or as a solo instrument using existing musical compositions. You are the chief engineer for a company whose present objective is to create and market such a new instrument. Design and build a prototype of a new instrument that would be salable. Prepare working drawings of your model together with cost estimates for volume production of the instrument.

16-116. Design some means of communicating with a deaf person who is elsewhere (such as by radio).

16-117. For a bicycle, design an automatic transmission that will change gears according to the force applied.

16-118. Design a "decommercializer" that will automatically cut out all TV commercial sounds for 60 sec.

16-119. Design a solar-powered refrigerator.

16-120. Design a small portable means for converting seawater to drinking water.

16-121. Design a fishing lure capable of staying at any preset depth.

16-122. Design an educational toy that may be used to aid small children in learning to read.

16-123. Design some device to help a handicapped person.

16-124. Design a heating and cooling blanket.

16-125. Design an automatic pulse-monitoring system for use in hospitals.

16-126. Design a portable solar cooker.

16-127. Design a carbon monoxide detector for automobiles.

16-128. Design a more effective method for prevention and/or removal of snow and ice from military aircraft.

16-129. Design a "practical" vehicle whose operation is based upon the "ground-effect" phenomenon.

16-130. Design a neuter (neither male nor female) connector for quick connect and disconnect that can be used on the end of flexible hose to transport liquids.

16-131. Design an electric space heater rated from 10,000 Btu/hr to 50,000 Btu/hr for military use in temporary huts and enclosures.

16-132. There is need for a system whereby one device emplaced in a hazardous area (minefield or other denial area) would interact with another device issued to each soldier, warn him of danger, and send guidance instructions for him to avoid or pass through the area of safety. Design such a system.

16-133. Develop some method to rate and/or identify the presence of rust spots when coatings fail to protect metal adequately. Present visual methods are unreliable and variable in results.

16-134. Develop a system whereby diseases of significance could be diagnosed rapidly and accurately.

16-135. Design a strong, flexible, lumpless, V-belt connector.

16-136. Design an inexpensive system for keeping birds out of ripening fruit trees.

16-137. Design a replacement for the paper stapler which will not puncture the paper.

17

The engineering design process

To many people engineering design means the making of engineering drawings, putting on paper ideas that have been developed by others, and perhaps supervising the construction of a working model. While engineers should possess the capability to do these things, the process of engineering design includes much more: the *formulation* of problems, the *development* of ideas, their *evaluation* through the use of models and analysis, the *testing* of the models, and the *description* of the design and its function in proposals and reports.

An engineering problem may appear in any size or complexity. It may be so small that an engineer can complete it in one day or so large that it will take a team of engineers many years to complete. It may call for the design of a tiny gear in a big machine, perhaps the whole machine, or an entire plant or process which would include the machine as one of its components. When the design project gets so big that its individual components can no longer be stored in one man's head, then special techniques are required to catalog all the details and to ensure that the components of the system work harmoniously as a coherent unit. The techniques which have been developed to ensure such coordination are called *systems design*.

Regardless of the complexity of a problem that might arise, the *method* for solving it follows a pattern similar to that represented in Figure 17-1. Each part of this "cyclic" process will be described in more detail, but first, two general characteristics of the process should be recognized:

1. Although the process conventionally moves in a circular direction, there is continuous "feedback" within the cycle.
2. The method of solution is a repetitious process that may be continuously refined through any desired number of cycles (Figure 17-2).

The concept of *feedback* is not new. For example, feedback is used by an individ-

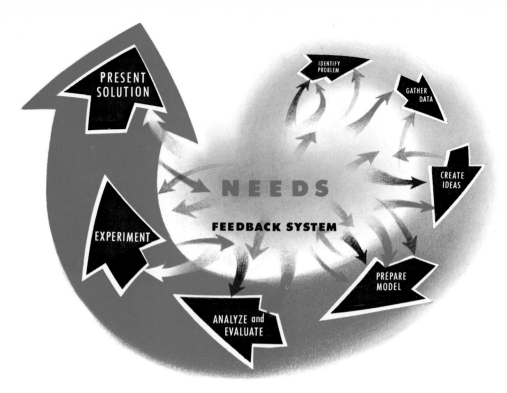

Figure 17-1 The design process.

ual to evaluate the results of actions that have been taken. The eye sees something bright that appears desirable and the brain sends a command to the hand and fingers to grasp it. However, if the bright object is also hot to the touch, the nerves in the fingers feed back information to the brain with the message that contact with this object will be injurious, and pain is registered to emphasize this fact. The brain reacts to this new information and sends another command to the fingers to release contact with the object. Upon completion of the feedback loop, the fingers release the object (Figure 17-3).

Another example is a thermostat. As part of a heating or cooling system, it is a feedback device. Changing temperature conditions produce a response from the thermostat to alter the heating or cooling rate.

The rate at which one proceeds through the problem-solving cycle is a function of many factors, and these factors change with each problem. Considerable time or very

Figure 17-2

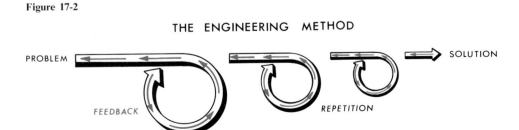

THE ENGINEERING METHOD

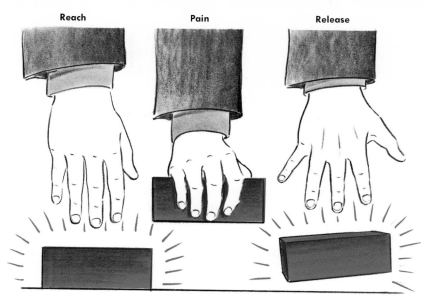

| Reach | Pain | Release |

Figure 17-3

little time may be spent at any point within the cycle, depending upon the situation.

Thus the problem-solving process is a dynamic and constantly changing process that provides allowances for the individuality and capability of the user.

The *design process* is used in each of the phases of design that were described in Chapter 16. Each phase starts with the identification of the problem and ends with a report. Some parts of the loop are more important in one phase than another. For example, the search for ideas is most important during the feasibility study and the preliminary design phases, as compared to analysis and experimentation, which tend to predominate in the preliminary and detail design phases. The *solution* of one phase often leads directly to the problem formulation for the next phase.

Identification of the problem

One of the biggest surprises that awaits the newly graduated engineer is the discovery that there is a significant difference between the classroom problems that he solved in school and the real-life problems that he is now asked to solve. This is true because problems encountered in real life are poorly defined. The individuals who propose such problems (whether they be commercial clients or the engineer's employer) rarely know or specify exactly what is wanted, and the engineer must decide for himself what information he needs to secure in order to solve the problem. In the classroom he was confronted with well-defined problems, and he usually was given most of the facts necessary to solve them in the problem statements. Now he finds that he has available insufficient data in some areas and an overabundance of data in others. In short, he must first find out what the problem *really* is. In this sense he is no different from the physician who must diagnose an illness or the attorney who must research a case before he appears in court. In fact, problem formulation is one of the most interesting and difficult tasks that the engineer faces. It is a necessary task, for one can arrive at a good and satisfactory solution only if the problem is fully understood. Many poor designs are the result of inadequate problem statements.

The ideal client who hires a designer to solve a problem will know what he wants the designer to accomplish; that is, he knows his problem. He will set up a list of limitations or restrictions that must be observed by the designer. He will know that an *absolute* design rarely exists—a *yes* or *no* type of situation—and that the designer usually has a number of choices available. The client can specify the most appropriate optimization criteria on which the final selection (among these choices) should be based. These criteria might be cost, or reliability, or beauty, or any of a number of other desirable results.

The engineer must determine many other basic components of the problem statement for himself. He must understand not only the task that the design is required to perform but what its range of performance characteristics is, how long it is expected to last in the job, and what demands will be placed on it one year, two years, or five years in the future. He must know the kind of an environment in which the design is to operate. Does it operate continuously or intermittently? Is it subject to high temperatures, or moisture, or corrosive chemicals? Does it create noise or fumes? Does it vibrate? In short, what type of design is best suited for the job?

For example, let us assume that the engineer has been asked by a physician to design a flow meter for blood. What does he need to know before he can begin his

Illustration 17-1
In the design process one of the most difficult tasks is to accurately define *the problem.*

design? Of course he should know the quantity of blood flow that will be involved. Does the physician want to measure the flow in a vein or in an artery? Does he want to measure the flow in the very small blood vessels near the skin or in the major blood vessels leading to and from the heart? Does he want to measure the average flow of blood or the way in which the blood flow varies with every pulse beat? How easy will it be to have access to the blood vessels to be tested? Will it be better to measure the blood flow without entering the vessel itself, or should a device be inserted directly into the vessel? One major problem in inserting any kind of material into the blood stream is a strong tendency to produce blood clots. In case an instrument can be inserted into the vessel, how small must it be so that it does not disturb the flow which it is to measure? How long a section of blood vessel is available, and how does the diameter of the blood vessel vary along its length and during the measurement? These and many more components of the problem statement must be determined by the engineer before an effective solution can be designed.

Another example of the importance and difficulty of problem definition is the urban transportation problem. Designers have proposed bigger and faster subways, monorails, and other technical devices because the problem was assumed to be simply one of transporting people faster from the suburbs into the city. In many cases, it was not questioned whether the problem that they were solving was *really* the problem that needed a solution.

Surely the suburbanite needs a rapid transportation system to get into the city, but the rapid transport train is not enough. He must also have "short haul" devices to take him from the train to his home or to his work with a minimum of walking and delay. Consequently, the typical rapid transit system must be coordinated with a citywide network of slower and shorter-distance transportation which permit the traveler to exit near his job, wherever it may be. For the suburbanite, speed is not nearly as important as frequent, convenient service on which he can rely and for which he need not wait.

Urbanites, particularly the poor, who generally live far from the places where they might find work, are also in need of better transportation. For these people, high speed again is not nearly as important as low cost and transportation routes and vehicles that provide access to the job market. Instead of placing emphasis on bigger and faster trains, designers should consider the *wants* and the *needs* of the people they are trying to serve and determine what these wants and needs really are.

How does the engineer find out? How does he define his problem and know that his definition is in fact what is needed? Of course the first step is to find out what is already known. He must study the literature. He must become thoroughly familiar with the problem, with the environments in which it operates, with similar machines or devices built elsewhere, and with peculiarities of the situation and the operators. *He must ask questions.*

It may be, after evaluating the available information, that the engineer will be convinced that the problem statement is unsatisfactory—just as today's statement of the transportation problem appears to be unsatisfactory. In that case he may suggest or perform additional studies—studies that involve the formulation of simulation models of the situation and the environment in which the machine is to be built. They

Illustration 17-2
The design of rapid transportation systems of the future must be carefully coordinated with the design of the city.

may include experiments with these models to show how this environment would react to various solutions of the problem.

The design engineer must work with many types of people. Some will be knowledgeable in engineering—others will not. His design considerations will involve many areas other than engineering, particularly during problem formulation. He must learn to work with physicists and physicians, with artists, architects, and city planners, with economists and sociologists—in short, with all those who may contribute useful information to a problem. He will find that these men have a technical vocabulary different from his. They look at the world through different eyes and approach the solution of problems in a different way. It is important for the engineer to have the experience of working with such people before he accepts a position in industry, and

Perhaps the most valuable result of all education is the ability to make yourself do the thing you have to do, when it ought to be done, whether you like it or not—however early a man's training begins, it is probably the last lesson that he learns thoroughly.
—Thomas Huxley

We can have facts without thinking but we cannot have thinking without facts.
—John Dewey

what better opportunity is there than to make their acquaintance during his college years. With the manifold problems that tomorrow's engineer will face—problems that involve human values as well as purely technical values—collaboration between the engineer and other professional people becomes increasingly important.

Collection of information

The amount of technical information available to today's scientists and engineers is prodigious and increasing daily. Two hundred years ago, during the time of Jefferson and Franklin, it was possible for an individual to have a fair grounding in all the social and physical sciences then known, including geography, history, medicine, physics, and chemistry, and to be an authority in several of these. Since the Industrial Revolution, or about the middle of the last century, the amount of knowledge in all the sciences has grown at such a rapid rate that no one can keep fully abreast of one major field, let alone more than one. It has been estimated that if a person, trained in speed reading, devoted 20 hours a day, seven days a week to nothing but study of the literature in a relatively specialized field, such as mechanical engineering, he would barely keep up with the current literature. He would not have time to go backward in time to study what has been published before or to consider developments in other fields of engineering. How then may one be able to find information that is available, or know what has been done concerning the solution of a particular problem? The answer is twofold: know *where* the information resources are located, and know *how* to retrieve information from a vast resource.

A typical technical library may contain from 10,000 to 200,000 books. It may subscribe to as many as 500 technical and scientific magazines, as well as a large store of technical reports published at irregular intervals by government agencies, universities, research institutes, and industrial organizations. The problem then is principally one of finding the proper books or articles.

Libraries have become quite efficient at cataloging books and major reports in their general catalog file. Usually these catalogs are arranged into three groups, one by author, one by title, and one by subject matter. Although the library catalog is an excellent source of book references, it does not contain any of the thousands of articles in magazines, technical journals, and special reports.

One's direction to these journal articles and special reports is through the refer-

Illustration 17-3
An overabundance of data does not necessarily guarantee that the engineer's task will be simplified.

ence section in the library. Here the abstract journals and books devoted to collecting and ordering all publications in a particular field are housed. For engineers, two of the most useful of these are the *Engineering Index* and the *Applied Science and Technology Index*. They appear annually and contain short abstracts of most important articles appearing in the engineering field. Articles are organized according to subject headings, so that all articles on a similar subject appear together. By looking under the appropriate heading, the searcher can discover the references of greatest interest to him or related headings where other references might be found. After satisfying himself that he has the correct references, the researcher then goes to the appropriate periodicals to find the full articles. There are many other indexes besides the two mentioned above, some more, some less specialized.

As an example, assume that we are concerned with the design of a pipeline to transport solid refuse (garbage) from the center of a large city to a disposal site where it may be processed, incinerated, or buried. Let us follow and observe an engineer making the required library search.

Literature search on "pipelining of refuse"

(The following "capsule narrative" indicates what actually happened during a quick noncomprehensive search conducted in an afternoon at a typical university library, and it is typical of the kind of search that an engineer might make for a brief study.)

> Fool me once, shame on you; fool me twice, shame on me.
> —Chinese proverb

Going first to the "subject" section of the catalog file, I could think of only three headings to look under: Refuse, Garbage, and Pipelines. There were 24 entries under "Refuse and Refuse Disposal." Some dealt with conveyors and trucking but only one with pipelining (not surprisingly since this is not a common way to convey garbage). I copied some of the titles and reference numbers because they might help to give me some idea of the composition and consistency of garbage and of the shredders and other devices used to make refuse more uniform in size and more capable of being transported in a pipeline.

The card under "Garbage" referred me right back to "Refuse and Refuse Disposal," a dead end.

Under "Pipe" there were some 75 entries under 23 different subheadings from "Pipe-Asbestos, Cement" to "Pipe-Welding." I copied the titles and numbers shown in Figure 17-4.

I wasn't satisfied that I had exhausted the subject file but could not think of any other pertinent headings. So I went next to the reference library and sat down in front of the shelf with the *Engineering Index*. The latest complete year was 1974. Looking under the headings like "Pipeline" and "Refuse Disposal" I paid particular attention to "See also" lists, Figure 17-5, and these eventually led me to a veritable gold mine of references under "Materials Handling." A few of these are shown in Figure 17-6. Notice that two of the most interesting articles are in German. If I find them, I will have to have them translated.

Figure 17-4

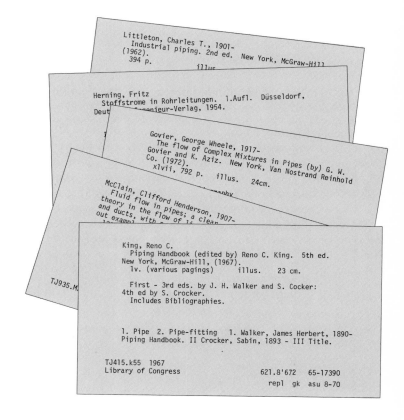

Littleton, Charles T., 1901-
  Industrial piping. 2nd ed. New York, McGraw-Hill
  (1962).
  394 p.   illus.

Herning, Fritz
  Stoffstrome in Rohrleitungen. 1.Aufl. Düsseldorf,
  Deut   ----nieur-Verlag, 1954.

Govier, George Wheele, 1917-
  The flow of Complex Mixtures in Pipes (by) G. W.
  Govier and K. Aziz. New York, Van Nostrand Reinhold
  Co. (1972).
  xlvii, 792 p. illus. 24cm.

McClain, Clifford Henderson, 1907-
  Fluid flow in pipes; a clea-
  theory in the flow of 1---
  and ducts,
  out exampl with -

King, Reno C.
  Piping Handbook (edited by) Reno C. King. 5th ed.
New York, McGraw-Hill, (1967).
  1v. (various pagings)   illus.   23 cm.

  First - 3rd eds. by J. H. Walker and S. Cocker:
4th ed by S. Crocker.
  Includes Bibliographies.

1. Pipe 2. Pipe-fitting   1. Walker, James Herbert, 1890-
Piping Handbook. II Crocker, Sabin, 1893 - III Title.

TJ415.k55 1967
Library of Congress   621.8'672   65-17390
                             repl gk asu 8-70

TJ935.M

REFUSE DISPOSAL

See Also Coal Mines and Mining–Waste Disposal; Coal Preparation–Refuse Disposal; Industrial Wastes.

TRANSPORTATION

See also Air Transportation; City Planning; Coal Transportation; Highway Engineering–Research; Highway Systems–Planning; Industrial Plants–Location; Mines and Mining–Surface Transportation; Motor Bus Transportation; Motor Truck Transportation; Noise–Control; Operations Research; Power Plant Equipment–Transportation; Railroads; Rapid Transit; Street Traffic Control; Traffic Surveys; Vehicles–Ground Effect; Waterway Transportation.

COAL HANDLING

See also Bins; Coal Transportation.

Hydraulic. See also Materials Handling–Pipe Lines; Ore Handling–Hydraulic; Pipe, Plastic.

MATERIALS HANDLING

See also Air Transportation; Aircraft Carriers; Barges; Bins; Bottle Caps; Cableways; Cargo Handling; Cement Handling; Chemicals–Transportation; Coal Handling; Construction Equipment; Containers; Conveyors; Cranes; Elevators; Flour Mills; Food Products–Packaging; Foundry Practice–Cleaning; Freight Handling; Furniture Manufacture–Finishing; Grinding Mills; Hoists; Industrial Engineering; Industrial Trucks; Logging; Machine Tools; Mail Handling; Mines and Mining–Surface Transportation; Nuclear Fuels–Safe Handling; Ore Handling; Packaging; Packaging Machines; Packaging Materials; Production Planning and Control; Rivers–Revetments; Salt–Storage; Scales and Weighing; Silos; Systems Engineering; Transportation; Warehouses.

Figure 17-5 Typical headings from *Engineering Index*.

Now that I know some of the best headings, I look through other years' editions of the *Index* and also study other indexes. I can also go back to the subject catalog and look under the headings that have been productive in the *Index,* headings like "Materials Handling," of which I did not think the first time.

In this way, in a short afternoon, one can assemble a reasonably good reference list on any subject one needs to study.

Next I have to obtain copies of those articles that I want to study in their entirety. If I found them as books in the subject catalog, I can ask for them or look for them myself in the stacks. For articles which have appeared in magazines, like the references from the *Engineering Index,* I first have to find out whether these references are available in the library and, if so, I'll find the appropriate order number and borrow them from the library. If they are particularly interesting, I may have the library make me a photostatic copy so that I can have a permanent record of the article.

If it is important for the searcher to find the very latest work done in his subject, the library will not be of much help. There is a time delay between the performance of a piece of research, its publication, and its appearance in any of the abstract

> Knowledge is of two kinds: we know a subject ourselves, or we know where we can find information upon it.
> —Samuel Johnson

042824 UMSCHLAG VON MASSENGUETERN SYSTEMATIC DER ANLAGEN-KOMPONENTEN. (Handling Bulk Goods, Systematics of the Installation Components.) Handling bulk goods is defined in all the various forms starting from the purpose of handling and storage. It is always resorted to where goods have to be transferred from one handling equipment to the other. The author describes the various methods of trans-shipment of bulk goods which can be classified according to the method of accumulation or to the time the goods remain in the rail and water termina. He also deals with the assessment of transshipment and distribution equipment and bulk goods transshipment installations in harbors. 10 refs. In German with English abstract.

Lubrich, W. S. W. Foerdern Heben v 23 n 14 Oct 1973 p 763-766.

042828 MATERIAL HANDLING AND THE IN-DUSTRIAL ENGINEER. The subjects discussed at this conference held in Philadelphia, Pa. from Jan 16-18 1974 can be broadly categorized as follows: analysis of materials handling problems which is a vital first step to efficient operation, selection criteria and determination of costs, computers in materials handling, automatic warehousing and retrieval systems and efficient plant layout and planned material flow. The 6 conference papers are indexed separately.

Semin, AIIE Mater Handl and the Ind Eng. Semin. 3rd Tech Pap. Philadelphia, Pa. Jan 16-18 1974 Publ by AIIE, New York, 1974, 44p.

042939 PROPOSED TRANS-ALASKA INTEGRATED PIPELINE TRANSPORTATION SYSTEM. The article describes the proposed Integrated Pipeline Transportation (IPT) system across Alaska (including its structural design features) and how it could be constructed, gives the advantages and obstacles to the plan, a breakdown of the cost, and discusses how the project could be financed.

Nasser, George D. Prestressed Concr Inst. Chicago, Ill. J Prestressed Concr Inst v 18 n Sep-Oct 1973 p 18-31.

042940 UEBERBLICK UEBER BETRIEBSAN-LAGEN ZUM HYDRAULISCHEN FESTSTOFF-TRANSPORT. (Survey of Installations for Pipelining of Solids.) Paper contains a world-wide survey of the most important installations for pipelining of solid materials with particular emphasis on installations located in Eastern Europe and Asia. Characteristic data related; throughput (slurry and solids); velocity of the flow, diameter of the pipeline; specific power requirements; screen analysis of the solids; distance involved; and type of solids are given in tabulated form. 21 refs. In German.

Boehme, Frank Ration Braunkohle, Grossaeschen, E Ger. Neue Bergbautech v 3n 8 Aug 1973 p 567-570.

061680 SIZE CHARACTERISTICS OF MUNICIPAL SOLID WASTE. A reclamation system currently under development at MIT seeks to separate automatically many items of refuse in the form in which they come from the packer truck. This system eliminates the unnecessary cost of pulverizing or pulping homogeneous refuse items such as bottles,cans and newspapers, and at the same time preserves them in the form in which their recycle value is a maximum. 5 refs.

Figure 17-6 Typical abstracts from *Engineering Index.*

journals. This delay is usually three years or more. The only source for the very latest materials is the expert himself. If one has made an exhaustive literature survey, he has usually found one or more researchers who are specialists and have published extensively in the field under investigation. These people are also the ones who probably can provide the latest technical information in the field. Often these men are happy to share their knowledge with the searcher in the field. However, it is customary to offer them a consultant's remuneration if a substantial amount of their time is required for this service.

Generation of ideas

It is incorrect to use the terms *synthesis, innovation,* and *creativity* interchangeably. They are not synonymous but all are used in the generation of ideas to solve engineering problems. *Synthesis* is the assembly of well-known components and parts to form a solution. *Innovation* is the discovery of a new, novel, or unusual idea or product by applying logic, experience, or artistry. *Creativity* originates an entirely new concept in response to a human need, a solution which is both satisfying and innovative. It presupposes an understanding of human experience and human values.

Problem solving does not necessarily require creative thought. Many kinds of problems can be solved by careful discriminating logic. An electronic computer can be programmed to perform synthesis—and perhaps even a certain degree of innovation—but it cannot create. Creativity is a *human* endeavor.

The engineer who redesigns a radio or improves an automobile engine uses established techniques and components; he synthesizes. Innovators are those who build something new, and who combine different ideas and facts with a purpose. Creativity is one of the rarest and highest forms of human activity. We only call those individuals "creative" who originate, make, or cause to come into existence an entirely new concept or principle. (Patents are mostly the result of clever innovation, rather than creative effort.) If we had to rely on creativity for patents, we would not have the nearly 4 million patents in the United States alone. All engineers must synthesize, some will innovate, but only a very few are able to be truly creative.

Since there is always a great demand for creative and innovative ideas, many attempts have been made to develop procedures for stimulating them. Certain of these procedures will work satisfactorily in one situation, yet at other times different methods may be needed.

There are many methods of stimulating ideas that are used in industry today: (*a*) the use of checklists and attribute lists, (*b*) reviewing of properties and alternatives, (*c*) systematically searching design parameters, (*d*) brainstorming, and (*e*) synectics. These methods will be discussed briefly.

Checklists and attribute lists

One of the simplest ways for an individual to originate a number of new ideas in a minimum amount of time is to make use of prepared lists of general questions to apply to the problem under consideration. A typical list of such questions might be the following:

1. In what ways can the idea be improved in quality, performance, and appearance?
2. To what other uses can the idea be put? Can it be modified, enlarged, or minified?
3. Can some other idea be substituted? Can it be combined with another idea?
4. What are the idea's advantages and disadvantages? Can the disadvantages be overcome? Can the advantages be improved?
5. What is the particular scientific basis for the idea? Are there other scientific bases that might work equally well?
6. What are the least desirable features of the idea? The most desirable?

Attribute listing is a technique of idea stimulation that has been most effective in improving tangible things—such as products. It is based upon the assumption that most ideas are merely extensions or combinations of previously recognized observations. Attribute listing involves the following:

1. Listing the key elements or parts of the product.
2. Listing the main features, qualities, or significant attributes of the product and of each of its key elements or parts.

3. Systematically modifying, changing, or eliminating each feature, quality, or attribute so that the *original purpose* is better satisfied, or perhaps a new need is fulfilled.

With both checklists and attribute lists one must be careful to recognize that these methods are merely "stimulators" and that they are not intended to replace original and intelligent thinking. They are certainly not intended to be used as crutches. Rather, like a wrench which extends the power or leverage of a man's fingers or arm, these ideation tools extend the power and effectiveness of the mind.

Reviewing the properties and alternatives

Another rather common procedure, somewhat similar to attribute listing, is to consider how all the various properties or qualities of a particular design might be changed, modified, or eliminated. This method lists the modifiable properties such as weight, size, color, odor, taste, shape, and texture. Functions that are desirable for the item's intended use may also be listed: automatic, strong, durable, or lightweight. After developing these lists, the engineer can consider and modify each property or function individually.

Imagine redesigning a lawn mower. The listed properties might include (1) metal, (2) two cycle, gasoline-powered, (3) four wheels, (4) rotary blade, (5) medium weight, (6) manually propelled, (7) chain driven, and (8) green in color. In beginning the design of an improved lawn mower, the engineer might first consider other possibilities for each property. What other materials could be used? Can the engine be improved—what about using electrical power? Should the mower operate automatically? Should the type of blade motion be changed? Questions like these may suggest how the design *could* be improved. The properties of lawn mowers have been changed many times, and these changes have presumably made lawn mowers more efficient and easier to use.

Again, besides considering the product's various properties the engineer must question, observe, and associate its functions. Can these functions be modified, rearranged, or combined? Can the product serve other functions or be adapted to other uses? Can we change the shape (magnify or minify parts of the design)? With this type of questioning we can stimulate ideas that will bring design improvements to the product.

Systematic search of design parameters

Frequently it is advisable to investigate alternatives more thoroughly. A systematic search considers all possible combinations of given conditions or design parameters. This type of search is frequently called a "matrix analysis" or a "morphological synthesis" of alternatives.[1] Its success in stimulating ideas depends upon the engi-

[1] K. W. Norris, "The Morphological Approach to Engineering Design," in J. Christopher Jones, *Conference on Design Methods,* Macmillan, New York (1963), p. 116.

> Use logic to decide between alternatives—not to initiate them.
>
> They can have any color they want . . . just as long as it's black.
> —Henry Ford

neer's ability to identify the significant parameters that affect the design. The necessary steps for implementing this type of idea search are the following:

1. **Describe the problem.** This description should be broad and general, so that it will not exclude possible solutions.
2. **Select the major independent-variable conditions** required in combination to describe the characteristics and functions of the problem under consideration.
3. **List the alternative methods** that satisfy each of the independent-variable conditions selected.
4. **Establish a matrix** with each of the independent-variable conditions as one axis of a rectangular array. Where more than three conditions are shown, the display can be presented in parallel columns.

Let us consider a specific example to see how this method can be applied.

1. *Problem Statement:* A continuous source of contaminate-free water is needed.
2. *Independent-Variable Conditions:*
   Energy
   Source
   Process
3. *Methods of Satisfying Each Condition:*
   Types of Energy:
   a. Solar
   b. Electrical
   c. Fossil
   d. Atomic
   e. Mechanical
   Types of Source:
   a. Underground
   b. Atmosphere
   c. Surface supply
   Types of Process:
   A. Distillation
   b. Transport
   c. Manufacture
4. *The Matrix* (Figure 17-7)
5. *Combinations*

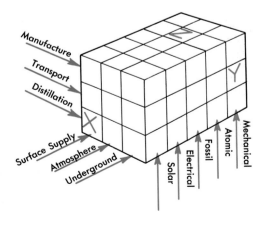

Figure 17-7

> Necessity may be the mother of invention, but imagination is its father.
>
> Whatever one man is capable of conceiving, other men will be able to achieve.
> —Jules Verne
>
> Originality is just a fresh pair of eyes.
> —W. Wilson

This particular matrix may be represented as an orderly arrangement of 45 small blocks stacked to form a rectangular parallelepiped. Every block will be labeled with the designations selected previously. Thus, block X in our preceding example suggests obtaining pure water by distilling a surface supply with a solar-energy power source, block Y means transporting water from an underground source by some mechanical means, and block Z recommends manufacturing water from the atmosphere using atomic power. Obviously, some of the blocks represent well-known solutions, and others suggest absurd or impractical possibilities. But, some represent untried combinations that deserve investigation.

Where more than three variables are involved, electronic computers may be used to excellent advantage. After the matrix has been programmed, the computer can print a list of all the alternative combinations. Use of the computer is especially helpful when considering a large number of parameters.

The preceding techniques of stimulating new design concepts are particularly useful for the individual engineer. But often several designers may be searching jointly for imaginative ideas about some particular product. Then it is advantageous to use one of the procedures, "brainstorming" or "synectics."

Brainstorming

The term "brainstorming" was coined by Alex F. Osborn[2] to describe an organized group effort aimed at solving a problem. The technique involves compiling all the ideas which the group can contribute and deferring judgment concerning their worth. This is accomplished (1) by releasing the imagination of the participants from restraints such as fear, conformity, and judgment; and (2) by providing a method to improve and combine ideas the moment an idea has been expressed. Osborn points out that this collaborative group effort does not replace individual ideative effort. Group brainstorming is used solely to supplement individual idea production and works very effectively for finding a large volume of alternative solutions or novel design approaches. It has been particularly useful for stimulating imaginative ideas for new products or product development. It is not recommended where the problem solution will depend primarily on judgment or where the problem is vast, complex, vague, or controversial. A homogeneous "status group" of six to twelve persons seems to be best for stimulating ideas with this method. However, the U.S. Armed Forces have used a hundred or more participants effectively. The typical brainstorming session has only two officials: a chairman and a recorder. The chairman's responsibility is to provide each panel member with a brief statement of the problem, preferably 24 hours prior to the meeting. He should make every effort to describe the problem in clear, concise terms. It should be *specific*, rather than *general*, in

[2] Alex F. Osborn, *Applied Imagination,* Scribner's, New York (1963), p. 151.

nature. Some examples of ideas that satisfy the problem statement may be included with the statement. Before beginning the session, the chairman should review the rules of brainstorming with the panel. These principles, although few, are very important and are summarized as follows:

1. **All ideas which come to mind are to be recorded.** No idea should be stifled. As Osborn says, "The wilder the idea, the better; it is easier to tame down than to think up." He recommends recording ideas on a chalkboard as they are suggested. Sometimes a tape recorder can be very valuable, especially when panel members suggest several different ideas in rapid succession.
2. **Suggested ideas must not be criticized or evaluated.** Judgments, whether adverse or laudatory, *must be withheld* until after the brainstorming session, because many ideas which are normally inhibited because of fear of ridicule and criticism are then brought out into the open. In many instances, ideas that would normally have been omitted turn out to be the best ideas.
3. **Combine, modify, alter, or add to ideas as they are suggested.** Participants should consciously attempt to improve on other people's ideas, as well as contributing their own imaginative ideas. Modifying a previously suggested idea will often lead to other entirely new ideas.
4. **The group should be encouraged to think up a large quantity of ideas.** Research at the State University of New York at Buffalo[3] seems to indicate that, when a brainstorming session produces more ideas, it will also produce higher-quality ideas.

The brainstorming chairman must always be alert to keep *evaluations* and *judgments* from creeping into the meeting. The spirit of enthusiasm that will permeate the group meeting is also very important to the success of the brainstorming session. The entire period should be conducted in a free and informal manner. It is most important to maintain, throughout the period, an environment where the group members are not afraid of seeming foolish. Both the speed of producing and recording ideas, and the number of ideas produced, help create this environment. Each panel member should bring to the meeting a list of new ideas that he has generated from the problem statement. These ideas help to get the session started. In general, the entire brainstorming period should not last more than 30 minutes to 1 hour.

The recorder keeps a stenographic account of all ideas presented and after the session lists them by type of solution without reference to their source. Team members may add ideas to the accumulated list for a 24-hour period. Later, the entire list of ideas should be rigorously evaluated, either by the original brainstorming group or, preferably, by a completely new team. Many of the ideas will be discarded quickly—others after some deliberation. Still others will likely show promise of success or at least suggest how the product can be improved.

Some specialists recommend that the brainstorming team include a few persons who are broadly educated and alert but who are amateurs in the particular topic to be discussed. Thus new points of view usually emerge for later consideration. Usually executives or other people mostly concerned with *evaluation* and *judgment* do not make good panel members. As suggested previously, particular care should be taken to confine the problem statement within a narrow or limited range to ensure that all

---

[3]Sidney J. Parnes and Arnold Meadow, "Effects of Brainstorming Instructions on Creative Problem Solving by Trained and Untrained Subjects," *Journal of Educational Psychology*, Vol. 50, No. 4 (1959), p. 176.

team members direct their ideas toward a common target. Brainstorming is no substitute for applying the fundamental mathematical and physical principles the engineer has at his command. It should be recognized that the objective of brainstorming is to stimulate ideas—not to effect a complete solution for a given problem.

Dr. William J. J. Gordon[4] has described a somewhat similar method of group therapy for stimulating imaginative ideas, which he calls "synectics."

Synectics

This group effort is particularly useful to the engineer in eliciting a radically new idea or in improving products or developing new products. Unlike brainstorming, this

[4] William J. J. Gordon, *Synectics,* Harper & Row, New York (1961).

> The person who is capable of producing a large number of ideas per unit of time, other things being equal, has a greater chance of having significant ideas.
> —J. P. Guilford
>
> What good is electricity, Madam? What good is a baby?
> —Michael Faraday
>
> He that answereth a matter before he heareth it, it is folly and shame unto him.
> —Proverbs xviii.13
>
> No idea is so outlandish that it should not be considered with a searching but at the same time with a steady eye.
> —Winston Churchill

technique does not aim at producing a large number of ideas. Rather, it attempts to bring about one or more solutions to a problem by drawing seemingly unrelated ideas together and forcing them to complement each other. The synectics participant tries to *imagine* himself as the "personality" of the inanimate object: "What would be my reaction *if I were that gear* (or drop of paint, or tank, or electron)?" Thus, familar objects take on strange appearances and actions, and strange concepts often become more comprehensible. A key part of this technique lies in the group leader's ability to make the team members "force-fit" or combine seemingly unrelated ideas into a new and useful solution. This is a difficult and time-consuming process. Synectics emphasizes the conscious, preconscious, and subconscious psychological states that are involved in all creative acts. In beginning, the group chairman leads the members to understand the problem and explore its *broad* aspects. For example, if a synectics group is seeking a better roofing material for traditional structures, the leader might begin a discussion on "coverings." He could also explore how the colors of coverings might enhance the overall efficiency (white in summer, black in winter). This might lead to a discussion of how colors are changed in nature. The group leader could then focus the group on more detailed discussion of how roofing materials could be made to change color automatically to correspond to different light intensities—like the biological action of a chameleon or a flounder. Similarly, the leader might approach the problem of devising a new type of can opener by first leading a group discussion of the word "opening," or he could begin considering a new type of lawn mower by first discussing the word "separation."

In general, synectics recommends viewing problems from various analogous situations. Paint that will not adhere to a surface might be viewed as analogous to water running off a duck's back. The earth's crust might be seen as analogous to the peel of an orange. The problem of enabling army tanks to cross a 40-ft-wide, bottomless crevass might be made analogous to the problem that two ants have in crossing chasms wider than their individual lengths.

Synectics has been used quite successfully in problem-solving situations in such diverse fields as military defense, the theater, manufacturing, public administration, and education. Where most members of the brainstorming team are very knowledgeable about the problem field, synectics frequently draws the team members from diverse fields of learning, so that the group spans many areas of knowledge. Philosophers, artists, psychologists, machinists, physicists, geologists, biologists, as well as engineers, might all serve equally well in a synectics group. Synectics assumes that someone who is imaginative but not experienced in that field may produce as many creative ideas as one who *is* experienced in that field. Unlike the expert, the novice can stretch his imagination. He approaches the problem with fewer preconceived ideas or theories, and he is thus freer from binding mental restrictions. (Obviously, this will not be true when the problem requires analysis or evaluation, where experience is a vital factor.) There is always present in the synectics conference an expert in the particular problem field. The expert can use his superior technical knowledge to give the team missing facts, or he may even assume the role of "devil's advocate," pointing out the weaknesses of an idea the group is considering. *All* synectics sessions are tape recorded for later review and to provide a permanent record.

Many believe that brainstorming comes to grips with the problem too abruptly while synectics delays too long. However, industry is using both methods successfully today.

Illustration 17-6

Preparation of a model

Thus far in analyzing the design process we have considered identification of the problem, collection of information, and generation of ideas. The next phase is one of bringing together all the diverse parts (ideas, data, parameters, etc.) into a meaningful whole. As suggested on page 466, this procedure is called *synthesis,* and it is best known to the engineer as the process of modeling.

Psychologists and others who study the workings of the human mind tell us that we can think effectively only about simple problems and small "bits" of information. They tell us that those who master complicated problems do so by reducing them to a series of simple problems which can be solved and synthesized to a final solution. This technique consists of forming a mental picture of the entire problem, and then simplifying and altering this picture until it can be taken apart into manageable components. These components must be simple and similar to concepts with which we are already familiar, to situations that we know. Such mental pictures are called *models.* They are simplified images of real things, or parts of real things—a special picture that permits us to relate it to something already known and to determine its behavior or suitability.

We are all familiar with models of sorts—with maps as models for a road system; with catalogs of merchandise as models of what is offered for sale. We have a model in our mind of the food we eat, the clothes we buy, and of the partner we want to marry.

We will often form judgments and make decisions on the basis of the model, even though the model may not be entirely appropriate. Thus the color of an apple may or may not be a sign of its ripeness, any more than the girl's apple-blossom cheeks and tip-tilted nose are the sign of a desirable girl-friend.

Engineering models are similar to sports diagrams that are composed of circles, squares, triangles, curved and straight lines, and other similar symbols which are used to represent a "play" in a football or basketball game (see Figure 17-8). Such geometrical models are limited because they are two dimensional and do not allow for the strengths, weaknesses, and imaginative decisions of the individual athletes. Their use, however, has proved to be quite valuable in simulating a brief action in the game and to suggest the best strategy for the player should he find himself in a similar situation.

An *idealized model* may emphasize the whole of the system and minimize its component parts, or it may be designed to represent only some particular part of the system. Its function is to make visualization, analysis, and testing more practical. The engineer must recognize that he is merely limiting the complexity of the problem in order to apply known principles. Often the model may deviate considerably from the true condition; and the engineer must, of necessity, select different models to represent the same real problem. Therefore, the engineer must view his answers with respect

> A damsel of high lineage and a brow
> May-blossom, and a cheek of apple-blossom
> Hawk-eyes; and lightly her slender nose
> Tip-tilted like the petal of a flower.
> —Tennyson

Illustration 17-7
Every football fan understands the value of diagrammed models in preparing for Saturday's "big game."

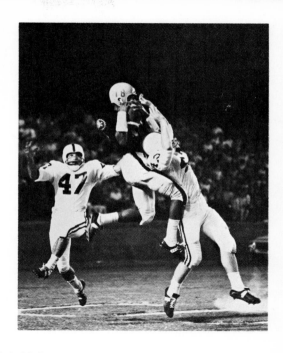

to the initial assumptions of the model. If the assumptions were in error, or if their importance was underestimated, then the engineer's analysis will not relate closely with the true conditions. The usefulness of the model to predict future actions must be verified by the engineer. This is accomplished by experimentation and testing. Refinement and verification by experimentation are continued until an acceptable model has been obtained.

Two characteristics, more than many others, determine an engineer's competence.

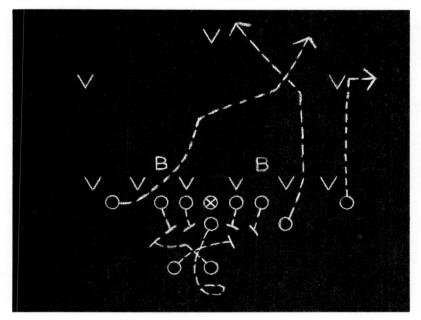

Figure 17-8

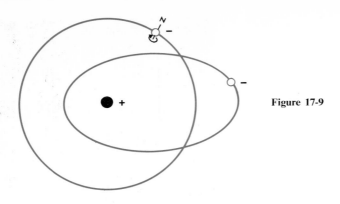

Figure 17-9

The first is his ability to devise simple, meaningful models; and second is the breadth of his knowledge and experience with examples with which he can compare his models. The simpler his models are, and the more generally applicable, the easier it is to predict the behavior and compute the performance of the design. *Yet models have value only to the engineer who can analyze them.* The beauty and simplicity of a model of the atom, Figure 17-9, will appeal particularly to someone familiar with astronomy. The free-body diagram of a wheelbarrow handle, Figure 17-10, has meaning only to someone who knows how such a diagram can be used to find the strength of the handle.

Aside from models for "things," we can make models of situations, environments, and events. The football or baseball diagram is such a model. Another familiar model of this type is the weather map, Figure 17-11, which depicts high- and low-pressure regions and other weather phenomena traveling across the country. Any meteorologist will tell you that the weather map is a very crude model for predicting weather, but that its simplicity makes the explanation of current weather trends more understandable for the layman. Models of situations and environmental conditions are particularly important in the analysis of large systems because they aid in predicting and analyzing the performance of the system before its actual implementation. Such models have been prepared for economic, military, and political situations and their preparation and testing is a science all its own.

Figure 17-10

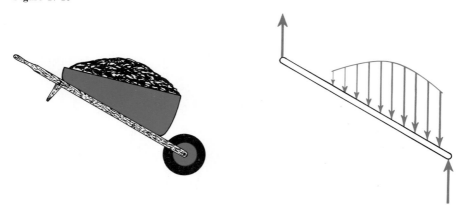

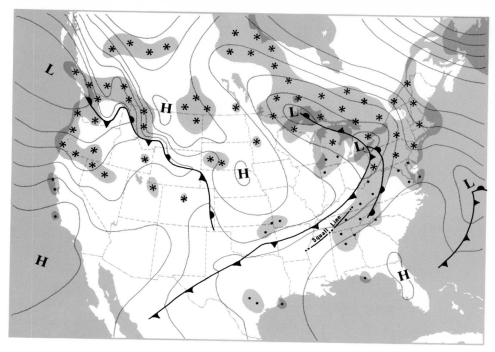

Figure 17-11

Charts and graphs as models

Charts and graphs are convenient ways to illustrate the relationship among several variables. We have all seen charts of the fluctuations of the stock market averages, Figure 17-12, in the newspaper from day to day, or you may have had your father plot your growth on the closet door. In these examples, *time* is one of the variables. The others, in the examples above, are the average value of the stock in dollars and your height in feet and inches, respectively. A chart or graph is not a model but

Figure 17-12

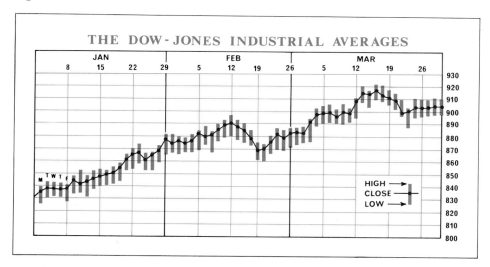

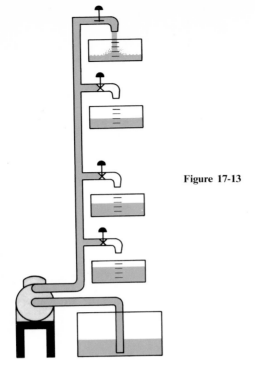

Figure 17-13

presents facts in a readily understandable manner. *It becomes a model only when used to predict, project, or draw generalized conclusions* about a certain set of conditions. Consider the following example of how facts can be used to develop a chart and a graphical model. An engineer may wish to test a pump and determine how much water it can deliver to different heights, Figure 17-13. Using a stopwatch and calibrated reservoirs at different heights, he measures the amount of water pumped to the different heights in a given time. His test results are plotted as crosses on a chart as shown in Figure 17-14. At this point, the plotted facts are a chart and not a model. Only when the engineer makes the assumption that the plotted points represent the typical performance of this or a similar pump under corresponding conditions can the chart be considered to be a graphical model. Once this assumption is made, he can draw a smooth curve through the points. With this performance curve as a model, the

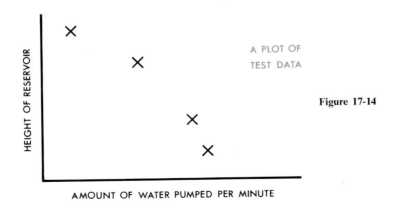

A PLOT OF
TEST DATA

Figure 17-14

engineer can predict that, if he put additional reservoirs between the actual ones, they would produce results much like those shown by the circles in Figure 17-15. He makes this assumption based on his experience that pumps are likely to behave in a "regular" way. *Now* he is using the graph as an engineering model of the performance of the pump.

The diagram

A model often used by the engineer is the *diagram*. Typical forms of diagrams are the *block diagram,* the *electrical diagram,* and the *free-body diagram.* Some attention should be given to each of these forms.

The *block diagram* is a generalized approach for examining the whole problem, identifying its main components, and describing their relationships and interdependencies. This type of diagram is particularly useful in the early stages of design work when representation by mathematical equations would be difficult to accomplish. Illustration 17-8 is an example of a block diagram in which components are drawn as blocks, and the connecting lines between blocks indicate the flow of information in the whole assembly. This type of presentation is widely used to lay out large or complicated systems—particularly those involving servoelectrical and mechanical devices. No attempt is made on the drawing to detail any of the components pictures. They are often referred to as "black boxes"—components whose *function* we know, but whose *details* are not yet designed.

The *energy diagram* is a special form of the block diagram and is used in the study of thermodynamic systems involving mass and energy flow. Some examples of the use of an *energy diagram* are given in Figures 17-16 and 17-17.

The *electrical diagram* is a specialized type of model used in the analysis of electrical problems. This form of *idealized model* represents the existence of particular electrical circuits by utilizing conventional symbols for brevity. These diagrams may be of the most elementary type, or they may be highly complicated and require many hours of engineering time to prepare. In any case, however, they are representations or models in symbolic language of an electrical assembly.

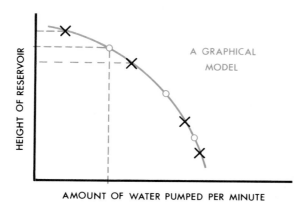

A GRAPHICAL MODEL

Figure 17-15

HEIGHT OF RESERVOIR

AMOUNT OF WATER PUMPED PER MINUTE

Illustration 17-8
The relation of
component parts of a
transistorized
telemetering system are
best shown by means of
a block diagram.

Figure 17-18 shows an electrical diagram of a photoelectric tube that is arranged to operate a relay. Notice that the diagram details only the essential parts in order to provide for electrical continuity, and thus is an idealization that has been selected for purposes of simplification.

The *free-body diagram,* Figure 17-19, is a diagramatic representation of a physical system which has been removed from all surrounding bodies or systems for purposes of examination and where the equivalent effect of the surrounding bodies is

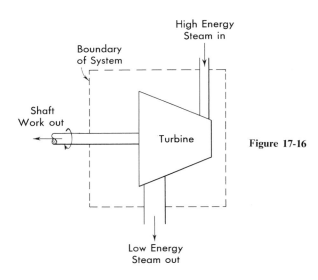

Figure 17-16

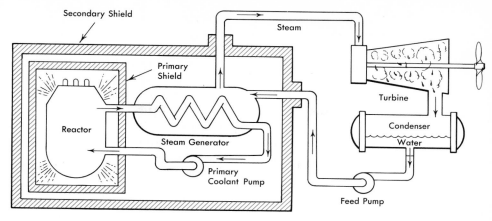

Courtesy: General Dynamics, Electric Boat Division

Figure 17-17 Diagramatic sketch showing how nuclear power can be used to operate a submarine.

shown as acting on the free body. Such a diagram may be drawn to represent a complex system or any smaller part of it. This form of *idealized model* is most useful in showing the effect of forces that can act upon a system. The free-body diagram was discussed more fully in Chapter 13.

The scale model

Scale models are used often in various problem-solving situations, especially when the system or product is very large and complex or very small and difficult to observe. A *scale model* is a replica, usually three dimensional, of the system, subsystem, or component being studied. It may be constructed to any desired scale relative to the

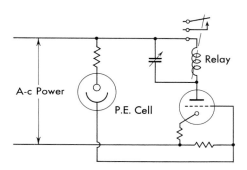

Figure 17-18 A simple photoelectric tube relay circuit.

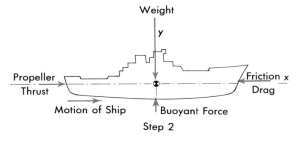

Figure 17-19 Free-body diagram of a ship.

actual design. Such projects as dam or reservoir construction, highway and freeway interchange design, factory layout, and aerodynamic investigations are particularly adaptable to study using scale models.

Scale models are useful for predicting performance because component parts of the model can be moved about to represent changing conditions within the system. Of considerably more usefulness are those scale models which are instrumented and subjected to environmental and load conditions that closely resemble reality. In such cases the models are tested and experimental data are recorded by an engineer. From an analysis of these data, predictions of the behavior of the real system can be made.

By using a scale model which can be constructed in a fraction of the time, a final design can be checked for accuracy prior to actual construction. Although scale models often cost many thousands of dollars, they are of relatively minor expense when compared with the total cost of a particular project.

Analog models

Analogs and similes are used to compare something that is unfamiliar to something else that is very familiar. Writers and teachers have found the *simile* to be a very effective way to describe an idea. Engineers use analogs in much the same way that teachers use similes. An analog, however, must provide more than a descriptive picture of what one wants to study; its action should correspond closely with the real thing. It should be *mathematically* similar to it, that is, the same type of mathematical expressions must describe well the action of both systems, the real and the analog.

A vibrating string is an analog of an organ pipe (Figure 17-20) because the sound in an organ pipe behaves quite similarly to the waves traveling along a vibrating

Figure 17-20

string. Under certain assumptions, similar mathematical equations can describe both systems. In other words, we can compare the corresponding actions of the *model* of the organ pipe with a *model* of the vibrating string. It is the *models* that behave exactly alike, *not the real systems*. If these models are "good" models, then, under certain conditions one can perform experiments with the string and draw valid conclusions concerning how the organ pipe would behave. Since one system may be much easier to experiment with than the other, one can work with the easier system and obtain results that are applicable to both.

An example of the use of a very successful analog is the electrical network that forms an analog for complete gas pipeline systems. Using such a model one can predict just what would happen if a lot of gas were suddenly needed at one point along the system. Experiments with the actual pipeline would be very costly and might disrupt service. The electrical network analog provides the answers faster, cheaper, and without disturbing anyone.

Analysis

One of the principal purposes of a model is to simplify the problem so that we can calculate the behavior, strength, and performance of the design. This is *analysis*.

Analysis is a mental process and, like any useful mental process, requires a store of basic knowledge and the ability to apply that knowledge. Since the amount of knowledge he possesses and his ability to use it are the major measures of a capable engineer, more time is spent at the university in studying analysis than any other subject. Just as one cannot solve a crossword puzzle without a knowledge of words, or make a medical diagnosis without a knowledge of the human body and its functions, so one cannot produce an acceptable engineering design without a basic knowledge of mathematics, physics, and chemistry, and their engineering relatives, such as stress analysis, heat transfer, electric network theory, and vibration. Nor can the engineer work effectively without an understanding of how his work affects man and his environment.

Analysis allows the engineer to "experiment on paper." For example, if he is concerned with the behavior of a wheel on a vehicle, his model might be a *rigid, perfectly round* wheel rolling on a *flat, unyielding* surface, Figure 17-21. (Words in italics indicate the assumptions made in the model.)

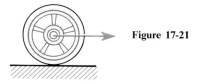

Figure 17-21

The motion of each point on the rim of the model wheel can be expressed by a well-known mathematical relationship called the *cycloid*. By knowing this motion the engineer can calculate the velocity of each point and determine how it varies with time, Figure 17-22. These calculations will enable him to solve for the centrifugal force on the wheel rim and to learn how fast the point makes contact with the ground.

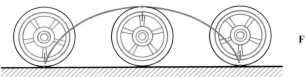

Figure 17-22

This elementary problem illustrates two important restrictions of engineering analysis: First, an equation usually describes only a very limited part of the function of the design, even in the case of a simple design such as a wheel. Second, an equation usually cannot describe *exactly* the action that takes place in the model. For example, we make the assumption that the wheel is perfectly rigid. This implies that it does not deform when it touches the ground. Such an assumption may be reasonably accurate in the case of a steel train wheel rolling on a steel rail, but it is probably a poor assumption to make for the rubber-tired wheels of an automobile.

Testing

The construction of a model and its analysis are based on assumptions, and these assumptions have to be verified.

One way in which the engineer can accomplish this is by the use of experiments. Experiments do not necessarily require construction of the entire design but only those portions of it that are important for the evaluation of the particular assumption. Components of a design are frequently tested instead of testing the entire design. If, in the design of a pipeline, the strength of the pipe is questionable, the engineer could obtain short sections of the pipe for testing. In the laboratory, liquid of appropriate density and pressure could be pumped through these sections of pipe to simulate the flowing fluid in the line. By measuring how much the pipe expands under the pressure and observing whether the results check with the calculations of his model, the engineer can verify if he has selected the proper pipe for his design.

It is important to remember that the value of the experiment is in the checking of the validity of the assumptions, not in checking the accuracy of the algebra. There is no need to run a rigid wheel on a rigid flat surface in order to prove the validity of the cycloidal motion of a point on the rim. If, however, the equation is to show the motion of a rubber-tired vehicle, then it may be well to run a rubber-tired wheel over a rigid surface to see how closely the cycloid does describe the motion of the nonrigid wheel.

For testing, one needs a model, a testing facility, and an arrangement of instruments suitable to measure what occurs during the test. Above all, one needs a test plan, just as a traveler needs directions in order to get to a desired destination. There is no sense in beginning a test without an objective and a plan for achieving the results necessary to satisfy that objective.

To test a completed design, the engineer should specify the characteristics that are most important and the instruments to be used for measuring these characteristics. In selecting the instruments the engineer must ask himself the question: "Does it provide the accuracy I need?" There being no such thing as *absolute* accuracy, the engineer must also know the probable error in the measurement. Only if that accuracy is

Illustration 17-9

This rare 1859 engraver's print shows "Watt's First Experiment." As with engineers of today, young James Watt relied upon the results of experimentation to verify his theories. Here Watt is shown in the act of placing a spoon on the spout of a steaming teakettle to verify his observations that steam exerts pressure when confined. *One can imagine that the steam pushed young Watt's spoon away and it fell again, in rapid succession, to produce a trip-hammer sound. The artist has centered interest on the young engineer, who, neglectful of his lunch, intently watches the steam's effect. So also do his father and his aunt, Mrs. Muirhead, with a degree of kindliness which evidences regard for the boy, however much she may chide him for "listless idleness."*

greater than his allowable error is the instrument suitable. He is concerned also with the effect of the measurement on the performance itself. This is very important in the case of small, intricate devices requiring great accuracy. According to Heisenberg,[5] we cannot measure any characteristic without affecting the system. (We all know this to be true from our experience in a physician's office. Our pulse rate and blood pressure quite often change as soon as the doctor starts to measure them. Whether this is psychologically or physically conditioned does not really matter; the fact is the measurement *does* influence the performance.)

The conditions under which tests are to be conducted must be defined; these must include all the important conditions of the design. A list of characteristics that might

[5] Werner Heisenberg (German physicist, 1901–   ) showed that, since observation must always necessarily affect the event being observed, this interference will lead to a fundamental limit on the accuracy of the observation. *Encyclopedia of Science,* Harper & Row, New York (1967).

be tested include start-up and shutdown conditions, operation under partial and full load,[6] operation under the failure of auxiliary equipment, operator errors, material selection, and many, many others.

There are five frequently used objectives for engineering tests. These objectives determine: (1) quality assurance of materials and subassemblies, (2) performance, (3) life, endurance, and safety, (4) human acceptance, and (5) effects of the environment. Some tests are required on every design, and in certain cases all the tests are needed. In general, when the analysis has been completed a prototype model of the design will be constructed. Usually this model is subjected to all the necessary tests. Once the prototype has passed the tests and the design has gone into the production stage, each final product may be subjected to selected tests. In this case, the tests are primarily for the purpose of assuring product uniformity and reliability.

Prototype testing generally applies only to products which go into mass production, such as the automobile wheel. When the design is for a "one-of-a-kind" item such as a pipeline, one will make as many tests as possible on raw materials and subassemblies to detect design errors before construction is completed. However, final tests on the complete design still will be necessary *to ensure* its safety and acceptability.

Quality assurance tests

Anyone who has selected wood at a lumber yard knows that the quality of raw material varies substantially from one piece to the next. It is less well known that such variations occur also in other materials, such as metals, ceramics, and polymers. Variation in these materials may be as great as the variation between the pieces of wood at the lumber yard and, just as the lumberman will provide more uniform wood at a higher price, so one can get a more uniform steel, aluminum, or Plexiglas at a higher price (to pay for preselection by the manufacturer). Typical variations in the strength of metals can be found in engineering handbooks.

The competent designer should account for this type of variation—either by designing the part so that it will perform satisfactorily with the least desirable (weakest) material or by prescribing tests that would ensure that only premium materials be used. Both approaches add to the cost. The conservative design may require more material and more weight; the testing process may require the use of more expensive material, or the extra cost may result from the testing and the discarding of unusable pieces.

Manufacturers, like lumber dealers, have realized the need for uniformity in engineering materials. For this reason, materials with more uniform properties than standard, or with guaranteed minimum properties, are available (at higher prices). For example, one can buy electrical carbon resistors in three ranges: The first grade, indicated by a gold band, varies a maximum of 5 per cent from its indicated value; the second grade (silver band) may vary as much as 10 per cent; and the standard product (no band) may vary as much as 20 per cent. Typically a silver-band resistor will cost twice as much as the standard, and the gold four times as much. Quality

---

[6] Whenever we talk about *load,* we mean this in the general sense to include such things as force, pressure, voltage, vibration, amplitude, temperature, and corrosive effects.

assurance includes the checking of dimensions of completed parts (a type of inspection routine in most modern machine shops), tests on the quality of joints between two members whether welded, brazed, soldered, riveted, or glued, and the continuity of electrical circuits.

Performance tests

What has been said about raw materials is also true for components and subassemblies which the designer may wish to include in his design. Electric motors, pumps, amplifiers, heat exchangers, pressure vessels, and similar items are designed to certain manufacturing standards. The products usually will be constructed at least as well as the manufacturer claims. However, if the quality of the total design depends critically upon the specifications of a subassembly, then it is best to inspect and test that subassembly separately before it is included in the construction. This is particularly true for one-of-a-kind designs, such as space capsules. Since performance of the capsule is critically dependent upon that of its components, the designer must specify a series of tests that will be made at the manufacturers' plants to assure performance. He may, in fact, personally supervise the testing.

A performance test simply shows whether a design does what it is supposed to do. It measures the skill of the engineer and the validity of the assumptions made in his analyses.

Performance testing generally does not wait until the design is completed. It follows step by step with the design. For example, the heat shield of the space capsule is tested in the supersonic wind tunnel to see if it can withstand the aerodynamic heating for which it was designed; the parachute is tested to see if it supports the capsule at just the right speed; structural members are tested for strength and stiffness; and instruments are checked to show if they indicate what they are supposed to measure.

Performance tests may require special testing apparatus, such as supersonic wind tunnels and space-simulation chambers. They always need careful planning and

Illustration 17-10
All new products, such as this improved golf club, should be tested in use and their performance measured.

instrumentation to assure that the tests measure what is really needed—a proof of the validity of the design.

Life, endurance, and safety tests

We know that machines, like people, age. One of the most important and most difficult tests to gain meaningful data from are the life tests, tests that tell how long a product will survive in service and whether it can take excessive loads, misoperations, and other punishment without failure. It is rarely possible to carry out life tests accurately, for one seldom has the time to subject the prototype to the same period of aging that the real part will experience in actual service. In some instances, accelerated "life" tests are used. For example, paints and other surface protections may be exposed to the actions of sunlight, wind, rain, snow, or salt-water spray for months or even years. In this way a body of knowledge relating to the "life" of these surface finishes slowly develops, but a final selection by the engineer may be delayed considerably. Therefore, these tests must be accelerated and the engineer does so by increasing the load, by applying the load more rapidly, or by subjecting the design to a more severe environment. However, he is never quite sure how accurately these short-term "life" tests *really* represent the effect of the aging conditions and how their results should be interpreted. Usually, this is done (in the case of mechanical tests) by making tests on several specimens, each at a different degree of overload. The length of life of each part tested is then plotted as a function of the applied load, and the resulting curve is extrapolated to the maximum load that the part is expected to endure in real life.

If different types of loading are applied to the part, such as pressure, temperature, and vibration, it may be necessary to make separate tests with overloads in each one of these areas to see how they extrapolate to "true life." It may be desirable to use overloads in combination and observe if the combined effect is different from the sum of the effects of the individual loads. Very often two combined loads have a much more serious effect on the part than the arithmetic sum of the separate effects of the two loads. This is called *synergistic behavior.*

We again use the automobile wheel for an example of "life" testing. It is reasonable to assume that a good automobile tire and wheel should survive without failure at least 50,000 mi under normal loads, at speeds of 50 mi/hr. This means that the tire must be tested at that load for at least 1000 hr—a period in excess of 40 days. Such a test may indeed be possible and, in fact, is often performed on new tire designs. The designer will want to know the effects of overload, of speeds higher than 50 mi/hr, of curves, rough roads, under- and overinflation, and of the effect of very high or very low temperatures. It is easy to see that one needs a battery of testing machines to find out all one wants to know within a reasonable period of time.

Even though the test part may pass the predicted life during the test, the test is usually not terminated, but is continued until the part actually fails. This then becomes an endurance test, and it determines the excess life of the part. Since the life of each part is likely to differ, and since not every part can be "life" tested, it is essential to know the excess "life" of the average part. Since there is a statistical

Illustration 17-11
*Some tests need to be
continued until the
part actually fails.*

variability between parts, the engineer will want to know not only the *average* excess life but also the range of *variation* in this lifetime.

Instead of measuring the endurance of a part under a constant load, the engineer may decide to increase the load until the part fails. This failure load will be higher than the design load if the part is properly designed. The ratio between the failure load and the design load is called the *factor of safety*. Factors of safety may be quite low where parts are very carefully manufactured, where excessive weight is undesirable, or where their failure causes no serious hardship. However, where human life is at stake, factors of safety must be so chosen that no variation in materials or workmanship, or simplified assumption in the designer's calculations, can possibly make the part unsafe and cause failure under normal operating loads.

Because the engineer has the responsibility to see that no one is injured as a consequence of his design, he must also consider the possibility of accidental or thoughtless misoperation of the design. For example, the automobile tire may be underinflated, it may be operated under too heavy a load, at too high a speed, or on a rough road, and yet it should not fail catastrophically.

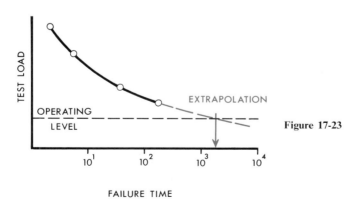

Figure 17-23

Human acceptance tests

For a long time the designers of consumer goods have been concerned with the appearance and acceptability of their product to the buyer. It is unfortunate that in many instances they have appealed more often to the buyer's baser instincts, such as pride, greed, and desire for status, rather than to his sense of quality and beauty. Engineers, on the other hand, have been concerned too little with the interaction of their designs with the people who buy or use them. They have often failed to ask themselves whether the physical, mental, and emotional needs and limitations of the human being permit him to operate the machine in the *best* possible way. In the past engineers have all too often assumed that the human body is sufficiently adaptable to operate any kind of control lever or wheel. Although the adaptability of people is truly astonishing, we now know that for maximum efficiency levers and wheels must be carefully designed, that the forces necessary to operate them must be neither too large nor too small, that the operator should be able to "feel" the effect that he is producing, and that the use of the device or design should not tire him physically. A new design, then, should be tested with real people. Such tests usually cannot be performed by an engineer alone, but require the aid of others, such as industrial psychologists and/or human factors engineers.

Those who have studied metal processes have found that a man's attention span is short, that he cannot be asked continually to peer at a gage or work piece unless things are "happening" to it. It is also known that warnings are better heeded if they are audible than if they are visual. If they are visual, a bright red light or a blinking light which draws the operator's attention is much better than the movement of a dial to some position that has been marked "unacceptable." In recent years we have learned that our emotional responses play a major part in our daily lives, that whether we are elated or depressed can affect the quality of our work and the attention we give to the instruments we are asked to observe, or the levers we are asked to operate. In turn, our emotions are influenced by color, beauty, or attractive design—in short, by our opinion of the design.

The purpose of acceptability testing is to see whether the design meets the physical, mental, and emotional requirements of the average person for whom it is designed. Since the "average" person can never be found, it is essential that a series of tests be made by and with different people, and that these observations be used to make whatever changes are necessary to make the device as acceptable as possible within the technical and economic constraints.

Environmental tests

The environment is the aggregate of all conditions that surround the design under operating conditions. It may be wind and weather; it may be the soil in which the design is buried or the chemicals in which it is immersed; it may be the vacuum of outer space or an intense field of nuclear or electromagnetic radiation. In general, the operating environment is different from the normal environment in the laboratory in which the tests are made. Sometimes the effects of the environment are not important,

Some products, such as these engines, must operate under conditions of extreme heat and cold. Environmental testing is, therefore, mandatory.

but more frequently the environment can strongly affect the functioning and life of the part. Therefore environmental testing has become an important part of the final testing of most new products.

We have already mentioned how paints and other surface finishes are exposed to sun, wind, and rain for long periods of time, thereby developing a considerable body of knowledge concerning how such surface finishes behave under both normal and unusual weather conditions. Similarly, an extensive body of knowledge exists on how chemicals deteriorate or "corrode" construction materials. These and other environmental factors are under continuing study. Therefore, unless the engineer is faced with an unusual environment, he can often (but not always) find pertinent information in the literature to predict how a particular environment is going to affect his design.

However, there are many instances where additional tests are needed. Tests are particularly important where two or more environmental effects work together, such as moisture and heat, or chemicals and vibrations. The result of such effects may not be predictable from either of the individual effects acting by themselves. Thus we know that a vibrating environment in a salt-spray atmosphere can cause corrosion fatigue at a rate far higher than that which might have been predicted from either the vibration or the saltwater corrosion taken independently.

With the advent of space travel, one of the most intensively studied types of environment is *space*. When away from the earth's atmosphere, a body in space will be in a nearly complete vacuum, but it will be exposed to a variety of types of radiation and to meteoric dust from which the earth's atmosphere normally protects it. The radiation effects may be severe enough to attack electronic circuits seriously and cause deterioration of transistors and other electronic devices. The meteoric dust, though generally quite fine, travels with speeds of 10,000 to 70,000 mi/hr and has sufficient energy to penetrate some of the strongest materials. Within the last few years some ways have been found to simulate in the laboratory both the high radiation and the presence of meteoric dust, and (on earth) to subject space equipment to these kinds of attack.

Safety and products liability

The engineer's role as a design specialist has always carried a sincere concern for the continued well-being of those who use his designs. In recent years the courts have taken a more positive role than heretofore in assessing the designer's responsibility concerning the health and safety of the user. Perhaps the most significant legislation pertaining to this matter to appear in modern times has been the Occupational Safety and Health Act (OSHA) which was signed into law on December 29, 1970.[10,11,12] This act sets forth the general conditions of safety which must be met for the manufacture or use of goods and services in this country. Basically it supplied the force of law to a host of previously defined "consensus standards" or codes that had been developed through the past 50 years by many professional, semi-professional, and governmental bodies. Although some of these codes, requirements, and recommended procedures were in keeping with modern engineering methods and materials, there were a host of others that were incorporated that were archaic and (although *technically* the latest printed standard available) were in need of modernization and change. Nevertheless the engineer-designer must exercise great care to be certain that his designs meet the legal requirements that pertain to the product or system that he has designed.

Products liability is civil liability to an ultimate consumer for injury to his person or property resulting from using a defective article that has been sold by a person in the business, or chain, of selling such articles.[13]

[10] R. J. Redding, *Intrinsic Safety,* McGraw-Hill, New York, 1976.
[11] Dan Petersen, *The OSHA Compliance Manual,* McGraw-Hill, New York, 1976.
[12] Peter S. Hopf, *Designer's Guide to OSHA,* McGraw-Hill, New York, 1976.
[13] Wyatt Jacobs, "Products Liability," *Mechanical Engineering,* November, 1972, p. 12.

Illustration 17-13
Communication is complex. An idea may be transmitted and acknowledged between earth and moon in a matter of seconds . . .

In the Middle Ages, the rule of "buyer beware" worked satisfactorily for all concerned because each manufacturer's products were used primarily in the local community. Thus, a swordsmaker, silversmith, or bootsmith was generally identified by the *quality* of his handicraft. Transactions were normally consummated in face-to-face encounters. This all changed with the advent of the industrial age, and today products made in a small factory in California may be used extensively in every part of the country. Also, the identity of the manufacturer and/or designers may be unknown.

For this reason those who engage in a particular field of manufacturing are held responsible for possessing the knowledge and skill of an expert in that field and they must keep reasonably abreast of methods and procedures used by practical men in the trade. This responsibility includes due care to conduct reasonable tests and to discover latent hazards.

It is incumbent upon the manufacturer to warn potential users or consumers when it is known that the use of a particular product in a certain way would be hazardous. The engineering field is complex and changes in available technical capability and materials selection occur rapidly. Because of this, design parameters may be adjudged as safe today, but unsafe tomorrow. The engineer must do his best to foresee conditions of possible failure or hazard. It is also very important that his records be accurate and complete, and that they reflect that he has designed the product to be as safe as possible, as the existing state of the art and his imagination will allow.[14]

[14] *Encyclopaedia of Occupational Health and Safety,* Vols. I and II, McGraw-Hill, 1976.

Illustration 17-14
. . . while a similar transmission and acknowledgment between individuals standing within arm's length may take years.

The report

Reporting is the process of information transfer. For the engineer the ability to write and speak clearly is most essential, for however good he may be as an analyst or experimenter, if he cannot convey his ideas clearly, concisely, and interestingly to others, then he is like a stranger in a foreign country whose people cannot understand him. The transmittal of information includes writing, drawing, and speaking. Refer to Chapter 8 for a more complete discussion of these processes.

Problems

17-1. After reviewing the ecological needs of your home town, state three problems that should be solved.

17-2. Give three examples of *feedback* that existed prior to A.D. 1800.

17-3. Talk to an engineer who is working in design or development in industry. Describe two situations in his work where he has not been able to rely on *theoretical textbook solutions* to solve his problems. Why was he forced to resort to other means to solve the problems?

17-4. Describe an incident where an individual or group abandoned their course of action because it was found that they were spending time working on the wrong problem.

17-5. List the properties of a kitchen electric mixer.

17-6. List the properties of the automobile that you would like to own.

17-7. Make a matrix analysis of the possible solutions to the problem of removing dirt from clothes.

17-8. For ten minutes solo brainstorm the problem of disposal of home wastepaper. List your ideas for solution.

17-9. List five types of models that are routinely used by the average American citizen.

17-10. Diagram the model of the football play that made the longest yardage gain for your team this year.

17-11. Draw an energy system of an ordinary gas-fired hot-water heater.

17-12. Draw an energy system representing a simple refrigeration cycle.

17-13. Draw an energy system representing a "perpetual motion" machine.

17-14. Draw an electrical circuit diagram containing two single-pole, double-throw switches in such a manner that a single light bulb may be turned on or off at either switch location.

17-15. Arrange three single-pole, single-throw switches in an electrical circuit containing three light bulbs in such a manner that one switch will turn on one of the bulbs, another switch will turn on two of the bulbs, and the third switch will turn on all three bulbs.

17-16. Describe three situations where a scale model would be the most appropriate kind of idealized model to use.

Appendixes

APPENDIX I

Trigonometry

Right triangles

It can be shown by measurements and by formal derivations that, for any given size of an angle at A or C, the ratio of the lengths of the sides to each other in a right triangle is a constant regardless of the numerical value of the lengths. In Figure AI-1, the sides of a right triangle are named in reference to the angle under consideration. In the cases, the angle is designated as θ (theta).

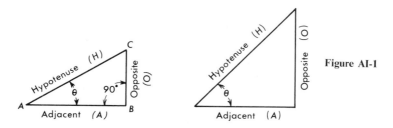

Figure AI-1

$$\frac{\text{Opposite Side}}{\text{Hypotenuse}} = \text{Sine } \theta \qquad \frac{O}{H} = \sin \theta$$

$$\frac{\text{Adjacent Side}}{\text{Hypotenuse}} = \text{Cosine } \theta \qquad \frac{A}{H} = \cos \theta$$

$$\frac{\text{Opposite Side}}{\text{Adjacent Side}} = \text{Tangent } \theta \qquad \frac{O}{A} = \tan \theta$$

$$\frac{\text{Adjacent Side}}{\text{Opposite Side}} = \text{Cotangent } \theta \qquad \frac{A}{O} = \cot \theta$$

$$\frac{\text{Hypotenuse}}{\text{Adjacent Side}} = \text{Secant } \theta \qquad \frac{H}{A} = \sec \theta$$

$$\frac{\text{Hypotenuse}}{\text{Opposite Side}} = \text{Cosecant } \theta \qquad \frac{H}{O} = \csc \theta$$

Methods of solving oblique triangle problems

In order to solve an oblique triangle problem, at least three of the six parts of the triangle must be known, and at least one of the known parts must be a side. In the suggested methods listed below, only the most effective methods are given.

1. Given: two sides and an angle opposite one of them:
   a. Law of sines.
   b. Right triangles.
2. Given: two angles and one side:
   a. Law of sines.
   b. Right triangles.
3. Given: two sides and the included angle:
   a. Law of cosines (answer is usually not dependable to more than three significant figures).
   b. Right triangles.
4. Given: three sides only:
   a. Tangent formula (half-angle solution).
   b. Sine formula (half-angle solution). This formula may not give exact results if the half-angle is near 90°.
   c. Cosine formula (half-angle solution). This formula may not give exact results if the half-angle is about 6° or less.
   d. Cosine formula (whole angle solution).
   e. Law of cosines (answer is usually not dependable to more than three significant figures).

Methods for finding areas of oblique triangles

The area of an oblique triangle may be found by any of several methods. Some of the more common methods are given below:

1. Area $= (\frac{1}{2})(\text{base})(\text{altitude})$.
2. Area $= \sqrt{(S)(S - AB)(S - BC)(S - AC)}$, where $S = \frac{1}{2}$ perimeter of the triangle.
3. Area $= \frac{1}{2}$ (product of two sides) (sine of the included angle).

Sine law

In any triangle the ratio of the length of a side to the sine of the angle opposite that side is the same as the ratio of any other side to the sine of the angle opposite it. In symbol form (see Figure AI-2):

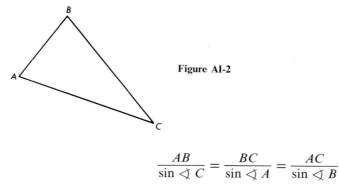

Figure AI-2

$$\frac{AB}{\sin \sphericalangle C} = \frac{BC}{\sin \sphericalangle A} = \frac{AC}{\sin \sphericalangle B}$$

This expression is called the *sine law*. The student is cautioned not to confuse the meanings of sine functions and sine law.

In the event one of the angles of a triangle is larger than 90°, a simple way to obtain the value of the sine of the angle is to subtract the angle from 180° and obtain the sine of this angle to use in the sine law expression.

The sine law can also be used if two sides and an angle of a triangle are known, provided the angle is not the one included between the sides. However, as explained in trigonometry texts, the product of the sine of the angle and the side adjacent must be equal to or less than the side opposite the angle; otherwise no solution is possible.

As an alternative method, the general triangle can be made into right triangles by adding construction lines. This method of using right triangle solutions is as exact as the sine law but usually will take more time than the sine law method.

Cosine law

In an oblique triangle, the square of any side is equal to the sum of the squares of the other two sides minus twice the product of the other two sides times the cosine of the included angle. In symbol form:

$$(AB)^2 = (AC)^2 + (BC)^2 - (2)(AC)(BC)(\cos \sphericalangle C)$$

This expression is called the *cosine law* and is useful in many problems, although it may not give an answer to the desired precision since we are adding and subtracting terms that may have only three significant figures.

After the side *AB* has been determined, the angles at *A* and *B* can be found by using the law of sines.

In the event that the angle used in the cosine law formula is larger than 90°, subtract the angle from 180°, and determine the cosine of this angle. Remember, however, that the cosine of an angle between 90 and 180° is negative. If the angle used in the formula is larger than 90°, the last term will add to the squared terms.

The problem above also can be solved by using construction lines and making right triangles from the figure (Figure AI-3). To do this, we construct the line BD

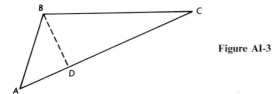

Figure AI-3

perpendicular to AC. This will form two right triangles, ABD and BCD. In triangle BCD, side BD may be found by using BD and the sine of $\angle C$. In a similar manner, by using the cosine of $\angle C$, side DC may be found. From this we can determine side AD in triangle ABD.

Using the tangent function, the angle at A can be found, and AB can be determined by the use of the sine or cosine function or the Pythagorean theorem $(AB)^2 = (BD)^2 + (AD)^2$. The right triangle method, while it may take longer to solve, will in general give a more accurate answer.

Three sides laws

There are a number of formulas derived in trigonometry that will give the angles of an oblique triangle when only the three sides are known. The formulas differ considerably in ease of application and precision, especially if logarithms are used. Of all the formulas available, in general the half-angle (tangent) formula is better than others. The formula (half-angle solution) is as follows:

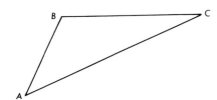

Figure AI-4

$$\tan \tfrac{1}{2} A = \frac{r}{S - BC}$$

where

$$r = \sqrt{\frac{(S - AB)(S - AC)(S - BC)}{S}}$$

and $\quad S = \tfrac{1}{2}$ perimeter of triangle

Other formulas that may be used are the following:

Sine formula (half-angle solution) $\sin \tfrac{1}{2} A = \sqrt{\dfrac{(S - AC)(S - AB)}{(AC)(AB)}}$

Cosine formula (half-angle solution) $\cos \tfrac{1}{2} A = \sqrt{\dfrac{(S)(S - BC)}{(AC)(AB)}}$

Cosine formula (whole angle solution) $\cos A = \dfrac{(2S)(S - BC)}{(AB)(AC)} - 1$

In the last formula, the quantity $(2S)(S - BC)/(AB)(AC)$ will usually be between 1 and 2 and should be calculated to four figures. Subtracting the 1 in the equation will leave the cosine of the angle correct to three figures. The formula has the advantage that it requires fewer operations.

After finding one angle, the remaining angles can be found by successive applications of the law, being careful to use the proper side of the triangle in the formula. The sine law can also be used after one angle is found. In order to have a check on the solution, it is better to solve for all three angles rather than solve for two angles and then subtract their sum from 180°. If each angle is computed separately, their sum should be within the allowable error range of 180°.

As an incidental item in the tangent formula, the constant r is equal to the length of the radius of a circle that can be inscribed in the triangle.

Geometric figures

Rectangle

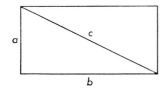

Area = (base)(altitude) = ab
Diagonal = $\sqrt{(\text{altitude})^2 + (\text{base})^2}$
$$C = \sqrt{a^2 + b^2}$$

Right triangle

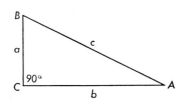

Angle A + angle B = angle C = $90°$
Area = $\frac{1}{2}$ (base)(altitude)
Hypotenuse = $\sqrt{(\text{altitude})^2 + (\text{base})^2}$
$$C = \sqrt{a^2 + b^2}$$

Any triangle

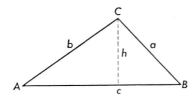

Angles $A + B + C = 180°$
(Altitude h is perpendicular to base c)
Area = $\frac{1}{2}$ (base)(altitude)

Parallelogram

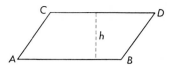

Area = (base)(altitude)
Altitude h is perpendicular to base AB
Angles $A + B + C + D = 360°$

Trapezoid

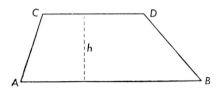

Area = ½ (altitude)(sum of bases)
(Altitude h is perpendicular to sides AB and CD. Side AB is parallel to side CD.)

Regular polygon

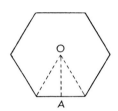

$$\text{Area} = \tfrac{1}{2} \begin{bmatrix} \text{length of} \\ \text{one side} \end{bmatrix} \begin{bmatrix} \text{number} \\ \text{of sides} \end{bmatrix} \begin{bmatrix} \text{distance} \\ OA \text{ to} \\ \text{center} \end{bmatrix}$$

A regular polygon has equal angles and equal sides and can be inscribed in or circumscribed about a circle.

Circle

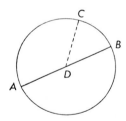

AB = diameter, CD = radius

$$\text{Area} = \pi(\text{radius})^2 = \frac{\pi(\text{diameter})^2}{4}$$

Circumference = π(diameter)

$$C = 2\pi(\text{radius})$$

$$\frac{\text{arc } BC}{\text{circumference}} = \frac{\text{angle } BDC}{360°}$$

$$1 \text{ radian} = \frac{180°}{\pi} = 57.2958°$$

Sector of a circle

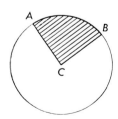

$$\text{Area} = \frac{(\text{arc } AB)(\text{radius})}{2}$$

$$= \pi \frac{(\text{radius})^2(\text{angle } ACB)}{360°}$$

$$= \frac{(\text{radius}^2)(\text{angle } ACB \text{ in radians})}{2}$$

Segment of a circle

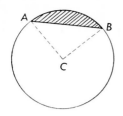

$$\text{Area} = \frac{(\text{radius})^2}{2}\left[\frac{\pi(\sphericalangle ACB^\circ)}{180} - \sin ACB^\circ\right]$$

$$\text{Area} = \frac{(\text{radius})^2}{2}[\sphericalangle ACB \text{ in radians} - \sin ACB^\circ]$$

Area = area of sector ACB − area of triangle ABC

Ellipse

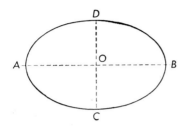

Area = π(long radius OA)(short radius OC)

Area = $\frac{\pi}{4}$(long diameter AB)(short diameter CD)

Volume and center of gravity equations for selected cases[1]

Volume equations are included for all cases. Where the equation for the CG (center of gravity) is not given, you can easily obtain it by looking up the volume and CG equations for portions of the shape and then combining values. For example, for the shape above, use the equations for a cylinder, case 1, and a truncated cylinder, case 10 (subscripts C and T, respectively, in the equations below). Hence taking moments,

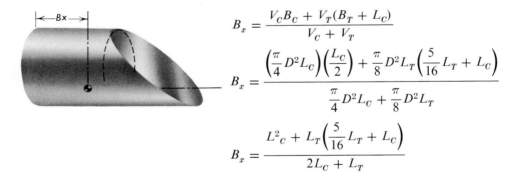

$$B_x = \frac{V_C B_C + V_T(B_T + L_C)}{V_C + V_T}$$

$$B_x = \frac{\left(\frac{\pi}{4}D^2 L_C\right)\left(\frac{L_C}{2}\right) + \frac{\pi}{8}D^2 L_T\left(\frac{5}{16}L_T + L_C\right)}{\frac{\pi}{4}D^2 L_C + \frac{\pi}{8}D^2 L_T}$$

$$B_x = \frac{L^2{}_C + L_T\left(\frac{5}{16}L_T + L_C\right)}{2L_C + L_T}$$

[1]Courtesy of Knoll Atomic Power Laboratory, Schenectady, N.Y., operated by the General Electric Company for the U.S. Atomic Energy Commission. Reprinted from *Products Engineering*—copyright owned by McGraw-Hill, New York.

In the equations to follow, angle θ can be either in degrees or in radians. Thus θ (rad) $= \pi\theta/180$ (deg) $= 0.01745\,\theta$ (deg).

For example, if $\theta = 30$ deg in Case 3, then $\sin\theta = 0.5$ and

$$B = \frac{2R(0.05)}{3(30)(0.01745)} = 0.637R$$

Symbols used are:

B = distance from CG to reference plane,
V = volume,
D and d = diameter,
R and r = radius,
H = height,
L = length.

---

 1. Cylinder

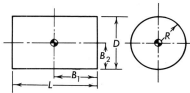

$$V = \frac{\pi}{4}D^2L = 0.7854D^2L \qquad B_1 = L/2$$

$$B_2 = R$$

Area of cylindrical surface = (perimeter of base)(perpendicular height)

---

 2. Half cylinder

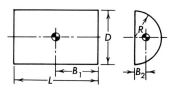

$$V = \frac{\pi}{8}D^2L = 0.3927D^2L$$

$$B_1 = L/2 \qquad B_2 = \frac{4R}{3\pi} = 0.4244R$$

 3. Sector of cylinder

$$V = \theta R^2 L \qquad B = \frac{2R\sin\theta}{3\theta}$$

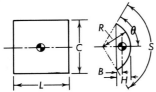

4. Segment of cylinder

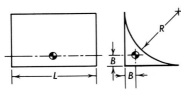

$$V = LR^2\left(\theta - \frac{1}{2}\sin 2\theta\right)$$

$$V = 0.5L[RS - C(R - H)]$$

$$B = \frac{4R \sin^3 \theta}{6\theta - 3\sin 2\theta}$$

$$S = 2R\theta$$
$$H = R(1 - \cos\theta)$$
$$C = 2R\sin\theta$$

 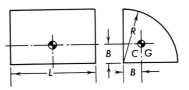

5. Quadrant of cylinder

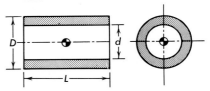

$$V = \frac{\pi}{4}R^2L = 0.7854R^2L$$

$$B = \frac{4R}{3\pi} = 0.4244R$$

6. Fillet or spandrel

$$V = \left(1 - \frac{\pi}{4}\right)R^2L = 0.2146R^2L$$

$$B = \frac{10 - 3\pi}{12 - 3\pi}R = 0.2234R$$

7. Hollow cylinder

$$V = \frac{\pi L}{4}(D^2 - d^2)$$

CG at center of part

8. Half-hollow cylinder

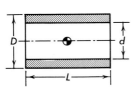

$$V = \frac{\pi L}{8}(D^2 - d^2)$$

$$B = \frac{4}{3\pi}\left[\frac{R^3 - r^3}{R^2 - r^2}\right]$$

9. Sector of hollow cylinder

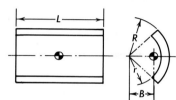

$$V = 0.01745(R^2 - r^2)\,\theta L$$

$$B = \frac{38.1972(R^3 - r^3)\sin\theta}{(R^2 - r^2)\,\theta}$$

10. Truncated cylinder (with full circle base)

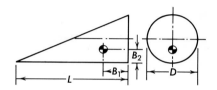

$$V = \frac{\pi}{8}D^2 L = 0.3927 D^2 L$$

$$B_1 = 0.3125L$$
$$B_2 = 0.375D$$

11. Truncated cylinder (with partial circle base)

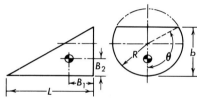

$$b = R(1 - \cos\theta)$$

$$V = \frac{R^3 L}{b}\left[\sin\theta - \frac{\sin^3\theta}{3} - \theta\cos\theta\right]$$

$$B_1 = \frac{L\left[\dfrac{\theta\cos^2\theta}{2} - \dfrac{5\sin\theta\cos\theta}{8} + \dfrac{\sin^3\theta\cos\theta}{12} + \dfrac{\theta}{8}\right]}{\left[1 - \cos\theta\right]\left[\sin\theta - \dfrac{\sin^3\theta}{3} - \theta\cos\theta\right]}$$

$$B_2 = \frac{2R\left[-\dfrac{\theta\cos\theta}{2} + \dfrac{\sin\theta}{2} - \dfrac{\theta}{8} + \dfrac{\sin\theta\cos\theta}{8} - N\right]}{\left[\sin\theta - \dfrac{\sin^3\theta}{3} - \theta\cos\theta\right]}$$

$$\text{where } N = \frac{\sin^3\theta}{6} - \frac{\sin^3\theta\cos\theta}{12}$$

12. Sphere

$$V = \frac{\pi D^3}{6} = 0.5236 D^3$$

Area of surface $= 4\pi(\text{radius})^2 = \pi D^2$

13. Hemisphere

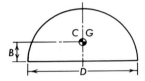

$$V = \frac{\pi D^3}{12} = 0.2618 D^3$$

$$B = 0.375R$$

14. Spherical segment

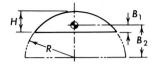

$$V = \pi H^2 \left(R - \frac{H}{3} \right)$$

$$B_1 = \frac{H(4R - H)}{4(3R - H)}$$

$$B_2 = \frac{3(2R - H)^2}{4(3R - H)}$$

15. Spherical sector

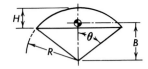

$$V = \frac{2\pi}{3} R^2 H = 2.0944 R^2 H$$

$$B = 0.375(1 + \cos\theta)$$

$$R = 0.375(2R - H)$$

16. Shell of hollow hemisphere

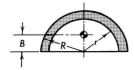

$$V = \frac{2\pi}{3}(R^3 - r^3)$$

$$B = 0.375 \left(\frac{R^4 - r^4}{R^3 - r^3} \right)$$

17. Hollow sphere

$$V = \frac{4\pi}{3}(R^3 - r^3)$$

18. Torus

$$V = \frac{1}{4}\pi^2 d^2 D = 2.467 d^2 D$$

19. Hollow torus

$$V = \frac{1}{4}\pi^2 D(d_1^2 - d_2^2)$$

 20. Bevel ring

 21. Bevel ring

$$V = \pi\left(R + \frac{1}{3}W\right)WH$$

$$B = H\left[\frac{\dfrac{R}{3} + \dfrac{W}{12}}{R + \dfrac{W}{3}}\right]$$

$$B > \frac{H}{3}$$

$$V = \pi\left(R - \frac{1}{3}W\right)WH$$

$$B = H\left[\frac{\dfrac{R}{3} - \dfrac{W}{12}}{R - \dfrac{W}{3}}\right]$$

 22. Curved-sector ring

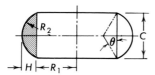

$$V = 2\pi R_2^2\left[R_1 + \left(\frac{4\sin 3\theta}{6\theta - 3\sin 2\theta} - \cos\theta\right)R_2\right][\theta - 0.5\sin 2\theta]$$

 23. Ellipsoidal cylinder

 24. Ellipsoid

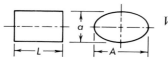

$$V = \frac{\pi}{4}AaL$$

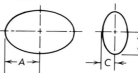

$$V = \frac{4}{3}\pi ACE$$

25. Paraboloid

$$V = \frac{\pi}{8} HD^2 \qquad B = \frac{1}{3} H$$

26. Pyramid (with base of any shape)

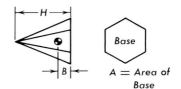

$A = $ Area of base

$A = $ Area of Base

$$V = \frac{1}{3} AH \qquad B = \frac{1}{4} H$$

27. Frustum of pyramid (with base of any shape)

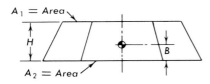

$$V = \frac{1}{3} H(A_1 + \sqrt{A_1 A_2} + A_2)$$

$$B = \frac{H(A_1 + 2\sqrt{A_1 A_2} + 3A_2)}{4(A_1 + \sqrt{A_1 A_2} + A_2)}$$

28. Cone

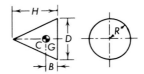

$$V = \frac{\pi}{12} D^2 H \qquad B = \frac{1}{4} H$$

Area of conical surface (right cone) = ½ (circumference of base) × (slant height)

29. Frustum of cone

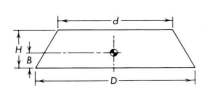

$$V = \frac{\pi}{12} H(D^2 + Dd + d^2)$$

$$B = \frac{H(D^2 + 2Dd + 3d^2)}{4(D^2 + Dd + d^2)}$$

30. Frustum of hollow cone

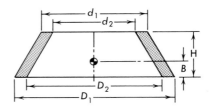

$$V = 0.2618H[(D_1^2 + D_1 d_1 + d_1^2) \\ -(D_2^2 + D_2 d_2 + d_2^2)]$$

31. Hexagon

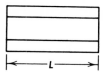

$$V = \frac{\sqrt{3}}{2}d^2L$$

$$V = 0.866d^2L$$

32. Closely packed helical springs

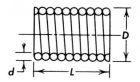

$$V = \frac{\pi^2 dL}{4}(D - d)$$

$$V = 2.4674(D - d)$$

33. Rectangular prism

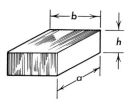

Volume = length × width × height
Volume = area of base × altitude

34. Any prism

(Axis either perpendicular or inclined to base)
Volume = (area of base)(perpendicular height)
Volume = (lateral length)(area of perpendicular cross section)

APPENDIX III

Tables

The Greek alphabet

A	α	Alpha	N	ν	Nu
B	β	Bēta	Ξ	ξ	Xī
Γ	γ	Gamma	O	o	Omicron
Δ	δ	Delta	Π	π	Pī
E	ϵ	Epsilon	P	ρ	Rhō
Z	ζ	Zēta	Σ	σ	Sigma
H	η	Eta	T	τ	Tau
Θ	θ	Thēta	Υ	υ	Upsilon
I	ι	Iōta	Φ	ϕ	Phī
K	κ	Kappa	X	χ	Chī
Λ	λ	Lambda	Ψ	ψ	Psī
M	μ	Mu	Ω	ω	Omega

Dimensional prefixes

Symbol	Prefix	Multiple
T	tera units	10^{12}
G	giga units	10^{9}
M	mega units	10^{6}
k	kilo units	10^{3}
h	hecto units	10^{2}
da	deca units	10^{1}
	units	10^{0}
d	deci units	10^{-1}
c	centi units	10^{-2}
m	milli units	10^{-3}
μ	micro units	10^{-6}
n	nano units	10^{-9}
p	pico units	10^{-12}
f	femto units	10^{-15}
a	atto units	10^{-18}

Conversion tables

The number in parentheses following a value in the table indicates the power of 10 by which this value should be multiplied. Thus, 6.214(-6) means 6.214×10^{-6}.

1. Length equivalents

cm	in.	ft	m	mi*	
1	3,937(−1)	3.281(−2)	1.0(−2)	6.214(−6)	cm
2.540	1	8.333(−2)	2.54(−2)	1.578(−5)	in.
3.048(+1)	1.2(+1)	1	3.048(−1)	1.894(−4)	ft
1.0(+2)	3.937(+1)	3.281	1	6.214(−4)	m
1.609(+5)	6.336(+4)	5.280(+3)	1.609(+3)	1	mi

* mile.

Additional Measures

Metric: 1 km = 10^3 m
1 mm = 10^{-3} m
1 μm = 10^{-6} m (micron)
1 Å = 10^{-10} m (angstrom)

English: 1 mil = 10^{-3} in.
1 yd = 3.0 ft
1 rod = 5.5 yd = 16.5 ft
1 furlong = 40 rod = 660 ft

2. Area equivalents

m²	in.²	ft²	acres	mi²	
1	1.55(+3)	1.076(+1)	2.471(−4)	3.861(−7)	m²
6.452(−4)	1	6.944(−3)	1.594(−7)	2.491(−10)	in.²
9.290(−2)	1.44(+2)	1	2.296(−5)	3.587(−8)	ft²
4.047(+3)	6.273(+6)	4.356(+4)	1	1.562(−3)	acres
2.590(+6)	4.018(+9)	2.788(+7)	6.40(+2)	1	mi²

Additional Measures

1 hectare = 10^4 m²

3. Volume equivalents

cm³	in.³	ft³	gal (U.S.)	
1	6.103(−2)	3.532(−5)	2.642(−4)	cm³
1.639(+1)	1	5.787(−4)	4.329(−3)	in.³
2.832(+4)	1.728(+3)	1	7.481	ft³
3.785(+3)	2.31(+2)	1.337(−1)	1	gal (U.S.)

Additional Measures

Metric: 1 liter = 10^3 cm³
1 m³ = 10^6 cm³

English: 1 quart = 0.250 gal (U.S.)
1 bushel = 9.309 gal (U.S.)
1 barrel = 42 gal (U.S.)
(petroleum measure only)
1 imperial gal = 1.20 gal (U.S.)
approx.
1 board foot (wood) = 144 in.³
1 chord (wood) = 128 ft³

4. Mass equivalents

kg	slug	lb$_m$*	g	
1	6.85(−2)	2.205	1.0(+3)	kg
1.46(+1)	1	3.22(+1)	1.46(+4)	slug
4.54(−1)	3.11(−2)	1	4.54(+2)	lb$_m$
1.0(−3)	6.85(−5)	2.205(−3)	1	g

* Not recommended.

5. Force equivalents

N*	lb$_f$†	dyn**	kg$_f$‡	g$_f$‡	Poundal‡	
1	2.248(−1)	1.0(+5)	1.019(−1)	1.019(+2)	7.234	N
4.448	1	4.448(+5)	4.54(−1)	4.54(+2)	3.217(+1)	lb$_f$
1.0(−5)	2.248(−6)	1	1.02(−6)	1.02(−3)	7.233(−5)	dyn
9.807	2.205	9.807(+5)	1	1.0(+3)	7.093(+1)	kg$_f$
9.807(−3)	2.205(−3)	9.807(+2)	1.0(−3)	1	7.093(−2)	g$_f$
1.382(−1)	3.108(−2)	1.383(+4)	1.410(−2)	1.410(+1)	1	poundal

* Newton.
† Avoirdupois.
** Dyne.
‡ Not recommended.

Additional Measures

1 metric ton = 10^3 kg$_f$ = 2.205 × 10^3 lb$_f$
1 pound troy = 0.8229 lb$_f$
1 oz† = 6.25 × 10^{-2} lb$_f$
1 oz troy = 6.857 × 10^{-2} lb$_f$

6. Velocity and acceleration equivalents

Velocity					Acceleration			
cm/s	ft/s	mi/h (mph)	km/h		cm/s^2	ft/s^2	ḡ*	
1	3.281(−2)	2.237(−2)	3.60(−2)	cm/s	1	3.281(−2)	1.019(−3)	cm/s^2
3.048(+1)	1	6.818(−1)	1.097	ft/s	3.048(+1)	1	3.109(−2)	ft/s^2
4.470(+1)	1.467	1	1.609	mi/h	9.807(+2)	3.217(+1)	1	ḡ
2.778(+1)	9.113(−1)	6.214(−1)	1	km/h				

* Standard acceleration of gravity.

Additional measures

1 knot = 1.152 mi/hr

7. Pressure equivalents

dyn/cm²	N/m²	lb/in.² (psi)	lb$_f$/ft² (psf)	atm*	Head†		
					in. (Hg)	ft (H$_2$O)	
1	1.0(−1)	1.45(−5)	2.089(−3)	9.869(−7)	2.953(−5)	3.349(−5)	dyn/cm²
1.0(+1)	1	1.45(−4)	2.089(−2)	9.869(−6)	2.953(−4)	3.349(−4)	N/m²
6.895(−4)	6.895(+3)	1	1.44(+2)	6.805(−2)	2.036	2.309	lb$_f$/in.²
4.788(+2)	4.788(+1)	6.944(−3)	1	4.725(−4)	1.414(−2)	1.603(−2)	lb/ft²
1.013(+6)	1.013(+5)	1.47(+1)	2.116(+3)	1	2.992(+1)	3.393(+1)	atm
3.336(+4)	3.386(+3)	4.912(−1)	7.073(+1)	3.342(−2)	1	1.134	in (Hg)
2.986(+4)	2.986(+3)	4.331(−1)	6.237(+1)	2.947(−2)	8.819(−1)	1	ft (H$_2$O)

*Standard atmospheric pressure.
† At std. gravity and 0°C for Hg, 15°C for H$_2$O.

Additional Measures

1 bar = 1 dyne/cm²
1 pascal = 1 N/m²

8. Work and energy equivalents

J*	ft-lb$_f$	W-h	Btu†	Kcal**	kg-m	
1	7.376(−1)	2.778(−4)	9.478(−4)	2.388(−4)	1.020(−1)	J
1.356	1	3.766(−4)	1.285(−3)	3.238(−4)	1.383(−1)	ft-lb$_f$
3.60(+3)	2.655(+3)	1	3.412	8.599(−1)	3.671(+2)	W-h
1.055(+3)	7.782(+2)	2.931(−1)	1	2.520(−1)	1.076(+2)	Btu
4.187(+3)	3.088(+3)	1.163	3.968	1	4.269(+2)	Kcal
9.807	7.233	2.724(−3)	9.295(−3)	2.342(−3)	1	kg-m

*Joule.
† British thermal unit.
** Kilocalorie.

Additional measures

1 Newton-meter = 1 J
1 erg = 1 dyne-cm = 10^{-7} J
1 cal = 10^{-3} kcal
1 therm = 10^{-5} Btu
1 million electron volts (Mev) = 1.602(10^{-13}) J

9. Power equivalents

J/s	ft-lb$_f$/s	hp*	kW	Btu/h	
1	7.376(−1)	1.341(−3)	1.0(−3)	3.412	J/s
1.356	1	1.818(−3)	1.356(−3)	4.626	ft-lb$_f$/s
7.457(+2)	5.50(+2)	1	7.457(−1)	2.545(+3)	hp
1.0(+3)	7.376(+2)	1.341	1	3.412(+3)	kW
2.931(−1)	2.162(−1)	3.930(−4)	2.931(−4)	1	Btu/h

* Horsepower.

Additional Measures

1 W = 10^{-3} kW
1 cal/s = 14.29 Btu/h
1 poncelet = 100 kg-m/sec = 0.9807 kW
1 ton of refrigeration = 1.2×10^4 Btu/h

Time

1 week		7 days	168 hours	10,080 minutes	604,800 seconds

1 week   7 days   168 hours   10,080 minutes   604,800 seconds
1 mean solar day   1440 minutes   86,400 seconds
1 calendar year   365 days   8760 hours   5.256(10)^5 minutes   3.1536(10)^7 seconds
1 tropical mean solar year   365.2422 days (basis of modern calendar)

Temperature

$\triangle 1°$ Celsius (formerly Centigrade)   (C) $= \triangle 1°$ Kelvin (K) $= 1.8°$ Fahrenheit (F)
$= 1.8°$ Rankine (R)

$0°C = 273.15°K = 32°F = 491.67°R = 0°R$
$0°K = -273.15°C = -459.67°F$

Electrical

1 coulomb	1.036(10)^5 faradays	0.1 abcoulomb	2.998(10)^9 statcoulombs
1 ampere		0.1 abampere	2.998(10)^9 statcoulombs
1 volt	10^3 millivolts	10^8 abvolts	3.335(10)$^{-3}$ statvolt
1 ohm	10^6 megohms	10^9 abohms	1.112(10)$^{-12}$ statohm
1 farad	10^6 microfarads	10^{-9} abfarads	8.987(10)^{11} statfarads

Trigonometric functions

$\sin(-\alpha) = -\sin\alpha$
$\cos(-\alpha) = \cos\alpha$
$\tan(-\alpha) = -\tan\alpha$
$\sin^2\alpha = \frac{1}{2} - \frac{1}{2}\cos 2\alpha$
$\cos^2\alpha = \frac{1}{2} + \frac{1}{2}\cos 2\alpha$
$\sin^2\alpha + \cos^2\alpha = 1$
$\sec^2\alpha = 1 + \tan^2\alpha$

$$\csc^2 \alpha = 1 + \operatorname{ctn}^2 \alpha$$

$$\sin 2\alpha = 2 \sin \alpha \cos \alpha$$

$$\cos 2\alpha = \cos^2 \alpha - \sin^2 \alpha = 1 - 2 \sin^2 \alpha = 2 \cos^2 \alpha - 1$$

$$\sin \alpha = \alpha - \frac{\alpha^3}{3!} + \frac{\alpha^5}{5!} - \frac{\alpha^7}{7!} + \frac{\alpha^9}{9!} \cdots$$

$$\cos \alpha = 1 - \frac{\alpha^2}{2!} + \frac{\alpha^4}{4!} - \frac{\alpha^6}{6!} + \frac{\alpha^8}{8!} \cdots$$

$$\sin (\alpha \pm \theta) = \sin \alpha \cos \theta \pm \cos \alpha \sin \theta$$

$$\cos (\alpha \pm \theta) = \cos \alpha \cos \theta \mp \sin \alpha \sin \theta$$

Differentials and integrals

$$\frac{dx^n}{dx} = nx^{n-1}$$

$$\frac{d(uv)}{dx} = U\frac{dv}{dx} + V\frac{du}{dx}$$

$$\frac{d(u/v)}{dx} = \frac{V(du/dx) - U(dv/dx)}{v^2}$$

$$\int x^n dx = \frac{x^{n+1}}{n+1} + C$$

$$\int u\, dv = uv - \int v\, du$$

$$\int \frac{dx}{x} = \log_\epsilon x + C$$

$$\int \sin x\, dx = -\cos x + C$$

$$\int \cos x\, dx = \sin x + C$$

$$\int \sin^2 x\, dx = \frac{x}{2} - \frac{\sin 2x}{4} + C$$

$$\int \cos^2 x\, dx = \frac{x}{2} + \frac{\sin 2x}{4} + C$$

Specific gravities and specific weights

Material	Average specific gravity	Average specific weight, $\mathrm{lb}_f/\mathrm{ft}^3$	Material	Average specific gravity	Average specific weight $\mathrm{lb}_f/\mathrm{ft}^3$
Acid, sulfuric, 87%	1.80	112	Iron, gray cast	7.10	450
Air, S.T.P.	0.001293	0.0806	Iron, wrought	7.75	480
Alcohol, ethyl	0.790	49			
Aluminum, cast	2.65	165	Kerosene	0.80	50
Asbestos	2.5	153			
Ash, white	0.67	42	Lead	11.34	710
Ashes, cinders	0.68	44	Leather	0.94	59
Asphaltum	1.3	81	Limestone, solid	2.70	168
			Limestone, crushed	1.50	95
Babbitt metal, soft	10.25	625			
Basalt, granite	1.50	96	Mahogany	0.70	44
Brass, cast-rolled	8.50	534	Manganese	7.42	475
Brick, common	1.90	119	Marble	2.70	166
Bronze, 7.9 to 14% S_n	8.1	509	Mercury	13.56	845
			Monel metal, rolled	8.97	555
Cedar, white, red	0.35	22			
Cement, portland, bags	1.44	90	Nickel	8.90	558
Chalk	2.25	140			
Clay, dry	1.00	63	Oak, white	0.77	48
Clay, loose, wet	1.75	110	Oil, lubricating	0.91	57
Coal, anthracite, solid	1.60	95			
Coal, bituminous, solid	1.35	85	Paper	0.92	58
Concrete, gravel, sand	2.3	142	Paraffin	0.90	56
Copper, cast, rolled	8.90	556	Petroleum, crude	0.88	55
Cork	0.24	15	Pine, white	0.43	27
Cotton, flax, hemp	1.48	93	Platinum	21.5	1330
Copper ore	4.2	262			
			Redwood, California	0.42	26
Earth	1.75	105	Rubber	1.25	78
Fir, Douglas	0.50	32	Sand, loose, wet	1.90	120
Flour, loose	0.45	28	Sandstone, solid	2.30	1444
			Seawater	1.03	64
Gasoline	0.70	44	Silver	10.5	655
Glass, crown	2.60	161	Steel, structural	7.90	490
Glass, flint	3.30	205	Sulfur	2.00	125
Glycerine	1.25	78	Teak, African	0.99	62
Gold, cast-hammered	19.3	1205	Tin	7.30	456
Granite, solid	2.70	172	Tungsten	19.22	1200
Graphite	1.67	135	Turpentine	0.865	54
Gravel, loose, wet	1.68	105			
			Water, 4°C	1.00	62.4
Hickory	0.77	48	Water, snow, fresh fallen	0.125	8.0
Ice	0.91	57			
			Zinc	7.14	445

Note: The value for the specific weight of water which is usually used in problem solutions is 62.4 $\mathrm{lb}_f/\mathrm{ft}^3$ or 8.34 $\mathrm{lb}_f/\mathrm{gal}$.

Coefficients of friction

Average values

Surfaces	Static	Kinetic
Metals on wood	0.4 –0.63	0.35–0.60
Wood on wood	0.3 –0.5	0.25–0.4
Leather on wood	0.38–0.45	0.3 –0.35
Iron on iron (*wrought*)	0.4 –0.5	0.4 –0.5
Glass on glass	0.23–0.25	0.20–0.25
Leather on glass	0.35–0.38	0.33–0.35
Wood on glass	0.35–0.40	0.28–0.31
Wood on sheet iron	0.43–0.50	0.38–0.45
Leather on sheet iron	0.45–0.50	0.35–0.40
Brass on wrought iron	0.35–0.45	0.30–0.35
Babbitt on steel	0.35–0.40	0.30–0.35
Steel on ice	0.03–0.04	0.03–0.04

Special-purpose formulas useful in solving uniform motion problems

Legend

V velocity	V_2 final velocity	S distance	a acceleration
V_1 initial velocity	V_{av} average velocity	t time	

Given	To find	Suggested formulas
V_1, V_2, t	S	$S = \left(\dfrac{V_1 + V_2}{2}\right)t$
V_1, V_2, a	S	$S = \dfrac{V_2^2 - V_1^2}{2a}$
V_1, a, t	S	$S = V_1 t + \dfrac{at^2}{2}$
V_1, V_2	V_{av}	$V_{av} = \dfrac{V_1 + V_2}{2}$
S, t	V_{av}	$V_{av} = \dfrac{S}{t}$
V_2, a, t	V_1	$V_1 = V_2 - at$
V_2, a, S	V_1	$V_1 = \sqrt{V_2^2 - 2aS}$
S, a, t	V_1	$V_1 = \dfrac{S}{t} - \dfrac{at}{2}$
V_1, a, t	V_2	$V_2 = V_1 + at$
V_1, a, S	V_2	$V_2 = \sqrt{V_1^2 + 2aS}$
V_1, S, t	V_2	$V_2 = \dfrac{2S}{t} - V_1$
V_1, V_2, S	t	$t = \dfrac{2S}{V_1 + V_2}$
V_1, a, S	t	$t = \dfrac{-V_1 \pm \sqrt{V_1^2 + 2aS}}{a}$
V_1, V_2, a	t	$t = \dfrac{V_2 - V_1}{a}$
V_1, V_2, t	a	$a = \dfrac{V_2 - V_1}{t}$
V_1, V_2, S	a	$a = \dfrac{V_2^2 - V_1^2}{2S}$
V_1, S, t	a	$a = \dfrac{2S}{t^2} - \dfrac{2V_1}{t}$

Anthropometric tables*

Anthropometric tables (*all dimensions in inches*)

Measurement	Range	Mean	Standard deviation	Percentiles				
				1st	5th	50th	95th	99th
Weight								
1. Weight (pounds)	104. –265.	163.66	20.86	123.1	132.5	161.9	200.8	215.9
Body Lengths								
2. Stature	59.45– 77.56	69.11	2.44	63.5	65.2	69.1	73.1	74.9
3. Nasal root height	56.30– 73.23	64.95	2.39	59.4	61.0	65.0	68.9	70.7
4. Eye height	56.30– 73.23	64.69	2.38	59.2	60.8	64.7	68.6	70.3
5. Tragion height	54.72– 74.41	63.92	2.39	58.4	60.0	64.0	67.8	69.6
6. Cervicale height	50.39– 66.93	59.08	2.31	53.7	55.3	59.2	62.9	64.6
7. Shoulder height	47.24– 64.17	56.50	2.28	51.2	52.8	56.6	60.2	61.9
8. Suprasternale height	48.03– 63.78	56.28	2.19	51.3	52.7	56.3	59.9	61.5
9. Nipple height	42.13– 57.09	50.41	2.08	45.6	47.0	50.4	53.9	55.3
10. Substernale height	41.34– 55.51	48.71	2.02	44.0	45.6	48.7	52.1	53.5
11. Elbow height	36.61– 49.21	43.50	1.77	39.5	40.6	43.5	46.4	47.7
12. Waist height	34.65– 48.82	42.02	1.81	37.7	39.1	42.1	45.0	46.4
13. Penale height	27.95– 41.34	34.52	1.75	30.6	31.6	34.5	37.4	38.7
14. Wrist height	27.56– 39.76	33.52	1.54	30.1	31.0	33.6	36.1	37.1
15. Crotch height (inseam)	26.77– 38.19	32.83	1.73	29.3	30.4	32.8	35.7	37.0
16. Gluteal furrow height	25.20– 37.01	31.57	1.62	27.9	29.0	31.6	34.3	35.5
17. Knuckle height	24.80– 35.04	30.04	1.45	26.7	27.7	30.0	32.4	33.5
18. Kneecap height	15.75– 23.23	20.22	1.03	17.9	18.4	20.2	21.9	22.7

* Adapted from H. T. E. Hertzberg, G. S. Daniels, and E. Churchill, *Anthropometry of Flying Personnel*—1950, WADC Technical Report 52–321, USAF, Wright Air Development Center, Wright-Patterson AFB, Ohio, September 1954. It should be noted that these data represent measurements made on approximately 4000 male USAF personnel and thus do not specifically represent the U.S. population at large.

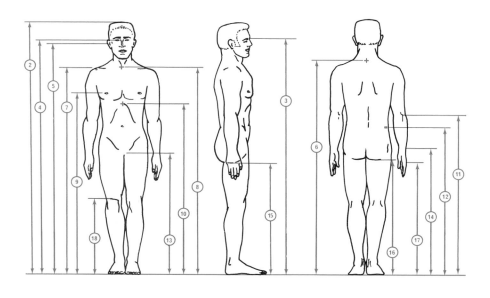

Measurement	Range	Mean	Standard deviation	Percentiles				
				1st	5th	50th	95th	99th
19. Sitting height	29.92–40.16	35.94	1.29	32.9	33.8	36.0	38.0	38.9
20. Eye	26.38–36.61	31.47	1.27	28.5	29.4	31.5	33.5	34.4
21. Shoulder	18.90–27.17	23.26	1.14	20.6	21.3	23.3	25.1	25.8
22. Waist height, sitting	6.30–12.99	9.24	0.76	7.4	7.9	9.3	10.4	10.9
23. Elbow rest height, sitting	4.33–12.99	9.12	1.04	6.6	7.4	9.1	10.8	11.5
24. Thigh clearance height	3.94– 7.09	5.61	0.52	4.5	4.8	5.6	6.5	6.8
25. Knee height, sitting	17.32–24.80	21.67	0.99	19.5	20.1	21.7	23.3	24.0
26. Popliteal height, sitting	14.17–19.29	16.97	0.77	15.3	15.7	17.0	18.2	18.8
27. Buttock–knee length	18.50–27.56	23.62	1.06	21.2	21.9	23.6	25.4	26.2
28. Buttock–leg length	35.43–50.00	42.70	2.04	38.2	39.4	42.7	46.1	47.7
29. Shoulder–elbow length	11.42–18.11	14.32	0.69	12.8	13.2	14.3	15.4	15.9
30. Forearm–hand length	15.35–22.05	18.86	0.81	17.0	17.6	18.9	20.2	20.7
31. Span	58.27–82.28	70.80	2.94	63.9	65.9	70.8	75.6	77.6
32. Arm reach from wall	27.56–39.76	34.59	1.65	30.9	31.9	34.6	37.3	38.6
33. Maximum reach from wall	31.10–46.06	38.59	1.90	34.1	35.4	38.6	41.7	43.2
34. Functional reach	26.77–40.55	32.33	1.63	28.8	29.7	32.3	35.0	36.4

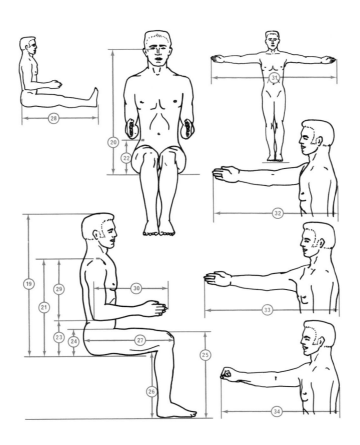

Measurement	Range	Mean	Standard deviation	Percentiles				
				1st	5th	50th	95th	99th
Body Breadths and Thicknesses								
35. Elbow to elbow breadth	11.42–23.62	17.28	1.42	14.5	15.2	17.2	19.8	20.9
36. Hip breadth, sitting	11.42–18.11	13.97	0.87	12.2	12.7	13.9	15.4	16.2
37. Knee-to-knee breadth	6.30–10.24	7.93	0.52	7.0	7.2	7.9	8.8	9.4
38. Biacromial diameter	12.60–18.50	15.75	0.74	14.0	14.6	15.8	16.9	17.4
39. Shoulder breadth	14.57–22.83	17.88	0.91	15.9	16.5	17.9	19.4	20.1
40. Chest breadth	9.45–15.35	12.03	0.80	10.4	10.8	12.0	13.4	14.1
41. Waist breadth	7.87–15.35	10.66	0.94	8.9	9.4	10.6	12.3	13.3
42. Hip breadth	8.27–15.75	13.17	0.73	11.3	12.1	13.2	14.4	15.2
43. Chest depth	6.69–12.99	9.06	0.75	7.6	8.0	9.0	10.4	11.1
44. Waist depth	5.51–11.81	7.94	0.88	6.3	6.7	7.9	9.5	10.3
45. Buttock depth	6.30–11.81	8.81	0.82	7.2	7.6	8.8	10.2	10.9
Circumferences and Body Surface Measurements								
46. Neck circumference	10.24–19.29	14.96	0.74	13.3	13.8	14.9	16.2	16.8
47. Shoulder circumference	35.43–56.69	45.25	2.43	40.2	41.6	45.1	49.4	51.5
48. Chest circumference	31.10–49.61	38.80	2.45	33.7	35.1	38.7	43.2	44.8
49. Waist circumference	24.41–47.24	32.04	3.02	26.5	27.8	31.7	37.5	40.1
50. Buttock circumference	29.92–46.85	37.78	2.29	33.0	34.3	37.7	41.8	43.5
51. Thigh circumference	14.57–28.74	22.39	1.74	18.3	19.6	22.4	25.3	26.4
52. Lower thigh circumference	11.81–23.23	17.33	1.41	14.2	15.1	17.3	19.6	20.9
53. Calf circumference	9.84–18.50	14.40	0.96	12.2	12.9	14.4	16.0	16.7
54. Ankle circumference	7.09–12.99	8.93	0.57	7.8	8.1	8.9	9.8	10.5

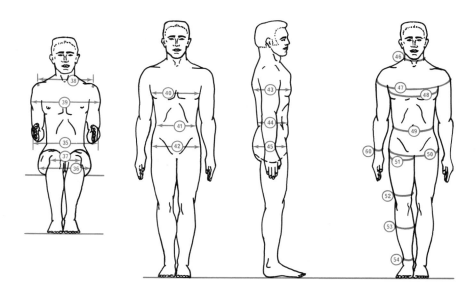

Measurement	Range	Mean	Standard deviation	Percentiles 1st	5th	50th	95th	99th
55. Scye circumference	11.02–22.83	18.09	1.38	15.1	16.1	18.0	20.5	21.8
56. Axillary arm circumference	7.87–16.54	12.54	1.10	10.2	10.9	12.4	14.4	15.2
57. Biceps circumference	8.27–16.93	12.79	1.07	10.5	11.2	12.8	14.6	15.4
58. Elbow circumference	8.27–15.35	12.26	0.80	10.7	11.1	12.2	13.6	14.3
59. Lower arm circumference	8.66–15.35	11.50	0.73	9.9	10.4	11.5	12.7	13.3
60. Wrist circumference	3.94– 8.27	6.85	0.40	6.0	6.3	6.8	7.5	7.8
61. Sleeve inseam	15.35–24.80	19.83	1.14	17.1	18.0	19.8	21.7	22.6
62. Sleeve length	27.56–38.98	33.64	1.50	30.2	31.3	33.7	36.0	37.3
63. Anterior neck length	1.38– 5.31	3.40	0.64	1.8	2.3	3.4	4.4	4.9
64. Posterior neck length	1.57– 6.10	3.64	0.61	2.3	2.7	3.6	4.7	5.2
65. Shoulder length	4.33– 8.66	6.77	0.56	5.5	5.9	6.8	7.7	8.1
66. Waist back	11.81–22.83	17.72	1.07	14.8	16.1	17.7	19.4	20.2
67. Waist front	10.63–21.26	15.24	1.12	12.3	13.5	15.2	17.0	18.1
68. Gluteal arc	7.87–17.32	11.71	0.92	9.7	10.4	11.7	13.1	14.8
69. Crotch length	20.08–38.19	28.20	2.00	23.7	25.1	28.2	31.6	33.5
70. Vertical trunk circumference	54.72–74.41	64.81	2.88	58.3	60.2	64.8	69.7	71.7
71. Interscye	12.20–24.41	19.62	1.40	16.3	17.3	19.6	22.0	22.9
72. Interscye maximum	17.72–27.17	22.85	1.33	19.8	20.7	22.9	25.1	26.0
73. Buttock circumference	33.46–52.36	41.74	2.82	36.1	37.4	41.5	46.7	49.3
74. Knee circumference	11.41–20.47	15.39	0.92	13.5	14.0	15.4	16.9	17.7

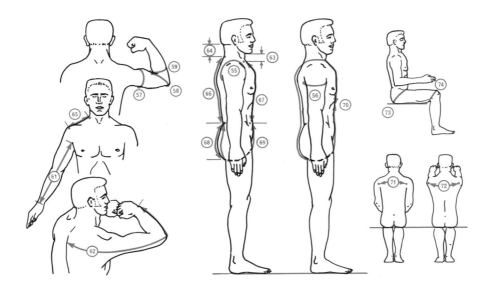

Measurement	Range	Mean	Standard deviation	Percentiles				
				1st	5th	50th	95th	99th
The Foot								
75. Foot length	8.86–12.24	10.50	0.45	9.5	9.8	10.5	11.3	11.6
76. Instep length	6.42– 8.86	7.64	0.34	6.9	7.1	7.6	8.2	8.4
77. Foot breadth	3.19– 4.65	3.80	0.19	3.40	3.50	3.78	4.10	4.36
78. Heel breadth	23.13– 3.27	2.64	0.15	2.30	2.40	2.63	2.87	3.01
79. Bimalleolar breadth	2.44– 3.58	2.95	0.15	2.61	2.70	2.95	3.19	3.32
80. Medial malleolus height	2.60– 4.29	3.45	0.21	3.0	3.1	3.5	3.8	4.0
81. Lateral malleolus height	2.01– 3.70	2.73	0.22	2.2	2.4	2.7	3.1	3.3
82. Ball of foot circumference	7.87–12.60	9.65	0.48	8.6	8.9	9.6	10.4	10.8
The Hand								
83. Hand length	5.87– 8.74	7.49	0.34	6.7	6.9	7.5	8.0	8.3
84. Palm length	3.39– 5.04	4.24	0.21	3.77	3.89	4.24	4.60	4.74
85. Hand breadth at thumb	3.23– 4.76	4.07	0.21	3.59	3.73	4.08	4.42	4.57
86. Hand breadth at metacarpale	2.99– 4.09	3.48	0.16	3.12	3.22	3.49	3.74	3.86
87. Thickness at metacarpale III	0.75– 1.54	1.17	0.07	1.00	1.05	1.17	1.28	1.35
88. First phalanx III length	2.21– 3.07	2.67	0.12	2.40	2.49	2.67	2.85	2.95
89. Finger diameter III	0.75– 1.00	0.86	0.05	0.77	0.79	0.85	0.93	0.96
90. Grip diameter (inside)	1.37– 2.63	1.90	0.14	1.52	1.62	1.83	2.05	2.16
91. Grip diameter (outside)	3.15– 4.72	4.09	0.21	3.58	3.72	4.09	4.44	4.57
92. Fist circumference	7.09–13.39	11.56	0.57	10.2	10.7	11.6	12.4	12.8

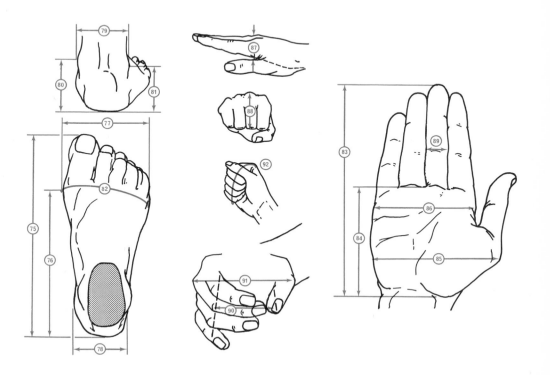

Measurement	Range	Mean	Standard deviation	Percentiles				
				1st	5th	50th	95th	99th
The Head and Face								
93. Head length	6.89–8.78	7.76	0.25	7.2	7.3	7.7	8.2	8.3
94. Head breadth	5.35–6.89	6.07	0.20	5.61	5.74	6.05	6.40	6.56
95. Minimum frontal diameter	3.54–5.00	4.35	0.19	3.88	4.04	4.35	4.68	4.80
96. Maximum frontal diameter	4.02–5.47	4.71	0.20	4.26	4.39	4.72	5.05	5.20
97. Bizygomatic diameter	4.72–6.22	5.55	0.20	5.07	5.21	5.54	5.88	6.02
98. Bigonial diameter	3.50–5.08	4.27	0.22	3.8	3.9	4.3	4.6	4.8
99. Bitragion diameter	4.76–6.30	5.60	0.21	5.1	5.3	5.6	5.9	6.1
100. Interocular diameter	0.87–1.65	1.25	0.10	1.03	1.09	1.25	1.42	1.50
101. Biocular diameter	3.19–4.45	3.78	0.17	3.38	3.48	3.78	4.06	4.19
102. Interpupiliary distance	2.01–2.99	2.49	0.14	2.19	2.27	2.49	2.74	2.84
103. Nose length	1.46–2.56	2.01	0.14	1.69	1.79	2.00	2.23	2.33
104. Nose breadth	0.91–1.85	1.31	0.11	1.09	1.16	1.31	1.49	1.58
105. Nasal root breadth	0.28–0.91	0.61	0.08	0.42	0.48	0.61	0.74	0.81
106. Nose protrusion	0.43–1.42	0.89	0.11	0.63	0.72	0.90	1.08	1.17
107. Philtrum length	0.35–1.46	0.77	0.14	0.48	0.54	0.76	0.98	1.09
108. Menton–subnasale length	1.81–3.54	2.63	0.27	2.05	2.19	2.62	3.07	3.28
109. Menton–Crinion length	6.18–8.58	7.36	0.36	6.6	6.8	7.4	8.0	8.2
110. Lip-to-lip distance	0.16–1.26	0.64	0.12	0.35	0.44	0.63	0.83	0.94
111. Lip length (bichelion dia.)	1.34–2.64	2.03	0.14	1.72	1.81	2.02	2.27	2.38
112. Ear length	1.69–3.15	2.47	0.16	2.08	2.21	2.47	2.73	2.85
113. Ear breadth	1.10–1.93	1.44	0.11	1.20	1.27	1.44	1.61	1.70
114. Ear length above tragion	0.79–1.61	1.17	0.11	0.92	0.99	1.17	1.35	1.42
115. Ear protrusion	0.31–1.54	0.84	0.14	0.55	0.63	0.83	1.10	1.23

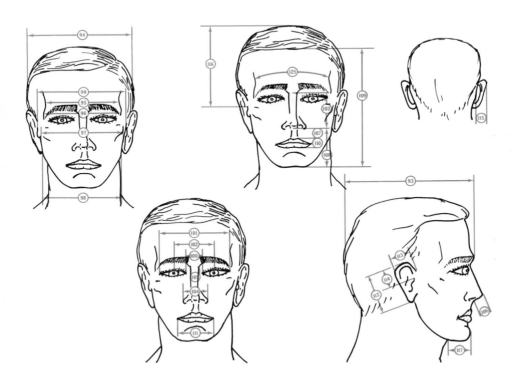

American national standard unit letter symbols

Unit	Symbol	Notes
ampere	A	SI unit of electric current
ampere per meter	A/m	SI unit of magnetic field strength
angstrom	Å	$1 \text{ Å} = 10^{-10} \text{ m}$
atmosphere, standard	atm	$1 \text{ atm} = 101{,}325 \text{ N/m}^2$
atto	a	SI prefix for 10^{-13}
barrel	bbl	$1 \text{ bbl} = 9702 \text{ in.}^3 = 0.15899 \text{ m}^3$
British thermal unit	Btu	
calorie	cal	$1 \text{ cal} = 4.1868 \text{ J}$
candela	cd	SI unit of luminous intensity
centi	c	SI prefix for 10^{-2}
centimeter	cm	
coulomb	C	SI unit of electric charge
cubic centimeter	cm³	
cubic foot	ft³	
cubic inch	in³	
cubic meter	m³	
curie	Ci	$1 \text{ Ci} = 3.7 \times 10^{10}$ disintegrations per second. Unit of activity in the field of radiation dosimetry.
cycle per second	Hz, c/s	See hertz. The name hertz is internationally accepted for this unit; the symbol Hz is preferred to c/s.
day	d	
decibel	dB	
degree (plane angle)	...°	
degree (temperature):		
degree Celsius	°C	Note that there is no space between the symbol ° and the letter. The use of the word *centigrade* for the Celsius temperature scale was abandoned by the Conférence Générale des Poids et Mesures in 1948.
degree Fahrenheit	°F	
degree Kelvin		See Kelvin
degree Rankine	°R	

Unit	Symbol	Notes
dyne	dyn	
electronvolt	eV	
erg	erg	
farad	F	SI unit of capacitance
foot	ft	
foot per second	ft/s	
foot pound-force	ft–lb$_f$	
gallon	gal	The gallon, quart, and pint differ in the US and the UK, and their use in science and technology is deprecated.
gauss	G	The gauss is the electromagnetic CGS unit of magnetic flux density. Use of SI unit, the tesla, is preferred.
gram	g	
henry	H	SI unit of inductance
hertz	Hz	SI unit of frequency
horsepower	hp	The horsepower is an anachronism in science and technology. Use of the SI unit of power (the watt) is preferred.
hour	h	
inch	in.	
joule	J	SI unit of energy
kelvin	K	In 1967 the CGPM gave the name *kelvin* to the SI unit of temperature which had formerly been called *degree Kelvin* and assigned it the symbol K (without the symbol °).
kilo	k	SI prefix for 10^3
kilogram	kg	SI unit of mass
kilogram-force	kg$_f$	In some countries the name *kilopond* (kp) has been adopted for this unit.
lambert	L	$1 \text{ L} = (1/\pi) \text{ cd/cm}^2$ A CGS unit of luminance. One lumen per square centimeter leaves a surface whose luminance is one lambert in all directions within a hemisphere. Use of the SI unit of luminance, the candela per square meter, is preferred.
liter	l	$1 \text{ l} = 10^{-3} \text{ m}^3$
lumen	lm	SI unit of luminous flux
mega	M	SI prefix for 10^6
megahertz	MHz	
meter	m	SI unit of length
mho	mho	CIPM has accepted the name *siemens* (S) for this unit and will submit it to the 14th CGPM for approval.
micro	μ	SI prefix for 10^{-6}
micron	μm	The name *micron* was abrogated by the Conférence Générale des Poids et Mesures, 1967.
mile (statute)	mi	$1 \text{ mi} = 5280 \text{ ft}$
mile per hour	mi/h	Although use of mph as an abbreviation is common, it should not be used as a symbol.
milli	m	SI prefix for 10^{-3}
minute (time)	min	Time may be designated by means of superscripts as in the following example: $9^h46^m30^s$.
mole	mol	SI unit of amount of substance
nano	n	SI prefix for 10^{-9}
newton	N	SI unit of force
newton per square meter	N/m^2	SI unit of pressure or stress; see pascal.
oersted	Oe	The oersted is the electromagnetic CGS unit of magnetic field strength. Use of the SI unit, the ampere per meter, is preferred.
ohm	Ω	SI unit of resistance

Unit	Symbol	Notes
pascal	Pa	$Pa = N/m^2$ SI unit of pressure or stress. This name accepted by the CIPM in 1969 for submission to the 14th CGPM.
pico	p	SI prefix for 10^{-12}
poise	P	SI unit of absolute viscosity
pound	lb	
pound-force	lb_f	The symbol lb, without a subscript, may be used for pound-force where no confusion is foreseen.
pound-force per square inch	lb_f/in^2	Although use of the abbreviation psi is common, it should not be used as a symbol. Refer to note on pound-force regarding subscript to the symbol.
radian	rad	SI unit of plane angle
revolution per minute	r/min	Although use of rpm as an abbreviation is common, it should not be used as a symbol.
revolution per second	r/s	
roentgen	R	Unit of exposure in the field of radiation dosimetry
second (time)	s	SI unit of time
siemens	S	$S = \Omega^{-1}$
slug	slug	1 slug = 14.5939 kg
steradian	sr	SI unit of solid angle
stokes	St	SI unit of dynamic viscosity
tesla	T	$T = N/(A-m) = Wb/m^2$ SI unit of magnetic flux density (magnetic induction)
ton	ton	1 ton = 2000 lb
volt	V	SI unit of voltage
voltampere	VA	IEC name and symbol for the SI unit of apparent power
watt	W	SI unit of power
watt-hour	Wh	
weber	Wb	$Wb = V-s$ SI unit of magnetic flux
yard	yd	

Code of ethics
for engineers[1]

The Fundamental Principles

Engineers uphold and advance the integrity, honor and dignity of the engineering profession by:

I. using their knowledge and skill for the enhancement of human welfare;
II. being honest and impartial, and serving with fidelity the public, their employers and clients;
III. striving to increase the competence and prestige of the engineering profession; and
IV. supporting the professional and technical societies of their disciplines.

The Fundamental Canons

1. Engineers shall hold paramount the safety, health and welfare of the public in the performance of their professional duties.
2. Engineers shall perform services only in the areas of their competence.
3. Engineers shall issue public statements only in an objective and truthful manner.
4. Engineers shall act in professional matters for each employer or client as faithful agents or trustees, and shall avoid conflicts of interest.
5. Engineers shall build their professional reputation on the merit of their services and shall not compete unfairly with others.
6. Engineers shall associate only with reputable persons or organizations.

[1] *Approved by the Board of Directors, October 1, 1974*

7. Engineers shall continue their professional development throughout their careers and shall provide opportunities for the professional development of those engineers under their supervision.

Suggested Guidelines for Use with the Fundamental Canons of Ethics

1. Engineers shall hold paramount the safety, health and welfare of the public in the performance of their professional duties.

   a. Engineers shall recognize that the lives, safety, health and welfare of the general public are dependent upon engineering judgments, decisions and practices incorporated into structures, machines, products, processes and devices.

   b. Engineers shall not approve nor seal plans and/or specifications that are not of a design safe to the public health and welfare and in conformity with accepted engineering standards.

   c. Should the Engineers' professional judgment be overruled under circumstances where the safety, health, and welfare of the public are endangered, the Engineers shall inform their clients or employers of the possible consequences and notify other proper authority of the situation, as may be appropriate.

      (c.1) Engineers shall do whatever possible to provide published standards, test codes and quality control procedures that will enable the public to understand the degree of safety or life expectancy associated with the use of the design, products and systems for which they are responsible.

      (c.2) Engineers will conduct reviews of the safety and reliability of the design, products or systems for which they are responsible before giving their approval to the plans for the design.

      (c.3) Should Engineers observe conditions which they believe will endanger public safety or health, they shall inform the proper authority of the situation.

   d. Should Engineers have knowledge or reason to believe that another person or firm may be in violation of any of the provisions of these Guidelines, they shall present such information to the proper authority in writing and shall cooperate with the proper authority in furnishing such further information or assistance as may be required.

      (d.1) They shall advise proper authority if an adequate review of the safety and reliability of the products or systems has not been made or when the design imposes hazards to the public through its use.

      (d.2) They shall withhold approval of products or systems when changes or modifications are made which would affect adversely its performance insofar as safety and reliability are concerned.

   e. Engineers should seek opportunities to be of constructive service in civic affairs and work for the advancement of the safety, health and well-being of their communities.

   f. Engineers should be commited to improving the environment to enhance the quality of life.

2. Engineers shall perform services only in areas of their competence.

   a. Engineers shall undertake to perform engineering assignments only when qualified by education or experience in the specific technical field of engineering involved.

b. Engineers may accept an assignment requiring education or experience outside of their own fields of competence, but only to the extent that their services are restricted to those phases of the project in which they are qualified. All other phases of such project shall be performed by qualified associates, consultants, or employees.

c. Engineers shall not affix their signatures and/or seals to any engineering plan or document dealing with subject matter in which they lack competence by virtue of education or experience, nor to any such plan or document not prepared under their direct supervisory control.

3. Engineers shall issue public statements only in an objective and truthful manner.

a. Engineers shall endeavor to extend public knowledge, and to prevent misunderstandings of the achievements of engineering.

b. Engineers shall be completely objective and truthful in all professional reports, statements, or testimony. They shall include all relevant and pertinent information in such reports, statements, or testimony.

c. Engineers, when serving as expert or technical witnesses before any court, commission, or other tribunal, shall express an engineering opinion only when it is founded upon adequate knowledge of the facts in issue, upon a background of technical competence in the subject matter, and upon honest conviction of the accuracy and propriety of their testimony.

d. Engineers shall issue no statements, criticisms, nor arguments on engineering matters, which are inspired or paid for by an interested party, or parties, unless they have prefaced their comments by explicitly identifying themselves, by disclosing the identities of the party or parties on whose behalf they are speaking, and by revealing the existence of any pecuniary interest they may have in the instant matters.

e. Engineers shall be dignified and modest in explaining their work and merit, and will avoid any act tending to promote their own interests at the expense of the integrity, honor and dignity of the profession.

4. Engineers shall act in professional matters for each employer or client as faithful agents or trustees, and shall avoid conflicts of interest.

a. Engineers shall avoid all known conflicts of interest with their employers or clients and shall promptly inform their employers or clients of any business association, interests, or circumstances which could influence their judgment or the quality of their services.

b. Engineers shall not knowingly undertake any assignments which would knowingly create a potential conflict of interest between themselves and their clients or their employers.

c. Engineers shall not accept compensation, financial or otherwise, from more than one party for services on the same project, nor for services pertaining to the same project, unless the circumstances are fully disclosed to, and agreed to, by all interested parties.

d. Engineers shall not solicit nor accept financial or other valuable considerations, including free engineering designs, from material or equipment suppliers for specifying their products.

e. Engineers shall not solicit nor accept gratuities, directly or indirectly, from contractors, their agents, or other parties dealing with their clients or employers in connection with work for which they are responsible.

f. When in public service as members, advisors, or employees of a governmental body or department, Engineers shall not participate in considerations or actions with respect to services provided by them or their organization in private or product engineering practice.

g. Engineers shall not solicit nor accept an engineering contract from a governmental body on which a principal, officer or employee of their organization serves as a member.

h. When, as a result of their studies, Engineers believe a project will not be successful, they shall so advise their employer or client.

i. Engineers shall treat information coming to them in the course of their assignments as confidential, and shall not use such information as a means of making personal profit if such action is adverse to the interests of their clients, their employers, or the public.

   (i.1) They will not disclose confidential information concerning the business affairs or technical processes or any present or former employer or client or bidder under evaluation, without his consent.

   (i.2) They shall not reveal confidential information nor findings of any commission or board of which they are members.

   (i.3) When they use designs supplied to them by clients, these designs shall not be duplicated by the Engineers for others without express permission.

   (i.4) While in the employ of others, Engineers will not enter promotional efforts or negotiations for work or make arrangements for other employment as principals or to practice in connection with specific projects for which they have gained particular and specialized knowledge without the consent of all interested parties.

j. The Engineer shall act with fairness and justice to all parties when administering a construction (or other) contract.

k. Before undertaking work for others in which Engineers may make improvements, plans, designs, inventions, or other records which may justify copyrights or patents, they shall enter into a positive agreement regarding ownership.

l. Engineers shall admit and accept their own errors when proven wrong and refrain from distorting or altering the facts to justify their decisions.

m. Engineers shall not accept professional employment outside of their regular work or interest without the knowledge of their employers.

n. Engineers shall not attempt to attract an employee from another employer by false or misleading representations.

o. Engineers shall not review the work of other Engineers except with the knowledge of such Engineers, or unless the assignments/or contractual agreements for the work have been terminated.

   (o.1) Engineers in governmental, industrial or educational employment are entitled to review and evaluate the work of other engineers when so required by their duties.

   (o.2) Engineers in sales or industrial employment are entitled to make engineering comparisons of their products with products of other suppliers.

   (o.3) Engineers in sales employment shall not offer nor give engineering consultation or designs or advice other than specifically applying to equipment, materials or systems being sold or offered for sale by them.

5. Engineers shall build their professional reputation on the merit of their services and shall not compete unfairly with others.

   a. Engineers shall not pay nor offer to pay, either directly or indirectly, any commission, political contribution, or a gift, or other consideration in order to secure work, exclusive of securing salaried position through employment agencies.

b. Engineers should negotiate contracts for professional services fairly and only on the basis of demonstrated competence and qualifications for the type of professional service required.

c. Engineers should negotiate a method and rate of compensation commensurate with the agreed upon scope of services. A meeting of the minds of the parties to the contract is essential to mutual confidence. The public interest requires that the cost of engineering services be fair and reasonable, but not the controlling consideration in selection of individuals or firms to provide these services.

(c.1) These principles shall be applied by Engineers in obtaining the services of other professionals.

d. Engineers shall not attempt to supplant other Engineers in a particular employment after becoming aware that definite steps have been taken toward the others' employment or after they have been employed.

(d.1) They shall not solicit employment from clients who already have Engineers under contract for the same work.

(d.2) They shall not accept employment from clients who already have Engineers for the same work not yet completed or not yet paid for unless the performance or payment requirements in the contract are being litigated or the contracted Engineers' services have been terminated in writing by either party.

(d.3) In case of termination of litigation, the prospective Engineers before accepting the assignment shall advise the Engineers being terminated or involved in litigation.

e. Engineers shall not request, propose nor accept professional commissions on a contigent basis under circumstances under which their professional judgments may be compromised, or when a contingency provision is used as a device for promoting or securing a professional commission.

f. Engineers shall not falsify nor permit misrepresentation of their, or their associates', academic or professional qualifications. They shall not misrepresent nor exaggerate their degree of responsibility in or for the subject matter of prior assignments. Brochures or other presentations incident to the solicitation of employment shall not misrepresent pertinent facts concerning employers, employees, associates, joint ventures, or their past accomplishments with the intent and purpose of enhancing their qualifications and work.

g. Engineers may advertise professional services only as a means of identification and limited to the following:

(g.1) Professional cards and listings in recognized and dignified publications, provided they are consistent in size and are in a section of the publication regularly devoted to such professional cards and listings. The information displayed must be restricted to firm name, address, telephone number, appropriate symbol, names of principal participants and the fields of practice in which the firm is qualified.

(g.2) Signs on equipment, offices and at the site of projects for which they render services, limited to firm name, address, telephone number and type of services, as appropriate.

(g.3) Brochures, business cards, letterheads and other factual representations of experience, facilities, personnel and capacity to render service, providing the same are not misleading relative to the extent of participation in the projects cited and are not indiscriminately distributed.

(g.4) Listings in the classified section of telephone directories, limited to name, address, telephone number and specialties in which the firm is qualified without resorting to special or bold type.

h. Engineers may use display advertising in recognized dignified business and professional publications, providing it is factual, and relates only to engineering, is free from ostentation, contains no laudatory expressions or implication, is not misleading with respect to the Engineer's extent of participation in the services or projects described.

i. Engineers may prepare articles for the lay or technical press which are factual, dignified and free from ostentations or laudatory implications. Such articles shall not imply other than their direct participation in the work described unless credit is given to others for their share of the work.

j. Engineers may extend permission for their names to be used in commercial advertisements, such as may be published by manufacturers, contractors, material suppliers, etc., only by means of a modest dignified notation acknowledging their participation and the scope thereof in the project or product described. Such permission shall not include public endorsement of proprietary products.

k. Engineers may advertise for recruitment of personnel in appropriate publications or by special distribution. The information presented must be displayed in a dignified manner, restricted to firm name, address, telephone number, appropriate symbol, names of principal participants, the fields of practice in which the firm is qualified and factual descriptions of positions available, qualifications required and benefits available.

l. Engineers shall not enter competitions for designs for the purpose of obtaining commissions for specific projects, unless provision is made for reasonable compensation for all designs submitted.

m. Engineers shall not maliciously or fasely, directly or indirectly, injure the professional reputation, prospects, practice or employment of another engineer, nor shall they indiscriminately criticize another's work.

n. Engineers shall not undertake nor agree to perform any engineering service on a free basis, except professional services which are advisory in nature for civic, charitable, religious or non-profit organizations. When serving as members of such organizations, engineers are entitled to utilize their personal engineering knowledge in the service of these organizations.

o. Engineers shall not use equipment, supplies, laboratory nor office facilities of their employers to carry on outside private practice without consent.

p. In case of tax-free or tax-aided facilities, engineers should not use student services at less than rates of other employees of comparable competence, including fringe benefits.

6. Engineers shall associate only with reputable persons or organizations.

a. Engineers shall not knowingly associate with nor permit the use of their names nor firm names in business ventures by any person or firm which they know, or have reason to believe, are engaging in business or professional practices of a fraudulent or dishonest nature.

b. Engineers shall not use association with non-engineers, corporations, nor partnerships as 'cloaks' for unethical acts.

7. Engineers shall continue their professional development throughout their careers,

and shall provide opportunities for the professional development of those engineers under their supervision.

a. Engineers shall encourage their engineering employees to further their education.

b. Engineers should encourage their engineering employees to become registered at the earliest possible date.

c. Engineers should encourage engineering employees to attend and present papers at professional and technical society meetings.

d. Engineers should support the professional and technical societies of their disciplines.

e. Engineers shall give proper credit for engineering work to those to whom credit is due, and recognize the proprietary interests of others. Whenever possible they shall name the person or persons who may be responsible for designs, inventions, writings or other accomplishments.

f. Engineers shall endeavor to extend the public knowledge of engineering, and shall not participate in the dissemination of untrue, unfair or exaggerated statements regarding engineering.

g. Engineers shall uphold the principle of appropriate and adequate compensation for those engaged in engineering work.

h. Engineers should assign professional engineers duties of a nature which will utilize their full training and experience insofar as possible, and delegate lesser functions to subprofessionals or to technicians.

i. Engineers shall provide prospective engineering employees with complete information on working conditions and their proposed status of employment, and after employment shall keep them informed of any changes.

Answers to selected problems

Chapter 9

9-1. *e.* $x = 5$
    j. $x = -3$

9-2. *e.* 318.3
    j. 702.25
9-3. *e.* -8.961
    j. 0.805

9-4. *e.* $5.6856(10)^4$
    j. $-5.345(10)^2$
    o. $5.2(10)^1$
9-5. *e.* 1.55
    j. $-3.064(10)^{-6}$
    o. 1.6
9-6. *e.* $7(10)^2\%$
    j. $1(10)^2\%$

9-7. *e.* $\pm 2(10)^{-4}$
    j. $\pm 10^{-1}$
9-10. *a.* 7.47 ft
    b. 1.89 ft
    c. 0.7% error

Chapter 10

10-65. $3.6379(10)^4$
10-70. $2.1764(10)^6$
10-75. $4.5745(10)^1$
10-80. $4.1886(10)^3$
10-85. $2.7321(10)^5$
10-90. $1.1866(10)^{-4}$
10-95. $3.9840(10)^{-3}$
10-100. $2.6588(10)^0$
10-105. $2.8661(10)^{-1}$
10-110. $9.6973(10)^2$
10-150. $1.9978(10)^0$
10-155. $2.0000(10)^3$
10-160. $1.0133(10)^{-4}$
10-165. $2.8421(10)^{-4}$
10-170. $4.9520(10)^3$

10-175. $1.5625(10)^0$
10-180. $1.3863(10)^0$
10-185. $5.9349(10)^{-5}$
10-190. $1.5943(10)^{23}$
10-235. $1.2264(10)^{-5}$
10-240. $1.8918(10)^3$
10-245. $3.5770(10)^{-1}$
10-250. $1.3349(10)^2$
10-255. $1.6428(10)^6$
10-260. $3.6281(10)^1$
10-265. $2.8306(10)^{-1}$
10-270. $1.1751(10)^0$
10-275. $2.5688(10)^1$
10-280. $8.0550(10)^{-13}$
10-335. $2.0070(10)^{-5}$

10-340. $2.8392(10)^2$
10-345. $4.2094(10)^3$
10-350. $3.7350(10)^1$
10-355. $3.6579(10)^0$
10-360. $1.7725(10)^0$
10-365. $2.5927(10)^1$
10-370. $3.1131(10)^2$
10-425. $1.2388(10)^0$
10-430. $7.6105(10)^5$
10-435. $1.6598(10)^{21}$
10-440. $4.0996(10)^{-1}$
10-445. $6.9306(10)^{-1}$
10-450. $2.6419(10)^{-2}$
10-455. $2.8048(10)^6$
10-460. $4.1209(10)^4$

10-465. $2.4831(10)^0$
10-515. $1.5814(10)^0$
10-520. $1.6801(10)^0$
10-525. $3.6915(10)^{-7}$
10-530. $2.2873(10)^3$
10-535. $6.7563(10)^{14}$
10-540. $8.9818(10)^{-1}$
10-545. $1.0181(10)^0$

10-550. $7.9671(10)^{-1}$
10-555. $6.4690(10)^{-2}$
10-560. $8.7288(10)^{-1}$
10-565. $4.2904(10)^{-1}$
10-570. $1.2472(10)^{-1}$
10-575. $5.7306(10)^{-1}$
10-580. $2.0879(10)^5$
10-585. $4.0486(10)^0$

10-655. $9.9956(10)^{-1}$
10-670. $9.8129(10)^{-1}$
10-675. $7.7289(10)^{-1}$
10-680. $5.7039(10)^{-1}$
10-685. $1.4410(10)^{-1}$
10-690. $1.5176(10)^1$
10-695. $4.5137(10)^0$
10-700. $1.0058(10)^0$

10-705. $1.4370(10)^0$
10-710. $7.2247(10)^1$
10-715. $6.0264(10)^1$
10-720. $6.3953(10)^{-1}$
10-725. $2.1405(10)^1$
10-730. $2.8508(10)^0$
10-735. $9.2958(10)^{-1}$
10-740. $2.5207(10)^{-1}$
10-745. $2.1032(10)^0$
10-750. $5.7588(10)^0$
10-755. $6.5559(10)^{-2}$
10-760. $3.6711(10)^{-2}$
10-765. a. $1.5865(10)^{-1}$
　　　b. $5.0052(10)^{-1}$
　　　c. $8.1441(10)^{-1}$
　　　d. $9.5953(10)^{-1}$
　　　e. $9.8161(10)^{-1}$
　　　f. $9.9170(10)^{-1}$
　　　g. $9.9991(10)^{-1}$
10-770. a. $-4.5539(10)^0 + j8.5646(10)^0$
　　　b. $-8.4106(10)^1 + j7.8430(10)^1$
　　　c. $-1.9640(10)^0 + j7.1482(10)^{-1}$
　　　d. $-1.9564(10)^0 + j5.3750(10)^0$

10-775. a. $1.7145(10)^1 \,\underline{/2.4427(10)^1}$
　　　b. $6.3474(10)^1 \,\underline{/2.5642(10)^2}$
　　　c. $1.1682(10)^0 \,\underline{/3.0188(10)^2}$
　　　d. $8.5834(10)^1 \,\underline{/1.1869(10)^2}$
10-780. $2.7803(10)^6$
10-785. $2.3909(10)^5$
10-790. $2.2684(10)^1$
10-795. $-5.9653(10)^{-3}$
10-800. $2.4621(10)^8$
10-805. $2.1956(10)^1$
10-810. $3.3279(10)^{-6}$
10-820. $5.6654(10)^4$
10-825. $2.6689(10)^1$
10-830. $4.0567(10)^{-3}$
10-835. $4.1181(10)^{-2}$
10-840. $2.9997(10)^3$
10-845. $1.4386(10)^2$
10-850. $7.9465(10)^{-1}$
10-855. $3.9216(10)^{-2}$
10-860. $x = 5.0274(10)^{-1}$
　　　$y = 8.3096(10)^5$

Chapter 11

11-1. $X = 2.01913$ g
　　$s = \pm 0.0016$ g
　　$s_m = \pm 4.25(10)^{-4}$ g
　　Wt. $= 2.0191 \pm 0.0004$ g

11-4. $\bar{X} = 50.6$
　　Median $= 51$
　　Mode $= 55$
　　Graph $=$ Slight tendency to skew

Chapter 12

12-5. $k = \dfrac{M^8 L^3}{F^4 \theta T}$

12-10. $k = \dfrac{FL^2}{Q^{1/2} M^3}$

12-15. (a) 1.076 dynes/cm^2: (b) $3.177(10)^{-5}$ in. Hg.
　　　$3.44(10)^8$ abhenries
12-25. (a) $6.97(10)^4$ ft^3/hr; (b) $5.48(10)^{-1}$ m^3/s
12-30. (a) $2.07(10)^{-4}$ acre; (b) $8.36(10)^{-1}$ m^2
12-35. $6.39(10)^{-2}$ ft^2
12-40. 1.274 in.
12-45. $M = FL$

12-50. (a) $3.93(10)^4$ lb_f; $1.748(10)^5$ N

(b) $5.68(10)^1$ ft/sec^2; $1.73(10)^1$ m/s^2

(c) $1.53(10)^5 \dfrac{lb_f \text{ sec}^2}{ft}$; $6.95(10)^{-1}$ kg

(d) $3.10(10)^5$ lb_f; $1.379(10)^5$ N

12-55. M = 1.953 lb_m

12-60. F = 1.473 lb_f; 6.55 N

12-65. F = $6.05(10)^1$ lb_f

12-70. g = 9.73 cm/sec^2

12-75. Invalid

12-80. F = $9.32(10)^1$ N

12-85. i = 3.52 amps

12-90. R = 1.58 ohms; power = 6.27 watts

Chapter 13

13-5. R = 360 lb @ N33.7°W

13-10. MA = 231 lb-ft

13-15. $R_R = 7.44(10)^3$N; $R_L = 3.06(10)^3$N

13-20. $R_R = 1.995(10)^4$ lb_f; $R_L = 1.355(10)^4$ lb_f

13-25. AB = $5.95(10)^1$ lb_f; C = $1.98(10)^1$ lb_f

13-30. P = $1.53(10)^2$ lb_f; N = $2.35(10)^2$ lb_f

13-35. CB = $1.597(10)^3$ lb_f; R = $1.984(10)^3$ lb_f @ 75.5°

13-45. $V_2 = 6.11(10)^1$ ft/sec; t = 7.11 sec

13-50. $V_2 = 2.87(10)^1$ ft/sec

13-55. $t_2 = 2.26$ min.; $1.784(10)^3$ m

13-60. 1st to 2nd: $4.70(10)^2$ in.; 2nd to 3rd: $4.33(10)^2$ in.

13-65. W = $8.90(10)^1$ rad/sec

13-70. V = $3.400(10)^3$ ft/min

13-75. (a) $-1.152 \dfrac{rev/min}{sec}$ (b) $\theta = 1.555(10)^5$ rad (c) $V_{av} = 1.57(10)^4$ ft/min

13-80. WK = $7.42(10)^4$ ft-lb

13-85. WK = $2.139(10)^3$ N-m

13-90. WK = $2.899(10)^5$ n-m

13-95. V = $2.88(10)^1$ ft/sec

13-100. hp = $1.865(10)^1$ hp

13-105. $P_{input} = 6.11(10)^{-1}$ hp

13-110. $KE_1 = 3.81(10)^1$ ft-lb

$KE_2 = 1.52(10)^2$ ft-lb

$KE_3 = 3.42(10)^2$ ft-lb

$KE_{10} = 3.810(10)^3$ ft-lb

13-115. t = 1.97 sec

13-120. $F_{total} = 1.43$ lb_f

13-125. WK = $1.8(10)^5$ N-m; P = $1.117(10)^1$ hp

13-130. F = $3.00(10)^2$ lb_f

13-135. P = $3.45(10)^2$ kw

13-140. Braking Force = $1.310(10)^3$ lb_f

Chapter 14

14-5. $a_{AV} = 9.8$ m/s^2

14-20. C = $1(10)^{-7}$ farads

14-25. $E_u = 1.8$ joules

14-40. $Q_c = 2.0(10)^5 \dfrac{N/m^2}{s} = $ pasquel/S

$\qquad$ $E = 2.4(10)^6$ joules

14-45. $I = 2.2(10)^7$ N–S^2/M^5

14-60. $qr = 5(10)^3$ watt

Index